The Origin and Evolution of Larval Forms

The Origin and Evolution of Larval Forms

Edited by
Brian K. Hall
Department of Biology
Dalhousie University
Halifax, Nova Scotia, Canada

Marvalee H. Wake
Department of Integrative Biology and
Museum of Vertebrate Zoology
University of California, Berkeley, California

ACADEMIC PRESS
San Diego   London   Boston   New York   Sydney   Tokyo   Toronto

Cover photo credit: Skulls of salamanders from exclusively larval reproducing families. For more information, see Figure 4 in Chapter 6, "Hormonal Control in Larval Development and Evolution—Amphibians," by Christopher Rose.

This book is printed on acid-free paper. ∞

Copyright © 1999 by ACADEMIC PRESS

All Rights Reserved.
No part of this publication may be reproduced or transmitted in any form or by any means, electronic or mechanical, including photocopy, recording, or any information storage and retrieval system, without permission in writing from the publisher.

Academic Press
a division of Harcourt Brace & Company
525 B Street, Suite 1900, San Diego, California 92101-4495, USA
http://www.apnet.com

Academic Press
24-28 Oval Road, London NW1 7DX, UK
http://www.hbuk.co.uk/ap/

Library of Congress Catalog Card Number: 98-86239

International Standard Book Number: 0-12-730935-7

PRINTED IN THE UNITED STATES OF AMERICA
98 99 00 01 02 03 EB 9 8 7 6 5 4 3 2 1

CONTENTS

PART I

Larval Types and Larval Evolution

PART **II**

Mechanisms of Larval Development and Evolution

8 Cell Lineages in Larval Development and Evolution of Echinoderms

Rudolf A. Raff

9 Cell Lineages in Larval Development and Evolution of Holometabolous Insects

Lisa M. Nagy and Miodrag Grbić

CONTRIBUTORS

MIODRAG GRBIĆ (275), Department of Zoology, University of Western Ontario, London, Ontario N6A 5B7, Canada

ERICK GREENE (379), Division of Biological Sciences, The University of Montana, Missoula, Montana 59812

BRIAN K. HALL (1, 411), Department of Biology, Dalhousie University, Halifax, Nova Scotia B3H 4J1, Canada

JAMES HANKEN (61), Department of Environmental, Population, and Organismic Biology, University of Colorado, Boulder, Colorado 80309

MICHAEL W. HART[1] (159), Section of Evolution and Ecology, University of California, Davis, California 95616

CAROLE S. HICKMAN (21), Department of Integrative Biology and Museum of Paleontology, University of California, Berkeley, California 94720

SARAH J. KUPFERBERG[2] (301), Department of Animal Ecology, Umeå University, S901-87, Umeå, Sweden; and Department of Integrative Biology, University of California, Berkeley, California 94720

LISA M. NAGY (275), Department of Molecular and Cellular Biology, University of Arizona, Tucson, Arizona 85475

H. FREDERIK NIJHOUT (217), Department of Zoology, Duke University, Durham, North Carolina 27708

[1] Present address: Department of Biology, Dalhousie University, Halifax, Nova Scotia B3H 4J1, Canada.
[2] Present address: Department of Integrative Biology, University of California, Berkeley, California 94720.

RUDOLF A. RAFF (255), Department of Biology and Indiana Molecular Biology Institute, Indiana University, Bloomington, Indiana 47405

CHRISTOPHER S. ROSE[3] (167), Department of Biology, Dalhousie University, Halifax, Nova Scotia B3H 3Y8, Canada

S. LAURIE SANDERSON (301), Department of Biology, College of William and Mary, Williamsburg, Virginia 23187

MARVALEE H. WAKE (1, 411), Department of Integrative Biology and Museum of Vertebrate Zoology, University of California, Berkeley, California 94720

JACQUELINE F. WEBB (109), Department of Biology, Villanova University, Villanova, Pennsylvania 19085

GREGORY A. WRAY (159), Department of Ecology and Evolution, State University of New York, Stony Brook, New York 11794

[3] Present address: Department of Biology, James Madison University, Harrisonburg, Virginia 22807.

PREFACE

If challenged by a 10-year-old, any biologist could confidently explain what a larva is, where larvae fit into the life cycle, and that larvae must metamorphose before they transform into, or are replaced by, adults. As documented in this volume, however, the answers to such fundamental questions as "What is a larva?" and "What is metamorphosis?" are not straightforward and the questions are not easy. The authors of the chapters in this volume discuss evidence bearing on the two questions "What is a larva?" and "What is metamorphosis?" in the twofold context of larval development and larval evolution. Four overarching themes—development, evolution, metamorphosis, and genetic mechanisms—inform the book. Morphological, functional, comparative, molecular, genetic, developmental, ecological, and evolutionary approaches are stressed.

The questions are old ones: The title of the book was the title of Walter Garstang's presidential address to Section D (Zoology) at the 96th Meeting of the British Association for the Advancement of Science in Glasgow in 1928. Garstang was especially concerned with whether larval evolution paralleled adult evolution. In his writings, he documented three important claims:

- ancestry should be sought in larvae and not in adults;
- ontogeny creates and does not recapitulate phylogeny; and
- many larval features are secondary adaptations to larval life.

Each of these claims is discussed in this volume to document the important conclusion that many organisms have separate developmental programs for larva and adult. Indeed, many organisms have separate developmental programs for embryo, larva, and adult. Such a conclusion raises important questions concerning the kinds of larvae that exist (primary–secondary; trochophore–dipleurula; feeding–nonfeeding; planktotrophic–lecithotrophic), how they arise developmentally, how they arose evolutionarily, and how they fit into and prepare adults for their environment. Many organisms have multiple larval stages in complex life histories. These are discussed. Larvae have been an important source of information on metazoan evolution for over 150 years. Whether larvae reveal anything about metazoan origins also is discussed. In short, the volume provides a comprehensive overview of larval origins, evolution, development, and function by contributors whom we thank with pleasure for their scholarly treatments of larval biology.

Brian K. Hall
Marvalee H. Wake

CHAPTER 1

Introduction: Larval Development, Evolution, and Ecology

BRIAN K. HALL* AND MARVALEE H. WAKE[†]

*Department of Biology, Dalhousie University, Halifax, Nova Scotia, Canada, [†]Department of Integrative Biology and Museum of Vertebrate Zoology, University of California, Berkeley, California

No questions . . . are of greater importance for the embryologist than . . . the secondary changes likely to occur . . . in the larval state.

(Balfour, 1880, p. 381)

I. INTRODUCTION

The life history of many organisms includes a larval stage that is morphologically distinct from the adult and/or that inhabits a different environment from the adult. The aquatic tadpole and the terrestrial frog are prime examples. Other examples are less familiar (Fig. 1). Such species display indirect development; larvae must metamorphose to transform into, or be replaced by, adults.

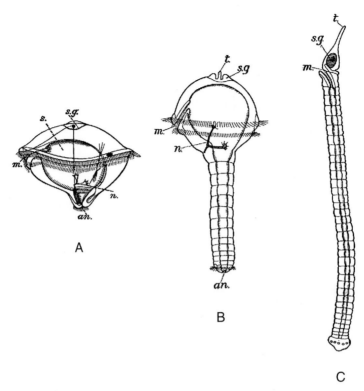

FIGURE 1   The life history of *Polygordius,* a paedomorphic annelid. (A) Trochophore larva with
the prominent ciliary band. (B) A later larval stage in which segments are beginning to form.
(C) A mature stage. Abbreviations: an, anal opening; m, mouth; n, nephridium; s, stomach; s.g.,
brain; t, tentacle. (From Kerr, 1926.)

Larvae and metamorphosis therefore are inextricably linked, a larva being the
"immature form of . . . animals that undergo some metamorphosis" (Oxford
English Dictionary). Despite a long history of research, however, even such
fundamental questions as "what is a larva?" and "what is metamorphosis?"
occupy us today as they occupied naturalists, zoologists, marine biologists,
and evolutionary biologists over a century and a half ago. It therefore seemed
to us that there were ample reasons for revisiting these topics. This volume,
our vehicle for that journey, deals with how vertebrate and invertebrate larvae
develop and have evolved. Each author was asked to address the issue "what is
a larva?" Comparative, functional morphological, physiological, and ecological
aspects are included where appropriate and where data allow. The chapters
are organized into sections that deal in turn with larval types and larval

evolution, mechanisms of larval development and evolution, and larval functional morphology, physiology, and ecology, highlighting four overarching themes: development, evolution, metamorphosis, and genetic mechanisms.

The title of the book is taken from the title of Walter Garstang's presidential address to Section D (Zoology) at the 96th Meeting of the British Association for the Advancement of Science, which was held in Glasgow, September 5–12, 1928. As Garstang reminded the audience in his opening statement, the evolution of larval transformations and metamorphoses had been featured in the presidential addresses of Milnes Marshall in 1890 and L. C. Miall in 1897 (Garstang, 1929). Garstang also could have mentioned the presidential address by Francis Balfour in 1880 on the same theme. During his long career, Garstang fought an energetic fight against Haeckel's biogenetic law—that ontogeny recapitulates phylogeny (Haeckel, 1866; Garstang, 1922)—and against contemporary recapitulationists such as Ernest MacBride, who was driven by notions of progress and for whom invertebrates were degenerate side branches from the central vertebrate line, which led inexorably to man. MacBride summed up his views on the proper ordering of life on earth in the following way: "It is, therefore, broadly speaking true that the invertebrates collectively represent those branches of the Vertebrate stock, which, at various times, have deserted their high vocation and fallen into lowlier habits of life" (1914, p. 662).

Garstang was especially concerned with how larval evolution paralleled adult evolution, "subject to conspicuous deviations" (1929, p. 77). These conspicuous deviations confound any simple parallel between larval and adult evolution and bedevil us still. He championed the view that

- ancestry should be sought in larvae rather than in adults;
- ontogeny creates rather than recapitulates phylogeny; and
- many larval features are secondary adaptations to larval life.

Many of us were introduced to Garstang's thinking through his delightful verses, *Larval Forms*, first published in 1951. A single Garstang verse contains as much erudition on larval development and evolution as is found in far lengthier (and loftier) treatises, as the following examples testify.

> Young Archi-mollusks went to sea with nothing but a velum—
> A sort of autocycling hoop, instead of pram—to wheel 'em;
> And, spinning round, they one by one acquired parental features,
> A shell above, a foot below—the queerest little creatures.
> (*The Ballard of the Veliger or How the Gastropod Got Its Twist*)

> MacBride was in his garden settling pedigrees,
> There came a baby Woodlouse and climbed upon his knees,
> And said: 'Sir, if our six legs have such an ancient air,
> Shall we be less ancestral when we've grown our mother's pair?'
> (*Isopod Phylogeny*)

> And newts Perennibranchiate have gone from bad to worse:
> They think aquatic life a bliss, terrestrial a curse.
> They do not even contemplate a change to suit the weather,
> But live as tadpoles, breed as tadpoles, tadpoles altogether!
> (*The Axolotl and the Ammocoete*)

Encapsulated in these verses are such issues as the relationship between ontogeny and phylogeny, larval adaptations, the transformation from larva to adult, retention of ancestral features, acquisition of new features, life history evolution, and neoteny. They raise major questions such as

- How far do larval and adult evolution run in parallel?
- How can we sort out secondary adaptations (caenogenesis) from primary features?
- How much of larval evolution has been an "escape from specialization," to use the title of a chapter by Sir Alister Hardy (1954) in a book devoted to evolution as a process?

In Garstang's day, many still regarded larval stages as foregone ancestors, evolution proceeding from adult to adult and/or through modifications at the end of ontogeny, a position Garstang roundly rejected. For Garstang, as with Wordsworth, "The Child is father of the Man." In adopting this view, Garstang was far ahead of his time; this was, after all, the man who coined the phrase "ontogeny does not recapitulate phylogeny, it creates it" (Garstang, 1922, pp. 21, 81). Garstang also coined the term "paedomorphosis," using it for the first time in his 1928 presidential address, with the origin of torsion in gastropods as his example. A paedomorphic annelid, *Polygordius,* is shown in Fig. 1.

Garstang saw larvae as a mechanism for dispersal, like the seeds of a plant. He saw secondary reduction of the free-swimming larval stage as a consequence of the adoption of an incubatory mode of development. Cephalopods lack larvae, bypassing the trochopore and veliger larvae and omitting metamorphosis, because the adult is locomotory—a dispersive larval stage is unnecessary. The relationship between larval and adult stages also is amply illustrated in organisms from groups that normally contain a larval stage as the primitive condition, but that have either modified or lost the larva. Such *direct-developing* organisms (e.g., many sea urchins, many frogs and salamanders, some ascidians) hatch as miniature adults, demonstrating the plasticity of early developmental processes and ontogenetic stages, as discussed by Rudy Raff in his chapter.

II. LARVAL ADAPTATIONS AND EVOLUTION

> Whole edifices of invertebrate relationships have been erected on the basis of larval similarities, and complex transformation sequences between dissimilar larvae and between similar larval and adult forms have been devised. (Willmer, 1990, p. 116)

Larvae may well be as ancient as the Metazoa; the oldest known fossil metazoan larvae are from early Cambrian deposits in China and Siberia (Bengtson and Zhao, 1997). Many early workers postulated that larval evolution provides the key to unlocking metazoan evolution and diversification [see Jägersten (1972) and Bowler (1996) for two monographic treatments on the origins of the Metazoa, and Wray (1992) for changes in larval morphology that accompanied echinoderm radiation after the Paleozoic era]. Yet because of larval adaptations (caenogenesis), only some larvae have provided useful information for reconstructing phylogenetic histories and evaluating evolutionary relationships (Strathmann, 1978, 1993). Ernst Haeckel introduced the concept of caenogenesis in 1866 to cover those situations in which recapitulation of phylogeny in ontogeny was obscured because of larval adaptations or the displacement of embryonic or larval stages in time or space during ontogeny. We have retained the term, but modern usage tends to follow de Beer (1958), who restricted caenogenesis to larval adaptations without any reference to recapitulation.

However, Smith (1997) has taken a new approach to larval characters in systematics. He has examined echinoderms using new phylogenies based on several data sets to trace the evolution of life-history strategies and to compare rates and patterns of larval and adult morphological change. Smith agrees that larval characters used exclusively can "mislead phylogenetic analysis" because of caenogenesis, expressed in the form of the extensive homoplasy (convergence) found in nonfeeding larvae through their loss of structures. His examination of larval and adult morphologies together indicates that larval morphology evolves independently of adult morphology, that larval morphology includes considerably more homoplasy than does adult morphology, and that patterns of early development are highly flexible.

Larvae came to prominence because Johannes Müller and other 19th century naturalists identified many types of marine larvae in plankton samples and raised major questions about animal diversity, relationships, and origins. Given that the larvae of marine invertebrates of all phyla display a profound tendency toward convergence, Müller's ability to recognize so many larval types was a staggering accomplishment.

The importance ascribed to larvae in the 19th century can be seen from Francis Balfour's treatment of larvae in a major review (Balfour, 1880), reproduced in his two-volume "Treatise on Comparative Embryology" (1880–1881). Balfour's definition of larvae and his view of their evolutionary importance were straightforward. Larvae are those animals "born in a condition differing to a greater or less extent from the adult" (Balfour, 1880, p. 381, and see the epigraph for this chapter). This line of thinking persisted for many decades.[1]

[1]Kerr (1921, p. 94) defined a larva as "a young developing individual, differing in form from the adult, but *not* contained within the body of the parent or other protective envelope," contrasting larvae with adults and embryos, the latter being "a young developing individual, which is contained within the body of the parent or within a protective shell or other envelope."

In the "Department of Phylogeny" Balfour saw the aim of embryological research as

> (1) To test how far Comparative Embryology brings to light ancestral forms
> common to the whole of the Metazoa.
> (2) How far some special embryonic larval form is constantly reproduced in
> the ontogeny of the members of one or more groups of the animal kingdom; and
> how such larval forms may be interpreted as the ancestral type of those groups.
> (3) How far such forms [larvae] agree with living or fossil forms in the adult
> state; such an agreement being held to imply that the living or fossil form in question
> is closely related to the parent stock of the group in which the larval form occurs.
> (4) How far organs appear in the embryo or larva which either atrophy or
> become functionless in the adult state, and which persist permanently in members
> of some other group or in lower members of the same group. (Balfour, 1880–1981,
> Vol. 1, pp. 4–5)

Balfour saw natural selection operating early in embryonic stages:

> I see no reason for doubting that the embryo in the earliest periods of develop-
> ment is as subject to the laws of natural selection as is the animal at any other
> period. Indeed, there appear to me grounds for the thinking that it is more so.
> (Balfour, 1874, p. 343)

> The principles which govern the perpetuation of variations which occur in
> either the larval or the foetal state are the same as those for the adult state. Variations
> favorable to the survival of the species are equally likely to be perpetuated, at
> whatever period of life they occur, prior to the loss of the reproductive powers.
> (Balfour, 1880, p. 381)

According to Balfour, development that included a larval stage was more
likely to repeat ancestral history than was direct development. Production of
a larva requires that organs be maintained without interruption of function to
allow independent larval existence. Even though secondary larval adaptations
occur, larval development is a closer representation of the evolutionary history
of the group than is the life cycle of a direct-developing species. Direct development appears simpler because it is abbreviated, but direct development is a
secondary modification of a primarily indirectly developing ontogeny, just as
the yolk-free mammalian egg is a secondary modification of a yolk-containing
egg. "There is a greater chance of the ancestral history being lost in forms
which develop in the egg; and masked in those which are hatched as larvae"
(Balfour, 1880, p. 383).

Adam Sedgwick, who wrote the extensive entry on larval forms for the 11th
edition of the "Encyclopædia Britannica" (1911) and who succeeded Balfour
at Cambridge, used Balfour's argument that direct development was more
likely to preserve ancestral features than development without a larval stage,
but came to quite different conclusions with respect to natural selection.
Balfour saw natural selection operating throughout development. For Sedgwick

"embryonic variations are not for the most part acted upon by natural selection, because they concern rudimentary organs only" (Sedgwick, 1894, p. 88).

We now know that natural selection can act separately on larval and adult stages. Consequently, larvae and adults can evolve on different schedules and with considerable independence from each other; witness direct development. Indeed, natural selection acts throughout ontogeny, as Charles Darwin discussed. The role played by natural selection on different parts of the life cycle has attracted considerable interest (Wilbur *et al.*, 1974; Calow, 1983; Mayo, 1983; Nielsen, 1995; Olive, 1985; Roff, 1992, 1996; Stearns, 1980, 1992; Ebenman, 1992; Williams, 1992).

III. LARVAL AND ADULT DEVELOPMENTAL PROGRAMS

Embryos contain cells and developmental programs for both larval and adult structures. These may be completely separate, as in insects (in which adult cells are set aside in imaginal disks within the larval body), or they may be admixed, as in amphibians. We know quite a bit about the former and astonishingly little about the latter. Specific larval structures can be retained as adult structures. As tadpoles metamorphose into frogs, larval jaw muscles are resorbed and new adult jaw muscles develop. In at least some frogs, however, adult muscles are innervated by larval nerves that persist through metamorphosis (Alley, 1989, 1990). Such equivalence or lack of equivalence between larval and adult body parts creates some interesting problems for the identification of homologous structures and for the analysis of serial homology (Cowley, 1991; Minelli and Peruffo, 1991; Minelli, 1996).

In Chapter 6 of this volume, Chris Rose calls attention to the paucity of information on mechanisms responsible for musculoskeletal remodeling during metamorphosis in amphibians. Although changes in rate of production of thyroid hormone have been implicated, we still know little of the hormonal changes associated with the evolution of larvae or of metamorphosis. Similarly, we currently are unable to explain the phylogenetic distribution of muscoskeletal remodeling within amphibians, why larval reproduction (neoteny) is found in salamanders but not in frogs, or how metamorphosis is linked with sexual maturation.

IV. KINDS OF LARVAE

Balfour summarized the evidence for two kinds of larvae: (1) primary larvae as modified ancestral forms that have existed as free larvae "from the time

when they constituted the adult form of the species" (1880, p. 383); and
(2) secondary larvae, introduced secondarily into the life history of a species
that previously developed directly (Fig. 2). Balfour set the primary larva, the
Planula (the ancestral form of coelenterates), apart from all other larval forms,
which he regarded as secondary. Secondary larval adaptations were thought
to arise from changes in larval life or changes in the order of appearance of
structures or to be related to the struggle for existence. Garstang saw secondary
larval characters as mainly anticipating adult characters. Do we still see them
in this light? He viewed primary larvae as "limited to the lower of more
primitive sections of the class" (1929, p. 77). Is this so?

Nowadays, zoologists recognize two fundamental types of invertebrate larvae, corresponding to protostome and deuterostome modes of development
and the superphyla that comprise protostomes and deuterostomes:

- The trochophore (trochosphere) found in animals with a protostome
  mode of embryonic development and spiral cleavage (Fig. 1A). Cladistic

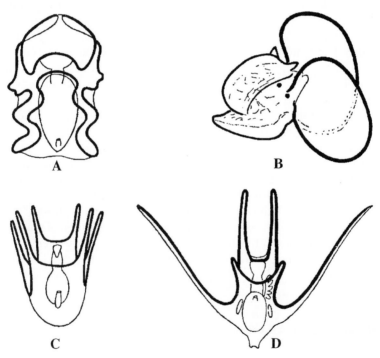

FIGURE 2   Representative larvae of echinoderms (A, C, D) and a gastropod (B). The thick black
line on each larva marks the location of the ciliary band. (Modified from Kerr, 1926.)

analyses of the distribution of larvae are consistent with the trochophore as the type of larva possessed by the last common ancestor of arthropods and chordates (Peterson *et al.*, 1997).

• The dipleurula (pluteus) found in animals with a deuterostome mode of embryonic development. The tornaria larvae found in hemichordates and the auricularia larvae of echinoderms are perhaps the least modified dipleurula larvae [Fig. 2; see Willmer (1990) for a discussion]. The divergence of the deuterostomes may well have started with the transformation of the protostome trochophore larva to the dipleurula larva (Peterson *et al.*, 1997).

The origin and evolution of vertebrate larvae have been considered by many authors, often in the context of the evolution of metamorphosis. A particularly cogent (and little cited) discussion is that of Szarski (1957). He thought it likely that a larval stage was present in the ancestor of all chordates and endorsed Garstang's (1928) view that chordates arose from sessile tunicate ancestors through neoteny of their larvae. Szarski compared fish and amphibian larvae, citing many similarities (e.g., external gills, lateral line system and large neurons in the central nervous system, pronephric kidney, and other features). He explored the significance of direct development in amphibians (but not in fishes, unfortunately) and questioned whether direct development was a "secondary adaptation" or ancestral for amphibians—a question that persists, although most workers maintain that direct development is a derived state (see Hanken's Chapter 3 in this volume). Szarski particularly was interested in the role of thyroxin primarily, but not only, in metamorphosis in fishes and amphibians; he also considered the role of thyroxin in development (presaging some of the work Rose reports in Chapter 6), in reproductive migrations, and in the maternal aspect of gestation in live-bearing taxa. He found that several assumptions about the origin of the Amphibia are not justified (e.g., that amphibians arose from a group of fishes that had increased thyroxin secretion and that larvae and metamorphosis in amphibians are "recent acquirements"). Though all larvae of the "Ichthyopsida" have many similarities, Szarski (1957) recognized that "amphibian larvae . . . have a considerable number of characteristics, which can be considered as comparatively new caenogenetic adaptations. They are most numerous in larvae of Anura."

Szarski (1957) concluded that the ancestors of Amphibia must have had a larval stage, that "a great evolutionary gain" was achieved by amphibians through prolongation of the larval period, acquisition of larval adaptations, and shortening and synchronizing metamorphosis. To this day, the variation in larval period, functions of many larval features (see Cohen, 1984; Kendall *et al.*, 1984), larval endocrinology, and modifications of metamorphosis are not well-understood in fishes or amphibians and provide an arena that might shed light on the origin of chordates.

V. LARVAE AND PHYLOGENY

Do the fundamental features of primary larvae provide any clues to the evolutionary origins of larvae, life histories that include larvae, the adults that derive from those larvae following metamorphosis, or even the Metazoa? Examples from studies on crustaceans and vertebrates illustrate how larvae have influenced approaches to relationships and origins.

Fritz Müller, who believed that he could identify a nauplius stage in all members of the Crustacea, identified the nauplius larva as the crustacean ancestor. He hypothesized that more advanced crustaceans, such as crabs, shrimps, and lobsters, had an additional larval stage that would be expected to appear later in ontogeny and be restricted to the life cycles of those higher forms. The zoea stage that he found in crabs and shrimp fit these expectations precisely: an additional stage inserted into the life cycle between nauplius and adult. Müller's book, "Für Darwin" (1864), which contained his studies on crustacean larvae, was the first explicit test of Darwin's theory of descent with modification. Darwin had the book translated in 1869 under the title "Facts and Arguments for Darwin" (Müller, 1869); see Bowler (1994, 1996) who has analyzed Müller's use of larvae to unravel crustacean relationships and therefore crustacean evolution.

Theories on the ancestors of vertebrates also bear witness to the importance attributed to larvae and larval evolution (Bowler, 1996; Gee, 1996). To cite only one instance: the gill slits of *Balanoglossus* were discovered by Kovalevsky in 1866 (Kovalevsky, 1866, 1867); Metschnikoff (1869) saw the resemblance of *Balanoglossus* and echinoderm larvae to the free-swimming Tornaria larvae, and by the early 1880s *Balanoglossus* was grouped with the echinoderms on the basis of the resemblance of their larvae, shared features of their embryonic development, and lack of segmentation. Garstang (1894, 1896) used these larval findings to propose an entirely novel theory for the origin of the chordates: they had an echinoderm ancestor and arose through neoteny, with the tornarian larva becoming an adult chordate.

Balfour ended his lengthy discussion of larval types with some phylogenetic conclusions. They hinge on the central notion that groups that share a common larval type are "descended from a common stem" (Balfour, 1880, p. 405). Common larval types can be used to deduce the form of the common ancestor for all triploblastic animals, which Balfour and others saw as a radially symmetrical, medusa-like organism.

More recently, Eric Davidson and colleagues have approached metazoan origins by searching for the evolutionary origins of the basic developmental regulatory mechanisms within embryonic and larval metazoans (Davidson, 1990, 1991, 1993; Davidson *et al.*, 1995; Peterson *et al.*, 1997). Davidson molded 19th century classification of larvae as primary and secondary into a modern approach to metazoan origins, because in his view, metazoans initially would have resembled modern, marine, microscopic invertebrate larvae.

Most invertebrate phyla have what Davidson calls type 1 embryogenesis and share common mechanisms for specification of cells and cell lineages (Table I). Clusters of genes encoding cell autonomous regions are expressed from cleavage onward. *lin* (cell *lineage* abnormal) genes in *Caenorhabditis elegans* (*lin-4, lin-14, lin-28,* and *lin-29*) control the "larva-to-adult switch" when adult genes are activated and larval genes repressed (Ambros, 1989). The importance of cell lineages in larval development and evolution is reflected in the two chapters devoted to them in this volume, in which Raff (Chapter 8) and Nagy and Grbic (Chapter 9) discuss cell lineages in echinoderms and in insects, respectively.

Cell lineages in indirect-developing sea urchins that specify only larval cell types are not needed in direct-developing embryos. Rather than being eliminated (a disruptive developmental event), they are respecified to form adult structures. In *Heliocidaris erythrogramma*, a direct-developing Australian sea urchin, for example, respecification occurs as early as the 16-cell stage. Cells that produce mesoderm in indirectly developing species form ectodermal and endodermal cells in the direct-developing species (Wray and Raff, 1990; Wray, 1994).

An interesting and possibly significant correlation is that no species with indirect development sets aside its germ line early in development (Ransick *et al.,* 1996). The germ line often only arises after both embryonic and larval structures are established.

Davidson and his colleagues regard type 1 embryogenesis (and therefore set-aside cells) as having arisen only once, i.e., as being homologous throughout the Metazoa (Peterson *et al.,* 1997). Primary larvae derived from type I embryogenesis are proposed to represent basic metazoan organization. (The embryo-

TABLE I   The Essential Features of Type 1 Embryogenesis as Enumerated by Davidson (1991)

- The type of embryogenesis characteristic of most invertebrate taxa
- Patterns of cleavage are invariant and specify cell lineages
- Spatial organization of embryo and larva primarily is specified by cell lineages
- Cell lineage plays an important role in specification of cell fate so that cell lineage normally is equivalent to cell fate
- Maternal cytoplasmic factors deposited in the ovum during oogenesis are important in specification of cell lineages and therefore cell fate
- Extensive movement of cells within the embryos is not required to specify cells
- One embryonic axis is specified during oogenesis and the second after fertilization
- Regulation (compensation for lost cells or portions of the embryos) is possible
- Selective gene expression occurs early in embryogenesis

genesis of advanced insects and vertebrates is derived and their larvae are secondary. Such larvae are unlikely to yield information pertinent to initial conditions.) The first essential element of Davidson's scenario therefore is that larvae of extant indirect-developing marine invertebrates with type 1 embryogenesis are surrogates of the Precambrian genetic regulatory systems that formed the basis for the evolution of metazoan body plans. The second and third elements are the evolution of set-aside cells and hierarchical, regulatory, developmental programs necessary for the diversification of metazoan body plans. Each lineage of set-aside cells evolved its own hierarchical program of gene expression. Elaboration of upstream regulatory processes within cell lineages or set-aside cells provides the basis for rapid specialization and evolution of novel cells and subsequent morphogenetic diversification. Such approaches demonstrate that the search for the origin and evolution of larval forms is as alive and active as it was 150 years ago.

VI. LARVAE IN CONTEXT

How are larvae to be viewed in functional, community, and ecological contexts?

A. FEEDING

Although most larvae feed, some coelenterate and sponge larvae and tunicate tadpoles are nonfeeding and therefore short-lived. They stand in sharp contrast to the ammocete larvae of lampreys, which can exist for 6 years or more buried in lake sediment during which time organs such as the lens grow exponentially (Hendrix and Rubinson, 1996). In Chapter 10 in this volume, Laurie Sanderson and Sarah Kupferberg discuss larval feeding mechanisms in amphibians and fishes, providing a physiological and ecological perspective to the chapters on the development and evolution of amphibian and fish larvae by Jim Hanken (Chapter 3) and Jackie Webb (Chapter 4), respectively. The authors of these three chapters call for studies that integrate development, physiology, ecology, and evolution and that are conducted against the background of a well-resolved phylogeny. Sanderson and Kupferberg especially emphasize approaches that take into account both ontogenetic changes in individual organs systems and the progressive integration between organ systems that occurs during ontogeny. Hunt von Herbing *et al.* (1996a,b) undertook an analysis of both aspects using larval cod; Hall (1998) discusses the importance of such studies for understanding embryonic and larval development and evolution.

Several authors have developed the theme that heterochrony, accompanied by developmental plasticity, was important in the evolution of nonfeeding larvae [Strathmann *et al.*, 1992; Ponder and Lindberg, 1997; see Hall (1998) for a discussion of heterochrony and the even more important mechanism of heterotopy]. In Chapter 5 in this volume, Mike Hart and Greg Wray critically evaluate heterochrony and question its role in larval evolution. They could find only one example in which heterochrony was responsible for the evolution of a derived larva—the study of Strathmann *et al.* (1992) of feeding sea urchin larvae—and conclude that heterochrony has not been important in larval origins or larval evolution. Their analysis indicates that we have a long way to go in understanding the generative mechanisms underlying larval evolution.

B. Multiple Stages and Plasticity

Whereas a life cycle may consist of a single larva that metamorphoses into the adult, many organisms have multiple larval stages, and some have multiple adult stages. The multiple larval stages of some parasitic insects each are specialized for different aspects of parasitic life. Barnacles may have as many as six nauplius feeding-stage larvae, the last of which transforms into the nonfeeding cypris larva, which settles and metamorphoses into the subadult (Walley, 1969). Insects that display larval heteromorphosis (different larval stages within the life cycle) include blister beetles (family Meloidae) and mantispids (Wilson, 1971). Other insects have different morphological types as adults—worker and soldier ants, for example. The two adult forms of the lancelet *Epigonichthys lucayanum* develop from two different larvae— amphioxides and amphioxus, originally thought to be distinct adult forms, even separate species—with larval features that are adapted for dispersal and nondispersal, respectively (Willey, 1894; Bone, 1957; Gibbs and Wickstead, 1996). Larval polymorphism therefore is an important topic (Hall, 1998), treated in depth by Erick Greene in Chapter 11 of this volume, in which he emphasizes evolutionary causes and consequences of phenotypic plasticity and phenotypic variation. As Greene notes and illustrates in his Fig. 1, colony mates of the Asian ant *Pheidologeton diversus* can vary by as much as 500-fold in dry weight!

One experimental approach to understanding phenotypic plasticity discussed by Greene is to combine experimental manipulation with field studies of natural patterns. Larval patterns in sea urchins can be altered by experimentally manipulating the volume of the egg (Sinervo and McEdward, 1988; McEdward, 1996). Consequences of egg cytoplasm reduction are the slowing of development and the production of smaller, simpler larvae. Egg size affects larval morphology, development and growth rate, feeding capacity, size, stage of maturation–size at metamorphosis, and timing of the metamorphic transition; see Hall (1998) for further discussion and examples.

Possession of a larval stage in the life cycle offers many organisms an opportunity for asexual reproduction. Alternatively, reproductive organs can mature precociously in the larva. When accompanied by failure of metamorphosis, the larva becomes the reproductive stage. The process, neoteny, is especially prevalent among salamanders, as Rose discusses in Chapter 6. Correlations between such life history traits as timing of metamorphosis, size at metamorphosis, attainment of sexual maturity, and development rate are discussed by McLaren (1965), Raff (1992), and McEdward and Janies (1997).

Complex life cycles—those that include multiple stages—have received considerable attention in studies of life history evolution. The heteromorphosis mentioned earlier is one example. Istock (1967) provided an ecological perspective on the evolution of complex life cycles. He characterized such cycles as those that include two or more ecologically distinct phases, without overlap among the factors that limit abundance in each phase. Consequently, each of the phases has its own set of interactions in terms of competition, predation, resource allocation, and environmental factors. Istock (1967) used several life table characters (e.g., age at first reproduction, age at change in life history phase) to evaluate the evolution and maintenance of complex life cycles in diverse invertebrate and vertebrate species, concluding that complex life cycles are inherently unstable over evolutionary time and that such life cycles lead to major changes if extinction does not occur first. He stressed that the evolution of the different phases of a complex life cycle are largely independent and that this independence would make complexity unstable by "moving the population away from maximum realization of the ecological advantages of such a life cycle" (Istock, 1967). Istock (1967) also stated that selective forces that promote the reduction or loss of one phase or another thus would be generated and that such selective forces have generated the diverse characteristics seen in many species that may once have had more complex life cycles (e.g., loss of feeding larvae in echinoderms, paedomorphic salamanders).

Istock's (1967) view of the instability of complex life cycles was challenged by several workers. In reviews of the maintenance of biphasic life cycles in frogs and the modification of the cycles in direct developers, Wassersug (1974, 1975) emphasized that larval ecology is the key to understanding these phenomena and that the specialized morphology and feeding behavior of tadpoles is the key to maintenance of their life cycles. Tadpoles are adapted to specific food resources that typically are temporary, and so they can deal with environmental fluctuations. Wassersug makes the case that living in an environment with marked changes in productivity makes it advantageous to have different morphological phases, so that different resources can be assimilated. He notes that environmental modification, temperature and moisture changes, and seasonality have characterized evolutionary time, so that biphasic life cycles are adaptive and provide stability for many species.

Wilbur *et al.* (1974), Werner (1988), and Ebenman (1992), among others, also challenged Istock's conclusions. Istock (1967) had focused on the several lineages that had secondary losses of phases of their life cycles; Werner (1988) noted that approximately 80% of extant animal species are bi- or multiphasic, suggesting long-term stability of the cycles. Further, the duration of different stages, the timing of metamorphosis, and the concomitant niche shift can be modified without complete loss of a life cycle phase. By using quantitative genetic models to assess the stability of complex life cycles, Ebenman (1992) found that trade-offs in the efficiency of resource utilization by larvae vs adults, caused by different selection on the two phases, select for a disruption of the genetic correlation between juvenile and adult traits so that they evolve independently. Ebenman asks for further research on the evolution of life cycle diversification and phenotypic evolution, especially using mechanistic and dynamic models that include quantitative genetic parameters.

C. LARVAL ECOLOGY

Clearly, larvae must also be viewed in ecological context. Take, for example, the production of soldier ants in *Pheidole bicarinata,* discussed by Fred Nijhout in Chapter 7. The number of soldiers is regulated by both interactions between individuals and the nutritional status of the colony. Adult soldiers release a pheromone that inhibits production of further soldiers by regulating the threshold of the larvae to juvenile hormone, the hormone that triggers development of soldiers. Size of individual soldiers and allometric relationships between soldiers are regulated by programming the growth patterns of imaginal disks. Hormone levels also are sensitive to colony nutritional status. Consequently, a balance of nutritional status and the number of soldiers already present determines whether additional soldiers will be produced (Wilson, 1971; Wheeler and Nijhout, 1983, 1986; Nijhout and Wheeler, 1996). In an analogous way—but where nutrition alone determines transformation from the larval phenotype—the development of larval bees into workers or queen depends on whether or not they are fed the "royal jelly" secreted by the workers' salivary glands. Hormonal control of larval development and evolution in both insects and amphibians is treated in the chapters by Nijhout (Chapter 7) and Rose (Chapter 6). Nijhout emphasizes the significance of the modularity of postembryonic insect development for the independent evolution of body parts and stages of the life cycle.

Some larvae, especially those of insects, can enter a phase of dormancy that permits them (and indeed the population or species) to survive what otherwise would be lethal environmental conditions. Dormancy is broken when conditions improve. Larval variation in many organisms is correlated with patterns of dispersal, involving a wonderfully complex interplay between seasonal signals that trigger larval development, physiological attributes of (marine) larvae such as positive or negative photo- and geotaxis, and larvae–environment interactions that determine both competence for site selection and the particular site(s) selected (Chia and Rice, 1978; Morse, 1991; Fell, 1997). The release, by females, of larvae that feed on phytoplankton within the water column is both coordinated with seasonal blooms of phytoplankton and controlled by phenolic compounds released by the phytoplankton (Starr *et al.*, 1990). This example of what one of us has called environmentally mediated induction (Hall, 1998) represents a major class of evidence in support of Van Valen's (1973) aphorism that evolution is the control of development by ecology and provides yet another reason for revisiting the origin and evolution of larval forms. Our journey ends with a final chapter including an inquiry into the accomplishments of recent research and the questions and research arenas that require attention if we are to more fully understand and appreciate the origin and evolution of larval forms.

ACKNOWLEDGMENTS

It is a pleasure to thank the contributors for their thorough, scholarly, and up-to-date contributions. Some of the work on the volume was undertaken during a sabbatical leave spent by B.K.H. with M.H.W. in Berkeley, CA. B.K.H. thanks Marvalee and the members of the Department of Integrative Biology for their warm and generous welcome and the Miller Institute for Basic Research for a Miller Visiting Research Professorship. Research support from NSERC (Canada) and the Killam Trust of Dalhousie University (B.K.H.) and from NSF (M.H.W.) is gratefully acknowledged.

REFERENCES

Alley, K. E. (1989). Myofiber turnover is used to retrofit frog jaw muscles during metamorphosis. *Am. J. Anat.* **184**, 1–12.

Alley, K. E. (1990). Retrofitting larval neuromuscular circuits in the metamorphosing frog. *J. Neurobiol.* **21**, 1092–1107.

Ambros, V. (1989). A hierarchy of regulatory genes controls a larva-to-adult developmental switch in *C. elegans. Cell (Cambridge, Mass.)* **57**, 49–57.

Balfour, F. M. (1874). A preliminary account of the development of the elasmobranch fishes. *Q. J. Microsc. Sci.* **14**, 323–364.

Balfour, F. M. (1880). Larval forms: Their nature, origin and affinities. *Q. J. Microsc. Sci.* **20**, 381–407.

Balfour, F. M. (1880-1881). "A Treatise on Comparative Embryology," 2 vols. Macmillan, London.

Bengtson, S., and Zhao, Y. (1997). Fossilized metazoan embryos from the earliest Cambrian. *Science* **277**, 1645–1648.

Bone, Q. (1957). The problem of the 'amphioxides' larva. *Nature (London)* **180**, 1462–1464.

Bowler, P. J. (1994). Are the Arthropoda a natural group? An episode in the history of evolutionary biology. *J. Hist. Biol.* **27**, 177–213.

Bowler, P. J. (1996). "Life's Splendid Drama. Evolutionary Biology and the Reconstruction of Life's Ancestry 1860–1940." University of Chicago Press, Chicago.

Calow, P. (1983). "Evolutionary Principles." Blackie, Glasgow and London.

Chia, F.-S., and Rice, M. E., eds. (1978). "Settlement and Metamorphosis of Marine Invertebrate Larvae." Elsevier/North-Holland Biomedical Press, New York.

Cohen, D. M. (1984). Ontogeny, systematics, and phylogeny. In "Ontogeny and Systematics of Fishes" (H. G. Moser, ed.), Am. Soc. Ichthyol. Herpetol., pp. 7–11. Allen Press, Lawrence, KA.

Cowley, D. E. (1991). Genetic prenatal maternal effects on organ size in mice and their potential contribution to evolution. J. Evol. Biol. 3, 363–381.

Davidson, E. H. (1990). How embryos work: A comparative view of diverse modes of cell fate specification. Development (Cambridge, UK) 109, 365–389.

Davidson, E. H. (1991). Spatial mechanisms of gene regulation in metazoan embryos. Development (Cambridge, UK) 113, 1–26.

Davidson, E. H. (1993). Later embryogenesis: Regulatory circuitry in morphogenetic fields. Development (Cambridge, UK) 118, 665–690.

Davidson, E. H., Peterson, K. J., and Cameron, R. A. (1995). Origin of Bilaterian body plans: Evolution of developmental regulatory mechanisms. Science 270, 1319–1325.

de Beer, G. R. (1958). "Embryos and Ancestors." Oxford University Press, Oxford.

Ebenman, B. (1992). Evolution in organisms that change their niches during the life cycle. Am. Nat. 139, 990–1021.

Fell, P. E. (1997). The concept of larvae. In "Embryology: Constructing the Organism" (S. F. Gilbert and A. M. Raunio, eds.), pp. 21–28. Sinauer Assoc., Sunderland, MA.

Garstang, W. (1894). Preliminary note on a new theory of the phylogeny of the Chordata. Zool. Anz. 17, 122–125.

Garstang, W. (1896). The origin of vertebrates. Proc. Cambridge Phil. S. Soc. 9, 19–47.

Garstang, W. (1922). The theory of recapitulation. A critical restatement of the biogenetic law. J. Linn. Soc. London, Zool. 35, 81–101.

Garstang, W. (1928). The morphology of the Tunicata and its bearing on the phylogeny of the Chordata. Q. J. Microsc. Sci. 72, 51–54.

Garstang, W. (1929). The origin and evolution of larval forms. British Association for the Advancement of Science Rep. 96th Meet. pp. 77–98 (reprinted in Garstang, 1985).

Garstang, W. (1951). "Larval Forms and Other Zoological Verses," with an Introduction by Sir Alister Hardy. Basil Blackwell, London.

Garstang, W. (1985), "Larval Forms and Other Zoological Verses," with a Foreward by M. La Barbara (reprint). University of Chicago Press, Chicago.

Gee, H. (1996). "Before the Backbone: Views on the Origin of the Vertebrates." Chapman & Hall, London.

Gibbs, P. E., and Wickstead, J. H. (1996). The myotome formula of the lancelet Epigonichthys lucayanum (Acrania): Can variations be related to larval dispersal patterns? J. Nat. Hist. 30, 615–627.

Haeckel, E. (1866). "Generelle Morphologie der Organismen: Allgemeine Grundzüge der organischen Formen-Wissenschaft, mechanisch begründet durch die von Charles Darwin reformite Descendenz-Theorie," 2 vols. Georg Reimer, Berlin.

Hall, B. K. (1998). "Evolutionary Developmental Biology: Embryos in Evolution and the Integration of Developmental and Evolutionary Biology," 2nd ed. Chapman & Hall, London.

Hardy, A. C. (1954). Escape from specialization. In "Evolution as a Process" (J. Huxley, A. C. Hardy, and E. B. Ford, eds.), pp. 122–142. Allen & Unwin, London.

Hendrix, R. W., and Rubinson, K. (1996). Cell growth patterns and lens geometry: A quantitative study from three-dimensional reconstructions. Tissue Cell 28, 473–484.

Hunt von Herbing, I., Miyake, T., Hall, B. K., and Boutilier, R. G. (1996a). The ontogeny of feeding and respiration in larval Atlantic cod, *Gadus morhua* (Teleostei; Gadiformes): (1) Morphology. *J. Morphol.* **227**, 15–35.

Hunt von Herbing, I., Miyake, T., Hall, B. K., and Boutilier, R. G. (1996b). The ontogeny of feeding and respiration in larval Atlantic cod, *Gadus morhua* (Teleostei; Gadiformes): (2) Function. *J. Morphol.* **227**, 37–50.

Istock, C. A. (1967). The evolution of complex life cycle phenomena: An ecological perspective. *Evolution (Lawrence, Kans.)* **21**, 592–605.

Jägersten, G. (1972). "Evolution of the Metazoan Life Cycle." Academic Press, London.

Kendall, A. W., Ahlstrom, E. H., and Moser, H. G. (1984). Early life history stages of fishes and their characters. *In* "Ontogeny and Systematics of Fishes" (H. G. Moser, ed.), Am. Soc. Ichthyol. Herpetol., pp. 11-22. Allen Press, Lawrence, KA.

Kerr, A. G. (1926). "Evolution." Macmillan, London.

Kerr, J. G. (1921). "Zoology for Medical Students." Macmillan, London.

Kovalevsky, A. O. (1866). Anatomie des Balanoglossus. *Mém. Acad. Sci. St. Petersbourg* **7**, 1–10.

Kovalevsky, A. O. (1867). Entwickelungsgeschichte des *Amphioxus lanceolatus*. *Mém. Acad. Sci. St. Petersbourg* **11**(4), 1–17.

MacBride, E. W. (1914). "Textbook of Embryology" Vol. 1. Macmillan, London.

Mayo, O. (1983). "Natural Selection and Its Constraints." Academic Press, London.

McEdward, L. R. (1996). Experimental manipulation of parental investment in echinoid echinoderms. *Am. Zool.* **36**, 169–179.

McEdward, L. R., and Janies, D. A. (1997). Relationships among development, ecology, and morphology in the evolution of echinoderm larvae and life cycles. *Biol. J. Linn. Soc.* **60**, 381–400.

McLaren, I. A. (1965). Temperature and frogs eggs. A reconsideration of metabolic control. *J. Genet. Physiol.* **48**, 1071–1079.

Metschnikoff, E. (1869). Uber ein Larvenstadium von *Euphasia*. *Z. Wiss. Zool.* **19**, 479–481.

Minelli, A. (1996). Some thoughts on homology 150 years after Owen's definition. *Mem. Soc. Ital. Sci. Nat. Mus. Civ. Storia Nat. Milan* **27**, 71–79.

Minelli, A., and Peruffo, B. (1991). Developmental pathways, homology and homonomy in metameric animals. *J. Evol. Biol.* **3**, 429–445.

Morse, A. N. L. (1991). How do planktonic larvae know where to settle? *Am. Sci.* **79**, 154–167.

Müller, F. (1864). "Für Darwin." Engelmann, Leipzig.

Müller, F. (1869). "Facts and Arguments for Darwin," with additions by the author, translated from the German by W. S. Dallas. John Murray, London.

Nielsen, C. (1995). "Animal Evolution: Interrelationships of the Living Phyla." Oxford University Press, Oxford.

Nijhout, H. F., and Wheeler, D. E. (1996). Growth models of complex allometries and evolution of form: An algorithmic approach. *Syst. Zool.* **35**, 445–457.

Olive, P. J. W. (1985). Covariability of reproductive traits in marine invertebrate, implications for the phylogeny of the lower invertebrates. *In* "The Origins and Relationships of Lower Invertebrates" (S. Conway Morris, J. D. George, R. Gibson, and H. M. Platt, eds.), pp. 42–59. Clarendon Press, London.

Peterson, K. J., Cameron, R. A., and Davidson, E. H. (1997). Set-aside cells in maximally indirect development: Evolutionary and developmental significance. *BioEssays* **19**, 623–631.

Ponder, W. F., and Lindberg, D. R. (1997). Towards a phylogeny of gastropod molluscs: An analysis using morphological characters. *Zool. J. Linn. Soc.* **119**, 83–265.

Raff, R. A. (1992). Direct-developing sea urchins and the evolutionary reorganization of early development. *BioEssays* **14**, 211–218.

Ransick, A., Cameron, R. A., and Davidson, E. H. (1996). Postembryonic segregation of the germ line in sea urchins in relation to indirect development. *Proc. Natl Acad. Sci. U.S.A.* **93**, 6759–6763.

Roff, D. A. (1992). "The Evolution of Life Histories: Theory and Analysis." Chapman & Hall, New York.

Roff, D. A. (1996). The evolution of threshold traits in animals. *Q. Rev. Biol.* **71**, 3–35.

Sedgwick, A. (1894). On the law of development commonly known as von Baer's law; and on the significance of ancestral rudiments in embryonic development. *Q. J. Microsc. Sci.* **36**, 35–52.

Sedgwick, A. (1911). Larval forms. In "Encyclopædia Britannica," Vol. 16, pp. 224–228. Encyclopædia Britannica, New York.

Sinervo, B., and McEdward, L. R. (1988). Developmental consequences of an evolutionary change in egg size: An experimental test. *Evolution (Lawrence, Kans.)* **42**, 885–899.

Smith, A. B. (1997). Echinoderm larvae and phylogeny. *Annu. Rev. Syst. Evol.* **28**, 219–241.

Starr, M., Himmelman, J. H., and Therriault, J. C. (1990). Direct coupling of marine invertebrate spawning with phytoplankton blooms. *Science* **247**, 1071–1074.

Stearns, S. C. (1980). A new view of life-history evolution. *Oikos* **35**, 266–281.

Stearns, S. C. (1992). "The Evolution of Life Histories." Oxford University Press, Oxford.

Strathmann, R. R. (1978). The evolution and loss of feeding larval stages of marine invertebrates. *Evolution (Lawrence, Kans.)* **32**, 894–906.

Strathmann, R. R. (1993). Hypotheses on the origins of marine larvae. *Annu. Rev. Ecol. Syst.* **24**, 89–117.

Strathmann, R. R., Fenaux, L., and Strathmann, M. F. (1992). Heterochronic developmental plasticity in larval sea urchins and its implications for evolution of nonfeeding larvae. *Evolution (Lawrence, Kans.)* **46**, 972–986.

Szarski, H. (1957). The origin of the larva and metamorphosis in Amphibia. *Am. Nat.* **91**, 283–301.

Van Valen, L. (1973). Festschrift. *Science* **180**, 488.

Walley, L. J. (1969). Studies on the larval structure and metamorphosis of *Balanus balanoides* (L.). *Philos. Trans. R. Soc. London, B Ser.* **256**, 237–280.

Wassersug, R. J. (1974). Evolution of anuran life cycles. *Science* **185**, 377–378.

Wassersug, R. J. (1975). The adaptive significance of the tadpole stage with comments on the maintenance of complex life cycles in anurans. *Am. Zool.* **15**, 405–417.

Werner, E. E. (1988). Size, scaling and the evolution of complex life cycles. In "Size-structured Populations: Ecology and Evolution" (B. Ebenman and L. Persson, eds.), pp. 68–81. Springer, Berlin.

Wheeler, D. E., and Nijhout, H. F. (1983). Soldier determination in ants: New role for juvenile hormone. *Science* **213**, 361–363.

Wheeler, D. E., and Nijhout, H. F. (1986). Soldier determination in the ant *Pheidole bicarinata:* Inhibition by adult soldiers. *J. Insect Physiol.* **30**, 127–135.

Wilbur, H. M., Tinkle, D. W., and Collins, J. P. (1974). Environmental certainty, trophic level, and resource availability in life history evolution. *Am. Nat.* **108**, 805–817.

Willey, A. (1894). "Amphioxus and the Ancestry of the Vertebrates." Macmillan, New York and London.

Williams, G. C. (1992). "Natural Selection: Domains, Levels and Challenges." Oxford University Press, Oxford.

Willmer, P. (1990). "Invertebrate Relationships: Patterns in Animal Evolution." Cambridge University Press, Cambridge, UK.

Wilson, E. O. (1971). "The Insect Societies." Harvard University Press, Cambridge, MA.

Wray, G. A. (1992). The evolution of larval morphology during the post-Paleozoic radiation of echinoids. *Paleobiology* **18**, 258–287.

Wray, G. A. (1994). The evolution of cell lineage in echinoderms. *Am. Zool.* **34**, 353–363.

Wray, G. A., and Raff, R. A. (1990). Novel origins of lineage founder cells in the direct-developing sea urchin *Heliocidaris erythrogramma. Dev. Biol.* **141**, 41–54.

PART **I**

Larval Types and Larval Evolution

CHAPTER 2

Larvae in Invertebrate Development and Evolution

CAROLE S. HICKMAN

Department of Integrative Biology and Museum of Paleontology, University of California, Berkeley, California

I. Introduction
  A. Invertebrate Larvae: A Structural Perspective
  B. Other Invertebrate Larval Perspectives
II. What Is a Larva?
  A. The Problem of Definition
  B. Three Conflicting Definitions and a Resolution
  C. Primary and Secondary Invertebrate Larvae
  D. Stages in Larval Development
III. The Concept of Larval Type and the Diversity of Marine Invertebrate Larvae
  A. Larval Types
  B. Structural Diversity of Larval Types
  C. Developmental and Ecological Diversity of Larval Types
IV. Polytypy and Inducible Traits in Larval Morphology
  A. Phenotypic Plasticity of Larval Morphology
  B. Plasticity and Flexibility in Developmental Programs
V. Origins of Marine Invertebrate Larvae
  A. The Nature of Questions
  B. Two Questions about Larval Origins
VI. Invertebrate Larvae and the Paleontological Record
  A. Evidence from First Appearances of Adults

The Origin and Evolution of Larval Forms

I. INTRODUCTION

A. INVERTEBRATE LARVAE:
A STRUCTURAL PERSPECTIVE

The larvae of marine invertebrates, like all larvae, are intricate functioning organisms in their own right. Their anatomical complexity and structural and functional diversity are all the more remarkable because marine larvae are so minute (typically ≤1 mm). Fascination with the beauty and diversity of larval forms is reflected in the broad range of empirical, comparative, and theoretical perspectives within the field of invertebrate larval biology.

The perspective of this chapter predominantly is structural and comparative. It asks what marine invertebrate larvae have to offer the ongoing synthesis of evolutionary and developmental biology when they are viewed as structurally and functionally complicated organisms, selected to operate efficiently in their own distinctive milieu. Other invertebrate larval perspectives, particularly those focusing on larvae as dispersive phases of life history, have generated an immense body of empirical data and theory, such that even a review of review articles and volumes would be a daunting task.

My interest primarily is in the staggering diversity and antiquity of invertebrate larval forms and the correspondingly considerable evolution of structure, function, and development that has occurred within the larval sphere of biphasic life history. The major themes of the chapter will emerge from a structural answer to the question "what is a larva?" They include the need for (1) more "new-fashioned" comparative structural data capable in their detail of unmasking convergence; (2) more rigorous phylogenetic analyses of the relationships of larval forms; (3) more full exploitation of the fossil record for direct evidence of the evolution of larval diversity; (4) more empirical and experimental studies of larval adaptations; and (5) greater attention to the phenotypic variability in populations and the plasticity and flexibility in individuals as factors promoting the evolution of larval diversity.

B. OTHER INVERTEBRATE LARVAL PERSPECTIVES

Depending upon the question, marine invertebrate larvae are defined differently and assume strikingly different roles in evolutionary and developmental studies. These include the idea that larvae exist for dispersal (Scheltema, 1977, 1986), aptly summarized by Hart and Strathmann (1995, p. 194) as "packages of genes dispersed away from each other and away from the parents." Alternatively, the larva is something waiting to happen, a sensory package designed to detect and respond to the cues that will insure recruitment at the right place at the right time. Larvae also can be characterized as "devices for turning smaller eggs into larger juveniles" (Hart and Strathmann, 1995, p. 194) or as "machines for feeding and growing" (Strathmann, 1987, p. 467). In the latter two perspectives, the role of the larva is to decrease parental investment per offspring or, alternatively, to increase its size or energy reserves at metamorphosis. For larval ecologists, the larva is a component of a larger life-history strategy. For developmental biologists, a contrasting view is that "the larva serves only as a life support system for the imaginal set-aside anlagen within which the adult body plan develops" (Davidson *et al.*, 1995, p. 1323).

The strict Haeckelian perspective that larvae are snapshots of ancestors has been laughed out of biology. But it has been variously resurrected and rephrased to suggest that larvae may retain in their development many of the features that once existed in the adults of the ancestors but subsequently were lost in the adults of the descendants. Indeed, one might suspect that it will be the kernel in the next iteration of the relationship between larval development and evolution.

Some larval perspectives, such as the notion that larvae are passively transported particles at the mercy of ocean currents, are correct at a large spatial scale, but not at the finer scale of vertical migration, orientation, and feeding within a water mass. Finally, from those perspectives that emphasize the adult, there is a mythology that larvae are all roughly spherical and lacking in complexity. The myth unfortunately is perpetuated in journalistic accounts of otherwise important discoveries in which larvae are characterized as "tiny, simple assemblages of cells" (Kerr, 1998, p. 804) or in which the first metazoans are characterized as "squishy larva-like things" (Li *et al.*, 1998, p. 809).

When is a larva more than a dispersal mechanism, a feeding machine, or a life-support system for set-aside cells? The next section sets up the structural definition essential to understanding larvae as organisms in their own right, with combinations of shared and uniquely derived traits and developmental patterns that have as much independent evolutionary history as their corresponding adults.

II. WHAT IS A LARVA?

A. THE PROBLEM OF DEFINITION

From an invertebrate perspective, there are two reasons to ask this question. The first has to do with the difficulty of achieving a precise definition, one that will permit unequivocal recognition across a broad structural diversity of invertebrate larval forms and varying degrees of discrete or gradual transformation during development. In addressing this question, McEdward and Janies (1993, pp. 259–260) stated that "in spite of numerous attempts, no one has provided a precise and generally accepted definition of the term 'larva'," and concluded that such a definition "is impossible." Likewise, Strathmann (1993, p. 90) considered the distinction between a larva and an embryo on one hand and a juvenile on the other to be "at some point arbitrary." The first problem is one of recognizing boundaries. The second reason for raising the question is that structural, ecological, and morphogenetic definitions do not coincide; a form may be a larva by one definition and not by the others.

B. THREE CONFLICTING DEFINITIONS AND A RESOLUTION

1. Structural Definition

In order to discuss the evolution of larvae and larval development, this chapter adopts a structural definition. The larva is a structural state or series of states that occurs between the onset of the divergent morphogenesis following embryonic development (cleavage, blastula, gastrula) and metamorphosis to the adult body plan. The larva is a mixture of transitory features and adaptations to larval life that will be lost at metamorphosis, larval features that will be remodeled for preadult and adult function at metamorphosis, and features of the adult that begin their development prior to metamorphosis. The specialized transitory features are crucial to the recognition of larvae, and recognition can be difficult in highly evolved developmental patterns in which larval adaptations have undergone considerable reduction. However, as long as morphogenesis continues to generate larval structure, and as long as that structure is lost during metamorphic transformation, there is a larva. Direct development becomes an evolutionarily derived reality when larval structures no longer are produced during ontogenesis.

2. Ecological Definition

The structural definition is in conflict with the standard ecological definition of a larva. In an ecological sense, the larva is a free-living life-history stage. The developing organism is a larva only if it hatches and passes through a pelagic phase. The larval ecologist views the larva as a dispersive phase of the life history. If the organism develops in a benthic egg mass or capsule, it is still an embryo *even if it has all the structural features of a larva*. One problem with this definition is that it does not differentiate between benthic development in which larval morphogenesis occurs and benthic development in which the organism never achieves any of the elements of the larval body plan because they have been eliminated (the strict definition of direct development). For the ecologist, a gastropod embryo with a fully developed ciliated velum is no different from a gastropod embryo that fails to develop a velum. For ecologists who violently object to this definition, I suggest at least modifying "embryo" by reference to its corresponding larval stage (e.g., "trochophore-stage embryo," "veliger-stage embryo").

In order to reconstruct the evolutionary history of larval forms, we need information about structures that are available as targets of selection during development. If a velum is present along with other transitory structures characteristic of a gastropod veliger, then it is a veliger larva, regardless of whether it is hatched and swimming in the plankton or "swimming" in intracapsular fluid on the bottom. An example developed in Section VII.B will illustrate how selection has modified planktonic larval structures for different intracapsular functions. The co-option of larval structures for novel functions is an important aspect of larval evolution that will be missed if what happens in the capsule is assumed to be direct development.

Larval ecologists have recognized that specialized structural features are essential to the dispersive and nutritional roles of both planktotrophic and pelagic lecithotrophic larvae. In Chia's (1974, p. 122) definition, "larva is a developmental stage, occupying the period from post embryonic stage to metamorphosis, and differs from the adult in morphology, nutrition, or habitat." Through the use of "or" we are given three alternative criteria. The difficulty with this definition for an evolutionary biologist, however, is not the use of "or" but rather the definition of postembryonic stage as the "time after emerging (hatching) from the primary egg membrane."

3. Morphogenetic Regulatory Definition

The third definition of a larva strictly is morphogenetic: it is the premetamorphic consequence of Type 1 embryogenesis, a type of specification of cell fates that is fundamentally different in invertebrate larvae until metamorphosis (Davidson, 1991; Davidson *et al.*, 1995; Hall and Wake, Chapter 1 of this volume, Table I). This larval developmental regulatory program builds larval structure as well as a field of cells that are set aside as the imaginal rudiment or progenitor field from which the adult body plan is generated. Like the structural definition, it is in conflict with the ecological definition if the adult hatches at metamorphosis and never experiences a pelagic phase. The morphogenetic regulatory definition requires two independent sets of developmental programs. It most readily applies to maximally indirect-developing larvae such as those of echinoderms, in which there is a massive collapse of larval tissues at metamorphosis. It is not equally applicable to all invertebrate larvae. It is in serious conflict with the structural definition if virtually all cells are involved in the production of adult structure, even if the structural reorganization at metamorphosis is profound. This would appear to be the case, for example, in gastropods, where only the velum is lost and all remaining cells of the larva are involved in the proliferation of adult structure. In the sense of Davidson *et al.* (1995), gastropods and many other invertebrate groups could be viewed as direct-developing (or perhaps "minimally indirect-developing"), even though there is a sudden and profound event that transforms the swimming veliger into a benthic juvenile snail.

C. PRIMARY AND SECONDARY INVERTEBRATE LARVAE

The concept of primary and secondary larvae (*sensu* Jägersten, 1972, pp. 8–9) is theoretical: a hypothetical explanation of two alternative evolutionary pathways to a biphasic life history. A secondary larva is a life stage that has assumed the character of a primary larva through modification of the juvenile or adult stage in a species after it has evolved direct development. In developmental terms, a secondary larva may be something else: an interpolated stage in development in which there is a second set of set-aside cells (Davidson *et al.*, 1995). The concept is not fruitful in generating a fundamental structural and comparative understanding of larval phenotypic and functional diversity and will not be discussed further.

D. Stages in Larval Development

The concept of *stage* in larval development is fruitful for organizing both information about structural traits and information about the order (sequence of events) and timing (onset and completion relative to a fixed point) of changes in organization. Many invertebrate groups have more than one kind of structurally distinct larva in their life history. For example, annelids, sipunculids, and mollusks all differentiate postembryonically into a trochophore larva. In annelids, prior to metamorphosis, ectodermal and mesodermal teloblasts divide to generate an elongate segmented trunk. In sipunculids, the characteristic pelagosphera larva is defined by Rice (1967) as a product of *metamorphosis* of the trochophore. In gastropods the trochophore is transformed gradually into the characteristic veliger larva. These later stage larvae are very much more complicated in their structure, function, and behavior than the preceding trochophores. Staging is rich with information content for phylogenetic analysis and understanding of evolutionary transformations (Hickman, 1992, 1995).

Stages exist in the mind of the biologist, not in the larva. All that is needed in order to recognize a stage is a fixed starting and finishing point. Naef (1919) articulated this clearly when he wrote, "We comprehend ontogenesis by fixing a series of momentary pictures or 'stages' out of an actually infinite number." If larval development were documented by time-lapse photography, there would be long series of frames that would be nearly identical, separated by short series of frames that are very different from one another. Hickman (1995) has shown that metamorphosis from larva to juvenile in gastropods can be staged if the events are recorded on an appropriate time scale. This is discussed further as an example in Section VIII.A.

The number of larval stages we can recognize and name is not of great significance. It is the ordering of events that is important, both for characterization of development and for comparative evolutionary studies, where this information can be treated as "sequence data" for phylogenetic reconstruction.

III. THE CONCEPT OF LARVAL TYPE AND THE DIVERSITY OF MARINE INVERTEBRATE LARVAE

A. Larval Types

Invertebrate larvae are notoriously diverse in their structure, development, and ecology. Classifications of larvae by structural, developmental, and ecological type can be dangerous in evolutionary studies for several reasons. First, similarity of type is not a statement of homology. Typology is useful in naming, recognizing, and sorting entities, but as a means of dealing with similarity it does not necessarily distinguish between homology and homoplasy. Second, typology idealizes structure and function and fails to capture the variability, flexibility, and plasticity in larval life histories (*sensu* Hadfield and Strathmann, 1996) that are crucial to an evolutionary perspective. Nonetheless, classifications of larval types figure prominently in developmental and evolutionary larval biology, both in the organization of comparative data and in the framing of hypotheses.

Structural, developmental, and ecological classifications of larval types all provide some measure of invertebrate larval diversity as a product of evolution, and all three have been sources of major evolutionary hypotheses and theories.

B. STRUCTURAL DIVERSITY OF LARVAL TYPES

There is no classification scheme that comes close to capturing the structural diversity of marine invertebrate larvae. A simple, dichotomous structural typology based on patterns of ciliary banding is useful for distinguishing planktotrophic protostomes (downstream-collecting bands with compound cilia on multiciliate cells) and deuterostomes (upstream-collecting bands with separate cilia on monociliate cells). However, these two conceptual types (*trochophora* and *dipleurula*, respectively) do not include letithotrophic (nonfeeding) larvae or forms such as the planktotrophic (feeding) ectoproct cyphonautes larva with its uniquely different ciliary bands. There is an immense and diffuse literature filled with narrative descriptions and illustrations of invertebrate larvae and larval types, but the names do not correspond to explicit sets of uniquely derived characters and character states. The names of invertebrate larvae primarily serve to call to mind images of disparate plans of organization. Some of the names refer to distinctive stages within the ontogeny of a single taxon, whereas others refer to a structural type that is relatively constant throughout larval ontogeny.

Perhaps the best way to convey the flavor of the diversity and disparity of invertebrate larval types is through a sampling of names, from A to Z: *actinotroch, actinula, amphiblastula, auricularia, benthosphaera, bipinnaria, brachiolaria, calyconula, cerinula, chaetosphera, conaria, coronate, cydippid, cyphonautes, cyprid, Desor's, doliolaria, echinopluteus, echinospira, Edwardsia, Götte's, halcampoides, Higgins', Iwata's, Lovén's, Luther's megalops, meraspis, metanauplius, metatrochophore, microniscid, mitraria, Müller's, nauplius, nectochaeta, ophiopluteus, parenchymella, pelagosphaera, pentacula, pericalymma, pilidium, planktosphaera, planula, pluteus, protaspis, protozoea, pseudocyphonautes, rataria, rostraria,*

Schmidt's, scyphistoma, siphonula, tornaria, trichimella, trochophore, veliconcha, veliger, vesiculariform, vitellaria, zoanthina, zoanthella, zoea. Only two of the names in this list are larval types known only from fossils: *meraspis* and *protaspis* are larval trilobite stages.

These names seriously underestimate larval diversity, because one name may apply to thousands of different combinations of uniquely derived character states. There is both room and need for more rigorous comparative study and documentation of larval structure and structural diversity.

Larval form and structure are variously illustrated in the primary literature and textbooks with a combination of simplified diagrams, rendered drawings, light micrographs, and scanning electron micrographs. Because biologists tend to specialize at the phylum level or lower in the taxonomic hierarchy, illustrations may be difficult to compare from one major group to another due to differences in orientation and vantage point as well as differences in the form of imagery. Some of the best comparisons can be made from drawings and scanning electron micrographs that show only exterior structure. Two sources of excellent drawings of a diversity of marine invertebrate larvae are Smith (1977) and Nielsen (1995). For the most extensive compilations of scanning electron micrographs of larvae, as well as excellent drawings and light micrographs, see Nielsen (1987) and Gilbert and Raunio (1997).

For most of the invertebrate phyla, larval structure is known primarily from intensive study of one or a few model species. One of the best systematic treatments of the range of structural variation within a phylum is that of Zimmer and Woollacott (1977) for gymnolaemate (Ctenostomta and Cheilostomata) bryozoans. Although the authors acknowledge that their larval categories may not be monophyletic and that knowledge of larval structure is limited to a small fraction of described species, they nonetheless provide detailed comparative structural data for many species. The format is one that can be modified and used to generate phylogenetic hypotheses once homoplasies are detected and eliminated. Their classification first distinguishes larvae according to presence or absence of a shell and then subdivides the shelled larvae into shelled planktotrophs (cyphonautes) and shelled lecithotrophs (unnamed larval type). They divide the unshelled, coronate larvae into five types. The other phyla for which there are similar levels of comparative data on larval structure, although less well-organized, are mollusks, annelids, and echinoderms.

Larval structural variation is relatively low in gastropod trochophore larvae, but incredible diversity is manifest in the succeeding veliger larval stage, both in the soft parts and in the features of the larval shell. Variation is especially apparent among the relatively long-lived planktotrophic veligers of the Caenogastropoda. Not only are there suites of structural features distinguishing veliger larvae at the level of family, but in most cases individual species have distinctive combinations of character states (C. S. Hickman, unpublished observation). Unique shell geometries and patterns of biomineralization, and the complex episodic patterns of reorganization of biomineralization at meta-morphosis, are preserved as detailed chronological records on larval shells (Hickman, 1995).

For some major taxa no name is assigned to the larval form, although it can be illustrated and recognized intuitively as distinct from the many named larval forms. For example, there is no name for the distinctive larval form of priapulids or for the equally distinctive larva of articulate brachiopods. There also is no name for the shelled larval form of inarticulate brachiopods, which bears little resemblance to the articulate brachiopod larva.

Different structural types may occur within larvae of a single species in different parts of its geographic range. Schroeder and Hermans (1975) discuss this phenomenon in the metatrochophore larvae of the polychaete *Polygordius lacteus* in the North Sea, where the forming segments are folded up within an enlarged region (the serosa) of the body posterior to the metatroch. This form is known as an *endolarva*. In contrast, metatrochophores of the same species in the Mediterranean form segments that trail posteriorly as a long appendage. This form is known as an *exolarva*.

Names are omitted from the preceding list for a variety of reasons. Some names (e.g., *megalops, copepodite* larva) actually are early juvenile stages. Still others (e.g., *phagocytella, trochophora, dipleurula*) are hypothetical constructs or archetypes—hopeful ancestors.

Finally, there are larval "types" that represent a grade of evolution. For example, the *atrochal* type is one that has uniform ciliation instead of distinct ciliary bands. Atrochal larvae are known from most of the invertebrate phyla, and the structural grade is correlated with a nonfeeding ecology (Nielsen, 1995). Another example is what Jägersten calls the *test-type* larva. This is a larval type that occurs in protobranch bivalves and some aplacophorans. In both of these groups the larva, which is planktic but nonfeeding, forms a temporary ectodermal envelope (the *test envelope*) around itself. In spite of overall structural similarities in the test envelope, there are striking differences. In protobranchs the envelope is shed at metamorphosis, and in aplacophorans it is incorporated into the anterior end of the juvenile body. The different patterns of locomotor ciliary bands on the test further suggest that the envelopes are not homologous.

The two expanded ciliated lobes comprising the mollusk velum usually are considered to be uniquely derived for larval swimming and feeding. However, the velum may represent a structural grade of organization within the phylum. A velum occurs only in larvae of the classes Gastropoda and Bivalvia, which are not sister taxa (Fig. 1). The alternative to two separate origins of this larval structure is the loss of a velum in the other three conchiferan classes, Monoplacophora, Cephalopoda, and Scaphopoda. The best resolution of these

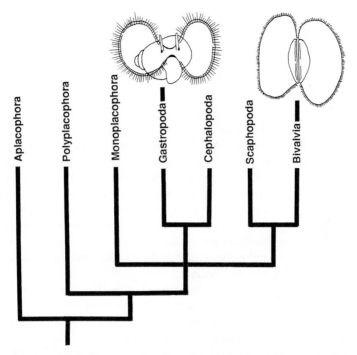

FIGURE 1 Phylogenetic hypothesis of the relationships of the major classes of Mollusca. Swimming larvae with a "velum" occur only in the Gastropoda and Bivalvia, suggesting that this structure is derived independently in the two classes or, alternatively, has been lost in other classes.

alternatives lies in more detailed comparison of velar structure in gastropods and bivalves.

The gastropod larval type known as the *sinusigera* has an interesting history. The name derives from what was once the genus *Sinusigera,* before it was recognized that the characteristic shell form on which the name was based was a larval shell and not the shell of an adult microgastropod. The sinusigera type of larval shell has a distinctive apertural margin, with two deep embayments separated medially by a projecting "beak." We now know that this larval shell form occurs in many caenogastropods and that the two sinusigeral notches accommodate the lobes of the velum, whereas the beak projects forward over the larval head (see Hickman, 1995). Careful examination of the position, shape, and other special features of the sinusigeral notches and beak and the distribution of these features on a phylogenetic tree derived from adult features suggests that sinusigeral apertures have evolved many times. The reward of more detailed comparative study of complex structure invariably is the unmasking of convergence.

C. Developmental and Ecological Diversity of Larval Types

Classification of invertebrate larvae by larval ecologists has focused on what are called "developmental modes." In a series of landmark papers, Thorson (1936, 1946, 1950) established the ecologically based classification and terminology of types of invertebrate larval development used by larval ecologists today. Thorson's classification has been amended, modified, reevaluated, and reviewed many times. Mileikovsky (1971) provides one of the most thoughtful reevaluations. Young (1990) places classificatory efforts in the larger perspective of the historical development of the field of larval ecology. The most comprehensive review and critique of classifications of invertebrate larval development is that of Levin and Bridges (1995).

The first criticism of these classifications is that they are cast in terms of simple dichotomies that underestimate developmental diversity (Hickman, 1992, p. 263). The main dichotomies include planktotroph vs nonplanktotroph, planktic vs benthic, planktic vs nonplanktic, and short-term vs long-term. Hadfield and Strathmann (1996) coined the term "polytypy" to include the different ways in which larval developmental features can be expressed in multiple states; the argument for greater attention to polytypic diversity is developed in Section IV.

A second criticism is that these schemes lack a single basis of organization. Levin and Bridges (1995) addressed this criticism with a new classification consisting of four categories: (1) mode of larval nutrition, (2) location or site of development, (3) dispersal potential, and (4) type of morphogenesis. This classification and its subdivisions, which are summarized in Table I, provide all the elements of a multidimensional ecodevelopmental design space (*sensu* Hickman, 1993) and might be used most profitably to examine patterns of occupation and/or to plot phylogenetic trajectories.

Larval developmental types may share suites of similar structural features as a consequence of convergence. For example, the *teleplanic* larval type (Scheltema, 1971; from the Greek "teleplanos" or "far-wandering") has evolved independently in a number of caenogastropod clades. Teleplanic larvae spend a very long time (6 months or more) in the plankton before settlement and metamorphosis. Patterns of expansion of the velar lobes, including elongation of individual lobes and formation of multiple lobes, represent alternative solutions to optimize feeding and swimming abilities at increased larval body size (see Section VII.A).

TABLE I   Classification of Invertebrate Larval
Development (Levin and Bridges, 1995)

Nutritional modes	Site of development
Planktotropy	Planktonic
Facultative planktotrophy	Demersal
Maternally derived	Benthic
Lecithotrophy	Aparental
Adelphophagy	Solitary
Translocation	Encapsulated
Osmotrophy	Parental
Autotrophy	Internal brooding
Photoautotrophy	External brooding
Chemoautotrophy	
Somatoautotrophy	

Dispersal potential	Morphogenesis
Teleplanic	Indirect
Actaeplanic	Free-living
Anchiplanic	Contained
Aplanic	Direct

IV. POLYTYPY AND INDUCIBLE TRAITS IN LARVAL MORPHOLOGY

The previous section considered the evolutionary diversity of larval forms from a higher taxonomic perspective, arriving at the conclusion that the terminology and classification of invertebrate larval types vastly underestimate their actual structural diversity. Similar typological rigidity at the species level has led to an assumption of invariant structure and fixed life-history and developmental characteristics. Accordingly, this section emphasizes the need for more data on differences between individual larvae in populations and species.

Evolutionary change depends upon the operation of selection on traits that vary in species populations. How much phenotypic variation is present within species? How plastic are the phenotypes of individual larvae? To what extent are individuals and entire populations flexible in the timing, rates, and ordering of events in development? Hadfield and Strathmann (1996) challenge the prevailing impression in the literature that most invertebrates are relatively fixed in their life-history parameters, which include the attributes of larval development. They provide examples across a range of invertebrate phyla to show that variability, flexibility, and plasticity (aspects they label collectively as *polytypy*) are not only common but probably much more widespread and

important in the evolution of invertebrate life histories than previously believed. Natural selection depends not only upon genetic variation among individuals within populations but also upon the extent to which each individual
genotype within a population is capable of environmental modulation of the
phenotype (Thompson, 1991).

One striking result of our progress in integrating developmental and evolutionary biology is increased appreciation of the importance of developmental
plasticity as a diversifying factor in evolution [see West-Eberhard (1989) for
a review]. How much do we know about polytypic structural and life-history
traits of marine invertebrate larvae? The following discussion draws examples
from structural traits, but the opportunities for major advances in understanding of the evolution of larval forms extend to the entire range of life-history
and developmental traits that may exhibit multiple states within species populations.

A. Phenotypic Plasticity of Larval Morphology

Quantitative expressions of larval shape and shape change during development
generally assume species specificity and a lack of environmental modulation of
individual development. But does the individual larva have a fixed morphology?
Although there are few studies to date, inducible plasticity may be characteristic
of many larvae and may play an important role in the evolution of larval forms.
There are theoretical grounds for expecting this to be the case. Larvae move
along environmental gradients in which continuous phenotypic modulation
(Smith-Gill, 1983) may be either highly adaptive, neutral, or highly nonadaptive, producing teratologies if a larva is unable to minimize the modulation
under severe perturbation.

Spatial and temporal variations in particulate food supply are environmental
variables that should induce phenotypic responses in the development of
ephemeral larval feeding structures. Data from invertebrate larvae as phylogenetically distant as echinoderms and mollusks suggest that phenotypic plasticity modulated by food abundance may be widespread in planktotrophic invertebrate larvae.

Echinoid larvae show induced phenotypic responses to low food levels,
both in laboratory culture (Boidron-Métairon, 1988; Strathmann et al., 1992)
and under natural conditions in the plankton (Fenaux et al., 1994). Under
food-limited conditions, echinopluteus larvae develop longer postoral arms
relative to body size, thereby increasing the length of the ciliary bands that
capture food from the plankton. Hart and Strathmann (1994) have shown that
this increase in ciliary band length enhances the maximum clearance rate (the
volume of water cleared of food per unit time).

Similar phenotypic responses have been observed in mollusk veliger larvae. Strathmann *et al.* (1993) reported larger velar lobes and longer prototrochal cilia relative to larval shell size in oyster larvae cultured at low food concentrations. Gastropod larvae, which develop more complicated multilobed velar shapes than bivalve larvae, may develop unusual velar morphologies when starved in culture (C. S. Hickman, unpublished observation). Gastropod larvae from plankton tows often show abrupt phenotypic shifts in shell form and microsculpture when placed in culture (Fig. 2). Such modulations of the phenotype are less likely to be adaptive, although survival of such larvae to metamorphosis indicates that they are not highly nonadaptive and the ability to modulate phenotype is adaptive.

B. PLASTICITY AND FLEXIBILITY IN DEVELOPMENTAL PROGRAMS

To what extent do conditions in the environment alter genetic programs of development? This aspect of larval developmental plasticity is even less well-understood than phenotypic modulation. The cueing of phenotypic conver-

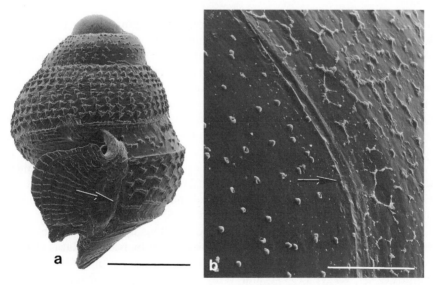

FIGURE 2 Discontinuities in secretory patterns in gastropod larval shells from the Hawaiian plankton. In each micrograph, the arrow indicates the transition from shell growth in the plankton to shell growth in culture. (a) A shell removed from culture showing a change in apertural form with a broadly expanded flange. Bar = 176 μm. (b) High magnification detail of a different shell, showing an abrup change in the pattern of microsculpture. Bar = 23 μm.

sions may show considerable variation in their timing along environmental gradients, but there is little evidence that differences in timing lead to phenotypic differences. However, Knowlton (1974) reported an effect of food supply on the number of zoeal instars in a decapod crustacean, suggesting that there can be a plasticity of staging that is of clear phenotypic consequence.

A poorly understood phenomenon that must involve major switches in the programming of development is asexual reproduction by echinoderm larvae. Bosh (1988) and Bosh *et al.* (1989) reported asexual cloning in sea star larvae in which secondary embryos resembling late stage gastrulae differentiated from posterolateral arms. Jaeckle (1994) subsequently reported two additional modes of asexual propagation in sea star larvae: autotomization of a portion of the preoral lobe and budding from the tips of the arms. The dedifferentiation and redifferentiation involved in this process must require switches in developmental program, but the literature reports do not explore or speculate on the nature of environmental cues.

V. ORIGINS OF MARINE INVERTEBRATE LARVAE

A. The Nature of Questions

When, where, and why did invertebrate larvae first originate? How many times have larvae been lost? How many times have larvae subsequently evolved? What are the primary selection pressures that have led to the appearance and maintenance larval forms? What are the ancestral and derived states of the major structural features and functional and behavioral attributes of larvae? How have novel larval forms arisen? How have novel larval traits arisen? Questions of origination are varied and usually not posed in the form of testable hypotheses. Many of the "answers" take the form of speculative scenarios involving hypothetical ancestral types and hypothetical transformations of types. Furthermore, the question of larval origins often gets mixed with questions of the origins of (1) multicellularity, (2) biphasic or complex life history, (3) metazoans, (4) contrasting nutritional modes, (6) contrasting developmental modes, (7) metamorphosis, and even (8) complexity.

Examples of the manner in which these questions become intertwined can be found in reviews and discussions of larval origins by Ivanova-Kazas (1985), Nielsen and Nørrevang (1985), Jägersten (1972), Rieger *et al.* (1991), Strathmann (1993), Rieger (1994), and Wray (1995). Progress in understanding various aspects of larval origins depends upon generating and testing new hypotheses, as advocated by Strathmann (1993) in a review that provides examples supporting the fruitfulness of this approach.

Questions about origins must distinguish clearly among questions of novel structure (features, measurable attributes of the larval phenotype), novel functions, novel developmental mechanisms, and novel taxa. The discussion that follows is limited to a brief summary of two contentious issues with respect to larval origins. In each case the objectives are to call attention to the pertinent literature, suggest how the question can be framed more clearly, and identify what must be done in order to obtain a resolution.

B. Two Questions about Larval Origins

1. How Many Times Have Invertebrate Larvae Arisen?

The disparity of larval forms among extant major invertebrate groups suggests that unique larval forms have arisen independently within at least 14 phyla and multiple times within many of those phyla. In reviewing the origin and early radiation of metazoan larvae, Wray (1995) discusses the historical evidence and phylogenetic inference of the antiquity and early evolutionary radiation of larvae. Although Wray remains cautious on the subject of multiple origins, he agrees with Strathmann (1993) that "multiple origins seem inescapable" (Wray, 1995, p. 415). Clearly, the answer to the question "how many times" is one that depends upon a strongly supported cladistic hypothesis of the relationships of major groups onto which we can map suites of larval characters.

2. Origins of Larval Feeding: Is Planktotrophy Primitive or Derived?

Much of the literature on larval origins focuses not on the question of structure or body plan, but upon the ancestral nutritional mode (feeding or nonfeeding) and habitat (pelagic or benthic) of marine invertebrate larvae. Although a few authors hold that nonfeeding is ancestral (von Salvini-Plawen, 1985; Chaffee and Lindberg, 1986; Haszprunar et al., 1995), the majority of authors conclude that feeding is the ancestral state and that nonfeeding larvae have evolved through the loss of feeding structures (Strathmann, 1978). There is more widespread agreement that the ancestral larva was pelagic. The exception is Chaffee and Lindberg (1986), who argue on the basis of small body size for a nonpelagic ancestral state.

Some confusion is introduced into this debate by the failure to recognize the distinction between the question of nutrition during the origins and early evolution of metazoans and questions of larval nutrition and switching of nutritional modes during the subsequent evolution of various metazoan clades.

This is another example of a question that requires robust hypotheses of relationships. Even if the ancestral form for all invertebrates was a feeding larva, it does not follow that feeding larvae are ancestral in all phyla or clades at various levels within those phyla. When nutritional modes are mapped onto phylogenetic trees, there are both losses and gains of feeding larvae. The cases with which I am most familiar are drawn from the Mollusca. For example, there are feeding larvae within both the Gastropoda and the Bivalvia; however, planktotrophy is absent from all the other mollusk classes, including the more basal ones. Therefore, lecithotrophy appears to be the ancestral mode for the phylum Mollusca. Because Gastropoda and Bivalvia are not sister taxa, it also appears that planktotrophy is derived independently in these two classes (Fig. 3). In fact, the basal clades in both classes have lecithotrophic larvae, and within the gastropods planktotrophy has been derived at least twice, once in the Neritomorpha and once in Caenogastropoda (Fig. 3). Reid (1989) shows

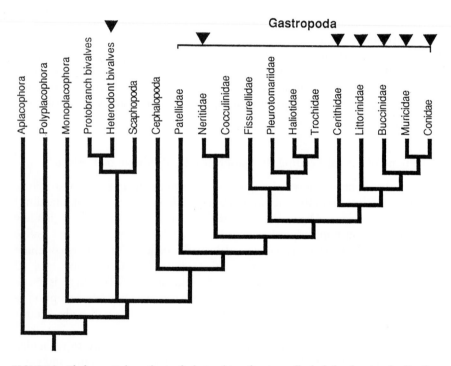

FIGURE 3   Phylogenetic hypothesis of relationships of major mollusk clades showing the distribution of planktotrophy. Feeding appears to have two independent origins in gastropods, as well as a third independent origin in one major group of bivalves. Alternatively, there is a much larger number of independent losses of feeding.

that the ancestral larva in the gastropod family Littorinidae is nonfeeding, although planktotrophy is the predominant mode within the family.

Outside of mollusks, it is more difficult to find examples in which planktotrophy is derived. Rouse and Fitzhugh (1994) present evidence that feeding is a derived condition in the family Serpulidae and that, although all larvae of Sabellidae are nonfeeding, pelagic larvae are derived within the family. The direction of change in nutritional mode within marine invertebrate clades, however, more frequently is in the direction of loss of feeding, and it often is accompanied by clear loss of structure (Jägersten, 1972; Strathmann, 1985; Hadfield and Iaea, 1989). The polarity of transformation in echinoids is to loss of feeding, with at least 14 echinoid lineages making the switch to lecithotrophic development (Emlet, 1990; Wray, 1992).

As ad hoc adaptive and functional arguments fall into increasing disfavor and as more rigorous phylogenetic hypotheses are generated, we will have a better basis for assessing ancestral larval states and the number of times that state changes have occurred.

VI. INVERTEBRATE LARVAE AND THE PALEONTOLOGICAL RECORD

Two kinds of paleontological data are used to argue that all of the most prominent and diverse invertebrate phyla had larvae at the beginning of the Phanerozoic eon. The first is indirect and based on the first appearances of fossil adults and the way that suites of larval synapomorphies of major groups map onto clades that had diverged by the beginning of the Cambrian period. The second is direct and based on the early stratigraphic appearances of larval skeletal remains and soft parts in the fossil record.

A. EVIDENCE FROM FIRST APPEARANCES OF ADULTS

Indirect evidence of the antiquity of larval diversity requires an adequate phylogenetic hypothesis for the major invertebrate groups and knowledge of larval synapomorphies shared by extant members of those groups. It assumes that basic larval features are conserved. If adult forms in two clades share a common set of homologous larval traits, then it is inferred that two Cambrian adults in the same two clades had larval traits in common with one another and with their common Precambrian ancestor. Wray (1995) provided an explicit hypothesis of the deep history of larval forms by mapping 16 larval features and developmental innovations onto a cladogram of 17 phyla that are preserved in the early metazoan radiation.

B. EVIDENCE FROM FOSSIL LARVAL SKELETAL REMAINS

There is some direct evidence of early diversification of larval forms from preserved skeletons. Larval shells of gastropods (Matthews and Missarzhevsky, 1975) and hyoliths (Dzik, 1978) occur in Cambrian and Ordovician sediments, and the shells of brachiopods and mollusks commonly retain the larval shell at the apex of the adult shell. The Upper Cambrian "Orsten" (limestone nodules in the Alum Shale of southern Sweden) preserve spectacular three-dimensional phosphatized nauplius and nauplius-like larvae of seven distinct kinds (Müller and Walossek, 1986; Walossek and Müller, 1989), as well as eight successive larval instars of an agnostid trilobite species (Walossek and Müller, 1990). The arm rods of echinoid and ophiuroid pluteus larvae occur as microfossils from the Oxfordian stage of the Jurassic (Deflande-Rigaud, 1946). The fossil record of larvae undoubtedly is much richer than previously appreciated. With intensified interest in fossil lagerstätten, new criteria for locating the appropriate preservational taphofacies, and a new focus on the microfossil component, we should expect much more direct evidence of invertebrate larval diversity and evolution.

C. EVIDENCE FROM FOSSIL LARVAL SOFT PARTS

Perhaps the most exciting paleontological evidence of a deep history of diversification of invertebrate early development is in phosphatized Cambrian and Precambrian taphofacies preserving embryos, larvae, and preadults or adults that originally were unmineralized. Phosphatized ovoid embryos (with blastomeres representing 64-cell and 128-cell cleavage stages) from the Middle Cambrian of China are of inferred arthropod affinity by virtue of their preservation with trilobite remains (Zhang and Pratt, 1994). Bengtson and Zhao (1997) reported eggs containing metazoan embryos of two types from Lower Cambrian rocks in China and Siberia. They inferred direct development from the morphology of "hatched" specimens. The antiquity of early multicellular developmental stages is pushed deeper by reports of spectacular three-dimensional embryos in Precambrian phosphorites of the Doushantuo formation (570 MYBP) in southern China (Xiao et al., 1998). It is unclear whether the absence of gastrulae or later developmental stages represents a taphonomic bias, a sampling bias, or a bias introduced by methods of fossil preparation and recovery. It is possible that it represents a stage in metazoan evolution preceding the developmental differentiation and radiation of both larva-like micrometazoans and adult metazoan body plans.

Although the Doushantuo fossils reported seem to be predominantly blastula stage embryos (Li *et al.*, 1998), there is encouraging evidence of possible parenchymella and amphiblastula sponge larvae. Perhaps it is too early in the investigation of these deposits to predict what they may hold in terms of metazoan larvae or micrometazoans of a grade of organization comparable to that of primary larvae.

D. Explaining the Missing Precambrian Record: A Developmental Regulatory Hypothesis with Paleontological Predictions

Davidson *et al.* (1995) have outlined an interesting argument for a Precambrian interval of evolution of microscopic (≤ 1 mm) multicellular organisms at an organizational grade equivalent to modern primary metazoan larvae. Their reasoning is based on the alleged universality of Type 1 embryogenesis (Davidson, 1991; Hall and Wake, Chapter 1 of this volume, Table I) in living invertebrates (both cnidarians and bilaterians) and the inability of this developmental mechanism to generate the adult body plans of these groups.

In their argument, maximal indirect development is regarded as the ancestral state, and Type 1 embryogenesis is the ancestral developmental regulatory mechanism. A qualitatively different regulatory innovation must be added in order to generate the adult phenotype. Because larval and postlarval developmental regulation are decoupled, the key evolutionary innovation is represented *morphologically* as the set-aside cells or imaginal rudiment in the primary larva and *developmentally* as the emplacement of the upstream set of genetic regulatory mechanisms that create these rudiments.

Comparative characterization of the upstream regulation of development is an exciting field in its own right, but it does not illuminate evolutionary history or generate paleontological predictions. However, the regulatory argument in this "micrometazoan hypothesis" does generate a predictive framework for paleontological investigation of the timing and origins of the metazoan radiation. Developmental biologists perhaps have dropped the ball by assuming that embryos and micrometazoan ancestors would not have left a fossil record either "because of their small size" or because of their "probable lack of skeletonization" (Davidson *et al.*, 1995) Paleontologists are beginning to realize unprecedented opportunities to analyze what is no longer a missing record of embryos. However, the furor over fossil embryos falls short of asking the key question. The next step must be an intensive search for minute larva-like forms that

test the micrometazoan hypothesis. The Late Precambrian fossil record potentially is the Rosetta stone for all of evolutionary developmental biology and
certainly is the final arbiter of debates over historical fact.

VII. EVIDENCE FOR LARVAL EVOLUTION: ADAPTATIONS OF LARVAE

> Did these planktonic larval stages represent a primitive ancestral pelagic adult
> type, or did they represent special larval adaptations? He left no one in doubt that
> the second view was correct. (Hardy, 1959, p. 180, speaking of Garstang)

Walter Garstang must have been an impressive lecturer. Sir Alister Hardy
(1959) based his chapter "Pelagic Larval Forms" upon a lecture by Garstang
and credited him not only with destroying his students' faith in recapitulation
but also with offering them the more exciting evolutionary concept of paedomorphosis. "He was one of the first to realize that selection acts just as power-
fully upon the young stages in development as upon the adult: that in fact
the developmental stages may—nay, must—become adapted to their particular
mode of life just as much as the adults are to theirs" (Hardy, 1959, p. 180).

Since Garstang's time, the study of larval adaptation has moved beyond the
telling of plausible stories to the analysis of form–function relationships in
terms of engineering optima and comparative measurements of performance.
This section highlights four aspects of the evolution of larval structure. It begins
with the evolution of larval ciliation, an area in which there is a substantial body
of literature, and moves on to "advertise" three aspects of larval structure
that are wide-open frontiers: modification of larval structure for encapsulated
functions, evolution of antipredatory structures, and, finally, the enigma of
gastropod torsion.

A. ADAPTATIONS FOR SWIMMING AND FEEDING

The diversity of patterns of epithelial ciliation is one of the most striking
features of marine invertebrate larvae. Nielsen (1987) provides a classification
of cilia according to their structure, arrangement, and functional coordination,
emphasizing the number of possibilities for creating distinctive combinations
of ciliary features. Nielsen and Nørrevang (1985) and Neilsen (1985, 1987)
emphasize the phylogenetic significance of both the structure and the function
of ciliary bands on invertebrate larvae.

Less attention has been directed to comparative adaptive performance of
cilia, although there is widespread recognition that different patterns of ciliation
represent alternative adaptive solutions to performing these functions effi-

ciently (e.g., Garstang, 1928; Strathmann, 1974, 1987; Nielsen, 1987; Emlet, 1994). A few examples of research investigating ciliary performance illustrate the opportunities for fruitful biomechanical studies of a diversity of features that have been selected for larval performance advantage.

The arrangement of ciliary bands and the form of the larva may reflect natural selection for high maximum rates of clearance (the volume of water from which food particles are removed per unit time) in feeding larvae. There are a many alternative bioengineering solutions to increasing clearance rates with different kinds of ciliary bands. Strathmann *et al.* (1972) have emphasized solutions based on (1) increasing ciliary length and velocity, (2) increasing the number of cilia per unit length of band, and (3) increasing the total length of the ciliary band. Emlet (1991) has shown that the amount of water moved can be increased by placing the cilia on ridges on the larval body.

Phylogenetic constraints may limit larvae within clades to subsets of the full range of theoretically possible solutions to increasing clearance rates. It therefore is instructive to examine solutions in taxa that have unique combinations of ciliary features. Noting that the mitraria larvae of oweniid polychaetes are unique in having simple cilia in their opposed band feeding apparatus, Emlet and Strathmann (1994) were able to explain the peculiar topology and enlarged body of the larvae as a functional solution to increasing band length in a taxon phylogenetically precluded from solutions open to other polychaetes, which have compound cilia.

Increases in ciliary band length do not require increases in larval body size. Much of the apparent diversity in form in the velar morphology of planktotrophic caenogastropods may represent alternative solutions to increasing band length within a compromise framework for maintaining complex locomotor functions (C. S. Hickman, unpublished observation). Very long narrow velar lobes are inferred to have evolved independently in the families Strombidae, Cypraeidae, and Conidae (Fig. 4). These families appear in different parts of the caenogastropod tree, and all grow to unusually large larval size (in the range of 2 mm) prior to metamorphosis. As the larva feeds and grows, the individual, roughly circular, lobes proliferate and elongate by patterns unique to each of the three clades. The resulting long, narrow lobes are, however, strikingly similar.

In another portion of the metazoan tree, the peculiar long tentaculate lobes that are the structural hallmark of the amphinomid polychaete rostraria larva are convergent on multiple elongate caenogastropod velar lobes. Jägersten (1972) interpreted the rostraria "tentacular apparatus" as a probable adaptation for feeding, in contrast to Häcker's (1898) earlier speculation that it was a propulsive structure ("Stossfühler"). Comparative study of the rostria tentacular apparatus, which appears to function only in feeding, and the multilobed forms of the gastropod velum, which function both in locomotion and in

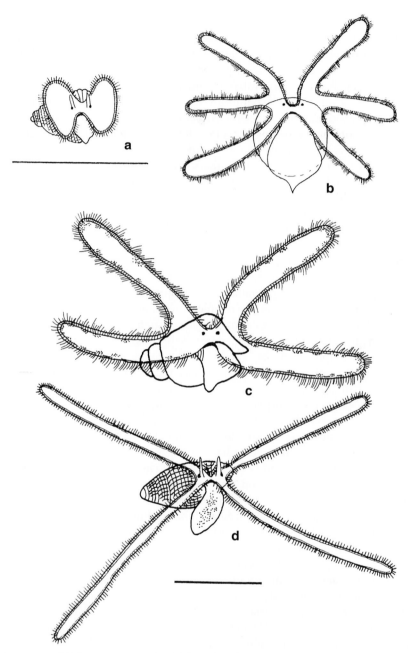

FIGURE 4   Velar shape in gastropods of differing larval size and duration of life in the plankton. (a) A representative small veliger with a short planktic life and simple bilobed velum. (b) Large veliger of *Strombus maculatus* with a six-lobed velum. (c) Large veliger of *Conus* sp. with a four-lobed velum. (d) Larva veliger of *Cypraea isabella* with a four-lobed velum. All drawings from video images were captured with the camera looking directly down on the swimming larva. Scale bars = 1 mm; the bottom scale bar applies to b–d.

feeding, could provide an excellent system for testing hypotheses of optimization and compromise.

B. Intracapsular Larval Adaptations

The importance of not restricting the term "larva" to hatched, planktic forms is illustrated in this section by larval adaptations to indirect development within the confines of intracapsular space. From an evolutionary perspective, one of the most compelling arguments for looking into the capsule is to document the modifications of larval structure that range from reduction and loss of function to fascinating adaptations that co-opt features for novel intracapsular functions. Hadfield and Iaea (1989) examined the encapsulated veligers of the vermetid gastropod *Petaloconchus montereyensis* Dall, and although larvae have a velum, it is a reduced feature in which both structure and function have been modified. Larvae no longer are able to swim or to feed in a normal manner because the pre- and postoral ciliary bands and food groove are missing. However, the entire ventral surface of the velum has become densely coated with motile cilia continuous with cilia surrounding the mouth, a novel modification for feeding on intracapsular nurse yolk.

Loss of swimming and feeding cilary bands is not always the case. It is common to observe encapsulated veligers "swimming" within the capsule, and Fretter and Graham (1994) speculate that this increased movement aids the larva both in uptake of nutrition and in respiration. The intracapsular larvae of a cassid gastropod observed by Fioroni and Sandmeier (1964) develop a fully functional velum in which the opposed band feeding system moves yolk grains from nurse eggs to the mouth. However, the anterior velar lobes are modified to fold back over nurse eggs and "grasp" them. Within these modified velar pouches, Fioroni and Sandmeier observed that the eggs actually are rotated by velar cilia to destroy the egg membrane and liberate yolk for ingestion. Natarajan (1957) provides a similar account in which a muricid gastropod grasps nurse eggs with the velum, but instead of destruction by abrasion, the eggs are collapsed by pressure from the velar lobes.

The velum is not the only structure modified for food uptake during an encapsulated stage of development. Rivest and Strathmann (1994) document the development of a transitory complex of unique large pedal cells at the tip of the foot of encapsulated larvae of the neritoidean gastropod *Nerita picea* (Recluz). The complex, experimentally implicated in the uptake of albumen proteins, begins to develop during the trochophore stage, reaches maximum development in the early veliger stage, and is resorbed by the time feeding veligers hatch.

The evolution and development of transitory larval features are poorly studied. Comparative study of the structure and function of such features could yield rich rewards, particularly in the encapsulated stage where the pressure presumably is great for efficient uptake within a nutritively rich environment.

C. ANTIPREDATORY ADAPTATIONS

Ever since Thorson (1936) there has been considerable interest in the very high rates of mortality during the planktic larval phase of invertebrate life histories. Thorson (1946, 1950, 1966) considered a number of sources of high "larval wastage" and concluded that predation undoubtedly was the most important single factor. High rates of mortality due to predation figure prominently in models of invertebrate reproductive strategies (e.g., Vance, 1973; Jackson and Strathmann, 1981), and reviews of larval mortality continue to uphold the traditional allegation that larval mortality primarily is a consequence of predation (Day and McEdward, 1984; Young and Chia, 1987; Rumrill, 1990). Morgan (1995) was more cautious in his conclusion that there are few accurate measurements of larval mortality, very few good data on the major predators of larvae, and even fewer data on feeding rates, prey preferences, and population densities of predators.

As a consequence of the assumption of high predation pressure on larvae, it should not be surprising that various aspects of larval morphology have been regarded as larval defense mechanisms. Young and Chia (1987) and Morgan (1955) review putative larval antipredatory adaptations that include structural, behavioral, and chemical defenses. Defensive structures include various mineralized structures (shells, spines, spicules), nematocysts, and erectile setae. Behavioral responses such as akenesis, spine flaring, negative rheotaxis, and vertical migration presumably also require the evolution of structural correlates. Many of the inferences of protective function are based on anecdotal observations of interactions: a predator spits out a larva with erected setae, is observed having difficulty handling a spiny prey, or is damaged by spines.

The study of larval antipredatory adaptations would benefit considerably from more data on predators and methods of prey capture, as well as from experimental studies. Who are the predators? It is curious that most of the accounts of predators on plankton, dating from the early studies of Lebour (1922, 1923), have focused on holoplanktonic prey rather than larvae. Reviews of the literature on larval predators (Young and Chia, 1987; Morgan, 1995) emphasize the importance of the most voracious and selective forms, including planktivorous fish, chaetognaths, and gelatinous zooplankton (hydromedusae, scyphomedusae, ctenophores, and siphonophores).

There are very few experimental studies of predation on larvae (Cowden *et al.*, 1984; Pennington and Chia, 1984, 1985; Rumrill *et al.*, 1985; Pennington *et al.*, 1986), and only one of these addresses a hypothesis of antipredatory adaptation (Pennington and Chia, 1985). The latter study, which is a putative test of Garstang's hypothesis of the adaptive significance of larval torsion, is addressed in Section VII.D.

The underlying assumptions in the experiments conducted to date are that predators "eat" larvae whole and that larval adaptations will take the form of structures and behaviors that prevent wholesale ingestion. However, observations of breakage and subsequent repair of the shells of larval gastropods (C. S. Hickman, unpublished observation) suggest that unsuccessful predatory attempts may be extremely common and that selection pressure may be very high for structures that aid larvae in surviving and repairing their shells after predatory encounters.

There are a number of ways to assess the potential protective function of larval shell features (C. S. Hickman, unpublished). One method is to propose engineering solutions that would retard mechanical breakage, followed by comparing engineering paradigms for goodness of fit with features actually observed on shells. Another method is experimental and involves simulated attacks on shells to determine whether there are features that consistently resist breakage. A third method involves the analysis of natural experiments in the plankton and examination of patterns of breakage and repair of shells that have survived predatory attack. The fourth method is direct experimentation with living larvae and zooplankton predators in culture.

An example of the use of the third method is illustrated in Fig. 5. Shells of larvae recovered from plankton tows frequently have damaged apertures as well as repaired breakages. Because the larvae are in a concentrate of zooplankton while the sorting is in progress, they are exposed to artificially high densities of potential predators. Although not a natural situation, it does demonstrate (1) how shells are damaged by real predators, (2) that larvae do survive this damage, and (3) that when placed in culture, damaged shells are repaired. One of the most commonly recurring structural features of larval shells in different major caenogastropod groups is a thickened peripheral keel or ridge on the shell. In the swimming larva, the growing margin of this ridge is positioned over the larval head and mouth. This ridge consistently resists mechanical breakage, and it is shell on either side of the ridge that is broken back (Fig. 5a,b). Repair of a broken aperture is illustrated in Fig. 5c.

D. Gastropod Larval Torsion: A Bizarre Twist in Development and Evolution

Gastropod torsion is one of the dramatic key innovations in the history of the invertebrates. It also is one of the few examples of a developmental novelty

FIGURE 5   Three larval shells prepared for SEM at the time they were sorted from a plankton tow, showing apertural breakage caused by unsuccessful predatory attempts. (a, b) Two unidentified species in which the thickened peripheral keel has remained intact and chipping is restricted to the unthickened shell margin on either side. Bar for a = 380 μm; bar for b = 350 μm. (c) *Nassarius dermestina* larval shell, with chipping most evident along the margin posterior to the beak. The beak is protected partially by heavier spiral ridges that retard breakage. Bar = 250 μm.

that occurs during larval life and persists in the adult body plan. During an extraordinary series of events, which may be completed in as little as 2 min (Garstang, 1928, 1929), the head and food are twisted by approximately 180° relative to the visceral mass, mantle, and shell. Recognition that torsion explains much of the complex morphology of the adult gastropod dates from Spengel (1881). Theories of the origin and adaptive significance of torsion have been argued for a century without resolution. If anything, the torsion debate becomes more rather than less befuddled as it continues.

The torsion story has the makings of a good book, but it cannot be written quite yet. The answer to "why not?" is the subject of this section. The short form of the answer is that we still do not have torsion characterized mechanically and ontogenetically. Many arguments about torsion are based on incorrect information and questionable assumptions. In the classical explanation, the first 90° of the twist is initiated by contraction of the right larval retractor muscle, which develops earlier than the left (Crofts, 1937, 1955). However, this explanation assumes that the larval shell on which the muscle is inserted is calcified and rigid at the time torsion occurs. C. S. Hickman and M. G. Hadfield (unpublished manuscript) have observed that muscle contractions produce transient dimpling rather than torsion in the flexible shells of pretorsional veligers of the "archaeogastropod" *Margarites pupillus* (Gould). Because the time required to complete torsion varies from 2 min to 9 days, it is likely that the torted end result now is generated by a variety of morphogenetic mechanisms. With growing recognition that development may be loosely constrained in its pathways to a more constrained end result (Wagner, 1994), the study of torsion should take on new interest.

Developmental explanations of torsion are crucial to an evolutionary understanding. Textbook accounts of torsion typically illustrate it as an innovation in a hypothetical ancestral adult gastropod: the twisted body plan of the nouveau gastropod is paired diagramatically with its untwisted hypothetical ancestral progenitor. There also is a common erroneous tendency to describe torsion as a twisting of the entire larval body relative to the shell, rather than a twisting of the head and foot (cephalopedal mass) relative to the visceral mass, shell, and mantle. Within this framework, which morphogenetic movements are common to all gastropods, and which are superimposed modifications of development? Page (1997) has shown that tissue movements during torsion in the abalone *Haliotis kamtschatkana* Jonas are much more complicated and that portions of the mantle epithelium and shell field accompany the head and foot. Her detailed computer-generated reconstructions provide an outstanding model for future comparative studies.

Torsion theories have regarded it alternatively as (1) originating in the adult to meet the needs of the adult, (2) originating in the larva, but to meet the needs of the adult, and (3) originating in the larva purely to meet the needs of the larva. Imbedded in these theories is the controversial component of whether torsion arose by a single genetic change of immediate functional utility.

Garstang (1928) was the first to propose that torsion not only originated in a veliger larva but also that it should have been more advantageous to the larva than to the adult. He argued that torsion would have had immediate protective value by enabling the larva to withdraw its head and velum into the shell. Many of Garstang's critics (e.g., Ghiselin, 1966) have been uncomfortable primarily with the macromutational implications: (1) torsion as a "sudden jump in the evolution of the veliger larva" and (2) ontogenetic torsion, as it occurs in living gastropods, as a conserved repetition in development of the original evolutionary event. Jägersten (1972, p. 124), for his part, did "not regard such sudden evolution as probable." However, his belief that torsion, as well as the shell itself, must have originated in the adult (and later shifted to the larva) appears to stem from his inability to see larvae as targets of selection for antipredatory modifications. This belief is reflected in his statement that "small planktonic creatures are as a rule swallowed whole" (Jägersten, 1972, p. 124).

Was Garstang correct that torted larvae are less subject to predation? This hypothesis can be tested. Pennington and Chia (1985) did so by exposing torted and untorted larvae of the abalone *H. kamtschatkana* Jonas to seven potential planktonic predator species. They concluded that torsion provided no significant protection, but what they measured was the number of larvae that "disappeared" and therefore were "swallowed whole." This may not have been the critical experiment. Can torted larvae survive injury, and can they repair damage better than untorted larvae? Incidences of shell breakage and repair argue strongly for an adaptive function of features that retard breakage. Under predatory attack from small zooplankton predators, many larval gastropods do withdraw the head and velum into the shell (C. S. Hickman, unpublished observation), and they do repair predator-inflicted damage to the growing margin of the shell aperture (Fig. 6).

In another variation on the theme of antipredatory adaptation, Stanley (1982) proposed that torsion not only permitted the evolution of the gastropod operculum but was "imperative" to the appearance of this evolutionary novelty. The operculum is another feature that develops in the planktic larva. As in the benthic adult, it serves to seal off the aperture when the animal withdraws into its shell. Its position is important. It develops on the dorsal posterior surface of the larval foot, which is the last portion of the torted larva to withdraw, following the velum and head.

Experimental approaches to the study of adaptation require considerable caution in extrapolating from what currently enhances survival to what enhanced the survival of ancestors. Thompson (1967) criticized Garstang's theory because he considered the foot to be as vulnerable to predatory attack as the velum and head. However, the question is not the tastiness of protruding parts but how well the larva can survive with various parts damaged. If larval torsion, the operculum, and features of the larval shell do enhance the survival of modern larvae in the plankton, there are significant research opportunities in comparative examination and measurement of performance directly in predatory confrontations.

VIII. LARVAL CHARACTERS IN PHYLOGENY RECONSTRUCTION

Adult structure is the traditional cornerstone of systematic biology. Invertebrate larval features have not been used extensively in phylogenetic reconstruction,

FIGURE 6   Three larval gastropod shells prepared for SEM at the time they were sorted from a plankton tow. All three shells show evidence of unsuccessful predation in the form of damage (at arrows) that was repaired in the plankton. Bar for a = 270 μm; bars for b and c = 380 μm.

primarily because of suspect rampant homoplasy. In spite of the recognition that larvae and adults live in separate milieu and are subject to different selection pressures, in spite of the recognition that larval and adult body plans and the morphogenetic programs that produce them are different, and in spite of increasing recognition of the decoupling (e.g., Wray, 1995) of larval and adult evolution, the bugabear of homoplasy looms larger in the larval realm than in the adult realm.

To what extent are the pronounced similarities of many larval forms indicative of shared ancestry, and to what extent are they a consequence of convergent modifications? What role, if any, does larval structure have to play in phylogenetic studies? There is strong divergence of opinion on this matter. The following brief discussion emphasizes (1) the potential role of substructural detail, particularly in structures that are not closely tied to major features of larval life histories and ecology, and (2) the contribution of data from the sequence and timing of events in larval development and metamorphosis.

A. STRUCTURAL CHARACTERS

Nielsen (1994, 1995) has been a leading advocate of the use of larval characters at higher taxonomic levels, and he has demonstrated their fruitfulness in analysis of the relationships of invertebrate phyla. At lower levels, however, the efficacy of larval characters is in question. From studies of echinoids, Smith (1997) argues that larval structure is more prone to homoplasy than adult structure. He also argues that simplicity of larval characters is insufficient to explain higher levels of larval homoplasy, concluding that larval structure is under stronger functional constraint. Nonetheless, larval characters have been used successfully in the analysis of echinoid relationships [Wray, 1992, 1996; see also Smith (1997) for a review].

The use of invertebrate larval data in cladistic analysis is in its infancy. The rest of this brief section is not so much a plea for phylogenetic analysis as it is a plea for more effort in the time-consuming but rewarding detailed examination of larval structure and for more emphasis on character development as the foundation for cladistic analysis.

Systematic biology needs more comparative morphologists who relish the unmasking of convergence. "Similarity" of structure is not the same as "identity" of structure. Homoplasy frequently disappears at deeper levels of structure. For example, the sharp boundary between the gastropod larval shell (proto-conch) and the juvenile shell (teleoconch) becomes indistinct and rich in structural detail when it is examined at higher magnification (Hickman, 1995) (Fig. 7). The reorganization of shell secretion begins prior to the onset of metamorphosis (as defined by settlement and loss of the velum) and leaves a rich record of "stages" in the incremental addition of shell material. Superficially identical protoconch–teleoconch boundaries record different patterns of secretory reorganization. Although these patterns can be analyzed comparatively after they have formed, the best way to determine the order of events (and the only way to determine timing and duration) is by following the process in populations of larvae in culture. Shell microstructure is a record of incremental additions of material and therefore is silent on aspects of reorganization that resorb or remodel.

Observation of larvae in culture allows not only direct observation of developmental events but also the opportunity to preserve individuals in various stages of transformation.

B. DEVELOPMENTAL SEQUENCE CHARACTERS

Changes in the order or sequences of structural change in developing larvae may be a richer source of data for phylogenetic reconstruction than structure

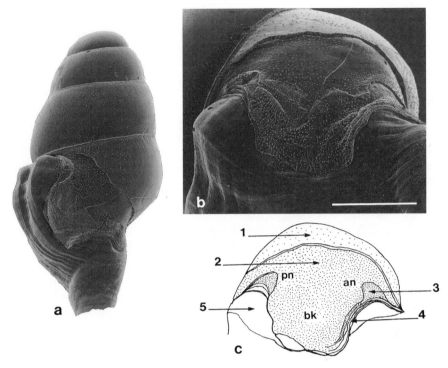

FIGURE 7   Shell of an unidentified gastropod removed from culture after metamorphosis, illustrating the complexity of the boundary between the larval and postlarval shell. (a) Entire shell in abapertural view, with the boundary at the arrow. Bar = 380 μm. (b) At higher magnification, many distinct episodes of incremental addition of shell material are visible at the boundary. Bar = 200 μm. (c) Simplified drawing of the boundary in b, noting five different stages in the transition (1–5). Note that 1 is terminated by a break that was repaired with growth of a new beak (bk) and that the anterior (an) and posterior (pn) notches were filled by younger layers secreted sequentially beneath older layers.

per se. Hickman (1992) established a basic chronological classification of developmental stages and events in the larval development of trochoidean gastropods. When sequences are compared from one species to another, there are some events that always occur in the same order and others that vary. For example, hatching, settlement, and the timing of torsion are highly variable within the sequence of events, as is the onset of calcification of the larval shell, the appearance of the foot rudiment, and the order of appearance, during the posttorsional veliger stage, of the diverse epithelial sensory structures that are hallmarks of the adult trochoidean body plan. Of particular interest are two events that appear to be consistent in their order: appearance of the shell gland

and the first shell material precede the appearance of the mantle fold. The mantle, as a uniquely mollusk secretory structure, clearly is not the structure that initiates formation of the mollusk larval shell.

IX. CONCLUSIONS

This chapter is presented from a predominantly structural perspective to highlight opportunities and questions in invertebrate larval evolution and development that accrue from a detailed comparative analysis of form and function. The staggering diversity of invertebrate larvae easily is masked by typology and classifications that are not concerned with structural, functional, and behavioral detail. Accordingly, the sections dealing with the question "what is a larva" and the concept of "larval type" emphasize the need for structural definition and a vocabulary that reflects the full information content of larval diversity.

Understanding of the origin and evolution of larval forms is advancing along many fronts. On the structural front, the greatest opportunities for major contributions will emerge from (1) moving to finer levels of comparative structure to distinguish between homology and analogy, (2) generating new character sets for phylogenetic analysis, (3) fuller exploration of the fossil record, especially in unusual taphofacies, for direct evidence of the antiquity and diversity of larvae, (4) more data on ranges of phenotypic variability, plasticity, and flexibility in larvae as factors contributing to the evolution of larval forms, and (5) more experimental and creative study of larval adaptation. In many instances, old "issues" such as "larval wastage" and "larval torsion" may find their resolution in reframing questions and hypotheses and through the use of new technology.

This chapter purposefully has not dealt with the embryo and early cleavage, with gene regulation and upstream development, or with metamorphosis. It has skirted the central questions of invertebrate larval ecology. There already are many excellent reviews of the issues in these areas, identifying questions that will fuel research programs well into the next century. As a morphologist, the theoretical questions I find most exciting and promising invariably emerge from empirical observation of the animals and details of structure. This is a time-consuming discipline, but advances in many forms of microscopy as well as new techniques for creating, manipulating, and analyzing images provide opportunities for studies that formerly were unimaginable.

The most serious threats to advances in understanding invertebrate larvae as living organisms are decreasing interest in this kind of research at marine laboratories, where it must be conducted, and increasing amounts of research time spent in front of computer screens. The strongest plea is for more time spent with the subjects of our inquiry.

ACKNOWLEDGMENTS

I thank Marvalee Wake and Brian Hall for the opportunity to contribute this chapter and for all of their encouragement and support as it went through its larval stages and metamorphosis. Original research reported herein was supported by NSF Grant DEB 9303169 and the Committee on Research and the Museum of Paleontology, University of California, Berkeley, CA.

REFERENCES

Bengston, S., and Zhao, Y. (1997). Fossilized metazoan embryos from the earliest Cambrian. *Science* 277, 1645–1648.

Boidron-Métairon, I. F. (1988). Morphological plasticity in laboratory-reared echinoplutei of *Dendraster excentricus* (Eschscholtz) and *Lytechinus variegatus* (Lamarck) in response to food conditions. *J. Exp. Mar. Biol. Ecol.* 119, 31–41.

Bosh, I. (1988). Replication by budding in natural populations of bipinnaria larvae of the sea star genus *Luidia*. *In* "Echinoderm Biology: Proceedings of the Sixth International Echinoderm Conference" (R. D. Burke, P. V. Mladenov, P. Lambert, and R. L. Parsley, eds.), p. 789. A. A. Balkema, Rotterdam.

Bosh, I., Rivkin, R. B., and Alexander, S. P. (1989). Asexual reproduction by oceanic planktotrophic echinoderm larvae. *Nature (London)* 337, 169–170.

Chaffee, C., and Lindberg, D. R. (1986). Larval biology of Early Cambrian molluscs: The implications of small body size. *Bull. Mar. Sci.* 39, 356–549.

Chia, F.-S. (1974). Classification and adaptive significance of developmental patterns in marine invertebrates. *Thalassia Jugos.* 10, 121–130.

Cowden, C., Young, C. M., and Chia, F.-S. (1984). Differential predation on marine invertebrate larvae by two benthic predators. *Mar. Ecol. Prog. Ser.* 14, 145–149.

Crofts, D. R. (1937). The development of *Haliotis tuberculata* with special reference to organogenesis during torsion. *Philos. Trans. R. Soc. London, Ser. B* 228, 219–268.

Crofts, D. R. (1955). Muscle morphogenesis and torsion. *Proc. Zool. Soc. London* 125, 711–750.

Davidson, E. H. (1991). Spatial mechanisms of gene regulation in metazoan embryos. *Development (Cambridge, UK)* 113, 1–26.

Davidson, E. H., Peterson, K., and Cameron, A. H. (1995). Origin of bilaterian body plans: Evolution of developmental regulatory mechanisms. *Science* 270, 1319–1325.

Day, R., and McEdward, L. (1984). Aspects of the physiology and ecology of pelagic larvae of marine benthic invertebrates. *In* "Marine Plankton Life Cycle Strategies" (K. A. Steidinger and L. M. Walker, eds.), pp. 93–120. CRC Press, Cleveland, OH.

Deflande-Rigaud, M. (1946). Vestiges microscopiques des larves d'Echinoderms de l'Oxfordien de Villers-sur Mer. *C. R. Hebd. Seances Acad. Sci.* 222, 908–910.

Dzik, J. (1978). Larval development of hyolithids. *Lethaia* 11, 293–299.

Emlet, R. B. (1990). World patterns of developmental mode in echinoderms. *Adv. Invertebr. Reprod.* 5, 329–335.

Emlet, R. B. (1991). Functional constraints on the evolution of larval forms of marine invertebrate larvae: Experimental and comparative evidence. *Am. Zool.* 31, 707–725.

Emlet, R. B. (1994). Body form and patterns of ciliation in nonfeeding larvae of echinoderms: Functional solutions to swimming in the plankton? *Am. Zool.* **34**, 570–585.

Emlet, R. B. and Strathmann, R. R. (1994). Functional consequences of simple cilia in the mitraria of oweniids (an anomalous larva of an anomalous polychaete) and comparisons with other larvae. *In* "Reproduction and Development of Marine Invertebrates" (W. H. Wilson, Jr., S. A. Stricker, and G. L. Shinn, eds.), pp. 143–157. Johns Hopkins University Press, Baltimore, Md.

Fenaux, L., Strathmann, M. F., and Strathmann, R. R. (1994). Five tests of food-limited growth of larvae in coastal waters by comparisons of rates of development and form of echinoplutei. *Limnol. Oceanogr.* **39**, 84–98.

Fioroni, P. and Sandmeier, E. (1964). Ueber eine neue Art der Naehereierbewaeltigung bei Prosobranchierveligern. *Vie Milieu* **17**, (Suppl.), 235–249.

Fretter, V., and Graham, A. (1994). "British Prosobranch Mollusks." Ray Society, London.

Garstang, W. (1928). The origin and evolution of larval forms. *Nature (London)* **122**, 366.

Garstang, W. (1929). The origin and evolution of larval forms. *Rep. Br. Assoc. Adv. Sci., Sect. D,* pp. 77–98.

Ghiselin, M. T. (1966). The adaptive significance of gastropod torsion. *Evolution (Lawrence, Kans.)* **20**, 337–348.

Gilbert, S. F., and Raunio, A. M., eds. (1997). "Embryology: Constructing the Organism." Sinauer, Sunderland, MA.

Häcker, V. (1898). Die pelagischen Polychaeten- und Achätenlarven der Plankton-Expedition. *Ergeb. Atlant. Ozean Planktonexp ed.* **2**, 2, 1–50.

Hadfield, M. G., and Iaea, D. K. (1989). Velum of encapsulated veligers of *Petaloconchus* (Gastropoda), and the problem of re-evolution of planktotrophic larvae. *Bull. Mar. Sci.* **45**, 377–386.

Hadfield, M. G., and Strathmann, M. F. (1996). Variability, flexibility and plasticity in life histories of marine invertebrates. *Oceanol. Acta* **19**, 323–334.

Hardy, A. C. (1959). "The Open Sea." Houghton Mifflin, Boston.

Hart, M. W., and Strathmann, R. R. (1994). Functional consequences of phenotypic plasticity in echinoid larvae. *Biol. Bull. (Woods Hole. Mass.)* **186**, 291–299.

Hart, M. W., and Strathmann, R. R. (1995). Mechanisms and rates of suspension feeding. *In* "Ecology of Invertebrate Larvae" (L. McEdward, ed.), pp. 193–221. CRC Press, Boca Raton, FL.

Haszprunar, G., von Salvini-Plawen, L., and Rieger, R. M. (1995). Larval planktotrophy—A primitive trait in the Bilateria? *Acta Zool. (Stockholm)* **76**, 141–154.

Hickman, C. S. (1992). Reproduction and development in trochacean gastropods. *Veliger* **35**, 245–272.

Hickman, C. S. (1993). Theoretical design space: A new program for the analysis of structural diversity. *Neues Jaheb. Geol. Paläontol. Abh.* **190**, 169–182.

Hickman, C. S. (1995). Asynchronous construction of the protoconch/teleoconch boundary: Evidence for staged metamorphosis in a marine gastropod larva. *Invertebr. Biol.* **114**, 295–306.

Ivanova-Kazas, O. M. (1985). On the evolution of polychaete larvae. *Proc. USSR Polychaete Conf., Leningrad, 1983,* pp. 40–45.

Jackson, G. A., and Strathmann, R. R. (1981). Larval mortality from offshore mixing as a link between precompetent and competent periods of development. *Am. Nat.* **118**, 16–26.

Jaeckle, W. B. (1994). Multiple modes of asexual reproduction by tropical and subtropical sea star larvae: An unusual adaptation for genet dispersal and survival. *Biol. Bull. (Woods Hole, Mass.)* **186**, 62–71.

Jägersten, G. (1972). "Evolution of the Metazoan Life Cycle: A Comprehensive Theory." Academic Press, London and New York.

Kerr, R. A. (1998). Pushing back the origins of animals. *Science* **279**, 803–804.

Knowlton, R. E. (1974). Larval developmental processes and controlling factors in decpod Crustacea, with emphasis on Caridea. *Thalassia Jugosl.* **10**, 138–158.

Lebour, M. V. (1922). The food of planktonic organisms. *J. Mar. Biol. Assoc. U.K.* **12**, 644–677.

Lebour, M. V. (1923). The food of planktonic organisms. II. *J. Mar. Biol. Assoc. U.K.* **13**, 70–92.

Levin, L. A. and Bridges, T. S. (1995). Pattern and diversity in reproduction and development. *In* "Ecology of Marine Invertebrate Larvae" (L. McEdward, ed.), pp. 1–48. CRC Press, Boca Raton, FL.

Li, C.-W., Chen, J.-Y., and Hua, T.-E. (1998). Precambrian sponges with cellular structures. *Science* **279**, 879–882.

Matthews, S. C., and Missarzhevsky, V. V. (1975). Small shelly fossils of late Precambrian and Early Cambrian age: A review of recent work. *J. Geol. Soc. London* **131**, 289–304.

McEdward, L. R., and Janies, D. A. (1993). Life cycle evolution in asteroids: What is a larva? *Biol. Bull. (Woods Hole, Mass.)* **184**, 255–268.

Mileikovsky, S. A. (1971). Types of larval development in marine bottom invertebrates, their distribution and ecological significance: A re-evaluation. *Mar. Biol.* **10**, 193–213.

Morgan, S. G. (1995). Life and death in the plankton: Larval mortality and adaptation. *In* "Ecology of Marine Invertebrate Larvae" (L. McEdward, ed.), pp. 279–321. CRC Press, Boca Raton, FL.

Müller, K. J., and Walossek, D. (1986). Arthropod larvae from the Upper Cambrian of Sweden. *Trans. R. Soc. Edinburg: Earth Sci.* **77**, 157–179.

Naef, A. (1919). "Idealistische Morphologie und Phylogenectik; zur Methodik der Systematischen Morphologie." Fischer, Jena.

Natarajan, A. V. (1957). Studies on the egg masses and larval development of some prosobranchs from the Gulf of Mannar and the Palk Bay. *Proc. Indian Acad. Sci. Sect. B* **46B**, 170–228.

Nielsen, C. (1985). Animal phylogeny in the light of the trochaea theory. *Biol. J. Linn. Soc.* **25**, 243–299.

Nielsen, C. (1987). Structure and function of metazoan ciliary bands and their phylogenetic significance. *Acta Zool. (Stockholm)* **68**, 205–262.

Nielsen, C. (1994). Larval and adult characters in animal phylogeny. *Am. Zool.* **34**, 492–501.

Nielsen, C. (1995). "Animal Evolution: Interrelationships of the Living Phyla." Oxford University Press, Oxford.

Nielsen, C., and Nørrevang, A. (1985). The trochaea theory: An example of life cycle phylogeny. *In* "The Origins and Relationships of Lower Invertebrate Groups" (S. Conway Morris, J. D. George, R. Gibson, and H. M. Platt, eds.), pp. 28–41. Oxford University Press, Oxford.

Page, L. R. (1997). Ontogenetic torsion and protoconch form in the archaeogastropod *Haliotis kamtschatkana:* Evolutionary implications. *Acta Zool. (Stockholm)* **78**, 227–245.

Pennington, J. T. and Chia, F.-S. (1984). Morphological and behavioral defenses of trochophore larvae of *Sabellaria cementarium* (Polychaeta) against four planktonic predators. *Biol. Bull. (Woods Hole, Mass.)* **167**, 168–175.

Pennington, J. T., and Chia, F.-S. (1985). Gastropod torsion: A test of Garstang's hypothesis. *Biol. Bull. (Woods Hole, Mass.)* **169**, 391–396.

Pennington, J. T., Rumrill, S. S., and Chia, F.-S. (1986). Stage-specific predation upon embryos and larvae of the Pacific sand dollar, *Dendraster excentricus*, by 11 species of common zooplankton predators. *Bull. Mar. Sci.* **39**, 234–240.

Reid, D. G. (1989). The comparative morphology, phylogeny and evolution of the gastropod family Littorinidae. *Philos. Trans. R. Soc. London, Ser. B* **234**, 1–110.

Rice, M. E. (1967). A comparative study of the development of *Phascolosoma agassizii, Golfingia pugettensis,* and *Themiste pyroides* with a discussion of developmental patterns in the Sipunculida. *Ophelia* **4**, 143–171.

Rieger, R. M. (1994). The biphasic life cycle—a central theme in metazoan evolution. *Am. Zool.* **34**, 484–491.

Rieger, R. M., Haszprunar, G., and Schuchert, P. (1991). On the origin of the Bilateria: Traditional views and alternative concepts. *In* "The Early Evolution of Metazoa and the Significance of

Problematical Taxa" (A. M. Simonetta and S. Conway Morris, eds.), pp. 107–112. Cambridge University Press, Cambridge, UK.

Rivest, B. R., and Strathmann, R. R. (1994). Uptake of protein by an independently evolved transitory cell complex in encapsulated embryos of neritoidean gastropods. In "Reproduction and Development of Marine Invertebrates" (W. H. Wilson, Jr., S. A. Stricker, and G. L. Shinn, eds.), pp. 166–176. Johns Hopkins University Press, Baltimore, MD.

Rouse, G. W. and Fitzhugh, K. (1994). Broadcasting fables: Is external fertilization really primitive? Sex, size, and larvae in sabellid polychaetes. Zool. Scr. 25, 271–312.

Rumrill, S. S. (1990). Natural mortality of marine invertebrate larvae. Ophelia 32, 163–198.

Rumrill, S. S., Pennington, T. J., and Chia, F.-S. (1985). Differential susceptibility of marine invertebrate larvae: Laboratory predation of sand dollar, Dendraster excentricus (Eschscholtz), embryos and larvae by zoeae of the red crab, Cancer productus (Randall). J. Exp. Mar. Biol. Ecol. 90, 193–208.

Scheltema, R. S. (1971). The dispersal of the larvae of the shoal-water benthic invertebrate species over long distances by ocean currents. In "Fourth European Marine Biology Symposium" (D. J. Crisp, ed.), pp. 7–28. Cambridge University Press, Cambridge, UK.

Scheltema, R. S. (1977). Dispersal of marine invertebrate organisms: Paleobiogeographic and biostratigraphic implications. In "Concepts and Methods of Biostratigraphy" (E. G. Kauffman and J. E. Hazel, eds.), pp. 73–108. Dowden, Hutchinson & Ross, Stroudsburg, PA.

Scheltema, R. S. (1986). Long-distance dispersal by planktonic larvae of shoal-water benthic invertebrates among central Pacific islands. Bull. Mar. Sci. 39, 241–256.

Schroeder, P. C., and Hermans, C. O. (1975). Annelida: Polychaeta. In "Reproduction of Marine Invertebrates" (A. C. Giese and J. S. Pearse, eds.), vol. 3, pp. 1–213. Academic Press, New York.

Smith, A. B. (1997). Echinoderm larvae and phylogeny. Annu. Rev. Ecol. Syst. 28, 219–241.

Smith, D. L. (1977). "A Guide of Marine Coastal Plankton and Marine Invertebrate Larvae." Kendall/Hunt, Dubuque, IA.

Smith-Gill, S. J. (1983). Developmental plasticity: Developmental conversion versus phenotypic modulation. Am. Zool. 23, 47–55.

Spengel, J. W. (1881). Die Geruchsorgane und das Nervensystem der Mollusken. Ein Beitrag zur Erkenntnis der Einheit der Molluskentypus. Z. Wiss. Zool. 35, 333–383.

Stanley, S. M. (1982). Gastropod torsion: Predation and the opercular imperative. Neues Jahrbu. Geol. Paläontol., Abh. 164, 95–107.

Strathmann, R. R. (1974). Introduction to function and adaptation in echinoderm larvae. Thalassia Jugosl. 10, 321–339.

Strathmann, R. R. (1978). The evolution and loss of feeding larval stages of marine invertebrates. Evolution (Lawrence, Kans.) 32, 894–906.

Strathmann, R. R. (1985). Feeding and non-feeding larval development and life history evolution in marine invertebrates. Annu. Rev. Ecol. Syst. 16, 339–361.

Strathmann, R. R. (1987). Larval feeding. In "Reproduction of Marine Invertebrates" (A. C. Giese, J. S. Pearse, and V. B. Pearse, eds.), vol. 9, pp. 465–550. Blackwell Scientific Publications, Palo Alto, CA, and Boxwood Press, Pacific Grove, CA.

Strathmann, R. R. (1993). Hypotheses on the origins of marine larvae. Annu. Rev. Ecol. Syst. 24, 89–117.

Strathmann, R. R., Jahn, T. L., and Fonseca, J. R. C. (1972). Suspension feeding by marine invertebrate larvae: Clearance of particles by ciliated bands of a rotifer, pluteus, and trochophore. Biol. Bull. (Woods Hole, Mass.) 142, 505–519.

Strathmann, R. R., Fenaux, L., and Strathmann, M. F. (1992). Heterochronic developmental plasticity in larval sea urchins and its implications for evolution of non-feeding larvae. Evolution (Lawrence, Kans.) 46, 972–986.

Strathmann, R. R., Fenaux, L. Sewell, A. T., and Strathmann, M. F. (1993). Abundance of food affects relative size of larval and postlarval structures of a molluscan veliger. *Biol. Bull.* (*Woods Hole, Mass.*) 185, 232–239.

Thompson, J. D. (1991). Phenotypic plasticity as a component of evolutionary change. *Trends Ecol. Evol.* 6, 246–249.

Thompson, T. E. (1967). Adaptive significance of gastropod torsion. *Malacologia* 5, 423–430.

Thorson, G. (1936). The larval development, growth and metabolism of Arctic marine bottom invertebrates compared with those of other seas. *Medd. Grønl.* 100, 1–155.

Thorson, G. (1946). Reproduction and larval development of Danish marine bottom invertebrates with special reference to planktonic larvae in the Sound (Øresund). *Medd. Komm. Dan. Fisk.-Havunders.* 4, 1–523.

Thorson, G. (1950). Reproductive and larval ecology of marine bottom invertebrates. *Biol. Rev. Cambridge Philos. Soc.* 25, 1–45.

Thorson, G. (1966). Some factors influencing the recruitment and establishment of marine benthic communities. *Neth. J. Sea Res.* 3, 267–293.

Vance, R. R. (1973). On reproductive strategies in marine benthic invertebrates. *Am. Nat.* 107, 339–352.

von Salvini-Plawen, L. (1985). Early evolution of the primitive groups. *In* "The Mollusca" (K. Wilber, ed.), Vol. 10, pp. 59–150. Academic Press, New York.

Wagner, G. P. (1994). Homology and the mechanisms of development. *In* "Homology: The Hierarchical Basis of Comparative Biology" (B. K. Hall, ed.), pp. 273–299. Academic Press, San Diego, CA.

Walossek, D., and Müller, K. J. (1989). A second type A-nauplius from the Upper Cambrian 'Orsten' of Sweden. *Lethaia* 22, 301–306.

Walossek, D. and Müller, K. J. (1990). Upper Cambrian stem-lineage crustaceans and their bearing upon the monophyletic origin of Crustacea and the position of Agnostus. *Lethaia* 23, 409–427.

West-Eberhard, M. J. (1989). Phenotypic plasticity and the origins of diversity. *Annu. Rev. Ecol. Syst.* 20, 249–278.

Wray, G. A. (1992). The evolution of larval morphology during the post-Paleozoic radiation of echinoids. *Paleobiology* 18, 258–287.

Wray, G. A. (1995). Evolution of larvae and developmental modes. *In* "Ecology of Marine Invertebrate Larvae" (L. McEdward, ed.), pp. 413–447. CRC Press, Boca Raton, FL.

Wray, G. A. (1996). Parallel evolution of nonfeeding larvae in echinoids. *Syst. Biol.* 45, 308–322.

Xiao, S., Zhang, Y., and Knoll, A. H. (1998). Three-dimensional preservation of algae and animal embryos in a Neoproterozoic phosphorite. *Nature* (*London*) 391, 553–558.

Young, C. M. (1990). Larval ecology of marine invertebrates: A sesquicentennial history. *Ophelia* 32, 1–48.

Young, C. M., and Chia, F.-S. (1987). Abundance and distribution of pelagic larvae as influenced by predation, behavior, and hydrographic factors. *In* "Reproduction of Marine Invertebrates" (A. C. Giese, J. S. Pearse, and V. B. Pearse, eds.), vol. 9, pp. 385–463. Blackwell Scientific Publications, Palo Alto, CA, and Boxwood Press, Pacific Grove, CA.

Zhang, X., and Pratt, B. R. (1994). Middle Cambrian arthropod embryos and blastomeres. *Science* 266, 637–639.

Zimmer, R. L. and Woollacott, R. M. (1977). Structure and classification of gymnolaemate larvae. *In* "Biology of Bryozoans" (R. M. Woollacott and R. L. Zimmer, eds.), pp. 57–89. Academic Press, New York.

CHAPTER 3

Larvae in Amphibian Development and Evolution

JAMES HANKEN

Department of Environmental, Population, and Organismic Biology, University of Colorado, Boulder, Colorado

I. INTRODUCTION

The tremendous intellectual excitement and activity that have attended the study of the relation between development and evolution over the past 20–25 years rival those characteristic of the "Golden Age" of zoology in the latter half of the 19th and early 20th centuries (Goldschmidt, 1956; Hall, 1992; Müller, 1991; Raff, 1996; D. B. Wake *et al.*, 1991). Then, as now, scientists strove to tease apart the complex relationship between ontogeny and phylogeny, including studies of both pattern and process (Churchill, 1997; Gould, 1977; Luckenbill-Edds, 1997). In both periods, amphibians—and especially amphibian larvae—occupy center stage. Evolutionary biologists have long appreciated that the presence of a free-living, aquatic larval stage, as well as the primitively complex (metamorphic) life history of which it is a part, confer on Recent amphibians a tremendous potential for adaptive diversification (Hanken, 1992; McDiarmid, 1978). This potential has been realized in the wide array of alternate life-history modes, developmental patterns, morphologies, and ecological relationships that are seen in both extinct and extant taxa (Duellman and Trueb, 1986; Lynn, 1961). At the same time, because of their relative accessibility, rapid external development, and ease of laboratory handling and experimental manipulation, amphibians offer many outstanding and practical opportunities to study animal development, including both features unique to amphibians as well as those characteristic of vertebrates in general (Armstrong and Malacinski, 1989; Cannatella and de Sá, 1993; Kay and Peng, 1991; Malacinski and Duhon, 1996; Shaffer, 1993). In short, amphibians and their larvae offer an excellent "system" to study the relation between development and evolution (Hanken, 1989).

The literature on larval amphibian biology is vast. It includes both classical and ongoing studies of morphology, physiology, ecology, behavior, genetics, and developmental biology. No attempt to offer a comprehensive review of this literature is made here. Instead, this chapter highlights several topics that are especially pertinent to contemporary studies of the development and evolution of amphibian larvae. In so doing, critical gaps in our knowledge and understanding of both the evolutionary history and developmental biology of modern amphibians, and especially their larvae, will be made evident. Hopefully, such an approach will help to guide or even initiate future research in these areas. The treatment begins with a short review of current ideas concerning the phylogenetic ancestry of amphibian larvae. This provides the historical context necessary for subsequent considerations of evolutionary trends in larval development and diversification involving Recent taxa. These are followed by a section that discusses two examples of the pervasive trend toward larval loss that is seen in many extant groups.

II. HISTORICAL CONTEXT: THE ANCESTRY OF AMPHIBIAN LARVAE AND THE COMPLEX, METAMORPHIC LIFE HISTORY

The free-living, aquatic larva is an ancient feature in amphibians. A complex, biphasic life history is phylogenetically widespread among Recent taxa and generally is accepted as the primitive condition for each of the three living orders (Duellman and Trueb, 1986). Interestingly, whereas a metamorphic ontogeny is characteristic of many species of frogs, salamanders, and caecilians, it is not the most frequent reproductive mode in all three groups. Most species of living salamanders, for example, have direct development: courtship, mating, and oviposition occur on land, and the terrestrial egg hatches as a fully formed, albeit miniature, adult; there is no free-living larva (D. B. Wake and Hanken, 1996). All of these species belong to a single family, the Plethodontidae, or lungless salamanders. Members of four additional families (Sirenidae, Proteidae, Cryptobranchidae, and Amphiumidae) display obligate loss of both a discrete metamorphosis and the subsequent terrestrial adult stage. Consequently, sexually mature adults reproduce essentially as modified larvae.

Paleontology provides an additional, valuable perspective regarding the ancestry of amphibian larvae. Extinct salamanders with a larval morphology are known from both the Cretaceous (Evans and Milner, 1996; Evans et al., 1996) and Triassic (Ivakhnenko, 1978, cited in Roček, 1996); some Paleocene fossils are assigned to extant genera (Naylor, 1978). Fossil tadpoles from the Cretaceous and Miocene are assignable to extant families (Estes et al., 1978) or even genera (Wassersug and Wake, 1995; Fig. 1) of frogs. Stem members

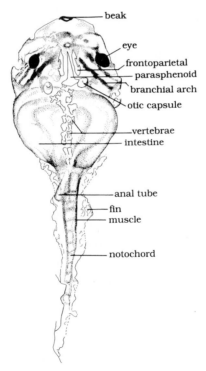

FIGURE 1   Reconstruction of a fossil tadpole from the Miocene of Turkey, which tentatively is assigned to the extant Eurasian genus *Pelobates* (Pelobatidae). Reprinted with permission from Wassersug and Wake (1995).

of the three Recent orders existed at least as far back as the early or middle Jurassic (Evans and Milner, 1996; Jenkins and Walsh, 1993; Shubin and Jenkins, 1995) or Triassic (Rage and Roček, 1989), but these clades likely had differentiated from one another by the end of the Permian, ca. 240 Myr BP (McGowan and Evans, 1995). If we accept the premise that a larval stage is a primitive characteristic in each order, then larvae also must date to at least this early.

The homology of amphibian larvae across all three Recent orders—that is, whether they represent just one versus two or more independent instances of the evolution of the metamorphic ontogeny among living taxa—is not fully resolved. Indeed, the topic has received surprisingly little attention. The most parsimonious interpretation of the presence of a larva (and metamorphosis) as a primitive trait in frogs, salamanders, and caecilians is that they represent the retention of this characteristic from the common ancestor of all three clades. This interpretation is supported by several lines of evidence. First, it avoids the vast amount of convergent evolution (homoplasy) of the many shared–derived, larva-specific traits that are common to all three groups that would be implied by positing two, or even three, independent origins of larvae. Second, morphological differences between larva and adult in the ancestral anuran likely were much less extreme—and metamorphosis much more gradual—than those typically seen in living frogs (Elinson, 1990; Fritzsch, 1990; Wassersug and Hoff, 1982). Thus, the ancestral anuran would have resembled urodeles much more closely than do living frogs. Finally, abundant fossil material documents the existence of a complex, biphasic life history, as adduced from larval and adult specimens, in many early tetrapod groups, including dissorophoid temnospondyls (Bolt, 1977, 1979; Boy, 1974; Carroll, 1986, 1988; Milner, 1982, 1990; Schoch, 1992, 1995; Werneburg, 1991; Fig. 2). Several dissorophoids have been variably proposed as outgroups to living amphibians (Bolt, 1969, 1977, 1991; McGowan and Evans, 1995; Milner, 1988, 1990, 1993; Schoch, 1995; Trueb and Cloutier, 1991), although this hypothesized relationship is not universally accepted (e.g., Carroll, 1988, 1992, 1995; Laurin and Reisz, 1997; Reisz, 1997).

Analysis of the larval homology problem is closely linked to ongoing debate regarding phylogenetic relationships among Recent amphibians. The debate centers around the claim that the three extant orders constitute a monophyletic group, Lissamphibia, which has a common ancestry distinct from that of all other living tetrapods and is derived from dissorophoid temnospondyls (see references in the preceding paragraph). New data bearing on these phylogenetic issues may have important implications for larval biology. For example, albanerpetontids are an additional group of salamander-like, "predominantly terrestrial," fossil amphibians (McGowan and Evans, 1995, p. 143). They have been proposed as a fourth lissamphibian lineage, closely allied with, but

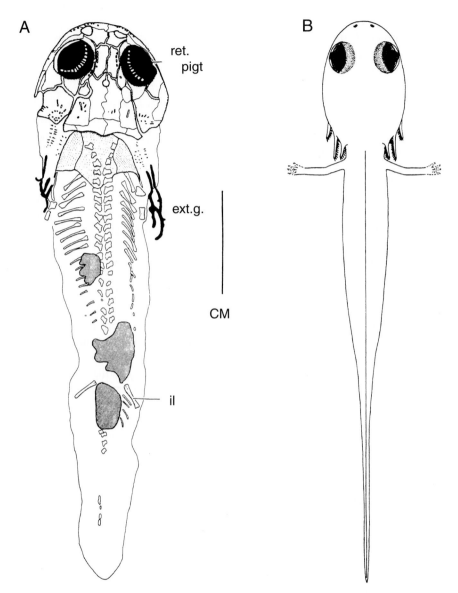

FIGURE 2 (A) *Saurerpeton* cf. *obtusum*, a basal temnospondyl amphibian (Saurerpetontidae) from the Middle Pennsylvanian of Illinois. The fossil is depicted in dorsal view; external gills (ext. g.) indicate a larval morphology. Abbreviations: il, ilium; ret, pigt, retinal pigment. Stippling denotes infilled intestines. Reprinted with permission from Milner (1982). (B) Larval specimen of *Branchiosaurus*, a temnospondyl amphibian (Branchiosauridae), reconstructed in dorsal view. Reprinted with permission from Boy (1974).

nevertheless distinct from, the three Recent orders. Larval albanerpetontids are unknown. If the presence of free-living larvae in living amphibians represents the retention of a primitive trait from dissorophoid ancestors, then larvae likely were either present in albanerpetontids as well or lost from this extinct lineage after it diverged from the three extant groups.

Finally, whereas there is widespread agreement that the presence of a free-living larva is a primitive life-history trait for living amphibians, alternate or divergent scenarios have been proposed. Bogart (1981) proposed that the primitive reproductive mode in anurans involved terrestrial, rather than aquatic, breeding. Although an aquatic larval stage still would have been present, it "probably had a relatively abbreviated free-swimming life" (Bogart, 1981, p. 34). This proposal has not been accepted by most subsequent authors (e.g., Duellman, 1989; Duellman and Trueb, 1986; Duellman *et al.*, 1988). Duellman and colleagues, however, have hypothesized that the free-living tadpole stage may have reevolved independently from a direct-developing ancestor as many as four times in a specialized group of South American marsupial frogs in the genera *Fritziana, Flectonotus,* and *Gastrotheca* (Duellman and Hillis, 1987; Duellman *et al.*, 1988; Wassersug and Duellman, 1984). Their hypothesis derives from a molecular phylogenetic analysis of these and related species, which was used to assess evolutionary trends in reproductive biology. Alternate hypotheses of life-history evolution in marsupial frogs have been offered by Haas (1996a,b) and Weygoldt and de Carvalho e Silva (1991). Similarly, the possibility of reevolution of the free-living, aquatic larval stage from direct-developing ancestors has been explored in plethodontid salamanders (Titus and Larson, 1996; D. B. Wake and Hanken, 1996; Fig. 3). In all of these studies, precise evolutionary sequences in relevant reproductive characters remain to be fully resolved.

III. EMBRYONIC DERIVATION OF LARVAL FEATURES

There is increased interest in resolving, with considerable precision, the embryonic derivation of both larval and adult (postmetamorphic) features in amphibians. These studies have been facilitated, if not promoted, by the advent of a variety of sophisticated molecular, experimental, and analytical tools for tracing embryonic cell and tissue lineages into postembryonic stages in animals generally (Collazo *et al.*, 1994; Gardner and Lawrence, 1986; Krotoski *et al.*, 1988; Le Douarin and McLaren, 1984; Thiébaud, 1983). Such techniques, although not infallible, offer considerable advantages over many comparable methods utilized by experimental embryologists earlier in this century to derive embryonic fate maps for various species, but especially amphibians, and to resolve

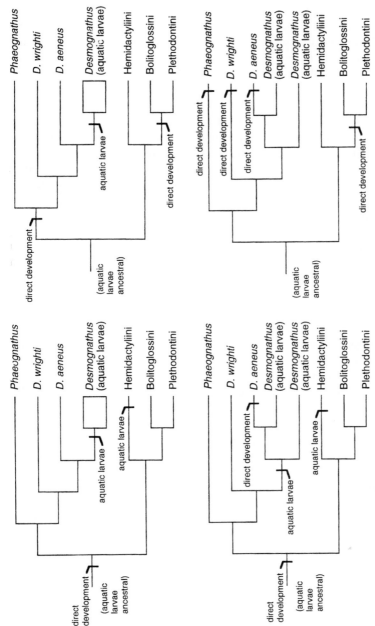

FIGURE 3 Four different scenarios for the evolution of aquatic larvae and direct development in lungless salamanders (Plethodontidae). Each scenario is a function of the particular scheme of phylogenetic relationships depicted and the parsimony method used to analyze life-history variation among the taxa. One consequence of the phylogenetic distribution of alternative life histories in these salamanders is that either aquatic larvae or direct development must have evolved more than once, and possibly several times, regardless of how the data are analyzed. Modified with permission from Titus and Larson (1996).

the embryonic derivation of particular organ systems (e.g., Hamburger, 1960). For example, a variety of fluorescent dyes will effectively label many cells and tissues for far longer than will traditional vital dyes, such as neutral red and Nile blue sulfate, and their use generally is not associated with many of the extreme developmental artifacts that are associated routinely with more invasive techniques, such as tissue ablation (e.g., Collazo et al., 1993; Eagleson et al., 1995; Olsson and Hanken, 1996).

In an evolutionary context, detailed knowledge of embryonic derivation can contribute much to our understanding of the developmental mechanisms that underlie the evolution of larval morphology per se, the extent to (and means by) which larval features mediate the evolution and development of adult features, and especially the reality and mode of operation of larval "constraints" on adult morphology. It can also contribute to more general discussions and controversies in evolutionary morphology, such as the developmental basis of homology (Abouheif, 1997; Hall, 1994, 1995). Nevertheless, most work on amphibians has been limited to the two standard, "model" amphibians, the clawed frog, Xenopus laevis (e.g., Collazo et al., 1993; Eagleson and Harris, 1990; Krotoski et al., 1988; Sadaghiani and Thiébaud, 1987), and the axolotl, Ambystoma mexicanum (e.g., Barlow and Northcutt, 1995, 1997; Northcutt et al., 1994, 1995; Smith, 1996; Smith et al., 1994); few studies focus on any of the large number of nonstandard species that display either the ancestral, biphasic ontogeny or some more derived reproductive mode. Moreover, most studies of Xenopus or the axolotl employ these species as model systems for understanding vertebrate development in general; they are not primarily interested in addressing larval specializations as such. One set of analyses that addresses the embryonic derivation of larval features in a comparative context focuses on the larval skull in anurans. These studies are reviewed briefly here.

Amphibians were the principal subjects of the large number of early studies that assessed the role of the embryonic neural crest in vertebrate development [reviewed in Hall and Horstädius (1988); Holtfreter, 1968]. These studies documented extensive contributions of the neural crest to many larval tissues and other features, especially the cartilaginous skull, spinal nerves, and pigmentation. In subsequent years, the focus of studies of neural crest biology moved to amniotes, and the use of amphibians was largely abandoned, at least for some organ systems such as the skull and skeleton. Embryonic derivation of the larval skull in anurans has been reinvestigated in several species of metamorphosing frogs. These studies provide an opportunity to confirm the results of earlier, classical studies by the use of alternate—and generally more sensitive and reliable—experimental and analytical methods. Moreover, by analyzing the data in a phylogenetic context, authors are beginning to assess various evolutionary topics, such as the relative lability versus conservatism of specific developmental characters.

For much of this century, the most comprehensive analysis of neural crest contribution to the tadpole skull was a study of the pickerel frog, *Rana palustris*, by Stone (1927, 1929). Phylogenetically, *R. palustris* (Ranidae) is a member of the so-called advanced frog clade, or Neobatrachia (Ford and Cannatella, 1993). Although *R. palustris* has a morphologically generalized tadpole, one can presume that it was chosen largely for reasons of practicality and availability; the species is widely distributed throughout much of eastern North America including New England, where Stone lived and worked (at Yale University). Stone assessed the extent of neural crest contribution to the exclusively cartilaginous larval skull of *R. palustris*, largely by ablating different portions of the neural crest before or soon after it began migrating from the neural folds and then identifying which cranial cartilages subsequently failed to form. By combining results from several ablated embryos, he inferred an extensive contribution of cranial neural crest to the larval skull. Indeed, all cranial cartilages were shown to be neural-crest-derived, except for posteroventral components of the skull proper and two small, median cartilages in the hyobranchial skeleton (Fig. 4).

Neural crest contribution to the tadpole skull has been assessed in three additional species of frogs. Sadaghiani and Thiébaud (1987) examined *Xenopus laevis* (Pipidae) by grafting cranial neural crest from a congeneric species,

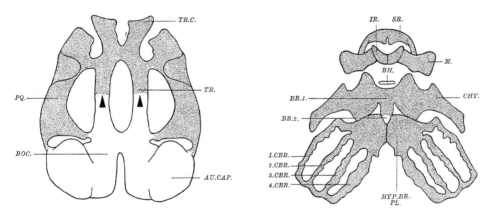

FIGURE 4 Neural crest derivation of the larval skull and hyobranchial skeleton of the pickerel frog, *Rana palustris* (Ranidae), seen in dorsal (left) and ventral (right) views. Crest-derived regions are stippled; arrows point to the boundary between crest- and non-crest-derived portions of the floor of the neurocranium. Anterior is at the top. Abbreviations: AU.CAP., auditory (otic) capsule; BB.1.–2 first and second basibranchials; BH., basihyal; BOC., basioccipital plate; 1.–4. CBR., ceratobranchials 1–4; CHY., ceratohyal; HYP.BR.PL., hypobranchial plate; IR., infrarostral; M., Meckel's cartilage; PQ., palatoquadrate; SR., suprarostral; TR., trabecular; TR.C., trabecular horn. After Stone (1929).

Xenopus borealis, to produce chimaeric embryos. Contrasting histological staining patterns by cell nuclei of these two species provide a permanent cell marker that can be used to distinguish donor from host cell lineages in differentiated tissues, such as the larval skull (Thiébaud, 1983; Fig. 5). Olsson and Hanken (1996) studied the Oriental fire-bellied toad, *Bombina orientalis* (Bombinatoridae), by labeling several different regions of premigratory neural crest with a fluorescent dye and following labeled cells into specific cranial cartilages (Fig. 6). Reiss (1997) mapped neural crest contributions in the tailed frog, *Ascaphus truei,* by following migrating crest cells, which can be distinguished histologically from non-crest (e.g., mesodermal) cell populations, at least during early developmental stages (Fig. 7). All three of these species are phylogenetically basal in comparison to all advanced frogs, including *R. palustris* (Ford and Cannatella, 1993). In this respect, each is more appropriate than any neobatrachian to serve as a source of "baseline" developmental data for evolutionary comparisons among anurans or between frogs and other vertebrates. Unlike *Bombina,* however, which has a morphologically generalized tadpole, larval cranial morphology in both *Xenopus* and *Ascaphus* is highly derived and unusual (Cannatella and Trueb, 1988; Gradwell, 1973; Trueb and Hanken, 1992). This at least potentially diminishes the suitability of *Xenopus* and *Ascaphus* as a baseline for anurans, in favor of *Bombina* (Cannatella and de Sá, 1993).

Despite the extensive phylogenetic and morphological diversity represented by these four species of anurans and the considerable differences among methods used to examine neural crest migration and fate, the general pattern of

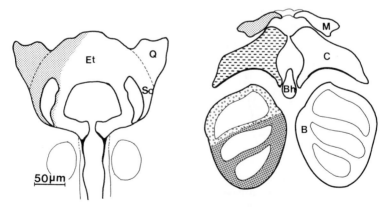

FIGURE 5   Neural crest derivation of the larval skull (left) and the lower jaw and hyobranchial skeleton (right) of the clawed frog, *Xenopus laevis* (Pipidae). Shading of crest-derived cartilages varies according to migratory stream; anterior is at the top. Abbreviations: B, ceratobranchials; Bh, basihyal; C, ceratohyal; Et, ethmoid–trabecular cartilage; M, Meckel's cartilage; Q, palatoquadrate; So, subocular. Modified with permission from Sadaghiani and Thiébaud (1987).

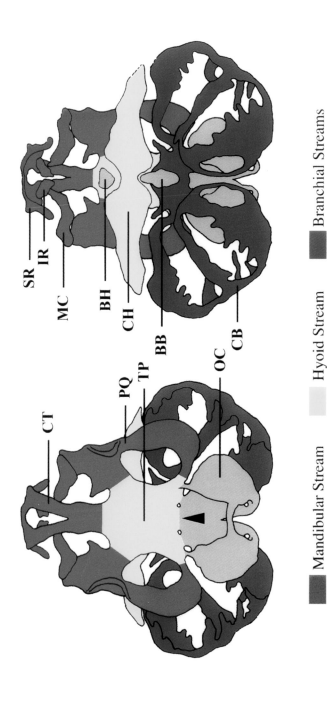

Mandibular Stream **Hyoid Stream** **Branchial Streams**

FIGURE 6 Neural crest derivation of the larval skull and hyobranchial skeleton of the Oriental fire-bellied toad, *Bombina orientalis* (Bombinatoridae), seen in dorsal (left) and ventral views. Cartilages are colored according to the cranial neural crest migratory streams(s) from which each is derived; non–crest-derived components are shaded gray. The arrow points to the boundary between crest- and non–crest-derived portions of the floor of the neurocranium. Anterior is at the top. Abbreviations: BB, basibranchial; BH, basihyal; CB, ceratobranchials I–IV; CH, certohyal; CT, trabecular horn; IR, infrarostral; MC, Meckel's cartilage; OC, otic capsule; PQ, palatoquadrate; TP, trabecular plate; SR, suprarostral. Olsson and Hanken (1996). Cranial neural crest migration and chondrogenic fate in the Oriental fire-bellied toad, *Bombina orientalis*: Defining the ancestral pattern of head development in anuran amphibians. *Journal of Morphology.* Copyright © 1996 Reprinted with permission from Wiley-Liss, Inc., a subsidiary of John Wiley & Sons, Inc.

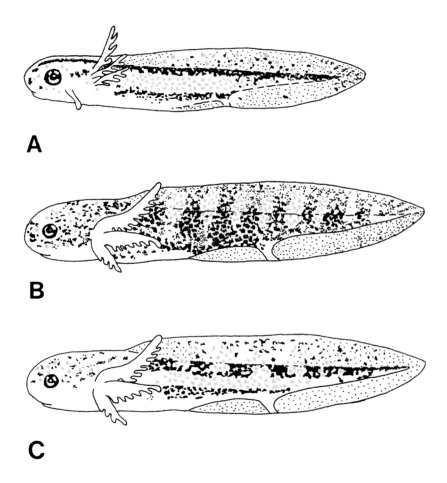

FIGURE 8 Variation in larval pigment patterning among three species of urodeles. (A) *Triturus alpestris* (Salamandridae) has a pattern of horizontal, lateral stripes above and below a melanophore-free region. (B) *Ambystoma mexicanum* has a series of vertical stripes or bars. (C) *Ambystoma t. tigrinum* has an intermediate pattern consisting of a series of vertical bars interrupted by a melanophore-free region. Reprinted with permission from Epperlein *et al.* (1996).

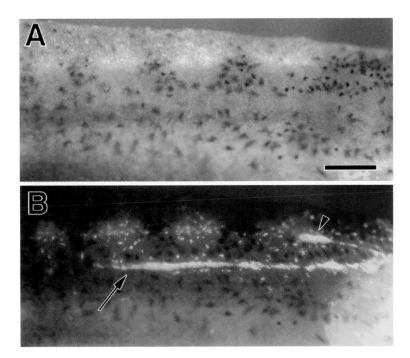

FIGURE 12 Interactions between pigment cells and the developing lateral line in the tiger salamander, *Ambystoma t. tigrinum*. (A) Brightfield photomicrograph of the lateral trunk region of an embryo. Dark melanophores are beginning to form a series of vertical bars, which are interrupted by a longitudinal, melanophore-free region. (B) Fluorescence double exposure of the same embryo shows dorsal aggregations of yellow xanthophores and dye-labeled, longitudinal lateral line primordia (arrow and arrowhead). The midbody lateral line primordium (arrow) lies within the melanophore-free region. Scale bar, 500 μm. Reprinted with permission from Parichy (1996a).

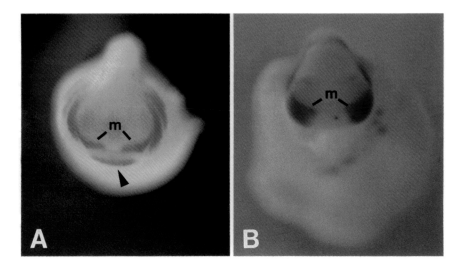

FIGURE 17   Distal-less gene expression in embryos of (A) the metamorphosing frog *Xenopus laevis* (Nieuwkoop–Faber stage 24) and (B) direct-developing *Eleutherodactylus coqui* (Townsend–Stewart stage 4) Both images are frontal views of the head; dorsal is at the top. Paired, dark areas on either side of the head in each embryo denote expression of homologous distal-less genes *X-dll4* (*Xenopus*) and *EcDlx2* (*Eleutherodactylus*) in migratory streams of cranial neural crest (m, mandibular stream). The arrow in A points to an additional, ventral site of distal-less expression in *Xenopus* that precedes embryonic differentiation of the larval cement gland in this species. The corresponding area of distal-less expression is absent from *Eleutherodactylus*, which also lacks a cement gland. Reprinted with permission from Fang and Elinson (1996).

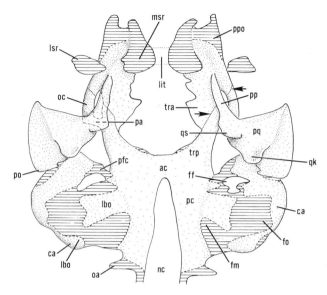

FIGURE 7 Larval skull of the tailed frog, *Ascaphus truei* (Ascaphidae), Gosner stage 23. Ventral view; anterior is at the top. Dotted lines (arrows) indicate the approximate boundary between neural crest and mesodermally derived tissues (all crest-derived tissues lie anterior to the boundary). Cartilage is stippled; undifferentiated mesenchyme is indicated by horizontal lines. Abbreviations: ac, acrochordal cartilage; ca, auditory capsule; ff, facial foramen; fm, mesotic fissure; fo, fenestra ovalis; bo, basiotic lamina; lit, intertrabecular ligament; sr, lateral suprarostral; msr, medial suprarostral; nc, notochord; oa, occipital arch; oc, orbital cartilage; pa, parachordal; pfc, prefacial commissure; po, otic process; pp, pterygoid process; ppo, pila preoptica; pq, palatoquadrate; qk, quadrate keel; qs, quadrate spur; tra, anterior trabecula; trp, posterior trabecula. Reiss (1997). Early development of chondrocranium in the tailed frog *Ascaphus truei* (Amphibia: Anura): Implications for anuran palatoquadrate homologies. *Journal of Morphology*, Copyright © 1997. Reprinted by permission of Wiley-Liss, Inc., a subsidiary of John Wiley & Sons, Inc.

neural crest derivation of larval cranial cartilages is highly concordant and consistent among them. Data for *Rana* and *Bombina* are the most extensive; these two species are compared in Table I (cf. Figs. 4 and 6). Patterns of derivation for these species are identical for 11 of the 12 discrete cartilages represented, which together constitute virtually the entire larval skull: seven are exclusively crest-derived, and four receive no contribution. Indeed, the sole apparent difference between these two species concerns the extent of neural crest contribution to the floor of the neurocranium (trabecular plate). In *Bombina,* the trabecular plate is entirely crest-derived. Consequently, the boundary between crest- and non-crest-derived regions of the cranial floor lies at the level of the anterior margin of the otic capsule (Fig. 6, arrow). In *Rana,* this boundary lies farther anterior, at a level corresponding to the middle of

TABLE I   Neural Crest Contribution to the Skull of
Larval Anurans[a]

Cartilage	Bombina	Rana
Cornua trabecula (trabecular cornu)	+	+
Suprarostral	+	+
Infrarostral	+	+
Meckel's	+	+
Palatoquadrate	+	+
Ceratohyal	+	+
Ceratobranchials I–IV	+	+
Trabecular plate (trabecular bar)	+	±
Basioccipital plate	−	−
Otic (auditory) capsule	−	−
Basihyal	−	−
Basibranchial (second basibranchium)	−	−

[a]A "plus" (+) denotes cartilages that are derived from neural
crest; the embryonic origin of remaining cartilages (−) remains
unknown, although it is presumed to be from cranial mesoderm.
A "±" denotes dual origin. Data for Bombina and Rana are based
on Olsson and Hanken (1996) and Stone (1929), respectively.

the orbit (Fig. 4, arrows). This results in a dual embryonic origin for the trabecular plate in this species (and in *Ascaphus;* Reiss, 1997). It should be noted, however, that accounts for these two species depict different larval stages: data for *Bombina* are mapped onto the skull of a "mature" larva (Gosner stage 36), whereas data for *Rana* depict a much smaller hatchling (ca. stage 23–25). Whereas the difference in derivation of the cranial floor noted previously may represent actual variation among species, it may also reflect ontogenetic change characteristic of both species.

Conserved patterns of neural crest derivation among anuran species also extend to the relative contributions of different migratory streams to specific cranial cartilages. The three principal cranial migratory streams—mandibular, hyoid, and branchial—exhibit the same basic pattern of cartilage derivation in both *Bombina* (Olsson and Hanken, 1996) and *Xenopus* (Sadaghiani and Thiébaud, 1987; cf. Figs. 4 and 5). The same general result even applies within streams. Thus, in both species the skeleton of the first branchial arch (ceratobranchial I) is derived exclusively from the anterior portion of the branchial crest stream, whereas posterior branchial arch elements (ceratobranchials II–IV) are derived principally from the posterior portion of the branchial crest stream.

Extensive similarity among anuran species in the preceding aspects of neural crest biology is of interest and significance in several respects. First, it likely betrays extreme evolutionary conservatism of these (and possibly other) under-lying features of cranial development. *Bombina* and *Rana,* for example, likely diverged from a common anuran ancestor no later than the Jurassic period, at least 144 Myr BP (Duellman and Trueb, 1986); the species *B. orientalis* may have differentiated as early as the Miocene (Maxson and Szymura, 1979). Yet, overall patterns of neural crest derivation of cranial cartilages are nearly identi-cal in these two taxa (Table I; cf. Figs. 4 and 5). Second, whereas these features of neural crest biology have changed little, many of the crest-derived larval cranial cartilages have evolved extensive morphological differences among species. This suggests that the developmental basis of larval cranial diversifica-tion lies not in gross aspects of neural crest biology, such as the timing of cell emergence or migration, the number or configuration of the principal migratory streams, or possibly even migration pathways, but rather in more subtle features of neural crest biology related to pattern formation, such as cell behavior and commitment or gene expression (Moury and Hanken, 1995; Hanken *et al.,* 1997a).

These and other results provide important baseline data that facilitate studies of larval amphibian development and evolution. At the same time they reveal conspicuous gaps in our knowledge that hamper subsequent analyses. There are, for example, no published studies that assess neural crest contribution to the cartilaginous or bony skull in either salamanders or caecilians using modern labeling techniques comparable to those reviewed earlier for anurans. For that matter, there is very little direct evidence of a neural crest contribution to the bony skull in *any* of the three living amphibian orders. Whereas the extent of neural crest contribution to the osteocranium has been documented convinc-ingly and precisely in at least some amniotes (Couly *et al.,* 1992, 1993; Köntges and Lumsden, 1996; Noden, 1986), comparable experimental data for amphibi-ans are extremely scarce (e.g., de Beer 1947; Wagner, 1949, 1955). Certainly no one has yet produced a detailed map between particular cranial neural crest streams and individual skull bones for amphibians as is available for the chicken (e.g., Köntges and Lumsden, 1996). Indeed, a neural crest derivation of the amphibian osteocranium can only be assumed by extrapolating to am-phibians the results obtained for other, distantly related taxa. Whereas general-ization across major taxa in this manner may be justified by the empirical observation that many such features of neural crest biology are highly conserved among vertebrates (see previous discussion; Hall and Horstädius, 1988), direct assessment in each extant amphibian order is urgently needed. Similarly, the embryonic derivation of non-crest-derived portions of either the larval or the adult skull remains to be assessed directly in any amphibian. Ignorance of the embryonic derivation of the bony skull in Recent amphibians is especially surprising in light of the large number of studies that assess various aspects of cranial ossification during larval development and metamorphosis in the

context of broader analyses of heterochrony, functional morphology, and endocrinology (e.g., Hanken and Hall, 1984, 1988a, b; Maglia and Púgener, 1998;
Púgener and Maglia, 1997; Reilly, 1986, 1994; Reilly and Altig, 1996; Rose,
1996; Trueb, 1985; D. B. Wake, 1980; Wild, 1997).

IV. HETEROCHRONY

The dominant paradigm for contemporary studies of the relation between
development and evolution has been heterochrony, or change in the relative
timing of developmental events (Gould, 1977; Hall, 1992; Raff, 1996).
Originally formulated in the 19th century by Ernst Haeckel as a means of
accounting for the many obvious (and to him, bothersome) exceptions to
his biogenetic law—"ontogeny recapitulates phylogeny"—heterochrony was
reformulated by Gavin de Beer in the 1930s in the context of then
contemporary interest in rates of development and relative growth, or
allometry [see reviews by Gould (1977) and Hall (1990, 1992)]. Widespread
interest in heterochrony exploded anew in the 1970s, following Gould's
(1977) "Ontogeny and Phylogeny," and this interest has continued virtually
unabated ever since. Studies range from several conceptual and theoretical
treatments that advocate one or another system of terminology, graphical
depiction, or other methods of analysis (e.g., Alberch *et al.*, 1979; McKinney,
1988b; McNamara, 1986; Pierce and Smith, 1979; Reilly *et al.*, 1997; Shea,
1983, 1988) to a large number of empirical analyses that document or
assess the existence and effects of heterochrony in particular groups of
both animals and plants (e.g., McKinney, 1988a; McKinney and McNamara,
1991; Hart and Wray, this volume, Chapter 5).

Whereas there is virtually unanimous agreement that heterochrony is a
prominent phenomenon in the evolution of organismal development, considerable disagreement remains regarding the extent to which heterochrony should
be emphasized at the expense of other potentially important factors or explanations. In a general treatment of heterochrony, Reilly *et al.* assert that "heterochrony may underlie all morphological variation and possibly is *the* developmental phenomenon producing all morphological change" (1997, p. 120).
Other authors take a more moderate view. Raff (1996), while conceding the
pervasiveness of heterochronic changes in ontogenetic pattern and agreeing
that such changes may offer important insights into underlying developmental
processes, regards such an extreme emphasis on heterochrony as undeserved
and, in some instances, even unhelpful. Indeed, "the uncritical attribution of
so many of the phenomena observed in the evolution of development to
heterochronic 'mechanisms' may be inhibiting a more penetrating investigation
of the subject" (Raff, 1996, p. 259). Thomson (1988) and Hall (1990, 1992)

express similar reservations. As pointed out by several authors, and as illustrated by the preceding quotations, at least part of the problem or disagreement may stem from the frequent failure to distinguish heterochrony as a *pattern* from heterochrony as a *mechanism* of evolutionary change (Hall, 1990, 1992; Raff, 1996). Others emphasize the important yet underappreciated role of changes in spatial patterning, or heterotopy, as a complement to changes in developmental timing (e.g., Hanken and Thorogood, 1993; Zelditch and Fink, 1996).

Amphibians have been central players in the heterochrony "industry." Indeed in the larval axolotl, amphibians present what arguably represents the paradigmatic example of heterochrony among animals (Gould, 1977). This distinction is embodied in Garstang's (1951) classic poem, "The *Axolotl* and the *Ammocoete*," which was first penned around 1922:

> *Amblystoma's* a giant newt who rears in swampy waters,
> As other newts are wont to do, a lot of fishy daughters:
> These *Axolotls*, having gills, pursue a life aquatic,
> But when they should transform to newts, are naughty and erratic.
>
> They change upon compulsion, if the water grows too foul,
> For then they have to use their lungs, and go ashore to prowl:
> But when a lake's attractive, nicely aired, and full of food,
> They cling to youth perpetual, and rear a tadpole brood.
>
> And newts Perennibranchiate have gone from bad to worse:
> They think aquatic life is bliss, terrestrial a curse.
> They do not even contemplate a change to suit the weather,
> But live as tadpoles, breed as tadpoles, tadpoles altogether!
>
> Now look at *Ammocoetes* there, reclining in the mud,
> Preparing thyroid-extract to secure his tiny food:
> If just a touch of sunshine more should make his gonads grow,
> The Lancelet's claims to ancestry would get a nasty blow!

Variation in developmental timing is a virtually ubiquitous phenomenon in amphibian evolution; it has played a pivotal role in the diversification of morphology, physiology, and ecology in both larvae and adults (Hanken, 1989, 1992). As stated by Milner, "The success of lissamphibians is very much a measure of the way in which adult and juvenile characters have been permutated by heterochrony and miniaturization to give new morphological combinations" (1993, p. 23). This variation has been couched explicitly in terms of heterochrony in numerous studies (Table II). Many of these studies principally are concerned with identifying changes in the timing or sequence of development of one or more individual characters, including those that underlie instances of evolutionary loss that are frequently associated with paedomorphosis (e.g., Hanken and Hall, 1984; Smirnov, 1989; D. B. Wake, 1980). Others constitute more comprehensive analyses that attempt to define, in terms of underlying developmental processes and mechanisms, particular pertubations

TABLE II Selected Examples of Studies of Heterochrony in Embryonic, Larval, and Adult Amphibians

Topic or feature addressed	Taxa[a]	Citation[b]
Origin and evolution of early and Recent amphibians	Temnospondyls, labyrinthodonts, lissamphibians (F, S, C)	9–11, 40, 41, 63
Skeletal morphology and dentition; progenesis and miniaturization, paedomorphosis, neoteny		
Paedomorphosis		
Phylogeny	S: *Ambystoma, Rhyacosiredon, Aneides, Batrachoseps*	34, 68, 84, 91
Genetic basis	S: *Ambystoma*	32, 65, 66, 70, 84, 85
Facultative	S: *Ambystoma, Notophthalmus*, others	6, 31, 33, 35, 64, 98, 99
Endocrine control		
Paedomorphosis, neoteny, perennibranchiation	S: *Ambystoma, Triturus, Proteus, Necturus, Pleurodeles, Eurycea*, others	54, 55
Direct development	F: *Eleutherodactylus*	30
Genome size and neoteny	S	83
Life-history evolution	C: *Siphonops, Gymnopis, Ichthyophis, Typhlonectes, Hypogeophis*	21
	S: *Salamandra*	17b
Population structure, diet, paedogenesis	S: *Eurycea, Gyrinophilus*	12, 13
Larval paedomorphosis	S: *Gyrinophilus*	15
Resource-based (trophic) polymorphism	S: *Ambystoma*	16, 95
Gonadal development, hemoglobins, nitrogen excretion, integument; neoteny	S: *Hynobius*, others	86–88
Embryonic development	C: *Typhlonectes*	60
	F: *Eleutherodactylus*	29, 61
	S: *Ambystoma, Gyrinophilus, Taricha*	14, 15, 44

Larval development and adult limb morphology	F: *Hyla, Rana, Bufo, Bombina*	8, 18–20
Larval pigment patterning	S: *Ambystoma*	37
	F: *Pseudis*	17a
Larval trophic morphology and behavior	F: *Lepidobatrachus*	27, 59
External morphology, size and shape; paedomorphosis	S: *Ambystoma, Rhyacosiredon, Notophthalmus, Bolitoglossa*	22, 31, 69
Cloacal anatomy; paedomorphosis	S: *Desmognathus, Phaeognathus, Leurognathus*	67
Neuroanatomy, sensory systems; paedomorphosis	C: *Epicrionops, Ichthyophis, Uraeotyphlus, others*	57, 62
	F: *Eleutherodactylus, Discoglossus, Bombina, Pipa, Ascaphus, others*	57, 61, 74
	S: *Batrachoseps, Plethodon, Thorius, Bolitoglossa, Hydromantes, Desmognathus, Ambystoma, others*	36, 44, 56–58, 77
Oral anatomy	F: *Gastrotheca, Ascaphus, Microhyla, Rhinophrynus, Scaphiopus, Alytes, Centrolenella, Colostethus, others*	96, 97
Larval and adult dentition	S: *Ambystoma, Rhyacosiredon, Hynobius, Eurycea, Gyrinophilus, others*	7, 42, 45
	F: *Pipa, Pyxicephalus, Pelobates, Uperoleia, Crinia, Pseudophryne*	17, 76
Cranial and hyobranchial morphology, ossification sequence and timing; paedomorphosis, neoteny, progenesis, peramorphosis, differential metamorphosis	S: *Batrachoseps, Notophthalmus, Thorius, Eurycea, Triturus, Aneides, Rhyacotriton, Siren, Cryptobranchus, Ambystoma, Rhyacosiredon, Necturus, Bolitoglossa, Salamandra, others*	1–5, 25, 39, 46–53, 89–92, 94
	F: *Rhinophrynus, Eleutherodactylus, Brachycephalus, Xenopus, Leiopelma, Uperoleia, Dendrobates, Ceratophrys, Bombina, Microhyla, Flectonotus, others*	17, 23, 24, 28, 29, 71–73, 75, 78–82, 100
	C: *Idiocranium*	93

(*Continues*)

TABLE II (Continued)

Topic or feature addressed	Taxa[a]	Citation[b]
Postcranial skeletal morphology, ossification sequence and timing; paedomorphosis, progenesis, neoteny	S: *Bolitoglossa, Thorius, Aneides, Necturus*	1–3, 26, 43, 94
	F: *Bombina, Leiopelma, Xenopus, Pseudophryne, Ceratophrys*, others	17, 38, 78, 79, 82, 100
	C: *Idiocranium*	93

[a]C, caecilians; F, frogs; S, salamanders.

[b]1, Alberch, 1980; 2, Alberch, 1983; 3, Alberch and Alberch, 1981; 4, Alberch and Blanco, 1996; 5, Alberch et al., 1979; 6, Begun and Collins, 1992; 7, Beneski and Larsen, 1989; 8, Blouin, 1991; 9, Bolt, 1977; 10, Bolt, 1979; 11, Bolt, 1991; 12, Bruce, 1979; 13, Bruce, 1976; 14, Collazo, 1994; 15, Collazo and Marks, 1994; 16, Collins et al., 1983; 17, Davies, 1989; 17a, de Sá and Lavilla, 1997; 17b, Dopazo and Alberch, 1994; 18, Emerson, 1986; 19, Emerson, 1987; 20, Emerson, 1988; 21, Exbrayat and Hraoui-Bloquet, 1994; 22, Green and Alberch, 1981; 23, Haas, 1995; 24, Haas, 1996a; 25, Hanken, 1982; 26, Hanken, 1984; 27, Hanken, 1993; 28, Hanken and Hall, 1984; 29, Hanken et al., 1992; 30, Hanken et al., 1997a; 31, Harris, 1989; 32, Harris et al., 1990; 33, Jackson and Semlitsch, 1993; 34, Larson, 1980; 35, Licht, 1992; 36, Linke and Roth, 1990; 37, Löfberg et al., 1989; 38, Maglia and Pügener, 1998; 39, Marconi and Simonetta, 1988; 40, Milner, 1988; 41, Milner, 1993; 42, Mutz and Clemen, 1992; 43, Naylor, 1978; 44, Northcutt et al., 1994; 45, Pedersen, 1991; 46, Reilly, 1986; 47, Reilly, 1987; 48, Reilly, 1994, 49, Reilly and Altig, 1996; 50, Reilly and Brandon, 1994; 51, Reilly et al., 1997; 52, Roček, 1996; 53, Rose, 1995; 54, Rose, 1996; 55, Rosenkilde and Ussing, 1996; 56, Roth and Schmidt, 1993; 57, Roth et al., 1992; 58, Roth et al., 1993; 59, Ruibal and Thomas, 1988; 60, Sammouri et al., 1990; 61, Schlosser and Roth, 1997; 62, Schmidt and Wake, 1997; 63, Schoch, 1995; 64, Scott, 1993; 65, Semlitsch and Wilbur, 1989; 66, Semlitsch et al., 1990; 67, Sever and Trauth, 1990; 68, Shaffer, 1984a; 69, Shaffer, 1984b; 70, Shaffer and Voss, 1996; 71, Smirnov, 1989; 72, Smirnov, 1990; 73, Smirnov, 1991; 74, Smirnov, 1993; 75, Smirnov, 1994; 76, Smirnov and Vasil'eva, 1995; 77, Smith et al., 1988; 78, Stephenson, 1960; 79, Stephenson, 1965; 80, Trueb, 1985; 81, Trueb and Alberch, 1985; 82, Trueb and Hanken, 1992; 83, Vignali and Nardi, 1996; 84, Voss, 1996; 85, Voss and Shaffer, 1996; 86, Wakahara, 1996; 87, Wakahara and Yamaguchi, 1996; 88, Wakahara et al., 1994; 89, D. B. Wake, 1966; 90, D. B. Wake, 1980; 91, D. B. Wake, 1989; 92, D. B. Wake and Larson, 1987; 93, M. H. Wake, 1986; 94, T. A. Wake et al., 1983; 95, Walls et al., 1993; 96, Wassersug, 1980; 97, Wassersug and Duellman, 1984; 98, Whiteman, 1994; 99, Whiteman et al., 1996; 100, Wild, 1997.

to the presumed ancestral ontogeny that are associated with specific hetero-chronic changes (e.g., Rose, 1996; Shaffer and Voss, 1996). Effects of phyloge-netic change in the relative timing or rate of character development in embryos and larvae range from subtle, yet functionally significant changes in adult body proportions (Blouin, 1991; Emerson, 1986, 1987, 1988) to large-scale shifts of entire suites of functionally integrated characters from one life-history stage to another (Carroll *et al.*, 1991; Hanken, 1993; Ruibal and Thomas, 1988). Heterochronic analysis has been especially important in suggesting the poten-tial for relatively small changes in developmental processes to effect large-scale changes in larval and adult morphology by shifting development along evolutionarily conserved ontogenetic trajectories (Alberch, 1980; Alberch *et al.*, 1979; D. B. Wake and Larson, 1987).

There are, nevertheless, several difficulties in assessing the heterochrony literature in amphibians. First, it is vast; the 100 studies listed in Table II are only representative of a much larger number of analyses that address evolutionary changes in developmental timing in larval and adult amphibians, both living and extinct. A more fundamental problem, however, is the large and to some extent unwieldy terminology that has characterized the field, especially the diverse and sometimes contradictory use of terms such as neo-teny, paedomorphosis, and paedogenesis. Frequently, the same term is used to describe vastly different evolutionary phenomena, which in turn likely reflect very different underlying developmental mechanisms. Conversely, sev-eral terms often are used in reference to the same phenomenon. Many endocri-nologists and molecular biologists, for example, employ the traditional defini-tion of neoteny as it applies to the numerous species of salamanders that typically retain a virtually intact larval morphology throughout life, such as the axolotl, *Ambystoma mexicanum,* and the mudpuppy, *Necturus maculosus* (e.g., Rosenkilde and Ussing, 1996; Vignali and Nardi, 1996; Wakahara, 1996). Whereas some contemporary comparative biologists endorse this definition (e.g., Reilly *et al.*, 1997), others favor a broader definition that emphasizes relative rates of development between ancestor and descendant taxa (Alberch *et al.*, 1979). They instead describe the extreme pattern of development in "neotenic" urodeles such as the axolotl and mudpuppy by the terms "larval reproduction" (Shaffer, 1984a), "paedomorphosis" (Voss and Shaffer, 1996), or "perennibranchiation" (Rose, 1996).

Such variable and inconsistent terminology complicates comparisons among studies; arguably it is retarding a synthesis of existing data. It also inhibits analyses of processes and mechanisms that underlie altered patterns of develop-ment. Although it may be unreasonable or even unnecessary to expect all authors to subscribe to and consistently employ the same, identical terminology for heterochrony, they should at least always define precisely how specific terms are being used.

V. DEVELOPMENTAL MECHANISMS OF MORPHOLOGICAL DIVERSIFICATION

Most studies of the evolution and development of amphibian larvae have focused on describing derived *patterns* of ontogeny in terms of the presence or absence of specific characters, or character states, and their developmental timing (i.e., heterochrony; see the previous discussion). Whereas such data have been used in some instances to generate hypotheses regarding the nature of specific perturbations to the ancestral ontogeny that underlie evolution of one or more derived patterns (e.g., Rose, 1996), there have been relatively few attempts to probe empirically and identify the actual developmental *mechanisms* involved. Among the notable exceptions are studies conducted over many years on the endocrinological basis of neoteny (larval reproduction) in the axolotl and other species of perennibranchiate urodeles (e.g., Rosenkilde and Ussing, 1996) and an embryological and molecular analysis of the evolutionary loss of the larval cement gland in the direct-developing frog, *Eleutherodactylus coqui* (Fang and Elinson, 1996). Many more such studies are needed.

Examples of the kinds of important insights into the developmental bases of evolutionary change that can be derived from such a mechanistic approach are provided by analyses of larval pigment patterning in urodeles. These analyses also demonstrate the additional benefits that accrue when comparative developmental studies are performed in an explicitly phylogenetic context. The studies address the extensive interspecific variation in external coloration that is found among urodele larvae. Notwithstanding the myriad subtle manifestations of pigment patterning in individual species, variation in larval coloration can be described largely in terms of two general pattern "elements:" vertical bars and horizontal stripes (Epperlein *et al.*, 1996; Olsson and Löfberg, 1992; Parichy, 1996a; Fig. 8). These pattern elements, in turn, form as a result of the migration and differentiation of three types of chromatophores, or pigment cells: black melanophores, yellow xanthophores, and, to a lesser extent, silvery iridophores. The two general pattern elements demonstrate a complex, overlapping distribution when superimposed on a scheme of phylogenetic relationships among urodele taxa (Parichy, 1996a; Fig. 9). One of the implications of the particular distribution of alternate character states (presence or absence) is that vertical barring has been acquired or lost two or more times independently in different lineages (Olsson, 1994; Fig. 10). Consequently, the absence of vertical barring that characterizes the salamandrid species *Triturus alpestris* and the ambystomatid species *Ambystoma maculatum* may not reflect simply the retention of a shared feature from the common ancestor of both species. Instead, it likely reflects the loss in *A. maculatum* of vertical barring, which evolved earlier in the common ancestor of all living ambystomatids (Olsson, 1993).

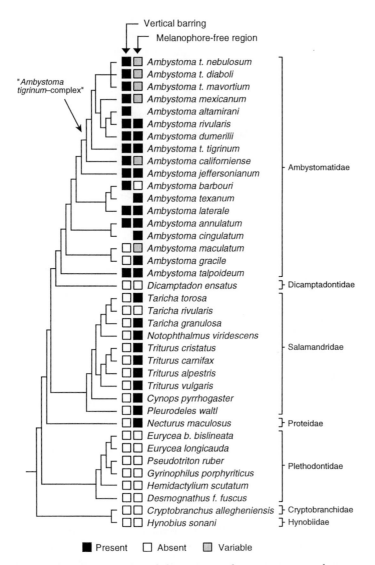

FIGURE 9 Phylogenetic distribution of the two general pigment–pattern elements—vertical barring and melanophore-free regions (horizontal stripes)—among urodele larvae representing several extant families. The particular scheme of phylogenetic relationships depicted is one of several alternate hypotheses derived from independent molecular and morphological analyses. Modified with permission from Parichy (1996a).

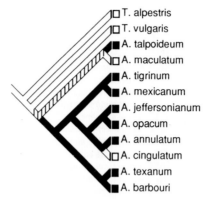

FIGURE 10   Evolution of vertical barring in the larval pigment pattern of urodeles. Black shading indicates that barring is present; in unshaded (white) areas barring is absent. Stripes indicate that the character state is equivocal (cannot be determined from existing data). T, *Triturus* (Salamandridae); A, *Ambystoma* (Ambystomatidae). The scheme of relationships among these 12 species is one of several possible phylogenetic hypotheses derived from consideration of both morphological and allozyme data. The particular distribution of character states shown here supports two equally parsimonious interpretations of the evolution of vertical barring: one gain and two losses, and two gains and one loss. Olsson (1994). Pigment pattern formation in larval ambystomatid salamanders: *Ambystoma talpoideum, Ambystoma barbouri,* and *Ambystoma annulatum. Journal of Morphology,* Copyright © 1994. Reprinted by permission of Wiley-Liss, Inc., subsidiary of John Wiley & Sons, Inc.

Whereas the mechanisms underlying pigment pattern formation in larval urodeles are far from understood completely, much fundamental information concerning the identity of and relationships among relevant components has been amassed in the last few years. In documenting the large number and diversity of developmental processes involved, these studies reveal the surprising complexity of mechanisms responsible for what might otherwise be regarded as a relatively simple patterning system. For example, development of vertical bars, one of the two basic patterning elements, occurs in two phases (Epperlein *et al.,* 1996; Parichy, 1996a). In the first phase, premigratory neural crest cells, from which pigment cells are derived, form a longitudinal series of cell aggregates dorsal to the neural tube (Fig. 11A). These aggregates eventually contain xanthophores, but not melanophores. While this "prepattern" is forming in the dorsal part of the embryo, melanophore precursors migrate out onto the dorsolateral portion of the flanks. In the second phase of vertical bar development, xanthophores migrate laterally from the cell aggregates atop the neural tube and invade regions of the flanks previously occupied by melanoblasts (Fig. 11B). Simultaneously, the melanophores aggregate among themselves and at a distance from the xanthophores. This results in a series of

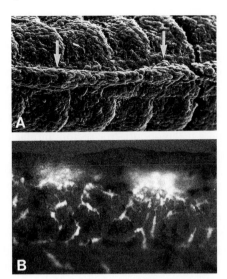

FIGURE 11   (A) Scanning electron micrograph of the trunk region of a stage 31 embryo of the ringed salamander, *Ambystoma annulatum* (Ambystomatidae). Dorsal view; surface ectoderm has been removed. Premigratory neural crest cells are beginning to form a longitudinal series of aggregates (arrows) dorsal to the neural tube and paired somites. (B) Lateral view of a slightly older, intact embryo (stage 38), viewed with fluorescence microscopy. Xanthophores (stained white with ammonia fluorescence) are beginning to migrate ventrolaterally from the crest cell aggregates. Olsson (1994). Pigment pattern formation in larval ambystomatid salamanders: *Ambystoma talpoideum, Ambystoma barbouri,* and *Ambystoma annulatum. Journal of Morphology,* Copyright © 1994. Reprinted by permission of Wiley-Liss, Inc., a subsidiary of John Wiley & Sons, Inc.

alternating bars, each principally containing one or the other of the two pigment cell types (Fig. 8). Factors governing the complex cell movements involved in these patterning events still are not well-understood, but likely include various kinds of cell–cell and cell–extracellular matrix interactions. Finally, there is compelling evidence that at least some of the preceding processes are directly involved in the evolution of pigment patterns (Epperlein *et al.,* 1996; Parichy, 1996a). In *A. maculatum,* for example, loss of vertical bars is correlated with the failure of premigratory neural crest cells to aggregate along the neural tube. Also, melanophores and xanthophores appear to migrate onto the flanks simultaneously in this species and not sequentially, as in species with vertical bars (Olsson, 1993).

Development of the second principal pattern element, horizontal stripes of melanophores on either side of a melanophore-free region, is mediated by processes that are both similar to and different from those involved in the development of vertical bars (Epperlein *et al.,* 1996; Parichy, 1996a). Especially important are interactions between pigment cells and the developing lateral lines, which comprise a neurosensory system that provides a mechanoreceptive sense important to aquatic vertebrates (Parichy, 1996a,b,c; Fig. 12). This too is an ancient component of the pigment-patterning mechanism in salamanders; it likely was present in the common ancestor of the families Ambystomatidae and Salamandridae and appears to be retained in most living species. Interestingly, however, at least one salamandrid species (*Taricha torosa*) possesses horizontal stripes and a melanophore-free region, but lacks the close coupling between pigment patterning and lateral-line formation. The appearance of novel, extracellular cues for melanophore localization accompanies the apparent loss of this component of the ancestral patterning mechanism (Parichy, 1996b; Fig. 13).

Horizontal stripe formation adds to the growing list of examples from evolutionary developmental biology in which the same or similar phenotype in different, often closely related, species is produced by different underlying mechanisms (Hall, 1995; Raff, 1996). Evolution of developmental mechanisms need not lead to evolution of phenotypes. This demonstrates the difficulty of inferring the nature of developmental mechanisms from their end products. It also cautions against strict use of features of development as unambiguous criteria for assessing the homology of larval or adult characters.

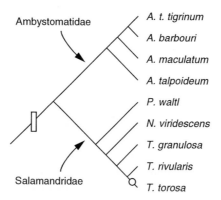

FIGURE 13 Hypothesis for the evolution of pattern-forming mechanisms underlying development of the horizontal stripe (melanophore-free region) in larval urodeles. This scenario posits the presence of a primitive, lateral-line-dependent pattern-forming mechanism in the common ancestor of the families Ambystomatidae and Salamandridae (open rectangle), which is retained in descendant taxa. A second, redundant, lateral-line-independent mechanism for stripe formation so far is known in only one species, *T. torosa*, and likely evolved much more recently (open circle). *A, Ambystoma; P., Pleurodeles; N., Notophthalmus; T., Taricha.* Reprinted with permission from Parichy (1996b).

VI. LOSS OF LARVAE

The presumed ancestral life history for all Recent amphibians comprises aquatic eggs and larvae and terrestrial adults. This biphasic, or metamorphic, ontogeny is retained in many living species of frogs, salamanders, and caecilians, although in many taxa embryonic and larval development has been altered to effect considerable morphological divergence among species. There are, however, a wide variety of alternate life histories and reproductive modes that have evolved independently within all three living orders (Altig and Johnston, 1989; Duellman and Trueb, 1986). Their distinguishing features range from virtually complete loss of the terrestrial adult (postmetamorphic) stage and obligate larval reproduction (e.g., the axolotl, *Ambystoma mexicanum*) to loss of the free-living, aquatic larva. Instances of larval loss are especially interesting, insofar as they can offer unique insights into the opportunities for adaptation and specialization that are conferred by the ancestral metamorphic ontogeny. They also provide opportunities to assess the possible role of larval features in mediating, or even constraining, the evolution of adult features. Two distinct examples of larval loss are discussed here: direct development and viviparity.

A. Direct Development

In direct development, most adult features form in the embryo and are present at hatching; there is no free-living larva. It is so named to distinguish it from the ancestral, "indirect" mode of development, in which most adult features form during the (posthatching) metamorphosis that follows the larval stage. Direct development has evolved independently within each of the three living amphibian orders and characterizes many hundreds of species of frogs, salamanders, and caecilians (Duellman and Trueb, 1986; M. H. Wake, 1989). The neotropical anuran genus *Eleutherodactylus* alone includes more than 500 species (Duellman, 1993), all of which are believed to display direct development or some even more extreme modification thereof (e.g., ovoviviparity in *E. jasperi;* M. H. Wake, 1978, 1993). Yet, the three orders differ considerably with respect to the phylogenetic distribution of direct development among their component taxa. At one extreme are the anurans. Direct development occurs in one or more species in each of at least 10 different families or family-level taxa of frogs (Fig. 14). Each of these lineages is believed to represent an independent acquisition of direct development as all also contain metamorphosing taxa, which represent the presumed ancestral life history. Accordingly, direct development must have evolved at least 10 times just in anurans. This number represents a conservative estimate, as direct development likely evolved more than once in at least some lineages (Duellman, 1989; Duellman

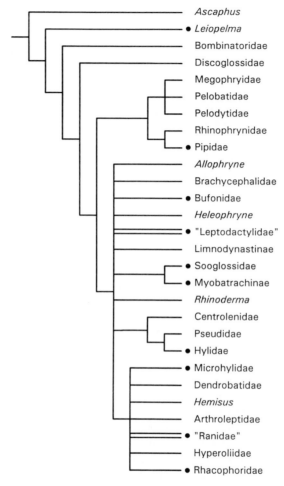

FIGURE 14   Phylogenetic distribution of direct development (●) among extant families and
family level taxa of frogs [based on Duellman and Trueb (1986)]. Relationships depicted are based
on Ford and Cannatella (1993). [Reprinted with permission from Ford and Cannatella (1993).]

and Trueb, 1986). Despite its repeated evolution and widespread distribution
among anurans, direct development nevertheless is not the predominant reproductive mode in living frogs (Duellman, 1989).

   At the other extreme are urodeles. Embryonic development outside the
reproductive tract of the female parent that proceeds directly to a terrestrial
hatchling without a free-living larval stage is restricted to just one of ten Recent
families, the Plethodontidae, or lungless salamanders (D. B. Wake and Hanken,

1996; D. B. Wake and Marks, 1993; Fig. 15). However, because the Plethodontidae account for nearly two-thirds of all species of living salamanders and because most plethodontids have direct development, it is the predominant reproductive mode in urodeles.

Caecilians represent an intermediate condition. The biphasic life history, including free-living, aquatic larvae, is retained by some or all species in four of the six extant families (Duellman and Trueb, 1986; Nussbaum and Wilkinson, 1989; M. H. Wake, 1993; Wilkinson and Nussbaum, 1996). Direct development occurs in at least three families and includes many instances of viviparity (see the following section). Whereas general trends in life-history evolution in caecilians have been defined, detailed knowledge of evolutionary patterns in most clades awaits more robust phylogenetic hypotheses for the group and reliable life-history data for many more species (M. H. Wake, 1977a, 1982, 1989, 1993; Wilkinson and Nussbaum, 1996).

Whereas there is widespread acceptance that direct development has evolved many times independently within and among the major groups of living amphibians, determination of the exact number of times has proven much more difficult. Phylogenetic analyses of plethodontid salamanders, for example, implicate as few as one and as many as five separate instances of the evolution of direct development in this clade (Collazo and Marks, 1994; D. B. Wake and Hanken, 1996). Determination of the exact number and phylogenetic position(s) of instances of the evolution of direct development are not trivial issues. In plethodontids, some plausible phylogenetic scenarios for the evolution of direct development necessarily would imply that a free-living aquatic larva must have reappeared one or more times in the history of the group (D. B. Wake and Hanken, 1996). Similar scenarios exist for the so-called

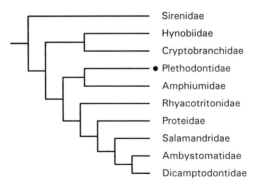

FIGURE 15   Phylogenetic distribution of direct development (●) among the 10 extant families of salamanders. Relationships depicted are based on Larson and Dimmick (1993). [Reprinted with permission from Larson and Dimmick (1993).]

marsupial frogs of South America (Hylidae: Hemiphractinae), which include
both direct-developing and metamorphosing taxa (Duellman and Hillis, 1987;
Duellman *et al.*, 1988; Wassersug and Duellman, 1984; but see Haas, 1996a,b).
Resolution of these problems awaits more robust phylogenetic hypotheses than
are available. Developmental data also may make an important contribution
if it can be shown that "direct development" is not a uniform developmental
mode wherever it exists within a particular clade, but instead comprises two
or more distinct ontogenetic patterns. For example, preliminary evidence that
direct development has evolved at least two or even three times within the
Plethodontidae is seen in the contrasting patterns of embryonic development
of several organ systems that distinguish several distantly related taxa (D. B.
Wake and Hanken, 1996).

    Almost by definition, direct development entails the precocious (embryonic)
formation of adult features, which typically form during (posthatching) metamorphosis in species that display the ancestral, biphasic life history. Most
studies of the embryology and developmental biology of direct-developing
amphibians have been concerned primarily with assessing the extent to which
larval features have been retained in or lost from this derived ontogeny (e.g.,
Orton, 1949; Wassersug and Duellman, 1984). As one might expect from a
characteristic that has evolved repeatedly, there is considerable variation among
direct-developing taxa with respect to the extent to which a given species or
members of a particular clade either recapitulate or lack larval features (Fig.
16). The eastern North American urodele *Desmognathus aeneus* lacks a free-
living, aquatic larva, but nevertheless it displays many larva-specific traits during
embryonic or immediate posthatching development (Marks, 1994; Marks and
Collazo, 1998). On the other hand, the many species of direct-developing
Eleutherodactylus retain relatively few larva-specific features, either externally
or internally (Hanken *et al.*, 1992; Hughes, 1959; Lynn, 1942). Indeed, *Eleutherodactylus* has long been regarded as among the most extreme examples of direct
development in amphibians in terms of the great extent to which the species
deviate from the ancestral, metamorphic ontogeny (Elinson, 1990; Orton,
1951).

    Relatively few studies have attempted to probe the developmental mecha-
nisms that underlie the evolution of direct development (Elinson, 1990). These
mechanisms, however, are the subject of several analyses, which address both
the loss of larval features and the embryonic formation of adult features.
Although our understanding of the developmental basis of direct development
is far from complete, the picture that is beginning to emerge is of a series
of perturbations to developmental processes at several levels of biological
organization, including gene expression, endocrine control, morphogenesis,
and pattern formation (Callery and Elinson, 1996; Elinson, 1994; Moury and
Hanken, 1995; Hanken *et al.*, 1997a,b; Jennings and Hanken, 1998; Richardson

Direct development in frogs

Simplification and loss of larval structures in embryos of some species that complete metamorphic changes before hatching

A typical aquatic tadpole for comparison

Shows typical structures and proportions: closed operculum, spiracle, internal foreleg buds, complex mouthparts, strong tail.

Pipa pipa (family Pipidae)

Yolk supply greatly increased, but embryo still develops all of the characters of a typical pipid tadpole.

Rhinoderma darwinii (family Atelopodidae)

Yolk supply moderately increased, but embryo is a typical tadpole in all important characters.

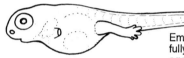

Nectophrynoides tornieri (family Bufonidae)

Embryo has many larval structures. Operculum and spiracle fully developed, foreleg buds internal, jaw muscles and cartilages larval, external mouthparts vestigial or absent, gill structures vestigial. Tail long and thin, consists chiefly of fins and notochord.

Hemiphractus divaricatus (family Hylidae)

Many larval structures are absent of vestigial. No spiracle, operculum remains open, foreleg bud exposed. Gills very specialized, have long gill stalks and sheet-like respiratory surfaces. Mouthparts greatly simplified, but still essentially larval. Tail rudimentary.

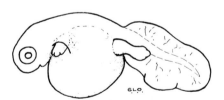

Eleutherodactylus cooki (family Leptodactylidae)

Most larval structures absent. Mouthparts embryonic, operculum reduced to vestige over base of exposed foreleg bud, no spiracle, gills vestigial or absent. Tail fins enlarged, vascular, function as a respiratory surface.

FIGURE 16   Direct development in frogs. Grace Orton's (1951) well-known illustration depicts the variable extent to which larval features have been lost during the evolution of direct development in different lineages of Recent anurans. Species are arranged in a cline from a typical tadpole (top), which represents taxa that retain the ancestral, biphasic life history, to *Eleutherodactylus* (bottom), which retains relatively few larval-specific features. Reprinted with permission from Orton (1951) courtesy of Ward's Natural Science Establishment, Inc.

et al., 1998). For example, evolutionary loss of the larval cement gland in direct-developing *Eleutherodactylus coqui*, a species native to Puerto Rico, is associated with altered spatial and temporal patterns of expression of distal-less genes during embryogenesis and modifications to the inductive interactions that mediate differentiation of the oral integument (Fig. 17; Fang and Elinson, 1996).

Direct development in plethodontid salamanders has been used to evaluate and support hypotheses regarding the existence of larval constraints on adult morphology in urodeles in general (D. B. Wake and Hanken, 1996; D. B. Wake and Marks, 1993). According to these hypotheses, the existence of fully differentiated, functionally specialized larval structures in metamorphosing salamanders limits, or constrains, the morphology of adult structures that can subsequently form at metamorphosis in these species (D. B. Wake, 1982; D. B. Wake and Roth, 1989). Because the evolution of direct development entails loss of the larva as a discrete, free-living, life-history stage, it also offers at least the possibility for loss of the larval constraint(s) on adult morphology. Direct-developing plethodontids offer at least circumstantial support for the larval constraint hypothesis, insofar as these species possess some of the most highly derived and complex functional systems found in any living salamanders. One example is the complex of muscular, skeletal, and nervous components that comprise the hyolingual apparatus, which mediates the tongue projection characteristic of many direct-developing species (Deban *et al.,* 1997; Lombard and Wake, 1977; Roth and Wake, 1985). If larval constraints have been relaxed or even lost in direct-developing plethontids, then we might expect to see the presence of derived adult morphologies correlated with novel developmental patterns of embryonic development that do more than simply recapitulate the ancestral ontogeny. Evidence for such "ontogenetic repatterning" in direct-developing plethodontids comes from several studies of the development of the limbs, the brain and cranial nerves, and the hyolingual skeleton (Alberch, 1987, 1989; Roth *et al.,* 1988; Shubin, 1995; Shubin and Wake, 1991; D. B. Wake and Roth, 1989; D. B. Wake and Shubin, 1994; D. B. Wake *et al.,* 1988).

Do larval features constrain adult morphology in other amphibians either in the same way or to the same extent that they appear to do so in many urodeles? Studies of at least some direct-developing frogs suggest that the answer is no. In *Eleutherodactylus,* for example, pronounced evolutionary modifications to embryonic development, including the loss of many larval components, are not associated with the origin of novel morphologies that characterizes many direct-developing (plethodontid) salamanders (Hanken, 1992; Hanken *et al.,* 1992). Moreover, the developmental mechanisms that mediate anuran metamorphosis appear to confer much greater developmental independence between larvae and adults than do comparable mechanisms in most metamorphosing urodeles (cf. Alberch, 1987, 1989; Elinson, 1990). Thus, in anurans, developmental mechanisms that mediate metamorphosis likely may facilitate, if not actually promote, the evolutionary dissociation of embryonic, larval, and adult features (Hanken *et al.*, 1997b).

B. VIVIPARITY

Viviparity, or live-bearing, refers to an array of reproductive mechanisms for embryonic or fetal maintenance by either the maternal or paternal parent. Viviparity has evolved independently in many groups of animals, especially vertebrates, and specific definitions of the term vary widely among both taxonomic groups and authors (Altig and Johnston, 1989; M. H. Wake, 1989, 1992). In amphibians, viviparity entails the retention of embryos in the maternal oviducts through development to the adult (postmetamorphic) stage, with maternally derived nutrition provided or consumed following exhaustion of the embryo's yolk supply (M. H. Wake, 1982). Viviparous amphibians lack a free-living larva.

Viviparity has evolved one or more times within each of the three Recent amphibian orders (M. H. Wake, 1977a, 1989, 1993). As with direct development, however, the orders differ considerably with respect to the prevalence of viviparity and its impact on the evolutionary history of each group. Obligate viviparity is very rare in frogs and salamanders, accounting for less than 1% of extant species. One species of salamander exhibits "optional viviparity" associated with intrauterine cannibalism, in which prenatal young feed on unfertilized eggs and developing siblings within the mother's reproductive tract (Dopazo and Alberch, 1994). A few additional species are ovoviviparous; embryos are retained but nutrition is exclusively yolk-dependent and (in salamanders) the young may be born as larvae (Alcobendas *et al.*, 1996; Joly, 1968). In caecilians, however, viviparity is the predominant reproductive mode among the approximately 170 described species; it has evolved at least two and as many as four times.

Several aspects of the developmental biology of amphibian viviparity have received considerable attention, especially those that concern fetal–maternal relations. These include studies of reproductive endocrinology (Exbrayat, 1992; Exbrayat and Morel, 1995; Xavier, 1977, 1986), urogenital anatomy (Exbrayat, 1983, 1984; Hraoui-Bloquet *et al.*, 1994; M. H. Wake, 1970, 1977b), reproductive physiology (Toews and Macintyre, 1977), fetal nutrition and growth rates (Dopazo and Alberch, 1994; Exbrayat and Hraoui-Bloquet, 1992; Hraoui-Bloquet and Exbrayat, 1992; M. H. Wake, 1980a), and organogenesis (Hraoui-Bloquet and Exbrayat, 1994; Sammouri *et al.*, 1990; M. H. Wake, 1976, 1980b,c; M. H. Wake and Hanken, 1982; M. H. Wake *et al.*, 1985). The taxonomic coverage of these treatments is spotty, however, and the comparative database for each topic is relatively small. Many do not concern larval biology per se and therefore are beyond the scope of this chapter. Interested readers are referred to several reviews and associated primary literature (see the references listed in this paragraph).

One important aspect of amphibian viviparity that does relate directly to larval biology and development is the evolutionary fate of larva-specific features. In general, there are two sharply contrasting fates. First, and as would be expected given that the neonate is born as a juvenile (postmetamorphic) adult, many larval features simply are lost. For example, several oral integumentary structures that contribute to the unique trophic apparatus that is characteristic of many anuran larvae are absent from embryos of the viviparous African frog, *Nectophrynoides occidentalis* (M. H. Wake, 1980c). These include the keratinized (horny) beaks and labial teeth. In lacking many larva-specific features, viviparous *N. occidentalis* exemplifies a predominant trend that is exhibited by several additional ovoviviparous or direct-developing species within the genus *Nectophrynoides,* all of which lack a free-living larva.

The second and arguably more interesting fate of larval features is their recruitment to perform new physiological functions in the developing embryo or fetus. Most of the examples offered to date center around apparent specializations for the transfer of nutrients and other substances from the mother to the fetus while it is still within the maternal oviduct. In *N. occidentalis,* for example, the fetal frog is believed to ingest uterine secretions orally as a supplemental food source following exhaustion of the embryonic yolk (Lamotte and Xavier, 1972; Vilter and Lugand, 1959; Xavier, 1973; cited in M. H. Wake, 1980c). Correlated with this novel feeding mode is the retention and, indeed, elaboration of the network of external buccal papillae, which surround the mouth and assist food gathering in typical tadpoles. In typlonectid caecilians, all of which are viviparous, embryos and larvae grow large, saclike, and highly vascularized external gills, which are resorbed or shed only at or soon after hatching (Hraoui-Bloquet and Exbrayat, 1994; Toews and Macintyre, 1977). The morphology and development of these gills differ considerably from those in viviparous species in other caecilian families, which are resorbed well before birth (M. H. Wake, 1993; Fig. 18). In free-living amphibian larvae, external gills provide sites for gas exchange that are used in aquatic respiration. In one typhlonectid species, *Typhlonectes compressicaudus,* the fetal gills and maternal oviduct have been posited to function as a "pseudoplacenta" that effects gas exchange and nutrient transfer between the mother and her offspring (Delsol *et al.,* 1981; Exbrayat and Hraoui-Bloquet, 1992; Hraoui-Bloquet and Exbrayat, 1994; Hraoui-Bloquet *et al.,* 1994; Sammouri *et al.,* 1990; Toews and Macintyre, 1977). However, despite extensive study of fetal–maternal relations in this species, evidence for this "placental" function is largely circumstantial. The

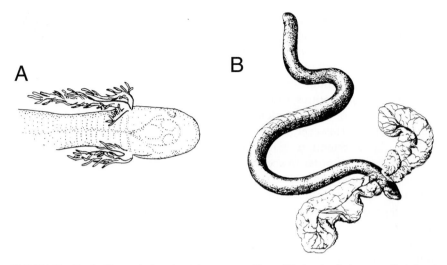

FIGURE 18   Fetal gill morphology in viviparous caecilians. (A) A pair of triramous, fimbriated, and relatively short gills are characteristic of nonviviparous and most viviparous species, such as *Dermophis mexicanus*. Reprinted with permission from M. H. Wake (1977a). (B) Viviparous typhlonectids, such as *Typhlonectes compressicaudus*, have a single, highly elongate, and saclike gill on each side of the head, which has been suggested to function as a "pseudoplacenta" (see text). Reprinted with permission from *Nature* (Toews and Macintyre, 1977) Copyright 1977. Macmillan Magazines Limited.

extent and mode of fetal–maternal exchange via the gills in this and other typhlonectid species await confirmation by direct observation and experiment (M. H. Wake, 1993).

Other features of the ancestral, complex life history may be recruited as well, including the biphasic pattern of development itself. For example, many viviparous caecilians possess a system for prenatal "food gathering" that is similar to but even more elaborate and complex than the one seen in *Nectophrynoides* described earlier; the fetus actively ingests secretions from the maternal oviducts (Exbrayat, 1984; Exbrayat and Hraoui-Bloquet, 1992; Hraoui-Bloquet and Exbrayat, 1992; M. H. Wake, 1977a,b). Associated with this novel feeding mode is the presence of highly specialized and unique fetal dentition (M. H. Wake, 1976, 1980b). The fetal teeth are shed at or shortly after birth, when they are replaced by the morphologically distinct and characteristic adult dentition. In effect, the biphasic pattern of tooth development and morphogenesis that is characteristic of the larval–adult transition in many metamorphosing amphibians is redeployed in the fetal–postnatal transition of viviparous caecilians.

Retention in viviparous amphibians of many larva-specific and related features of the ancestral, complex life history, despite loss of the corresponding, free-living larval stage, provides excellent examples of exaptation, i.e., the evolutionary cooption of preexisting adaptations in novel functional and ecological contexts (Gould and Vrba, 1982). Indeed, it seems likely that the existence of larval features, discrete from those in the adult, may have allowed, or even facilitated, the evolution of this novel reproductive mode in at least some instances. However, much greater knowledge of phylogenetic trends involving larval retention or loss, as well as their underlying developmental bases, is needed in order to accurately and reliably define the nature and role of such biases and constraints in the evolution of amphibian viviparity.

It also should be emphasized that the evolution of viviparity and other, related reproductive specializations in Recent amphibians involves much more than the simple retention, or even redeployment, of preexisting features of the ancestral ontogeny. For example, the morphology of individual teeth that comprise the fetal dentition in viviparous caecilians is unlike that seen in either the larval or the adult dentition of metamorphosing species (M. H. Wake, 1976). Many other aspects of early ontogeny, such as cranial ossification, similarly are highly derived in viviparous caecilians (M. H. Wake and Hanken, 1982) and salamanders (Alberch and Blanco, 1996) in comparison to the respective presumed ancestral ontogenies. In both instances, the sequence and timing of bone formation have been altered in similar fashion, apparently in response to similar functional demands imposed by precocious use of the jaws in feeding before birth. As with the evolution of most adaptations, viviparity comprises a complex mix of conserved and novel traits.

VII. CONCLUSION

The broad array of reproductive modes and life histories displayed by Recent amphibians constitutes a series of variations on the ancestral theme of metamorphosis (Hanken, 1992). These variations have allowed frogs, salamanders, and caecilians to diversify and adapt to a wide range of aquatic and terrestrial environments, as well as other ecological opportunities. A central element in this evolutionary success story is the free-living larva—retained or even amplified in many contexts and abandoned in others.

The evolution of larvae fundamentally involves modifications in development. Changes in developmental patterns, especially those concerned with the timing of developmental events, i.e., heterochrony, are well-documented. The mechanistic bases of these and other evolutionary changes in developmental pattern, however, generally remain poorly known. Both kinds of information are needed to reliably define the range of evolutionary opportunities and constraints conferred by the ancestral, metamorphic life history, as well as to assess the extent to which these opportunities and constraints vary among different amphibian lineages.

Many of the most important evolutionary changes in larval development likely involve pertubations to earlier ontogenetic stages, especially the embryo. Yet, despite its long history, the comparative embryology of amphibians is relatively poorly known. Consequently, renewed interest in comparative studies of development has revealed an unexpected diversity of molecular, cellular, and developmental patterns and mechanisms among amphibian embryos (e.g., Del Pino, 1989; Del Pino and Elinson, 1983; Elinson and Del Pino, 1985; Novoselov, 1995; Purcell and Keller, 1993; Tiedemann *et al.*, 1995), including comparisons of close phylogenetic relatives (Minsuk and Keller, 1996). Comparative embryological studies need to be extended to define the consequences of these and other evolutionary modifications of early development for larval, as well as adult, features.

Finally, studies of larval development and evolution will be most effective and insightful when data are analyzed in a rigorous phylogenetic context, both as a means of deriving the likely sequences of character transformation and to most effectively integrate these studies with comparable and complementary analyses of amphibian phylogeny and larval biology.

ACKNOWLEDGMENTS

I thank Tim Carl and John Foskett for providing valuable comments on earlier drafts. Research support is provided by the U.S. National Science Foundation (IBN 94-19407 to J. H.).

REFERENCES

Abouheif, E. (1997). Developmental genetics and homology: A hierarchical approach. *Trends Ecol. Evol.* **12**, 405–408.

Alberch, P. (1980). Ontogenesis and morphological diversification. *Am. Zool.* **20**, 653–667.

Alberch, P. (1983). Morphological variation in the neotropical salamander genus *Bolitoglossa*. *Evolution (Lawrence, Kans.)* **37**, 906–919.

Alberch, P. (1987). Evolution of a developmental process—Irreversibility and redundancy in amphibian metamorphosis. In "Development as an Evolutionary Process" (R. A. Raff and E. C. Raff, eds.), pp. 23–46. Liss, New York.

Alberch, P. (1989). Development and the evolution of amphibian metamorphosis. In "Trends in Vertebrate Morphology" (H. Splechtna and H. Hilgers, eds.), pp. 163–173. Fischer, Stuttgart.

Alberch, P., and Alberch, J. (1981). Heterochronic mechanisms of morphological diversification and evolutionary change in the neotropical salamander, *Bolitoglossa occidentalis* (Amphibia: Plethodontidae). *J. Morphol.* **167**, 249–264.

Alberch, P., and Blanco, M. J. (1996). Evolutionary patterns in ontogenetic transformation: From laws to regularities. *Int. J. Dev. Biol.* **40**, 845–858.

Alberch, P., Gould, S. J., Oster, G. F., and Wake, D. B. (1979). Size and shape in ontogeny and phylogeny. *Paleobiology* 5, 296–317.

Alcobendas, M., Dopazo, H., and Alberch, P. (1996). Geographic variation in allozymes of populations of *Salamandra salamandra* (Amphibia: Urodela) exhibiting distinct reproductive modes. *J. Evol. Biol.* 9, 83–102.

Altig, R., and Johnston, G. F. (1989). Guilds of anuran larvae: Relationships among developmental modes, morphologies, and habitats. *Herpetol. Monogr.* 3, 81–109.

Armstrong, J. B., and Malacinski, G. M., eds. (1989). "Developmental Biology of the Axolotl." Oxford University Press, New York.

Barlow, L. A., and Northcutt, R. G. (1995). Embryonic origin of amphibian taste buds. *Dev. Biol.* 169, 273–285.

Barlow, L. A., and Northcutt, R. G. (1997). Taste buds develop autonomously from endoderm without induction by cephalic neural crest or paraxial mesoderm. *Development (Cambridge, UK)* 124, 949–957.

Begun, D. J., and Collins, J. P. (1992). Biochemical plasticity in the Arizona tiger salamander *Ambystoma tigrinum nebulosum*. *J. Hered.* 83, 224–227.

Beneski, J. T., Jr., and Larsen, J. H., Jr. (1989). Interspecific, ontogenetic, and life history variation in the tooth morphology of mole salamanders (Amphibia, Urodela, and Ambystomatidae). *J. Morphol.* 199, 53–69.

Blouin, M. S. (1991). Proximate developmental causes of limb length variation between *Hyla cinerea* and *Hyla gratiosa* (Anura: Hylidae). *J. Morphol.* 209, 305–310.

Bogart, J. P. (1981). How many times has terrestrial breeding evolved in anuran amphibians? *Monit. Zool. Ital.* [N.S.] 40, Suppl. XV, 29–40.

Bolt, J. R. (1969). Lissamphibian origins: Possible protolissamphibian from the Lower Permian of Oklahoma. *Science* 166, 888–891.

Bolt, J. R. (1977). Dissorophoid relationships and ontogeny, and the origin of the lissamphibia. *J. Paleontol.* 51, 235–249.

Bolt, J. R. (1979). *Amphibamus grandiceps* as a juvenile dissorophid: Evidence and implications. *In* "Mazon Creek Fossils" (M. H. Nitecki, ed.), pp. 529–563. Academic Press, New York.

Bolt, J. R. (1991). Lissamphibian origins. *In* "Origins of the Higher Groups of Tetrapods: Controversy and Consensus" (H. P. Schultze and L. Trueb, eds.), pp. 194–222. Comstock, Ithaca, NY.

Boy, J. A. (1974). Die larven der rhachitomen Amphibien (Amphibia: Temnospondyli; Karbon— Trias.). *Paläontol. Z.* 48, 236–268.

Bruce, R. R. (1976). Population structure, life history and evolution of paedogenesis in the salamander *Eurycea neotenes*. *Copeia*, pp. 242–249.

Bruce, R. C. (1979). Evolution of paedomorphosis in salamanders of the genus *Gyrinophilus*. *Evolution (Lawrence, Kans.)* 33, 998–1000.

Callery, E. M., and Elinson, R. P. (1996). Developmental regulation of the urea-cycle enzyme arginase in the direct developing frog *Eleutherodactylus coqui*. *J. Exp. Zool.* 275, 61–66.

Cannatella, D. C., and de Sá, R. O. (1993). *Xenopus laevis* as a model organism. *Syst. Biol.* 42, 476–507.

Cannatella, D. C., and Trueb, L. (1988). Evolution of pipoid frogs: Intergeneric relationships of the aquatic frog family Pipidae (Anura). *Zool. J. Linn. Soc.* 94, 1–38.

Carroll, E. J., Jr., Seneviratne, A. M., and Ruibal, R. (1991). Gastric pepsin in an anuran larva. *Dev., Growth Differ.* 33, 499–507.

Carroll, R. L. (1986). Developmental processes and the origin of lepospondyls. *In* "Studies in Herpetology" (Z. Roček, ed.), pp. 45–48. Eur. Herpetol. Prague.

Carroll, R. L. (1988). "Vertebrate Paleontology and Evolution." Freeman, New York.

Carroll, R. L. (1992). The primary radiation of terrestrial vertebrates. *Annu. Rev. Earth Planet. Sci.* 20, 45–84.

Carroll, R. L. (1995). Problems of the phylogenetic analysis of paleozoic choanates. *Bull. Mus. Natl. Hist. Nat.,* [4] **17**, 389–445.

Churchill, F. B. (1997). Life before model systems: General zoology at August Weismann's institute. *Am. Zool.* **37**, 260–268.

Collazo, A. (1994). Molecular heterochrony in the pattern of fibronectin expression during gastrulation in amphibians. *Evolution (Lawrence, Kans.)* **48**, 2037–2045.

Collazo, A., and Marks, S. B. (1994). Development of *Gyrinophilus porphyriticus:* Identification of the ancestral developmental pattern in the salamander family Plethodontidae. *J. Exp. Zool.* **268**, 239–258.

Collazo, A., Bronner-Fraser, M., and Fraser, S. E. (1993). Vital dye labeling of *Xenopus laevis* trunk neural crest reveals multipotency and novel pathways of migration. *Development (Cambridge, UK)* **118**, 363–376.

Collazo, A., Mabee, P. M., and Fraser, S. E. (1994). A dual embryonic origin for vertebrate mechanoreceptors. *Science* **264**, 426–430.

Collins, J. P., Zerba, K. E., and Sredl, M. J. (1993). Shaping intraspecific variation: Development, ecology and the evolution of morphology and life history variation in tiger salamanders. *Genetica* **89**, 167–183.

Couly, G. F., Coltey, P. M., and Le Douarin, N. M. (1992). The developmental fate of the cephalic mesoderm in quail-chick chimeras. *Development (Cambridge, UK)* **114**, 1–15.

Couly, G. F., Coltey, P. M., and Le Douarin, N. M. (1993). The triple origin of skull in higher vertebrates: A study in quail-chick chimeras. *Development (Cambridge, UK)* **117**, 409–429.

Davies, M. (1989). Ontogeny of bone and the role of heterochrony in the myobatrachine genera *Uperoleia, Crinia,* and *Pseudophryne* (Anura: Leptodactylidae: Myobatrachinae). *J. Morphol.* **200**, 269–300.

Deban, S. M., Wake, D. B., and Roth, G. (1997). Salamander with a ballistic tongue. *Nature (London)* **389**, 27.

de Beer, G. R. (1947). The differentiation of neural crest cells into visceral cartilages and odontoblasts in *Ambystoma* and a re-examination of the germ-layer theory. *Proc. R. Soc. London, Ser. B* **134**, 377–398.

Del Pino, E. M. (1989). Modifications of oogenesis and development in marsupial frogs. *Development (Cambridge, UK)* **107**, 169–187.

Del Pino, E. M., and Elinson, R. P. (1983). A novel developmental pattern for frogs: Gastrulation produces an embryonic disk. *Nature (London)* **306**, 589–591.

Delsol, M., Flatin, J., Exbrayat, J.-M., and Bons, J. (1981). Développement de *Typhlonectes compressicaudus,* Amphibien apode vivipare. Hypothèses sur sa nutrition embryonnaire et larvaire par un ectotrophoblaste. *C. R. Seances Acad. Sci., Sér. 3* **293**, 281–285.

de Sá, R. O., and Lavilla, E. O. (1997). The tadpole of *Pseudis minuta* (Anura: Pseudidae), an apparent case of heterochrony. *Amphibia-Reptilia* **18**, 229–240.

Dopazo, H., and Alberch, P. (1994). Preliminary results on optional viviparity and intrauterine siblicide in *Salamandra salamandra* populations from northern Spain. *Mertensiella* **4**, 125–137.

Duellman, W. E. (1989). Alternative life-history styles in anuran amphibians: Evolutionary and ecological implications. *In* "Alternative Life-History Styles of Animals" (M. N. Bruton, ed.), pp. 101–126. Kluwer Academic Publishers, Dordrecht, The Netherlands.

Duellman, W. E. (1993). "Amphibian Species of the World: Additions and Corrections," Mus. Nat. Hist., University of Kansas, Spec. Publ. No. 21. Lawrence, Kansas.

Duellman, W. E., and Hillis, D. M. (1987). Marsupial frogs (Anura: Hylidae: *Gastrotheca*) of the Ecuadorian Andes: Resolution of taxonomic problems and phylogenetic relationships. *Herpetologica* **43**, 141–173.

Duellman, W. E., and Trueb, L. (1986). "Biology of Amphibians." McGraw-Hill, New York.

Duellman, W. E., Maxson, L. R., and Jesiolowski, C. A. (1988). Evolution of marsuapial frogs (Hylidae: Hemiphractinae): Immunological evidence. *Copeia,* pp. 527–543.

Eagleson, G., Ferreiro, B., and Harris, W. A. (1995). Fate of the anterior neural ridge and the morphogenesis of the *Xenopus* forebrain. *J. Neurobiol.* **28,** 146–158.

Eagleson, G. W., and Harris, W. (1990). Mapping of the presumptive brain regions in the neural plate of *Xenopus laevis. J. Neurobiol.* **21,** 427–440.

Elinson, R. P. (1990). Direct development in frogs: Wiping the recapitulationist slate clean. *Semin. Dev. Biol.* **1,** 263–270.

Elinson, R. P. (1994). Leg development in a frog without a tadpole *(Eleutherodactylus coqui). J. Exp. Zool.* **270,** 202–210.

Elinson, R. P., and Del Pino, E. M. (1985). Cleavage and gastrulation in the egg-brooding marsupial frog, *Gastrotheca riobambae. J. Embryol. Exp. Morphol.* **90,** 223–232.

Emerson, S. B. (1986). Heterochrony and frogs: The relationship of a life history trait to morphological form. *Am. Nat.* **127,** 167–183.

Emerson, S. B. (1987). The effect of chemically produced shifts in developmental timing on postmetamorphic morphology in *Bombina orientalis. Exp. Biol.* **47,** 105–109.

Emerson, S. B. (1988). Evaluating a hypothesis about heterochrony: Larval life-history traits and juvenile hind-limb morphology in *Hyla crucifer. Evolution (Lawrence, Kans.)* **42,** 68–78.

Epperlein, H.-H., Löfberg, J., and Olsson, L. (1996). Neural crest cell migration and pigment pattern formation in urodele amphibians. *Int. J. Dev. Biol.* **40,** 229–238.

Estes, R., Špinar, Z. V., and Nevo, E. (1978). Early Cretaceous pipid tadpoles from Israel (Amphibia: Anura). *Herpetologica* **34,** 374–375.

Evans, S. E., and Milner, A. R. (1996). A metamorphosed salamander from the early Cretaceous of Las Hoyas, Spain. *Philos. Trans. R. Soc. London, Ser. B* **351,** 627–646.

Evans, S. E., Milner, A. R., and Werner, C. (1996). Sirenid salamanders and a gymnophionan amphibian from the Cretaceous of the Sudan. *Palaeontology* **39,** 77–95.

Exbrayat, J.-M. (1983). Premières observations sur le cycle annuel de l'ovaire de *Typhlonectes compressicaudus* (Duméril et Bibron, 1841), Batracien Apode vivipare. *C. R. Seances Acad. Sci., Ser. 3* **296,** 493–498.

Exbrayat, J.-M. (1984). Quelques observations sur l'évolution des voies génitales femelles de *Typhlonectes compressicaudus* (Duméril et Bibron, 1841), Amphibien Apode vivipare, au cours du cycle de reproduction. *C. R. Seances Acad. Sci., Ser. 3* **298,** 13–18.

Exbrayat, J.-M. (1992). Reproduction et organes endocrines chez les femelles d'un amphibien gymnophione vivipare *Typhlonectes compressicaudus. Bull. Soc. Herpetol. Fr.* **64,** 37–50.

Exbrayat, J.-M., and Hraoui-Bloquet, S. (1992). La nutrition embryonnaire et les relations foeto-maternelles chez *Typhlonectes compressicaudus* amphibien gymnophionen vivipare. *Bull. Soc. Herpetol. Fr.* **61,** 53–61.

Exbrayat, J.-M., and Hraoui-Bloquet, S. (1994). An example of heterochrony: The metamorphosis in Gymnophiona. *Bull. Soc. Zool. Fr.* **119,** 117–126.

Exbrayat, J.-M., and Morel, G. (1995). Prolactin (PRL)-coding mRNA in *Typhlonectes compressicaudus,* a viviparous gymnophionan amphibian: An in situ hybridization study. *Cell Tissue Res.* **280,** 133–138.

Fang, H., and Elinson, R. P. (1996). Patterns of distal-less gene expression and inductive interactions in the head of the direct developing frog *Eleutherodactylus coqui. Dev. Biol.* **179,** 160–172.

Ford, L. S., and Cannatella, D. C. (1993). The major clades of frogs. *Herpetol. Monogr.* **7,** 94–117.

Fritzsch, B. (1990). The evolution of metamorphosis in amphibans. *J. Neurobiol.* **21,** 1011–1021.

Gardner, R. L., and Lawrence, P. A. (1986). "Single Cell Marking and Cell Lineage in Animal Development." Royal Society, London.

Garstang, W. (1951). "Larval Forms and Other Zoological Verses." Basil Blackwell, Oxford.

Goldschmidt, R. B. (1956). "The Golden Age of Zoology: Portraits from Memory." University of Washington Press, Seattle.

Gould, S. J. (1977). "Ontogeny and Phylogeny." The Belknap Press of Harvard University Press, Cambridge, MA.

Gould, S. J., and Vrba, E. (1982). Exaptation—A missing term in the science of form. *Paleobiology* **8**, 4–15.

Gradwell, N, (1973). On the functional morphology of suction and gill irrigation in the tadpole of *Ascaphus,* and notes on hibernation. *Herpetologica* **29**, 84–93.

Green, D. M., and Alberch, P. (1981). Interdigital webbing and skin morphology in the neotropical salamander genus *Bolitoglossa* (Amphibia; Plethodontidae). *J. Morphol.* **170**, 273–282.

Haas, A. (1995). Cranial features of dendrobatid larvae (Amphibia: Anura: Dendrobatidae). *J. Morphol.* **224**, 241–264.

Haas, A. (1996a). Non-feeding and feeding tadpoles in hemiphractine frogs: Larval head morphology, heterochrony, and systematics of *Flectonotus goeldii* (Amphibia: Anura: Hylidae). *J. Zool. Syst. Evol. Res.* **34**, 163–171.

Haas, A. (1996b). Das larvale Cranium von *Gastrotheca riobombae* und seine Metamorphose (Amphibia, Anura, Hylidae). *Verh. Naturwiss. Ver. Hamburg* [NS] **36**, 33–162.

Hall, B. K. (1990). Heterochronic change in vertebrate development. *Semin. Dev. Biol.* **1**, 237–243.

Hall, B. K. (1992). "Evolutionary Developmental Biology." Chapman & Hall, London.

Hall, B. K., ed. (1994). "Homology: the Hierarchical Basis of Comparative Biology." Academic Press, San Diego, CA.

Hall, B. K. (1995). Homology and embryonic development. *Evol. Biol.* **28**, 1–38.

Hall, B. K., and Horstãdius, S. (1988). "The Neural Crest." Oxford University Press, Oxford.

Hamburger, V. (1960). "A Manual of Experimental Embryology," rev. ed. University of Chicago Press, Chicago.

Hanken, J. (1982). Appendicular skeletal morphology in minute salamanders, genus *Thorius* (Amphibia: Plethodontidae): Growth regulation, adult size determination, and natural variation. *J. Morphol.* **174**, 57–77.

Hanken, J. (1984). Miniaturization and its effects on cranial morphology in plethodontid salamanders, genus *Thorius* (Amphibia: Plethodontidae). I. Osteological variation. *Biol. J. Linn. Soc.* **23**, 55–75.

Hanken, J. (1989). Development and evolution in amphibians. *Am. Sci.* **77**, 336–343.

Hanken, J. (1992). Life history and morphological evolution. *J. Evol. Biol.* **5**, 549–557.

Hanken, J. (1993). Model systems versus outgroups: Alternative approaches to the study of head development and evolution. *Am. Zool.* **33**, 448–456.

Hanken, J., and Hall, B. K. (1984). Variation and timing of the cranial ossification sequence of the Oriental Fire-bellied Toad, *Bombina orientalis* (Amphibia, Discoglossidae). *J. Morphol.* **182**, 245–255.

Hanken, J., and Hall, B. K. (1988a). Skull development during anuran metamorphosis. I. Early development of the first three bones to form—the exoccipital, the parasphenoid, and the frontoparietal. *J. Morphol.* **195**, 247–256.

Hanken, J., and Hall, B. K. (1988b). Skull development during anuran metamorphosis. II. Role of thyroid hormone in osteogenesis. *Anat. Embryol.* **178**, 219–227.

Hanken, J., and Thorogood, P. (1993). Evolution and development of the vertebrate skull: The role of pattern formation. *Trends Ecol. Evol.* **8**, 9–15.

Hanken, J., Klymkowsky, M. W., Summers, C. H., Seufert, D. W., and Ingebrigtsen, N. (1992). Cranial ontogeny in the direct-developing frog, *Eleutherodactylus coqui* (Anura: Leptodactylidae), analyzed using whole-mount immunohistochemistry. *J. Morphol.* **211**, 95–118.

Hanken, J., Jennings, D. H., and Olsson, L. (1997a). Mechanistic basis of life-history evolution in anuran amphibians: Direct development. *Am. Zool.* **37**, 160–171.

Hanken, J., Klymkowsky, M. W., Alley, K. E., and Jennings, D. H. (1997b). Jaw muscle development as evidence for embryonic repatterning in direct-developing frogs. *Proc. R. Soc. London, Ser. B* **264**, 1349–1354.

Harris, R. N. (1989). Ontogenetic changes in size and shape of the facultatively paedomorphic salamander *Notophthalmus viridescens dorsalis. Copeia,* pp. 35–42.

Harris, R. N., Semlitsch, R. D., Wilbur, H. M., and Fauth, J. E. (1990). Local variation in the genetic basis of paedomorphosis in the salamander *Ambystoma talpoideum. Evolution (Lawrence, Kans.)* **44**, 1588–1603.

Holtfreter, J. (1968). On mesenchyme and epithelia in inductive and morphogenetic processes. *In* "Epithelial-Mesenchymal Interactions" (R. Fleischmajer and R. E. Billingham, eds.), pp. 1–30. Williams & Wilkins, Baltimore, MD.

Hraoui-Bloquet, S., and Exbrayat, J.-M. (1992). Développement embryonnaire du tube digestiv chez *Typhlonectes compressicaudus* (Dumeril et Bibron 1841), Amphibien Gymnophione vivipare. *Ann. Sci. Nat., Zool. Biol. Anim.* [13] **13**, 11–23.

Hraoui-Bloquet, S., and Exbrayat, J.-M. (1994). Développement des branchies chez les embryons de *Typhlonectes compressicaudus,* amphibien gymnophione vivipare. *Ann. Sci. Nat., Zool. Biol. Anim.* [13] **15**, 33–46.

Hraoui-Bloquet, S., Escudie, G., and Exbrayat, J.-M. (1994). Aspects ultrastructuraux de l'évolution de la muqueuse utérine au cours de la gestation chez *Typhlonectes compressicaudus* Amphibien Gymnophione Vivipare. *Bull. Soc. Zool. Fr.* **119**, 237–242.

Hughes, A. (1959). Studies in embryonic and larval development in Amphibia. I. The embryology of *Eleutherodactylus ricordii,* with special reference to the spinal cord. *J. Embryol. Exp. Morphol.* **7**, 22–38.

Ivakhnenko, M. F. (1978). [Tailed amphibians from the Triassic and Jurassic of Central Asia]. *Paleontol. Zh.* **12**(3), 84–88 (in Russian).

Jackson, M. E., and Semlitsch, R. D. (1993). Paedomorphosis in the salamander *Ambystoma talpoideum:* Effects of a fish predator. *Ecology* **74**, 342–350.

Jenkins, F. A., and Walsh, D. M. (1993). An early Jurassic caecilian with limbs. *Nature (London)* **365**, 246–248.

Jennings, D. H., and Hanken, J. (1998). Mechanistic basis of life history evolution in anuran amphibians: thyroid gland development in the direct-developing frog, *Eleutherodactylus coqui. Gen. Comp. Endocrinol.* **111**, 225–232.

Joly, J. (1968). Données écologiques sur la salamandre tachetée *Salamandra salamandra* (L.). *Ann. Sci. Nat., Zool. Biol. Anim.* [12] **10**, 301–366.

Kay, B. K., and Peng, H. B., eds. (1991). "*Xenopus laevis:* Practical Uses in Cell and Molecular Biology," Methods Cell Biol., Vol. 36. Academic Press, San Diego, CA.

Köntges, G., and Lumsden, A. (1996). Rhombencephalic neural crest segmentation is preserved throughout craniofacial ontogeny. *Development (Cambridge, UK)* **122**, 3229–3242.

Krotoski, D. M., Fraser, S. E., and Bronner-Fraser, M. (1988). Mapping of neural crest pathways in *Xenopus laevis* using inter- and intra-specific cell markers. *Dev. Biol.* **127**, 119–132.

Lamotte, M., and Xavier, F. (1972). Recherches sur le développement embryonnaire de *Nectophrynoides occidentalis* Angel, amphibien anoure vivipare. I. Les principaux traits morphologiques et biometriques du développement. *Ann. Embryol. Morphog.* **5**, 315–340.

Larson, A. (1980). Paedomorphosis in relation to rates of morphological and molecular evolution in the salamander *Aneides flavipunctatus* (Amphibia, Plethodontidae). *Evolution (Lawrence, Kans.)* **34**, 1–17.

Larson, A., and Dimmick, W. W. (1993). Phylogenetic relationships of the salamander families: An analysis of congruence among morphological and molecular characters. *Herpetol. Monogr.* **7**, 77–93.

Laurin, M., and Reisz, R. R. (1997). A new perspective on tetrapod phylogeny. *In* "Amniote Origins: Completing the Transition to Land" (S. S. Sumida and K. L. M. Martin, eds.), pp. 9–59. Academic Press, San Diego, CA.

Le Douarin, N. M., and McLaren, A., eds. (1984). "Chimeras in Developmental Biology." Academic Press, London.

Licht, L. E. (1992). The effect of food level on growth rate and frequency of metamorphosis and paedomorphosis in *Ambystoma gracile*. *Can. J. Zool.* **70**, 87–93.

Linke, R., and Roth, G. (1990). Optic nerves in plethodontid salamanders (Amphibia, Urodela): Neuroglia, fiber spectrum and myelination. *Anat. Embryol.* **181**, 37–48.

Löfberg, J., Perris, R., and Epperlein, H. H. (1989). Timing in the regulation of neural crest cell migration: Retarded "maturation" of regional extracellular matrix inhibits pigment cell migration in embryos of the white axolotl mutant. *Dev. Biol.* **131**, 168–181.

Lombard, R. E., and Wake, D. B. (1977). Tongue evolution in the lungless salamanders, family Plethodontidae. II. Function and evolutionary diversity. *J. Morphol.* **153**, 39–80.

Luckenbill-Edds, L. (1997). Introduction: Research news in developmental biology in 1895 and 1995. *Am. Zool.* **37**, 213–219.

Lynn, W. G. (1942). The embryology of *Eleutherodactylus nubicola*, an anuran which has no tadpole stage. *Contrib. Embryol. Carnegie Inst. Washington Publ.* **541**, 27–62.

Lynn, W. G. (1961). Types of amphibian metamorphosis. *Am. Zool.* **1**, 151–161.

Maglia, A. M., and Púgener, L. A. (1998). Skeletal development and adult osteology of *Bombina orientalis* (Anura: Bombinatoridae). *Herpetologica* **54**, 344–363.

Malacinski, G. M., and Duhon, S. T. (1996). Developmental biology of urodeles. *Int. J. Dev. Biol.* **40**, 617–916.

Marconi, M., and Simonetta, A. M. (1988). The morphology of the skull in neotenic and normal *Triturus vulgaris meridionalis* (Boulenger) (Amphibia Caudata Salamandridae). *Monit. Zool. Ital.* [N. S.] **22**, 365–396.

Marks, S. B. (1994). Development of the hyobranchial apparatus in *Desmognathus aeneus*, a direct-developing salamander. *J. Morphol.* **220**, 371.

Marks, S. B., and Collazo, A. (1998). Direct development in *Desmognathus aeneus* (Caudata: Plethodontidae): A staging table. *Copeia*, 637–648.

Maxson, L. R., and Szymura, J. M. (1979). Quantitative immunological studies of the albumins of several species of fire bellied toads, genus *Bombina*. *Comp. Biochem. Physiol.* **63B**, 517–519.

McDiarmid, R. W. (1978). Evolution of parental care in frogs. *In* "The Development of Behavior: Comparative and Evolutionary Aspects" (G. M. Burghardt and M. Bekoff, eds.), pp. 127–147. Garland STPM Press, New York.

McGowan, G., and Evans, S. E. (1995). Albanerpetontid amphibians from the Cretaceous of Spain. *Nature (London)* **373**, 143–145.

McKinney, M. L., ed. (1988a). "Heterochrony in Evolution." Plenum, New York.

McKinney, M. L. (1988b). Classifying heterochrony: Allometry, size, and time. *In* "Heterochrony in Evolution" (M. L. McKinney, ed.), pp. 17–34. Plenum, New York.

McKinney, M. L., and McNamara, K. J., eds. (1991). "Heterochrony: The Evolution of Ontogeny." Plenum, New York.

McNamara, K. J. (1986). A guide to the nomenclature of heterochrony. *J. Paleontol.* **60**, 4–13.

Milner, A. R. (1982). Small temnospondyl amphibians from the Middle Pennsylvanian of Illinois. *Palaeontology* **25**, 635–664.

Milner, A. R. (1988). The relationships and origin of living amphibians. *In* "The Phylogeny and Classification of the Tetrapods" (M. J. Benton, ed.), Vol. 1, pp. 59–102. Clarendon Press, Oxford.

Milner, A. R. (1990). The radiation of temnospondyl amphibians. *In* "Major Evolutionary Radiations" (P. D. Taylor and G. P. Larwood, eds.), Syst. Assoc. Spec. Vol. No. 42, pp. 321–349. Clarendon Press, Oxford.

Milner, A. R. (1993). The Paleozoic relatives of lissamphibians. *Herpetol. Monogr.* **7**, 8–27.

Minsuk, S. B., and Keller, R. A. (1996). Dorsal mesoderm has a dual origin and forms by a novel mechanism in *Hymenochirus,* a relative of *Xenopus. Dev. Biol.* **174**, 92–103.

Moury, J. D., and Hanken, J. (1995). Early cranial neural crest migration in the direct-developing frog, *Eleutherodactylus coqui. Acta Anat.* **153**, 243–253.

Müller, G. B. (1991). Experimental strategies in evolutionary embryology. *Am. Zool.* **33**, 605–615.

Mutz, T., and Clemen, G. (1992). Development and dynamics of the tooth-systems in *Eurycea* and comparison of the definitive dentition of the palate of *Gyrinophilus* (Urodela: Plethdontidae). *Z. zool. Syst. Evolut.-forsch.* **30**, 220–233.

Naylor, B. G. (1978). The earliest known *Necturus* (Amphibia, Urodela), from the Paleocene Ravenscrag formation of Saskatchewan. *J. Herpetol.* **12**, 565–569.

Noden, D. M. (1986). Origins and patterning of craniofacial mesenchymal tissues. *J. Craniofacial Genet. Dev. Biol., Suppl.* **2**, 15–31.

Northcutt, R. G., Catania, K. C., and Criley, B. B. (1994). Development of lateral line organs in the axolotl. *J. Comp. Neurol.* **340**, 480–514.

Northcutt, R. G., Brändle, K., and Fritzsch, B. (1995). Electroreceptors and mechanosensory lateral line organs arise from single placodes in axolotls. *Dev. Biol.* **168**, 358–373.

Novoselov, V. V. (1995). Notochord formation in amphibians: Two directions and two ways. *J. Exp. Zool.* **271**, 296–306.

Nussbaum, R. A., and Wilkinson, M. (1989). On the classification and phylogeny of caecilians (Amphibia: Gymnophiona), a critical review. *Herpetol. Monogr.* **3**, 1–42.

Olsson, L. (1993). Pigment pattern formation in the larval salamander *Ambystoma maculatum. J. Morphol.* **215**, 151–163.

Olsson, L. (1994). Pigment pattern formation in larval ambystomatid salamanders: *Ambystoma talpoideum, Ambystoma barbouri,* and *Ambystoma annulatum. J. Morphol.* **220**, 123–138.

Olsson, L., and Hanken, J. (1996). Cranial neural crest migration and chondrogenic fate in the Oriental fire-bellied toad, *Bombina orientalis:* Defining the ancestral pattern of head development in anuran amphibians. *J. Morphol.* **229**, 105–120.

Olsson, L., and Löfberg, J. (1992). Pigment pattern formation in larval ambystomatid salamanders: *Ambystoma tigrinum tigrinum. J. Morphol.* **211**, 73–85.

Orton, G. L. (1949). Larval development of *Nectophrynoides tornieri* (Roux) with comments on direct development in frogs. *Ann. Carnegie Mus.* **31**, 257–277.

Orton, G. L. (1951). Direct development in frogs. *Turtox News* **29**, 2–6.

Parichy, D. M. (1996a). Salamander pigment patterns: How can they be used to study developmental mechanisms and their evolutionary transformation? *Int. J. Dev. Biol.* **40**, 871–884.

Parichy, D. M. (1996b). Pigment patterns of larval salamanders (Ambystomatidae, Salamandridae): The role of the lateral line sensory system and the evolution of pattern-forming mechanisms. *Dev. Biol.* **175**, 265–282.

Parichy, D. M. (1996c). When neural crest and placodes collide: Interactions between melanophores and the lateral lines that generate stripes in the salamander *Ambystoma tigrinum tigrinum* (Ambystomatidae). *Dev. Biol.* **175**, 283–300.

Pedersen, S. C. (1991). Dental morphology of the cannibal morph in the tiger salamander, *Ambystoma tigrinum. Amphibia-Reptilia* **12**, 1–14.

Pierce, B. A., and Smith, H. M. (1979). Neoteny vs. paedogenesis. *J. Herpetol.* **13**, 119–120.

Púgener L. A., and Maglia, A. M. (1997). Osteology and skeletal development of *Discoglossus sardus* (Anura: Discoglossidae). *J. Morphol.* **233**, 267–286.

Purcell, S. M., and Keller, R. (1993). A different type of amphibian mesoderm morphogenesis in *Ceratophrys ornata. Development, (Cambridge, UK)* **117**, 307–317.

Raff, R. A. (1996). "The Shape of Life." University of Chicago Press, Chicago.

Rage, J.-C., and Roček, Z. (1989). Redescription of *Triadobatrachus massionti* (Piveteau, 1936), an anuran amphibian from the early Triassic. *Palaeontographica A,* **206**, 1–16.

Reilly, S. M. (1986). Ontogeny of cranial ossification in the eastern newt, *Notophthalmus viridescens* (Caudata: Salamandridae) and its relationship to metamorphosis and neoteny. *J. Morphol.* **188**, 315–326.

Reilly, S. M. (1987). Ontogeny of the hyobranchial apparatus in the salamanders *Ambystoma talpoideum* (Ambystomatidae) and *Notophthalmus viridescens* (Salamandridae): The ecological morphology of two neotenic strategies. *J. Morphol.* **191**, 205–214.

Reilly, S. M. (1994). The ecological morphology of metamorphosis: Heterochrony and the evolution of feeding mechanisms in salamanders. *In* "Ecological Morphology: Integrative Organismal Biology" (P. C. Wainwright and S. M. Reilly, eds.), pp. 319–338. University of Chicago Press, Chicago.

Reilly, S. M., and Altig, R. (1996). Cranial ontogeny in *Siren intermedia* (Caudata: Sirenidae): Paedomorphic, metamorphic, and novel patterns of heterochrony. *Copeia,* pp. 29–41.

Reilly, S. M., and Brandon, R. A. (1994). Partial paedomorphosis in the Mexican stream ambystomatids and the taxonomic status of the genus *Rhyacosiredon* Dunn. *Copeia,* pp. 656–662.

Reilly, S. M., Wiley, E. O., and Meinhardt, D. J. (1997). An integrative approach to heterochrony: The distinction between interspecific and intraspecific phenomena. *Biol. J. Linn. Soc.* **60**, 119–143.

Reiss, J. O. (1997). Early development of chondrocranium in the tailed frog *Ascaphus truei* (Amphibia: Anura): Implications for anuran palatoquadrate homologies. *J. Morphol.* **231**, 63–100.

Reisz, R. R. (1997). The origin and early evolutionary history of amniotes. *Trends Ecol. Evol.* **12**, 218–222.

Richardson, M. K., Carl, T. F., Hanken, J., Elinson, R. P., Cope, C., and Bagley, P. (1998). Limb development and evolution: A frog embryo with no apical ectodermal ridge (AER). *J. Anat.* **192**, 379–390.

Roček, Z. (1996). Skull of the neotenic salamandrid amphibian *Triturus alpestris* and abbreviated development in the Tertiary Salamandridae. *J. Morphol.* **230**, 187–197.

Rose, C. S. (1995). Skeletal morphogenesis in the urodele skull: I. Postembryonic development in the Hemidactyliini (Amphibia: Plethodontidae). *J. Morphol.* **223**, 125–148.

Rose, C. S. (1996). An endocrine-based model for developmental and morphogenetic diversification in metamorphic and paedomorphic urodeles. *J. Zool.* **239**, 253–284.

Rosenkilde, P., and Ussing, A. P. (1996). What mechanisms control neoteny and regulate induced metamorphosis in urodeles? *Int. J. Dev. Biol.* **40**, 665–673.

Roth, G., and Schmidt, A. (1993). The nervous system of plethodontid salamanders: Insight into the interplay between genome, organism, behavior, and ecology. *Herpetologica* **49**, 185–194.

Roth, G., and Wake, D. B. (1985). Trends in the functional morphology and sensorimotor control of feeding behavior in salamanders: An example of internal dynamics in evolution. *Acta Biotheor.* **34**, 175–192.

Roth, G., Nishikawa, K., Dicke, U., and Wake, D. B. (1988). Topography and cytoarchitecture of the motor nuclei in the brainstem of salamanders. *J. Comp. Neurol.* **278**, 181–194.

Roth, G., Dicke, U., and Nishikawa, K. (1992). How do ontogeny, morphology, and physiology of sensory systems constrain and direct the evolution of amphibians? *Am. Nat.* **139**, S105–S124.

Roth, G., Nishikawa, K. C., Naujoks-Manteuffel, C., Schmidt, A., and Wake, D. B. (1993). Paedomorphosis and simplification in the nervous system of salamanders. *Brain, Behav. Evol.* **42**, 137–170.

Ruibal, R., and Thomas, E. (1988). The obligate carnivorous larvae of the frog, *Lepidobatrachus laevis* (Leptodactylidae). *Copeia,* pp. 591–604.

Sadaghiani, B., and Thiébaud, C. H. (1987). Neural crest development in the *Xenopus laevis* embryo studied by interspecific transplantation and scanning electron microscopy. *Dev. Biol.* **124**, 91–110.

Sammouri, R., Renous, S., Exbrayat, J.-M., and Lescure, J. (1990). Développement embryonnaire de *Typhlonectes compressicaudus* (Amphibia, Gymnophiona). *Ann. Sci. Nat., Zool. Biol. Anim.* [13] **11**, 135–163.

Schlosser, G., and Roth, G. (1997). Evolution of nerve development in frogs. II. Modified development of the peripheral nervous system in the direct-developing frog *Eleutherodactylus coqui* (Leptodactylidae). *Brain, Behav. Evol.* **50**, 94–128.

Schmidt, A., and Wake, M. H. (1997). Cellular migration and morphological complexity in the caecilian brain. *J. Morphol.* **231**, 11–27.

Schoch, R. R. (1992). Comparative ontogeny of early Permian branchiosaurid amphibians from southwestern Germany. *Palaeontographica A,* **222**, 43–83.

Schoch, R. (1995). Heterochrony in the development of the amphibian head. *In* "Evolutionary Change and Heterochrony" (K. J. McNamara, ed.), pp. 107–124. Wiley, Chichester.

Scott, D. E. (1993). Timing of reproduction of paedomorphic and metamorphic *Ambystoma talpoideum. Am. Midl. Nat.* **129**, 397–402.

Semlitsch, R. D., and Wilbur, H. M. (1989). Artificial selection for paedomorphosis in the salamander *Ambystoma talpoideum. Evolution (Lawrence, Kans.)* **43**, 105–112.

Semlitsch, R. D., Harris, R. N., and Wilbur, H. M. (1990). Paedomorphosis in *Ambystoma talpoideum:* Maintenance of population variation and alternative life-history pathways. *Evolution (Lawrence, Kans.)* **44**, 1604–1613.

Sever, D. M., and Trauth, S. E. (1990). Cloacal anatomy of female salamanders of the plethodontid subfamily Desmognathinae Amphibia Urodela. *Trans. Am. Microsc. Soc.* **109**, 193–204.

Shaffer, H. B. (1984a). Evolution in a paedomorphic lineage. I. An electrophoretic analysis of the Mexican ambystomatid salamanders. *Evolution (Lawrence, Kans.)* **38**, 1194–1206.

Shaffer, H. B. (1984b). Evolution in a paedomorphic lineage. II. Allometry and form in the Mexican ambystomatid salamanders. *Evolution (Lawrence, Kans.)* **38**, 1207–1218.

Shaffer, H. B. (1993). Phylogenetics of model organisms: The laboratory axolotl, *Ambystoma mexicanum. Syst. Biol.* **42**, 508–522.

Shaffer, H. B., and Voss, S. R. (1996). Phylogenetic and mechanistic analysis of a developmentally integrated character complex: Alternative life history modes in ambystomatid salamanders. *Am. Zool.* **36**, 24–35.

Shea, B. T. (1983). Allometry and heterochrony in the African apes. *Amer. J. Phys. Anthropol.* **62**, 275–289.

Shea, B. T. (1988). Heterochrony in primates. *In* "Heterochrony in Evolution" (M. L. McKinney, Ed.), pp. 237–266. Plenum Publ. Corp., New York.

Shubin, N. (1995). The evolution of paired fins and the origin of tetrapod limbs: Phylogenetic and transformational approaches. *Evol. Biol.* **28**, 39–86.

Shubin, N. H., and Jenkins, F. A., Jr. (1995). An Early Jurassic jumping frog. *Nature (London)* **377**, 49–52.

Shubin, N. H., and Wake, D. B. (1991). Implications of direct development for the tetrapod limb bauplan. *Am. Zool.* **31**, 8A.

Smirnov, S. V. (1989). Postmetamorphic skull development in *Bombina orientalis* (Amphibia, Discoglossidae), with comments on neoteny. *Zool. Anz.* **233**, 91–99.

Smirnov, S. V. (1990). Evidence of neoteny: A paedomorphic morphology and retarded development in *Bombina orientalis* (Anura, Discoglossidae). *Zool. Anz.* **225**, 324–332.

Smirnov, S. V. (1991). The anuran middle ear: Developmental heterochronies and adult morphology diversification. *Belg. J. Zool.* **121**, 99–110.

Smirnov, S. V. (1993). The anuran amphibian papilla development, with comments on its timing, rate, and influence on adult papilla morphology. *Zool. Jahrb. Abt. Anat. Ontog. Tiere* **123**, 273–289.

Smirnov, S. V. (1994). Postmaturation skull development in *Xenopus laevis* (Anura, Pipidae): Late-appearing bones and their bearing on the pipid ancestral morphology. *Russ. J. Herpetol.* **1**, 21–29.

Smirnov, S. V., and Vasil'eva, A. B. (1995). Anuran dentition: Development and evolution. *Russ. J. Herpetol.* **2**, 120–128.

Smith, S. C. (1996). Pattern formation in the urodele mechanoreceptive lateral line: What features can be exploited for the study of development and evolution? *Int. J. Dev. Biol.* **40**, 727–733.

Smith, S. C., Lannoo, M. J., and Armstrong, J. B. (1988). Lateral-line neuromast development in *Ambystoma mexicanum* and a comparison with *Rana pipiens. J. Morphol.* **198**, 367–379.

Smith, S. C., Graveson, A. C., and Hall, B. K. (1994). Evidence for a developmental and evolutionary link between placodal ectoderm and neural crest. *J. Exp. Zool.* **270**, 292–301.

Stephenson, E. M. (1960). The skeletal characters of *Leiopelma hamiltoni* McCulloch, with particular reference to the effects of heterochrony on the genus. *Trans. R. Soc. N. Z.* **88**, 473–488.

Stephenson, N. G. (1965). Heterochronous changes among Australian leptodactylid frogs. *Proc. Zool. Soc. London* **144**, 339–350.

Stone, L. S. (1927). Further experiments on the transplantation of neural crest (mesectoderm) in amphibians. *Proc. Soc. Exp. Biol. Med.* **24**, 945–948.

Stone, L. S. (1929). Experiments showing the role of migrating neural crest (mesectoderm) in the formation of head skeleton and loose connective tissue in *Rana palustris. Wilhelm Roux' Arch. Entwickungs mech. Org.* **118**, 40–77.

Thiébaud, C. H. (1983). A reliable new cell marker in *Xenopus. Dev. Biol.* **98**, 245–249.

Thomson, K. S. (1988). "Morphogenesis and Evolution." Oxford University Press, Oxford.

Tiedemann, H., Tiedemann, H., Grunz, H., and Knöchel, W. (1995). Molecular mechanisms of tissue determination and pattern formation in amphibian embryos. *Naturwissenschaften* **82**, 123–134.

Titus, T. A., and Larson, A. (1996). Molecular phylogenetics of desmognathine salamanders (Caudata: Plethodontidae): A reevaluation of evolution in ecology, life history, and morphology. *Syst. Biol.* **45**, 451–472.

Toews, D., and Macintyre, D. (1977). Blood respiratory properties of a viviparous amphibian. *Nature (London)* **266**, 464–465.

Trueb, L. (1985). A summary of osteocranial development in anurans with notes on the sequence of cranial ossification in *Rhinophrynus dorsalis* (Anura: Pipoidea: Rhinophrynidae). *S. Afr. J. Sci.* **81**, 181–185.

Trueb, L., and Alberch, P. (1985). Miniaturization and the anuran skull: A case study of heterochrony. *In* "Functional Morphology of Vertebrates" (H. R. Duncker and G. Fleischer, eds.), pp. 113–121. Fischer, Stuttgart.

Trueb, L., and Cloutier, R. (1991). A phylogenetic investigation of the inter- and intrarelationships of the lissamphibia (Amphibia: Temnospondyli). *In* "Origins of the Higher Groups of Tetrapods" (H. P. Schultze and L. Trueb, eds.), pp. 223–313. Comstock, Ithaca, NY.

Trueb, L., and Hanken, J. (1992). Skeletal development in *Xenopus laevis* (Anura: Pipidae). *J. Morphol.* **214**, 1–41.

Vignali, R., and Nardi, I. (1996). Unusual features of the urodele genome: Do they have a role in evolution and development? *Int. J. Dev. Biol.* **40**, 637–643.

Vilter, V., and Lugand, A. (1959). Trophisme intrautérin et croissance embryonnaire chez le *Nectophrynoides occidentalis* Ang., crapaud totalement vivipare du Mont Nimba (Haute Guinée). *C. R. Seances Soc. Biol. Ses Fil.* **153**, 29–32.

Voss S. R. (1996). Genetic basis of paedomorphosis in the axolotl, *Ambystoma mexicanum:* A test of the single gene hypothesis. *J. Hered.* **86**, 441–447.

Voss, S. R., and Shaffer, H. B. (1996). What insights into the developmental traits of urodeles does the study of interspecific hybrids provide? *Int. J. Dev. Biol.* **40**, 885–893.

Wagner, G. (1949). Die Bedeutung der Neuralleiste für die Kopfgestaltung der Amphibienlarven: Untersuchungen an Chimaeren von *Triton* und *Bombinator. Rev. Suisse Zool.* **56**, 519–620.

Wagner, G. (1955). Chimaerische Zahnanlagen aus *Triton*-Schmelzorgan und *Bombinator*-Papille. Mit Beobachtungen über die Entwicklung von Kiemenzahnchen und Mundsinnesknospen in den *Triton*-Larven. *J. Embryol. Exp. Morphol.* **3**, 160–188.

Wakahara, M. (1996). Heterochrony and neotenic salamanders: Possible clues for understanding the animal development and evolution. *Zool. Sci.* **13**, 765–776.

Wakahara, M., and Yamaguchi, M. (1996). Heterochronic expression of several adult phenotypes in normally metamorphosing and metamorphosis-arrested larvae of a salamander *Hynobius retardatus. Zool. Sci.* **13**, 483–488.

Wakahara, M., Miyashita, N., Sakamoto, A., and Arai, T. (1994). Several biochemical alterations from larval to adult types are independent on morphological metamorphosis in a salamander, *Hynobius retardatus. Zool. Sci.* **11**, 583–588.

Wake, D. B. (1966). Comparative osteology and evolution of the lungless salamanders, family Plethodontidae. *Mem. S. Calif. Acad. Sci.* **4**, 1–111.

Wake, D. B. (1980). Evidence of heterochronic evolution: A nasal bone in the Olympic salamander, *Rhyacotriton olympicus. J. Herpetol.* **14**, 292–295.

Wake, D. B. (1982). Functional and developmental constraints and opportunities in the evolution of feeding systems in urodeles. *In* "Environmental Adaptation and Evolution" (D. Mossakowski and G. Roth, eds.), pp. 51–66. Fischer, Stuttgart.

Wake, D. B. (1989). Phylogenetic implications of ontogenetic data. *Geobios, Mem. Spec.* **12**, 369–378.

Wake, D. B., and Hanken, J. (1996). Direct development in the lungless salamanders: What are the consequences for developmental biology, evolution and phylogenesis? *Int. J. Dev. Biol.* **40**, 859–869.

Wake, D. B., and Larson, A. (1987). Multidimensional analysis of an evolving lineage. *Science* **238**, 42–48.

Wake, D. B., and Marks, S. B. (1993). Development and evolution of plethodontid salamanders: A review of prior studies and a prospectus for future research. *Herpetologica* **49**, 194–203.

Wake, D. B., and Roth, G. (1989). The linkage between ontogeny and phylogeny in the evolution of complex systems. *In* "Complex Organismal Functions: Integration and Evolution in Vertebrates" (D. B. Wake and G. Roth, eds.), pp. 361–377. Wiley, Chichester.

Wake, D. B., and Shubin, N. (1994). Urodele limb development in relation to phylogeny and life history. *J. Morphol.* **220**, 407–408.

Wake, D. B., Nishikawa, K. C., Dicke, U., and Roth, G. (1988). Organization of the motor nuclei in the cervical spinal cord of salamanders. *J. Comp. Neurol.* **278**, 195–208.

Wake, D. B., Mabee, P. M., Hanken, J., and Wagner, G. (1991). Development and evolution—The emergence of a new field. *In* "The Unity of Evolutionary Biology: The Proceedings of the Fourth International Congress of Systematic and Evolutionary Biology" (E. C. Dudley, ed.), pp. 582–588. Dioscorides Press, Portland, OR.

Wake, M. H. (1970). Evolutionary morphology of the caecilian urogenital system. Part II. The kidneys are urogenital ducts. *Acta Anat.* **75**, 321–358.

Wake, M. H. (1976). The development and replacement of teeth in viviparous caecilians. *J. Morphol.* **148**, 33–64.

Wake, M. H. (1977a). The reproductive biology of caecilians: An evolutionary perspective. *In* "The Reproductive Biology of Amphibians" (D. H. Taylor and S. I. Guttman, eds.), pp. 73–101. Plenum, New York.

Wake, M. H. (1977b). Fetal maintenance and its evolutionary significance in the Amphibia: Gymnophiona. *J. Herpetol.* 11, 379–386.

Wake, M. H. (1978). The reproductive biology of *Eleutherodactylus jasperi* (Amphibia, Anura, Leptodactylidae) with comments on the evolution of live-bearing systems. *J. Herpetol.* 12, 121–133.

Wake, M. H. (1980a). Reproduction, growth, and population structure of the Central American caecilian *Dermophis mexicanus*. *Herpetologica* 36, 244–256.

Wake, M. H. (1980b). Fetal tooth development and adult replacement in *Dermophis mexicanus* (Amphibia: Gymnophiona): Fields versus clones. *J. Morphol.* 166, 203–216.

Wake, M. H. (1980c). The reproductive biology of *Nectophrynoides malcolmi* (Amphibia: Bufonidae), with comments on the evolution of reproductive modes in the genus *Nectophrynoides*. *Copeia*, pp. 193–209.

Wake, M. H. (1982). Diversity within a framework of constraints. Amphibian reproductive modes. *In* "Environmental Adaptation and Evolution" (D. Mossakowski and G. Roth, eds.), pp. 87–106. Fischer, Stuttgart.

Wake, M. H. (1986). The morphology of *Idiocranium russeli* (Amphibia: Gymnophiona), with comments on miniaturization through heterochrony. *J. Morphol.* 189, 1–16.

Wake, M. H. (1989). Phylogenesis of direct development and viviparity in vertebrates. *In* "Complex Organismal Functions: Integration and Evolution in Vertebrates" (D. B. Wake and G. Roth, eds.), pp. 235–250. Wiley, Chichester.

Wake, M. H. (1992). Evolutionary scenarios, homology and convergence of structural specializations for vertebrate viviparity. *Am. Zool.* 32, 256–263.

Wake, M. H. (1993). Evolution of oviductal gestation in amphibians. *J. Exp. Zool.* 266, 394–413.

Wake, M. H., and Hanken, J. (1982). The development of the skull of *Dermophis mexicanus* (Amphibia: Gymnophiona), with comments on skull kinesis and amphibian relationships. *J. Morphol.* 173, 203–233.

Wake, M. H., Exbrayat, J.-M., and Delsol, M. (1985). The development of the chondrocranium of *Typhlonectes compressicaudus* (Gymnophiona), with comparison to other species. *J. Herpetol.* 19, 68–77.

Wake, T. A., Wake, D. B., and Wake, M. H. (1983). The ossification sequence of *Aneides lugubris*, with comments on heterochrony. *J. Herpetol.* 17, 10–22.

Walls, S. C., Beatty, J. J., Tissot, B. N., Hokit, D. G., and Blaustein, A. R. (1993). Morphological variation and cannibalism in a larval salamander (*Ambystoma macrodactylum columbianum*). *Can. J. Zool.* 71, 1543–1551.

Wassersug, R. J. (1980). Internal oral features of larvae from eight anuran families; Functional, systematic, evolutionary and ecological considerations. *Misc. Publ. Mus. Nat. Hist. Univ. Kansas* 68, 1–146.

Wassersug, R. J., and Duellman, W. E. (1984). Oral structures and their development in egg-brooding hylid frog embryos and larvae: Evolutionary and ecological implications. *J. Morphol.* 182, 1–37.

Wassersug, R. J., and Hoff, K. (1982). Developmental changes in the orientation of the anuran jaw suspension. *Evol. Biol.* 15, 223–246.

Wassersug, R. J., and Wake, D. B. (1995). Fossil tadpoles from the Miocene of Turkey. *Alytes* 12, 145–157.

Werneburg, R. (1991). Die Branchiosaurier aus dem Unterrotliegend des Döhlener Beckens bei Dresden. *Veröff. Naturhist. Mus. Schleusingen* 6, 75–99.

Weygoldt, P., and de Carvalho e Silva, P. S. (1991). Observations on mating, oviposition, egg sac formation and development in the egg-brooding frog, *Fritziana goeldii. Amphibia-Reptilia* 12, 67–80.

108   James Hanken

Whiteman, H. H. (1994). Evolution of facultative paedomorphosis in salamanders. *Q. Rev. Biol.* **69**, 205–221.

Whiteman, H. H., Wissinger, S. A., and Brown, W. S. (1996). Growth and foraging consequences of facultative paedomorphosis in the tiger salamander, *Ambystoma tigrinum nebulosum. Evol. Ecol.* **10**, 433–446.

Wild, E. R. (1997). Description of the adult skeleton and developmental osteology of the hyperossified horned frog, *Ceratophrys cornuta* (Anura: Leptodactylidae). *J. Morphol.* **232**, 169–206.

Wilkinson, M., and Nussbaum, R. A. (1996). On the phylogenetic arrangement of the Uraeotyphlidae (Amphibia: Gymnophiona). *Copeia,* pp. 550–562.

Xavier, F. (1973). Le cycle des voies génitales femelées de *Nectophrynoides occidentalis* Angel, amphibien anoure vivipare. *Z. Zellforsch. Mikrosk. Anat.* **140**, 509–534.

Xavier, F. (1977). An exceptional reproductive strategy in Anura: *Nectophrynoides occidentalis* Angel (Bufonidae), an example of adaptation to terrestrial life by viviparity. *In* "Major Patterns in Vertebrate Evolution" (M. K. Hecht, P. C. Goody, and B. M. Hecht, eds.), pp. 545–552. Plenum, New York.

Xavier, F. (1986). La réproduction des *Nectophrynoides. In* "Traité de Zoologie Amphibiens" (P. P. Grasse and M. Delsol, eds.), Vol. 14, pp. 497–513. Masson, Paris.

Zelditch, M. L., and Fink, W. L. (1996). Heterochrony and heterotopy: Stability and innovation in the evolution of form. *Paleobiology* **22**, 241–254.

CHAPTER 4

Larvae in Fish Development and Evolution

JACQUELINE F. WEBB

Department of Biology, Villanova University, Villanova, Pennsylvania

I. INTRODUCTION

The literature on larval fishes is vast and distributed among the fields of systematic ichthyology, comparative functional morphology and physiology, limnology, biological oceanography, fisheries science, aquaculture, applied ecology, and behavioral ecology. Data on larval morphology, growth, development, behavior, and ecology are distributed among these diverse disciplines and clearly are in need of synthesis. There are over 24,000 valid species of fishes in 482 families (Nelson, 1994) and, thus, at least a comparable number of early life-history stages. Most of these have not been studied and in many cases have not even been described. In several instances, the larval development and larval biology of even the most conspicuous taxa are not well-known (e.g., Leis, 1989, 1991). This chapter presents a sampling of the literature and is intended to be an introduction to fish larvae, especially larvae of marine teleost fishes, as potential or realized subjects of developmental and evolutionary studies. Data on early life-history stages of freshwater fishes are included for additional clarification or where data on marine fishes are not available.

The most comprehensive source of gross morphological data on the early life-history stages of marine teleost fishes is found in the taxonomic literature (Kendall and Matarese, 1994). Larval fish identification guides describe and illustrate species assemblages from specific geographic regions and are arranged taxonomically (e.g., Russell, 1976; Fahay, 1983; Leis and Rennis, 1983; Ozawa, 1986; Leis and Trnski, 1989; Matarese et al., 1989; Moser, 1996a; Able and Fahay, 1998). Larval characters [especially osteological characters, reviewed by Dunn (1983, 1984)] have been employed in systematic analyses at specific, familial, and ordinal levels [see Moser et al. (1984) and works cited therein]. Specialized larval characters have been especially useful in establishing mono-phyly and identifying interrelationships within several taxa (e.g., Leis, et al., 1997), but the utility of larval characters in resolving systematic problems has not yet been fully realized (Johnson, 1993).

In addition to being the subject of systematic studies, fish larvae also are the subject of functional morphological, behavioral, and ecological investigations. Feeding strategies and sensory capabilities of larvae have been analyzed to deter-mine larval behavioral strategies and their potential for survival in the field (see Section III). Larval fishes are studied as important components of marine communities [e.g., pelagic larvae of coral reef fishes, reviewed by Leis (1991)] and of the plankton (ichthyoplankton) of the world's oceans, and as such are an important subject of study in biological oceanography and plankton ecology. The study of the determinants of larval distribution, the physiological condition of fish larvae (Ferron and Leggett, 1994), and predation on and by fish larvae (Hunter, 1981; see Section III) provides data that are essential for an understanding of population dynamics, recruitment, and fisheries resources (Lasker, 1981; Blax-ter, 1984; Heath, 1992; Moser and Smith, 1993; Chambers and Trippel, 1997).

The growth, morphological development, and behavioral ontogeny of fish larvae have been studied in the context of the commercial aquaculture industry (e.g., Kuronuma, 1984). Hundreds of species of marine and freshwater fishes are raised worldwide for direct commercial harvest or, to a lesser degree, for reintroduction into habitats where naturally occurring stocks have been depleted. The development of reliable, large-scale closed or semi-enclosed rearing systems, a knowledge of conditions required for successful fertilization, hatching, and proper behavioral development, and the ability to rear preferred larval food sources consistently and in large quantities are required for the growth and success of the commercial fish aquaculture industry.

II. GENERAL CHARACTERISTICS OF FISH LARVAE

The development of fishes from fertilization to sexual maturity is a continuum that is punctuated by developmental events and transitions, which may be either gradual and unremarkable (Copp and Kovač, 1996) or abrupt and quite dramatic [reviewed by Youson (1988)]. The nature and organization of these events and transitions, including the definition of metamorphosis (see Rose and Reiss, 1993), are the bases for extensive and ongoing debate (e.g., Balon, 1981, 1990). The result has been the formulation of several terminological schemes that define and describe developmental stages found among fishes (Kendall *et al.*, 1984). For the purposes of discussion here, a larva will be defined simply as the stage between hatching and metamorphosis (transformation to the juvenile stage) (Fig. 1).

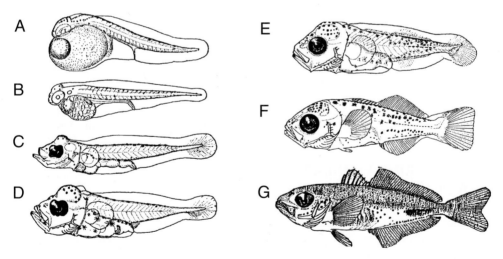

FIGURE 1 Early life-history stages of *Trachurus symmetricus* [reprinted with permission from Ahlstrom and Ball (1954)]. (A, B) Yolk sac larvae; (C, D) preflexion larvae; (E) flexion larva; (F) postflexion larva; (G) juvenile. Flexion refers to the flexing of the caudal end of the notochord, an ontogenetic marker, which is followed by the morphogenesis of the hypural plate, fin rays, and supporting elements typical of the caudal fin of teleost fishes.

Freshwater teleost fishes are generally characterized by unspecialized larvae with a gradual transition to the juvenile stage. Some marine teleost fishes develop from demersal eggs and larvae, and these also tend to be unremarkable and have a gradual transition to the juvenile stage. In contrast, the majority of marine teleost fishes have planktonic eggs and larvae, which may have extensive specializations and whose transition to the juvenile stage generally is dramatic (Matarese *et al.*, 1989). Planktonic marine fish larvae generally are 2–3 mm long at hatching and are characterized by a large yolk sac ("yolk sac larva," Fig. 1), poorly developed mouth, unpigmented eyes, pectoral fin buds, ' and the adult number of myomeres (Moser, 1996b). By the time yolk sac resources are exhausted, the larva is 4–5 mm long and the mouth, digestive system, sensory systems, and pectoral fins are evident. Caudal fin development, involving the flexion of the end of the notochord, is gradual and continues well into the larval period (Moser, 1996b) (Fig. 1). The larval period may last from several days to several months. Transformation to the juvenile stage typically occurs at a length of 10–30 mm, although a prolonged larval period and giant larvae do occur (see Section III). Metamorphosis is characterized by the loss of larval features and transient larval specializations (see Section IV.B) and the acquisition of a new set of juvenile–adult features, including thicker pigmented skin, silvery guanine pigment, scales, and the characteristic number of fin rays [reviewed by Youson (1988)]. In addition, most transforming marine fish larvae experience overall rapid growth, significant changes in body shape (Moser, 1996b), and changes in metabolism and physiology [reviewed by Youson (1988)]. A distinct transformation stage may be present before a distinct juvenile stage can be clearly identified (Kendall *et al.*, 1984). Some species have transitional juveniles (Kaufman *et al.*, 1991) or pelagic juveniles (e.g., *Sebastes*) that continue to occupy the larval habitat (Moser and Boehlert, 1991). Like many invertebrates, larval fishes may delay metamorphosis and/or settlement until a suitable settlement site is found, thus resulting in intraspecific variation in the length of the larval period (Markle *et al.*, 1992).

III. MARINE FISH LARVAE: CHALLENGES OF A PLANKTONIC LIFESTYLE

Like the planktonic larvae of marine invertebrates, planktonic fish larvae are faced with a series of physiological, behavioral, and ecological challenges in the face of chemical and physical stresses, the patchy distribution of food resources, the threat of predation, and the probability of dispersal in the plankton.

Planktonic larvae must maintain their position in the upper layers of the ocean in order to exploit planktonic food resources. Persistent fin folds, gelatinous body envelopes, and elaborated fins may all provide buoyancy (Moser, 1981), in addition to that provided by the swim bladder (see Section IV.B). Larvae are capable of active locomotion through undulatory body movements or the movement of fins, which allow them to actively change their vertical and horizontal position. Whereas the larvae of some rocky intertidal species remain inshore (Marliave, 1986), other larvae may actively move into estuaries or offshore into open water (e.g., Able et al., 1989; Able and Kaiser, 1994; Keefe and Able, 1994). Many marine larvae exhibit diurnal vertical migration and active movements, which allow these fishes to exploit food resources at varying depths during a 24-hr period (Röpke, 1993). Oceanic currents play a critical role in establishing planktonic fish egg and larval distributions (Kleckner and McCleave, 1985; Lobel, 1989; Appeldoorn et al., 1994; Hensley et al., 1994). Such processes are thought to account for biogeographic patterns, including the limited distribution of individual species on isolated oceanic islands (Schultz and Cowen, 1994) and reefs (Leis, 1991), and for patterns of endemism (Lobel and Robinson, 1986).

Vertical and horizontal movements subject larvae to significant metabolic, osmoregulatory, and behavioral challenges. Temperature and salinity will vary dramatically when moving between estuarine and nearshore marine habitats over time or through the thermocline on a daily basis. Variation in light quality and quantity due to differences in turbidity in coastal and estuarine waters and open water and diurnal changes in light with depth experienced during diel vertical migration have significant implications for visual function and, thus, behavioral capabilities (Connaughton et al., 1994; Burke et al., 1995; see Section IV.E.1).

Larvae metabolize their yolk sac resources and, thus, may not need to be dependent upon exogenous food sources at hatching. Lecithotrophic larvae use their extensive yolk reserves for an extended period of time, whereas planktotrophic larvae quickly exhaust their limited yolk supply after hatching. The trophic niche of actively feeding planktonic larvae is limited by the capabilities of developing sensory systems used in prey detection (see Section IV.E), gape size and developmental condition of the feeding apparatus (see Chapter 10 in this volume), and hydrodynamic constraints on prey capture at low Reynolds numbers. The hydrodynamics of prey capture is dependent on the locomotory ability of both the larva (Hunter, 1972) and the prey (Buskey et al., 1993), larval body shape (Webb and Weihs, 1986), and temperature (Fuiman and Batty, 1994). Prey capture success also is dependent upon the ability of a larva to change its behavior where the density of food particles varies (Hunter and Thomas, 1973; Coughlin et al., 1992; Davis and Olla, 1995). Marine fish larvae experience high rates of predation by planktivorous fishes (Fuiman, 1994), cnidarians (Duffy et al., 1997), ctenophores (Cowan and Houde, 1993; Elliott and Leggett, 1996), and other planktivorous invertebrates (Hunter, 1981; Leis,

1991). Therefore, there must be strong selection for predator escape mechanisms in addition to morphological deterrents to predation in larval fishes (see Section IV.B), especially in those with extended larval stages.

Larval fishes are capable of rapid escape responses known as fast starts or c-starts (Webb, 1981; Eaton and DiDomenico, 1986), which generally are followed by burst swimming during which the Reynolds number is no different from that for adult fishes (Webb and Corolla, 1981; Webb and Weihs, 1986). Large Mauthner neurons in the hindbrain of larval fishes have been shown, in a small number of experimental species, to generate the c-start escape response (Eaton and DiDomenico, 1986; Williams *et al.*, 1996). Sensory input to Mauthner cells is established within the first 4 days posthatch in zebrafish larvae (Kimmel, 1982), and a c-start can be elicited in embryos released from their chorionic membrane 2 days before normal hatching (Eaton and DiDomenico, 1986). This suggests that the development of the c-start in embryos could facilitate the process of hatching itself. Selection for an effective escape response in small larvae that need to overcome the constraints of viscous forces encountered at low Reynolds numbers could account for the unusually early development of this neural pathway. The speed of c-start escape responses is correlated with larval body size (Fuiman, 1994; Williams *et al.*, 1996) and the developmental state of the musculoskeletal system (Kohno *et al.*, 1984). However, the timing of the initiation of the escape response appears to be due to the relative state of the development of the visual system, lateral line system, and components of the central nervous system mediating the response (Blaxter and Fuiman, 1989). Temperature also has a significant metabolic and hydrodynamic effect on the escape response and, thus, on susceptibility to predation (Fuiman, 1986; Fuiman and Batty, 1994).

IV. DEVELOPMENTAL MORPHOLOGY OF MARINE FISH LARVAE

The planktonic lifestyle of larval marine fishes clearly is distinct from that of the juvenile and adult fishes into which they develop. These larvae generally occupy a different habitat, encounter a different physical environment, exploit different food resources, and experience different predator pressures than juvenile or adult fishes. Therefore, it is not surprising that planktonic larvae exhibit a wide range of structural specializations.

In some species, larval morphology is so dramatically different from that of more familiar adult forms that field-collected larvae have had to be raised through metamorphosis in order just to identify them. In several notable cases, the degree of morphological specialization found in marine fish larvae resulted in their initial identification as distinct species. For example, "leptocephalus" larvae of several hundred species of eels originally were identified as members of the single genus *Leptocephalus*. In addition, the genus *Caulolepis* has been correctly identified as the larval form of *Anoplogaster,* the "Kasidoron" larva is now known to be the larval form of *Gibberichthys,* and the "Rosaura" larva has been identified as a larval giganturid (Cohen, 1984; Johnson, 1984; Johnson and Bertelsen, 1991).

A. GROWTH AND ALLOMETRY

Marine fish larvae generally range in size from 2.5–3.0 mm at hatching to 10–30 mm at transformation, a process that occurs from several days to several months after hatching, depending on the species (Ahlstrom *et al.,* 1984; Victor, 1986, 1991; Moser, 1996a,b). The largest marine fish larvae are the unusual leptocephalus larvae of elopomorph fishes (eels and relatives; Smith, 1979), which are known to reach lengths of more than 1 m (Fig. 2; Nielsen and Larsen, 1970). The flatfishes (Order Pleuronectiformes) exhibit the widest range of sizes of larvae among teleost fishes (see Fig. 3; Ahlstrom *et al.,* 1984), which generally does not correlate with adult size. For instance, the lined sole (*Achirus lineatus*) metamorphoses at 5.0 mm, whereas the rex sole (*Glyptocephalus zachirus*) may metamorphose at >90 mm. Some bothid larvae may reach lengths of more than 100 mm at transformation (Ahlstrom *et al.,* 1984), but their adult sizes are not extraordinary when compared to other species of flatfishes. The leaflike shape of the larvae of bothid flatfishes is the result of lengthening of the epaxial and hypaxial musculature, pterygiophores, and dorsal and ventral fin rays (Fig. 3; Moser, 1981; Ahlstrom *et al.,* 1984). The disk-shaped larvae of the Dover sole (*Microstomus pacificus*), which metamorphose at 60 mm, typically live in the plankton for 1 year but may remain there for more than 2 years (Markle *et al.,* 1992; Butler *et al.,* 1996).

Standard morphometric measurements are used to identify closely related species (e.g., Leis and Rennis, 1983). A comprehensive review of larval fishes in the California current region (Moser, 1996a) includes comparable morphometric data for more than 500 species, a database ripe for analysis in a comparative–phylogenetic or ecological context (H. G. Moser, personal communication). Thus, a vast but unexploited database concerning the comparative morphometry of larval body form is available in the taxonomic literature. Fishes provide an excellent opportunity to examine patterns of growth and allometry and the effects of scaling on a wide spectrum of biomechanical and physiological problems (see Chapter 10 in this volume).

B. LARVAL SKELETAL SPECIALIZATIONS

Of all of the morphological features of marine fish larvae, the skeleton exhibits the most impressive range of specializations, especially among "higher teleosts"

A

B

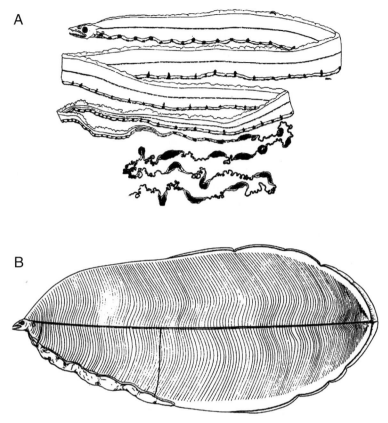

FIGURE 2   Giant leptocephalus larvae of elopiform fishes. (A) *Leptocephalus giganteus,* midstage larva, 314 mm (reprinted with permission from Moser and Charter, 1996). (B) *Thalassenchelys coheni,* 246 mm (reprinted with permission from Charter, 1996).

[the Acanthopterygii, reviewed by Moser (1981); Leis and Trnski, 1989; Kendall *et al.,* 1984; Moser, 1996b] in which the diversity of fin and head spine specializations in acanthopterygian larvae exceeds that among the adults into which they develop (Moser, 1981).

Head spines may be found on many of the cranial bones, including the frontal, opercular, preopercular, and infraorbitals, in a wide variety of species (Fig. 4). The highly specialized larvae of butterflyfishes ("tholichthys" larva; Leis, 1989) and the larvae of surgeonfishes (Leis and Rennis, 1983) are characterized by a solid, spiny head shield. Head spines may be a transient larval feature lost at transformation (Matarese *et al.,* 1989; Moser, 1996b), or they may be retained in juvenile and adult stages where they are characteristic of

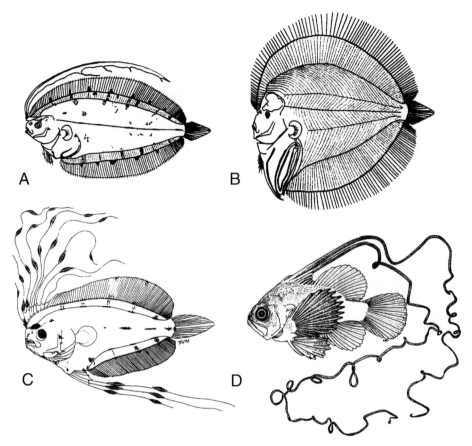

FIGURE 3   Larvae of fishes with elaborated fins and fin rays. (A) Bothid flatfish, *Arnoglossus yamanakai*, 49.0 mm SL (reprinted with permission from Fukui, 1997); (B) bothid flatfish, *Bothus pantherinus*, 32 mm SL (reprinted with permission from Fukui, 1997); (C) *Cyclopsetta panamensis*, God's flounder, 17.8 mm SL, postflexion, post-eye migration larva (reprinted with permission from Moser and Sumida, 1996); (D) the serranid, *Diploprion bifasciatum*, 16.2 mm SL, 24 day posthatch, postflexion larva (reprinted with permission from Baldwin *et al.*, 1991).

these groups (e.g., some scorpaeniforms; Moser, 1981). The presence of head spines at the widest or deepest point on the head may make a larva "effectively larger, painful to ingest, thus more resistant to predation" (Moser, 1981). Head spines are common in some taxa with extended larval periods (Moser, 1981), further suggesting a role in predator defense. The mass imposed by increased spination increases specific gravity, which generally is compensated by the presence of a larger, more anteriorly placed swim bladder in order to maintain

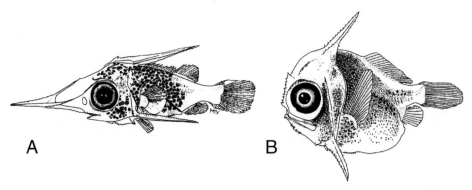

FIGURE 4 Cranial spines in teleost larvae. (A) a holocentrid, *Sargocentron suborbitalis*, 5.8 mm, postflexion larva (reprinted with permission from Watson, 1996b); (B) an anoplogastrid, *Anoplogaster cornuta,* 6 mm SL, (reprinted with permission from Watson, 1996a).

buoyancy in these planktonic animals (Moser, 1981). Given the rapid growth and drastic morphological changes experienced by transforming larvae, head spines also may serve as calcium stores that can be mobilized when both the cranial and postcranial skeletons undergo rapid ossification and growth.

Elongation of fin rays and fin spines and elaboration of the epithelium that covers them are also common among marine fish larvae (e.g., Baldwin *et al.,* 1991; Masuma *et al.,* 1993; Kohno *et al.,* 1993; Brogan, 1996); they generally are lost at the end of the larval period (Figs. 3, 5, and 6). Elongated fin spines, fin rays, or caudal fin filaments may increase surface area to reduce the rate of sinking, or they may provide predator defense in these planktonic larvae and probably are due to a simple change in the pattern of allometric growth (Fig. 5). The elongated single fin ray in some bothid flatfishes (Fig. 3A) and the dorsal fin crest in paralichthyid flatfishes may serve as a rudder during undulatory locomotion, act as a stabilizing vane during feeding strikes, or increase apparent size and thus dissuade predators (Moser, 1981). The larvae of *Diploprion* (Fig. 3D) and *Liapropoma* and the larvae of carapid fishes ("vexillifer larvae") (Fig. 6) have a peculiarly exaggerated elaboration of the dorsal fin. A "vexillum," or similar structures, may be many times longer than body length, bears fleshy tabs, and is both vascularized and innervated. It is presumed to serve a sensory function, and it may serve to distract predators or mimic siphonophores or salps (Govoni *et al.,* 1984; Baldwin *et al.,* 1991).

Medial fin folds are present in yolk sac and preflexion larvae and generally regress as fin spines and rays develop. Persistent "voluminous fin folds" are an interesting feature of some larvae, including those of some myctophids and argentinoids (Fig. 7B; Moser and Ahlstrom, 1970; Moser, 1981). The additional

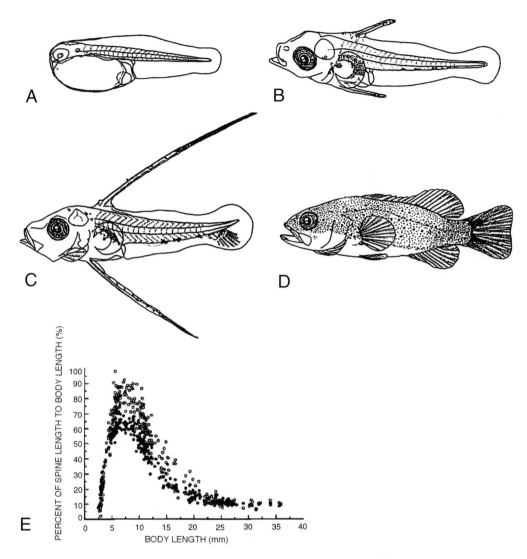

FIGURE 5   Fin spine allometry in reared larvae and juveniles of the coral trout, *Plectropomus leopardus.* (A) Newly hatched larva 1.62 mm; (B) postlarva, 7 days posthatch, 3.22 mm; (C) postlarva, 13 days posthatch, 6.10 mm; (D) juvenile, 65 days posthatch, 35 mm; (E) changes in length of second dorsal spine (○) and pelvic spine (●) as percentage of body length during larval and juvenile phases (reprinted with permission from Masuma *et al.,* 1993).

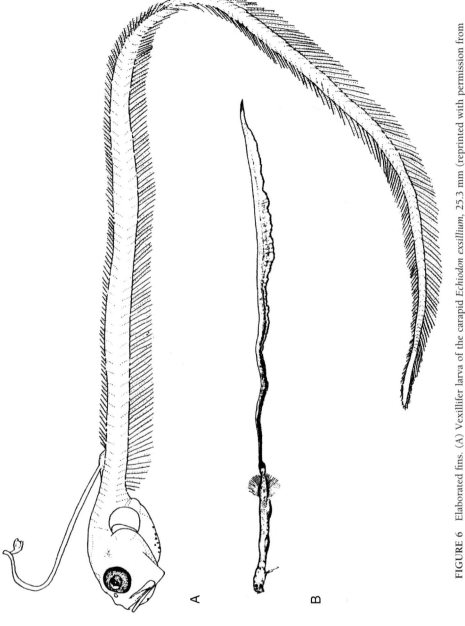

FIGURE 6 Elaborated fins. (A) Vexillifer larva of the carapid *Echiodon exsillium*, 25.3 mm (reprinted with permission from Ambrose, 1996a); (B) postflexion larva of the festive ribbontail, *Eutaeniophorus festivus*, Mirrapinnidae (reprinted with permission from Charter and Moser, 1996).

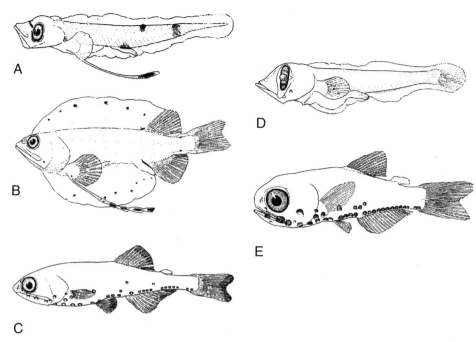

FIGURE 7   Specializations in the larvae and juveniles of lanternfishes (Myctophidae). Dwarf lanternfish, *Loweina rara:* (A) preflexion larva, 4.8 mm, showing elongation of pectoral fin ray; (B) postflexion larva, 17.6 mm, with persistent voluminous fin folds; (C) juvenile, 24.5 mm with photophores. *Electrona risso:* (D) oval eyes, larva, 6.3 mm; (E) juvenile, 9.9 mm (reprinted with permission from Moser and Ahlstrom, 1996b).

surface area that they provide may promote buoyancy, assist in locomotion, or enhance apparent size and thus deter predators (Moser, 1981). Alternatively, they could promote cutaneous gas exchange and/or the absorption of dissolved organic nutrients (see Section IV.C).

One other unusual feature of unpaired fins is the fin migration characteristic of clupeiform and elopiform larvae, in which the dorsal and anal fins and internal fin supports migrate anteriorly relative to the underlying myomeres. Moser (1981) suggests that the more posterior placement during the larval stage may assist in the tail beat thrust needed for a fast start.

Perhaps the most radical modification of the skeleton found among larval fishes is associated with the unique eye migration characteristic of larval flatfishes (Order Pleuronectiformes, the flounders, halibuts, and soles). Eye migration is characterized by the movement of one eye over the middorsal line to the other side of the head during the planktonic larval period. This results in perturbation of bilaterial symmetry in the neurocranium and dermatocranium, especially in the vicinity of the orbits. This morphological modification is accompanied by a 90° rotation in body orientation, which allows these fishes to rest on their "eyeless" side when they eventually settle to the benthos upon completion of transformation. These unique behavioral modifications are followed by the development of bilateral asymmetry in pigmentation, paired fins, teeth, olfactory organs (Webb, 1993), neuromast receptors, and lateral line canals (e.g., Webb, 1988, 1995). The result is a redefinition of the functional body axes such that the eyed side (e.g., the right side in pleuronectids) becomes the functional dorsal side of the body in adult flatfishes. Eye migration is a synapomorphy of the Order Pleuronectiformes (Chapleau, 1993). Although studied for more than a century (see Williams, 1902; Kyle, 1921; Berrill, 1925), the genetic, physiological, and biomechanical mechanisms responsible for the generation of this unique developmental pattern still are not fully understood (see Brewster, 1987; Youson, 1988; Rose and Reiss, 1993).

C. SPECIALIZATIONS OF THE LARVAL DIGESTIVE SYSTEM

The ability of a larva to process exogenous food resources is essential when yolk sac reserves are exhausted. In most marine fish larvae, a foregut, midgut, and hindgut are evident at first feeding, and then additional specializations of the digestive system occur at transformation (J. J. Govoni, personal communication) and continue throughout the juvenile period (Govoni et al., 1986; Bisbal and Bengtson, 1995; Tanaka et al., 1996).

The most unusual features of the digestive system among larval marine fishes are the trailing gut and external intestinal loops, in which the terminal or midportion of the digestive tract extends beyond the ventral boundary of the trunk (Fig. 8; Moser, 1981). A trailing gut is found in mesopelagic families such as myctophids, melanostomids, malacosteids, and idiacanthids (Kawaguchi and Moser, 1984; Matarese et al., 1989) and probably arises as the simple result of positive allometry. For example, in one unidentified stomatioid species, the gut may be up to 5 times the length of the body and exceed the body in mass (Moser, 1981). It has been suggested that by increasing the length of the digestive system, the trailing guts increase the surface area across which nutrients may be absorbed. This could increase the range of prey types that may be eaten (Moser, 1981) during periods of rapid growth in environments in which food resources may be patchy. Certain bythitid larvae ("exterilium larvae") have an ornate, free-trailing intestinal loop, which may serve to mimic gelatinous zooplankton (Moser, 1981). Other larvae have a saclike gut (e.g., *Bathophilus*), a straight gut with distinctly visible rugae (some myctophids), or a coiled gut (found only in

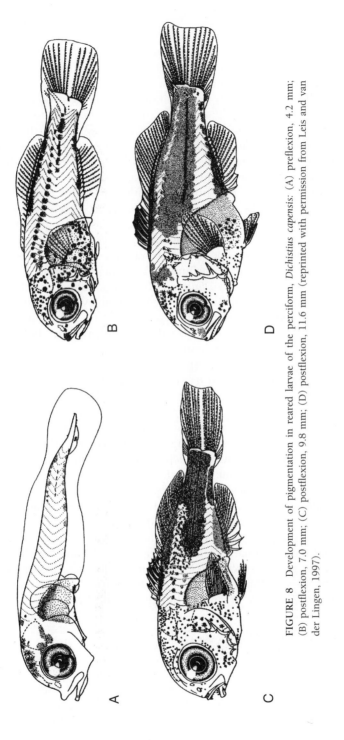

FIGURE 8 Development of pigmentation in reared larvae of the perciform, *Dichistius capensis*: (A) preflexion, 4.2 mm; (B) postflexion, 7.0 mm; (C) postflexion, 9.8 mm; (D) postflexion, 11.6 mm (reprinted with permission from Leis and van der Lingen, 1997).

gadiforms and "higher teleosts"), which may extend beyond the body margin (some flatfishes; see Ahlstrom *et al.*, 1984) to further increase the surface area of the digestive tube for absorption of nutrients (Moser, 1981).

Marine fish larvae generally are planktivorous and feed on a wide variety of invertebrates and other fish larvae (e.g., Leis, 1991). Planktivory is considered by some to be a specialization of marine fish larvae (if the adult form occupies a different trophic niche), but given the size of marine fish larvae and the hydrodynamic constraints with which they are faced, planktivory most likely is the only option for larvae exploiting particulate food sources. However, particulate food resources may not be the only exploitable nutritional source available to larvae (Otake, *et al.*, 1993). Pfeiler (1986) presents an argument that suggests that, like invertebrate larvae, the unusual leptocephalus larvae of elopomorph fishes (see Fig. 2) obtain nutrition through the cutaneous absorption of dissolved organic nutrients found in seawater. These larvae are covered by a thin external epithelium with filamentous projections thought to be analogous to microvilli (Hulet, 1978). In addition, chloride cells are found in the external epithelium (Pfeiler and Lindley, 1989), suggesting an osmoregulatory role. Biochemical analyses suggest that larvae are capable of water and salt loading, which is also thought to promote the transport of amino acids across the skin epithelium. In addition, these larvae are reported to have a poorly developed digestive system, which in some species lacks a lumen, and in which solid food has never been found (Pfeiler, 1986). These data suggest that the ability to absorb dissolved organic nutrients could allow leptocephalus larvae, which have a low metabolic rate (Pfeiler and Govoni, 1993), to live in the plankton for long periods of time (Pfeiler, 1986). J. J. Govoni (personal communication) suggests, however, that these larvae live in oligotrophic, open ocean waters, which do not have sufficient nutrient levels to warrant adaptations for the absorption of dissolved organic nutrients. Instead, he suggests that they probably are eating prey items such as aloricate (naked) ciliates and tintinnids that do not leave a digestive residue. Further, the presence of a formidable set of fanglike teeth, which are lost at metamorphosis, suggests that these larvae may eat particulate food. Leptocephalus larvae will feed on solid food in aquaria (Mochioka *et al.*, 1993), and zooplankton fecal pellets and larvacean houses have been found in leptoceophalus guts. These observations suggest that the fanglike teeth are used to pierce and capture gelatinous zooplanktonic organisms (Mochioka and Iwamizu, 1996).

D. Pigmentation, Squamation, and Photophores

Melanophores in marine fish larvae are thought to serve several purposes, including protection from ultraviolet (UV) radiation, masking of the light emitted by gut contents, masking refracted light emanating from the swim bladder, and concealment from predators (Moser, 1981). Most interestingly, Moser (1981) suggests that pigmentation patterns in larvae may facilitate intraspecific recognition, which would promote attraction to a food patch and, thus, may give clues to the origins of schooling behavior. Pigmentation develops in characteristic and, often, diagnostic patterns in larvae. Melanophore size, shape, pattern, and sequence of development are of taxonomic importance (Matarese *et al.*, 1989; Fig. 9). The diversity of pigment patterns among freshwater centrarchids (Mabee, 1995) and zebrafishes (*Danio*, M. M. McClure, personal communication) is the result of variation in patterns of pigment distribution that arise in early life-history stages. Parichy (1996a,b) has shown that the presence of the lateral line (superficial neuromasts) on the trunk plays a role in the development of pigment patterns of ambystomatid salamanders. The nature of developmental interactions between pigment cells and neuromast precursor cells has not yet been explored in fishes, although interesting correlations have been noted (Milos and Dingel, 1978).

Scale development generally commences at the end of the larval period, so that only the precocious development of squamation is of interest among fish larvae. Scales clearly are present in the larvae of many taxa, including trachichythiform fishes (Fig. 10) and swordfishes (where adults have either reduced or lost squamation).

The serial, neurogenic photophores of mesopelagic fishes (e.g., myctophids) may appear during the larval phase (Moser, 1996b), but generally develop in juveniles (see Fig. 7). Their distribution is of taxonomic value at the specific, generic, and higher levels among myctophids (Moser, 1981, 1996b). The presence of photophores in shallow water fishes is unusual, but their occurrence in the shallow water midshipman, *Porichthys* has allowed studies of their development. *Porichthys* photophores contain lenses that, like the lens of the eye, are transparent and transmit light from inside the photophore to the external environment (Ancil, 1977). Dove *et al.* (1993) have shown that although the biochemical mechanism for the development of transparency of the photophore lens is similar to that in the eye lens, there is no evidence that the photophore lens protein is similar to the eye lens protein. The nature of the mechanisms underlying developmental patterning and morphogenesis of photophores is an area in need of experimental investigation.

E. CRANIAL SENSORY SYSTEMS

Larval fishes are equipped with several sensory systems that are functional at or soon after hatching and whose function is modified further throughout the larval and juvenile periods. The timing of the development of different features and functional attributes of larval sensory systems generally is correlated with changes in the physical and chemical characteristics of the environment occupied, and

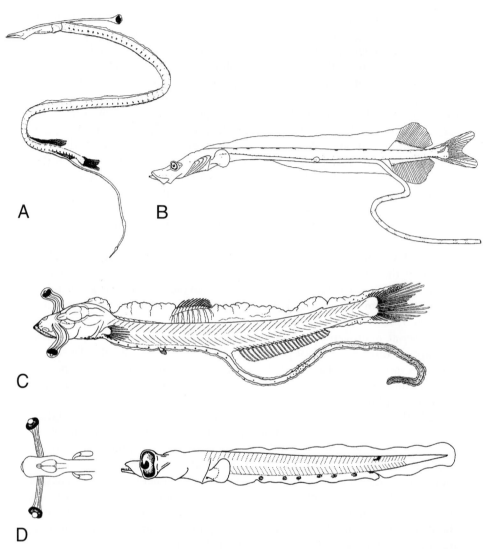

FIGURE 9 Visual system specializations and trailing guts in mesopelagic fish larvae. (A) *Idiacanthus antrostomus,* (Idiacanthidae), postflexion larva, 55 mm (reprinted with permission from Moser, 1996d); (B) *Aristostomias scintillans* (Malacosteidae), postflexion larva, 34.7 mm (reprinted with permission from Moser, 1996c); (C) *Myctophum aurolaternatum* (Myctophidae), postflexion larva, 25.8 mm (reprinted with permission from Moser and Ahlstrom, 1996b); (D) lateral view of the stalked eyes of the preflexion larva of the bathylagid, *Bathylagus bericoides,* 7.3 mm (reprinted with permission from Moser and Alhstrom, 1996a).

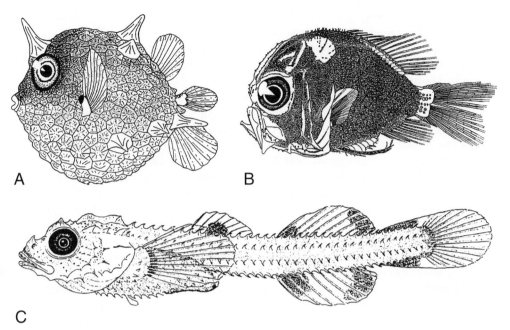

FIGURE 10   Squamation in larvae and pelagic juveniles. (A) Pelagic juvenile of the ostraciid, *Lactoria diaphana,* 6.8 mm (reprinted with permission from Watson, 1996c); (B) larva of *Aulotrachichthys,* 7.4 mm, showing heavy cranial spination and spiny scales on trunk (reprinted with permission from Konishi and Okiyama, 1997); (C) postflexion larva of *Agonopsis sterletus,* 14.5 mm (reprinted with permission from Ambrose, 1996b).

significant transitional events in life history and behavior [reviewed by Blaxter (1986, 1991); Noakes and Godin, 1988]. Ontogenetic changes in the input from different sensory systems are reflected in the ontogeny of different regions of the brain associated with these different sensory modalities (e.g., Branstätter and Kotrschal, 1989; Tomoda and Uematsu, 1996). In general, our knowledge of the sensory biology of larval fishes is limited largely to commercially important marine and freshwater species (e.g., clupeoids, pleuronectiforms, salmonids); there is little information about the sensory biology of the larvae of coral reef or of tropical marine fishes in general (Leis, 1991), and few if any comparisons of sensory development have been carried out in a formal phylogenetic context.

1. Vision

The development of functional eyes generally is correlated with the onset of feeding (Blaxter, 1988a,b; Moser, 1996b; Pankhurst and Eagar, 1996). The retina of larval fishes generally is distinct from that of juvenile and adult stages (Schmitt and Kunz, 1989; Powers and Raymond, 1990; Pankhurst and

Montgomery, 1994). Larval fishes first develop a pure cone retina; a duplex retina containing rods as well as several types of cones (Bowmaker, 1990) develops later during development (e.g., Sandy and Blaxter, 1980). Major events in the functional ontogeny of the visual system are closely correlated with life-history events [reviewed by Noakes and Godin (1988)] where the fish experiences changes in the "photic environment" (Boehlert, 1979; Loew and McFarland, 1990) due to a change in vertical or horizontal position or changes in behavioral repertoire (see references in Table I). The configuration of the retinal receptor cell mosaic and the density of receptor cells change during ontogeny (Pankhurst, 1984; Miller *et al.*, 1993; Pankhurst, 1994; van der Meer, 1995), thus contributing to ontogenetic changes in visual acuity, which is of great importance in the localization of prey (Miller *et al.*, 1993). Distinct changes in retinal morphology occur with a shift from a pelagic (planktonic) to a benthic habitat at transformation (Boehlert, 1979; Kawamura *et al.*, 1984, 1989; Evans and Fernald, 1990; Shand, 1993, 1994; Kvenseth *et al.*, 1996).

The ability to detect ultraviolet light is a transient feature of the eyes of the early life-history stages of salmonids (Browman and Hawryshyn, 1994), juvenile perch (Loew and Wahl, 1991), and *Lepomis* (Loew and McFarland, 1990). Ultraviolet light reception has been observed in several species of adult freshwater fishes (e.g., Douglas, 1989) and has also been recorded in some adult (pomacentrids; McFarland and Loew, 1994) and larval (topsmelt and grunion; Loew *et al.*, 1997) marine fishes, but its distribution among marine fishes as a group has yet to be determined. Ultraviolet light reception is a capability of single cones in the retina (McFarland and Loew, 1994), and is thought to enhance ability to feed on zooplankton (Loew and Sillman, 1993). In salmonids, the loss of ultraviolet light reception appears to be due to changes in thyroxine levels associated with normal development (Browman and Hawryshyn, 1994) and is accomplished through apoptosis of UV receptors (Kunz *et al.*, 1994).

The most dramatic morphological specializations of the visual system are found in the larvae of mesopelagic fishes. Whereas most fishes have round eyes, some marine fish larvae have narrowed, elliptical eyes (Figs. 7 and 8; e.g., argentinoids, notosudids, scopelarchids, some myctophids; Moser and Ahlstrom, 1970; Moser, 1981). In addition, the larvae of many midwater fishes have eyes on short or long stalks (Fig. 8; e.g., anguilliforms, argentinoids, stomatioids, myctophids; Moser, 1981; Matarese *et al.*, 1989). Weihs and Moser (1981) suggest that an elliptical eye may permit increased rotational ability, which would increase the volume of the visual field, and that placing an eye at the end of a stalk would allow the visual field to approximate the volume of a sphere. They further suggest that each of these features would provide a large increase in the volume of the visual field thus allowing these fishes to detect prey from a greater distance with minimal body movement. This would minimize expenditure of energy and the production of predator cues (Moser,

1981). Stalked eyes are a transient larval feature that disappears as the eyes return to a "normal" position, a process that simply may be due to the relative growth rates of the head and stalk (H. G. Moser, personal communication) or may be the result of the reabsorption of the cartilaginous part of the stalk (Beebe, 1934).

2. Chemosensory Systems

Fishes have two major chemosensory systems that are present in all vertebrates, the olfactory (smell) system and the gustatory (taste) system. Whereas olfaction has been implicated in habitat selection, migration, predator avoidance, and social communication, gustation appears to play a role in feeding behavior [reviewed by Noakes and Godin (1988); Hara, 1993, 1994]. Fishes also have several other chemosensory systems, including a "common chemical sense," solitary chemoreceptor cells, and spinal nerve chemoreceptors [reviewed by Finger (1988); Hara, 1992], whose distribution, function, and development are not well-known.

As in all vertebrates, the olfactory system of fishes develops from a pair of olfactory placodes during embryogenesis and is evident as a pair of patches of ciliated epithelial receptor cells in larvae [reviewed by Hara and Zielinski (1989); Blaxter, 1988a; Noakes and Godin, 1988]. During the larval stage, the epithelium invaginates to form the nasal sacs, leaving one or two nares open to the outside (Applebaum *et al.*, 1983; Hara and Zielinski, 1989; Olsen, 1993; Hansen and Zeiske, 1993; Harvey, 1996). The olfactory epithelium folds to form olfactory lamellae, which increase in number as the fish grows, providing additional surface area for ciliated olfactory receptor cells. Whereas the relative size of the olfactory organs appears to reflect the importance of olfaction in different species, attempts to correlate the size and number of lamellae with olfactory function as reflected by feeding behavior and ecology have not been successful [reviewed by Noakes and Godin (1988)]. Accessory olfactory sacs, described in a wide variety of teleost taxa (e.g., Verraes, 1976; Applebaum *et al.*, 1983; Webb, 1993), are nonsensory diverticula of the olfactory sacs and facilitate active olfactory ventilation, which presumably permits chemical sampling of the aquatic environment.

The gustatory system of fishes generally is composed of a population of taste buds in the epithelial lining of the oral cavity and on the gill arches. In some fishes, a second population of taste buds is present on the external epithelia of the face and body and on maxillary and mandibular barbels [reviewed by Hara (1992)]. For instance, in catfishes, external taste buds function in food localization, in contrast to the internal taste buds, which function in final food acceptance and initiation of the swallowing reflex [reviewed by Finger (1988)]. In the goatfish, *Upeneus tragula,* barbels are present in larvae but are used only in juveniles and adults for food localization. At settlement, barbels increase in length by up to 52% and the gustatory receptor cells that make up the external taste buds on the barbels increase in size by up to 100% (McCormick, 1993). Development of the barbels is influenced by food availability, indicating the presence of complex interactions between environmental factors and sensory development (McCormick, 1993). Among fishes, taste buds generally appear before hatching (Kawamura et al., 1990), at or after first feeding [reviewed by Noakes and Godin (1988)], or within 2 weeks of hatching in many species (Blaxter, 1988a). Their appearance has been correlated with behavioral evidence of food preference (Kawamura and Ishida, 1985).

Analysis of the functional development of the olfactory and gustatory systems in larval marine fishes has centered on correlative ontogenetic studies of sensory morphology and behavior (see Table I). Some studies have examined the developmental morphology of the olfactory system in economically important species with contrasting feeding strategies in order to argue for adaptive sensory morphology as an explanation for feeding behavior (e.g., Harvey et al., 1992; Harvey, 1996). There has been considerable interest in the functional development of the olfactory system in the context of its role in the migratory and homing behavior of salmonids (e.g., Olsen, 1993; Werner and Lannoo, 1994) and other anadromous species (e.g., VanDenBossche et al., 1995).

A few experimental behavioral studies have examined the role of chemical stimuli in the initiation of feeding behavior in larvae (Sweatman, 1988) and the use of chemical cues to identify suitable postlarval settlements sites (Knutsen, 1992). Because of the diversity of chemical stimuli and the diversity of chemosensory systems that can potentially respond to them at similar concentrations, the identity of a system mediating a particular behavioral response cannot be identified clearly without direct physiological recording or selective ablation of individual systems. The taxonomic distribution and the relative importance of the chemosensory systems other than olfaction and gustation have not been assessed adequately in fishes.

3. Auditory System

The inner ear of teleosts generally is composed of three otolithic organs and three semicircular canals, each equipped with a population of innervated sensory hair cells. In addition, a small macula neglecta of unknown function may be present. The otolithic organs are responsible for detection of sound (pressure) and gravity, whereas the semicircular canals allow the determination of movement in 3D space (Platt, 1988; Schellart and Popper, 1992).

Both the otoliths and the semicircular canals generally are present at hatching, although ossification of the neurocranium in which they are contained is not complete. The swim bladder develops during the larval stage (Allen et al.,

TABLE I Selected Studies of Sensory Ontogeny and Behavior in Clupeiform, Pleuronectiform, and Other Teleost Fishes

Species	Vision	Chemoreception	Mechanoreception	Correlation with behavior?	Source
Clupea harengus	XX		XX		Blaxter and Fuiman (1990)
Engraulis mordax	XX		XX	Yes	O'Connell (1981)
Brevoortia tyrannus	XX		XX	Yes	Higgs and Fuiman (1996a,b)
Paralichthys olivaceus	XX	XX	XX	Yes	Kawamura and Ishida (1985)
Paralichthys californicus	XX	XX	XX	Yes	Lara (1992)
Rhombosolea tapirina	XX	XX	XX	Yes	Pankhurst and Butler (1996)
Hippoglossus hippoglossus	XX				Pittman et al. (1990)
Hypseleotris galii	XX		XX	Yes	Konagai and Rimmer (1985)
Astractoscion nobilis	XX		XX	Yes	Margulies (1989)
Micropterus salmoides	XX	XX	XX	Yes	Kawamura and Washiyama (1989)
Tilapia nilotica	XX	XX	XX	Yes	Kawamura and Washiyama (1989)
Takifugu rubripes	XX		X	Yes	Nobuhiro et al. (1995)
Oxyeleotris marmoratus	XX	XX	XX	Yes	Senoo et al. (1994)
Leptocephalus larvae	XX	XX	XX	Yes	Pfeiler (1989)
Trichiurus lepturus	XX	XX	XX	Yes	Kawamura and Munekiyo (1989)
Hemiramphus sajori	XX	XX	XX	Yes	Kawamura et al. (1990)
Cynoscion regalis	XX		XX	Yes	Connaughton et al. (1994)
Micropogonias undulatus	XX		XX	Yes	Poling and Fuiman (1997)
Theragra chalcogramma	XX	XX		Yes	Davis and Olla (1995)

1976; Hoss and Blaxter, 1982; O'Connell, 1981). Several specializations of the auditory system that enhance pressure sensitivity develop during the larval period. These include the recessus lateralis of clupeomorph fishes, which provides a three-way connection between the swim bladder, the inner ear, and the lateral line canals of the head (Allen *et al.*, 1976; O'Connell, 1981; Hoss and Blaxter, 1982), and the Weberian apparatus of ostariophysan fishes (Vandewalle *et al.*, 1989; Radermaker *et al.*, 1989), which links the swim bladder to the inner ear via modified vertebral bones. The development of otophysic connections present in other marine and freshwater teleosts [reviewed by Schellart and Popper (1992)], which link the swim bladder and the inner ear, is not well-known.

The comparative functional morphology of otoliths as components of the auditory system of fishes is not well-understood, but these structures are the subject of an extensive literature because they are exploited as tools (see Secor *et al.*, 1995) for the determination of growth rates in fishes and the duration of the larval stage (Brothers, 1984; Victor, 1986, 1991). In addition, changes in the composition of the otolith or patterns of otolith growth are used as markers for significant events during ontogeny (Volk *et al.*, 1984; Toole *et al.*, 1993; Tsukamoto and Okiyama, 1993; Otake *et al.*, 1994; Modin *et al.*, 1996). Such determinations allow the assessment of age structure within a population and, thus, provide data for the analysis of population dynamics (e.g., Lecomtefiniger, 1992, 1994).

4. Mechanosensory Lateral Line System

The mechanosensory lateral line system is a primitive craniate feature (Northcutt, 1992) composed of hair-cell-based neuromast receptors. The lateral line system allows fishes to respond to unidirectional or oscillatory (vibratory) water movement at relatively short distances. Effective lateral line stimuli may arise from prey, predators, environmental obstacles, and neighbors in a school (Bleckmann, 1993; Montgomery *et al.*, 1995) and facilitate rheotaxis (Montgomery *et al.*, 1997). Functional superficial neuromasts generally are present in the epithelium on both the head and trunk of teleost fishes at hatching (Fig. 11; Blaxter *et al.*, 1983). A kinocilium and several sterocilia are present on the apical surface of each hair cell in neuromasts (e.g., Iwai, 1983). The cilia of all the hair cells in a neuromast are embedded in a prominent gelatinous cupula (Fig. 11; Mukai and Kobayashi, 1991), which grows continuously (e.g., Iwai, 1964, 1965, 1967; Yamashita, 1982; Mukai and Kobayashi, 1992). Each hair cell is directionally sensitive as a result of the placement of the kinocilium and stereocilia on their apical surface (e.g., Mukai *et al.*, 1992). Hair cells in a neuromast are oriented 180° to one another (Nobuhiro *et al.*, 1996) and hair cell orientation does not change through ontogeny (Münz, 1989; Webb, 1989b; Vischer, 1989), thus preserving the orientation of the axis of best physiological response

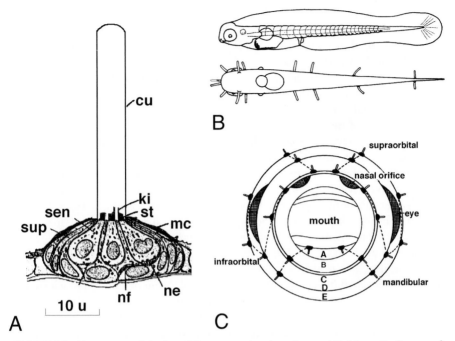

FIGURE 11   Neuromasts of the lateral line system in teleost larvae. (A) Schematic diagram of
a section of a superficial neuromast: cu, cupula; ki, kinocilium; mc, mantle cell; ne, nerve ending;
nf, nerve fiber; sen, sensory cell; st, stereocilium; sup, supporting cell (reprinted with permission
from Iwai, 1967 © University of Indiana Press). (B) Dorsal view of a larva of rabbitfish, *Siganus
guttatus* (2.8 mm TL), indicating the location of cupulae of superficial neuromasts (reprinted with
permission from Hara *et al.*, 1986 with kind permission from Elsevier Science-NL). (C) Anterior
view of the arrangement of superficial neuromasts on the head of a goldfish larvae: a, rostral to
olfactory sac; b, at level of olfactory sac; c, rostal to eye; d, caudal to eye; e, halfway between eye
and auditory vesicle (reprinted with permission from Iwai, 1967 © University of Indiana Press).

of a given neuromast. The number of hair cells within each neuromast increase
during ontogeny (e.g., Münz, 1989; Harvey *et al.*, 1992; Wonsettler and Webb,
1997). Hair cell density, number, neuromast shape (Münz, 1989; Webb, 1989c;
Wonsettler and Webb, 1997), and cupular shape (e.g., Blaxter *et al.*, 1983) also
change during ontogeny, and this variation is predicted to have functional consequences (Denton and Gray, 1989; Van Netten and Kroese, 1989).

   Late during the larval period, or at transformation, a subset of neuromasts
(presumptive canal neuromasts) becomes enclosed in pored lateral line canals;
other neuromasts remain superficial throughout life and may increase in number. Presumptive canal neuromasts become enclosed in epithelial grooves and
then canals, which ossify intramembranously, to form the pored lateral line

canals (Webb, 1989a). These canals are incorporated into a subset of dermal bones to form "lateral line bones." On the trunk of teleost fishes with a single lateral line canal, a neuromast develops in the epithelium overlying each of the scales of the lateral line series, and individual canal segments form around each neuromast to form the pored trunk canal ("the lateral line;" see Webb, 1989a,c). Wonsettler and Webb (1997) have shown, however, that the multiple lateral line canals on the trunk of two species of hexagrammid fishes develop in the absence of neuromasts.

The morphology of the lateral line canals and the bones in which they are contained, and the number and distribution of canal pores have been used extensively as characters in species descriptions and in phylogenetic reconstructions (see Webb, 1989a). The morphological diversity of the lateral line canal system in bony fishes has been reviewed (Coombs *et al.,* 1988; Webb, 1989a). The developmental osteology of the lateral line canals of the head has been described in several nonteleost actinopterygians (e.g., Allis, 1889; Holmgren and Pehrson, 1949; Disler, 1960), clupeoids (e.g., Blaxer, 1987), and ostariophysans (Lekander, 1949; Disler, 1960). However, detailed studies of lateral line canal development in a formal phylogenetic context at a specific or generic level are lacking. Further, the mechanisms underlying the distribution of neuromasts, including the relationship between neuromasts and dermal bone, remain unresolved (see Section V.C). Both canal and superficial neuromasts are innervated by a well-defined series of cranial nerves: the anterior, middle, and posterior lateral line nerves, *not* branches of cranial nerves V, VII, IX, and X [reviewed by Northcutt (1989)] as still indicated in most textbooks. The distribution and innervation of neuromasts have been used to reconstruct the morphotype of the ancestral lateral line system (Northcutt, 1989), and heterochrony in lateral line development has been used to explain the evolution of the system among major vertebrate clades (Northcutt, 1992). The role of developmental patterns in explaining the evolution of the system in teleost fishes has been discussed (e.g., Webb, 1989a,b; Wonsettler and Webb, 1997), but we still do not understand the mechanisms that may explain the spatial relationships of neuromast receptor organs and dermal bone ossifications on the head and trunk of bony fishes (see Section V.C).

5. Electrosensory System

Electroreception is a vertebrate synapomorphy (Northcutt, 1992; New, 1997) that was lost with the origin of the Neopterygii (*Amia* and teleost fishes), but has re-evolved convergently in two monophyletic lineages of teleosts (the mormyriform osteoglossomorphs and the common ancestor of siluriform and gymnotiform ostariophysans). Among non-teleost marine fishes, the lampreys, elasmobranchiomorphs, and the living coelocanth, *Latimeria chalumnae,* are electroreceptive (Northcutt, 1986), but none have a larval phase. Among marine teleosts, catfishes of the families Ariidae and Plotocidae are electroreceptive (J. G. New, personal communication). *Arias felis* has a reduced electrosensory system, whereas marine plotocids have a well-developed system (Obara, 1976; J. G. New, personal communication) but their development presently is unknown. Neither group has planktonic larvae; the marine ariids are oral incubators whose young emerge as juveniles, and the marine plotocids have demersal larvae (C. Ferraris, personal communication).

In elasmobranchiomorphs and nonteleost actinopterygians, ampullary electroreceptors detect electrical fields, which are used in prey detection and navigation (New and Tricas, 1998). In freshwater mormyriform and gymnotiform teleosts, a new class of electroreceptors, the tuberous organs, evolved convergently to detect electrical signals generated by their electric organs, which are used for social communication [Moller, 1995; reviewed by New (1997)]. The development of ampullary and tuberous electroreceptors, the larval and adult electric organs, in the relatively few species that have been studied has been reviewed briefly by Kirschbaum (1995). The development of the electrosensory system has been described in larval *Eigenmannia* (Vischer, 1989) and *Sternopygus* (Zakon, 1984, 1986).

The larvae of many weakly electric fishes (mormyriforms and gymnotiforms) have a morphologically distinct larval electric organ, which generates a characteristic larval electric organ discharge (EOD; Kirschbaum, 1995). The adult electric organ develops later and only becomes active after the first month of larval life. The larval and adult organs function in tandem, and then the larval organ becomes silent (C. Hopkins, personal communication). The comparison of the electric organ discharges of these weakly electric fishes, their behavioral significance, and their utility as a model for the study of neurophysiological principles has been the subject of intense study (Bullock and Heiligenberg, 1986; Moller, 1995). The morphological basis for variation in functional attributes of electric organs in mormyriform and gymnotiform fishes has been reviewed by Hopkins (1995).

6. Multiple Sensory Modalities and the Ontogeny of Behavior

Behavior is likely the result of the integration of information from multiple sensory systems (Blaxter, 1988a,b). The functional role of individual sensory systems in the production of behavior is difficult to establish, especially in small larvae, because mechanical experimental manipulations generally are difficult if not impossible to carry out. Thus, many studies have focused on the correlation of sensory system development and the onset of behaviors [see reviews by Blaxter (1986, 1991); Noakes and Godin, 1988; see Table I]. Such studies generally can only infer the function of a single sensory system through an examination of the gross morphology of receptors and/or receptor organs (generally with scanning electron microscopy and less often with histological analysis), and with comparison to other species in which the function of a system with similar morphology has been determined experimentally.

Experiments in which one sensory system has been rendered ineffective by selective ablation have allowed the determination of the role of that sensory system in the production of a particular behavior. Most commonly, the role of visual stimuli in the production of a behavior is determined by conducting experiments under different light conditions or in complete darkness. Under these circumstances, the relative contributions of the lateral line, auditory, and chemosensory systems cannot be established. However, methods have been developed in which the hair cells of neuromasts can be reversibly inactivated using streptomycin (see Blaxter and Fuiman, 1989) to determine the role of mechanoreception in the initiation of feeding behavior (e.g., Batty and Hoyt, 1995) and predator avoidance (Blaxter and Fuiman, 1990). The species that have been used for these interesting morphological and functional studies span the phylogenetic spectrum (Table I). However, questions pertaining to the comparative functional and developmental sensory morphology of the major sensory systems have not yet been placed in an appropriate phylogenetic context.

V. WHAT CAN FISH LARVAE TEACH US ABOUT THE DEVELOPMENTAL BASIS FOR EVOLUTIONARY CHANGE?

The establishment of the developmental basis for evolutionary change requires comparable developmental data on related species for which there is a good hypothesis of phylogenetic relationships. The development of individual species must be defined carefully as a series of stages (e.g., Gorodilov, 1996), and standard and reliable methods of interspecific comparison of development must be established (e.g., relative order of development; see Cubbage and Mabee, 1996; Mabee and Trendler, 1996). Finally, comparative developmental studies are essential for the identification of homologies among related species (Mabee, 1988; Johnson, 1993) upon which phylogenetic hypotheses are based, especially in species where adult morphologies are widely divergent.

Heterochrony has been suggested as the mechanism for evolution in several highly modified deep sea fish taxa (Moore, 1994), including deep water anglerfishes, which may be neotenic derivatives of shallow water benthic forms (Moser, 1981). Johnson and Brothers (1993) have determined that *Schindleria*, a taxon whose phylogenetic position had long been debated, is a radically paedomorphic goby. Heterochrony also has been implicated in the morphological evolution of the jaw and teeth (Fink, 1981; Meyer, 1987; Boughton *et al.*, 1991), the lateral line system (Webb, 1989b, 1990; Northcutt, 1992), and miniaturization (Weitzman and Vari, 1988; Johnson and Brothers, 1993) in teleost fishes.

The development of species with interesting larval attributes can be used to address broader issues, which address not only intraspecific variation but morphological novelty among species and among higher taxa. These issues include the definition of adaptations in larvae and the mechanisms underlying patterns of developmental plasticity, developmental integration, and metamorphosis.

A. Character Evolution in Larval Fishes

The presence of distinctive features in marine fish larvae (or any group of organisms, for that matter) would lead one to expect that they are functionally important, that they are the result of natural selection, and that they are thus adaptive [see Rose and Lauder (1996) for an extensive discussion of this issue]. However, unique or unusual features in larvae must be interpreted in the context of the comparative development of the larval, juvenile, and adult stages in order to be considered properly. Further, the adaptive significance of such specializations cannot be determined until structure–function relationships are better understood.

A larval feature may be a "normal" stage in the ontogenetic trajectory of an adult feature, but may not necessarily be an adaptation or a result of natural selection. For instance, superficial neuromasts in larvae function differently from canal neuromasts, but may not be an adaptation because superficial neuromasts are the normal ontogenetic precursor of the canal neuromasts present in juveniles and adults.

Alternatively, a larval feature may be a transient specialization atypical of a simple ontogenetic trajectory, as described earlier. For instance, in the lateral line system, a proliferation then regression of neuromasts or the development of a transient, specialized neuromast or cupular morphology that is functionally distinct from, and not a simple ontogenetic precursor of, adult morphology should be considered a larval specialization. Other specializations such as exaggerated fin rays, spines, or epithelial elaborations of the medial or paired fins may arise as the result of a shift to positive allometry and may be lost due to a shift to negative allometry at metamorphosis. The evolution of a larval specialization also may be the result of a shift in developmental onset. For example, if a feature that normally develops in juveniles develops early, during the larval stage, then this is a larval specialization (e.g., squamation).

A specialization in the larva of one taxon may be similar to a specialization in the adults of other taxa, but in order to interpret these, a hypothesis of phylogenetic relationships must be available so that the pattern of character evolution can be determined. In other words, it is essential that both the pattern of ontogenetic appearance and phylogenetic distribution of morphological features be established (Alberch, 1985; Mabee, 1989). For instance, if a feature in a larva is actually the primitive adult condition for the taxon to which it belongs, then the ontogenetic loss of such a feature at metamorphosis will result in the adult possessing the derived condition. The ontogenetic loss of a primitive feature during development, not its presence in the larval stage, should be considered to be a specialization or potential adaptation in that species (e.g., loss of swim bladder).

Larval characters may evolve as nonterminal or terminal additions or as nonterminal or terminal deletions (e.g., Mabee, 1993). Elongated head or fin spines may be a nonterminal addition, whereas the development of elongated spines that persist into the juvenile and adult stages could be described as a terminal addition. For example, loss of the swim bladder is a terminal deletion in a taxon that otherwise has a swim bladder in adults. Among percomorph fishes, eye migration in flatfishes is a terminal addition.

It is a difficult task to define larval features, such as those described previously, as adaptations. Larval features may or may not be functional and may or may not be adaptive during the larval stage. Does a particular feature appear during the larval stage as the result of adaptation, or because that is the time during development when that feature normally develops (e.g., fin rays, spines) or can develop? Is eye migration in flatfishes, for example, an adaptation to a planktonic lifestyle or to the benthic habit of juveniles and adults but can only develop during the larval stage while the skull bones are still flexible and capable of modification? The identification of novel characteristics of larvae as adaptations requires that their functional significance be demonstrated using comparative, functional, behavioral, and/or physiological approaches, which are no different from those used to evaluate such features in adult organisms (Rose and Lauder, 1996).

B. DEVELOPMENTAL PLASTICITY

Among vertebrates, teleost fishes exhibit an unparalleled amount of intraspecific variation in both meristics and morphometric features, which may be attributed to either genetic variation or epigenetic effects of environmental factors. Developmental morphology (Suda et al., 1987), rate of development of sensory organs (Kawamura et al., 1989), and development of behavior (Kawamura et al., 1989; Pittman et al., 1990; Coughlin, 1991) vary among wild and reared populations. These are phenomena that can be attributed to the effects of differences in unidentified environmental factors. Many environmental factors, including photoperiod (Tandler and Helps, 1985), space (Suda et al., 1987), food quality (Gatten et al., 1983), and temperature (Beacham and Murray, 1986), have been implicated as the cause of variation in vertebral and fin ray meristics [reviewed by Tåning (1952); Lindsey, 1988].

Experimental work has shown that temperature can affect the timing of development (Hunt von Herbing et al., 1996a), the development of otoliths (Hoff and Fuiman, 1993), and variation in meristics [reviewed by Blaxter (1988a); Cubbage and Mabee, 1996]. Sensitive periods during which temperature affects meristics have been identified (e.g., Pavlov and Moksness, 1996). Tooth polymorphism is correlated with variation in diet in the field (Meyer, 1990) and can be induced experimentally (Liem and Kaufman, 1984). Most interestingly, body shape may change during ontogeny in response to the presence of predators, thus offering individuals a refuge from predation and presumably increasing their survivorship and potentially their fitness (Brönmark and Miner, 1992).

Different cranial sensory systems have different capacities to be plastic and apparently are affected by different environmental factors. The size of external taste buds and the barbels on which they are found is affected by food availability, but not temperature, in goatfish (McCormick, 1993). The development of the lateral line system in cichlids is correlated with size, but not age (Münz, 1986). In contrast, the growth of eyes differs in wild and laboratory-reared fishes (Boehlert, 1979) and is correlated with light levels experienced during captive rearing (Pankhurst, 1992; Helvik and Karlsen, 1996). However, growth of eyes appears to be rather constant, despite food shortages that cause a decrease in somatic growth rate (Pankhurst and Montgomery, 1994).

The variation in juvenile and adult morphology that arises as a result of developmental plasticity during the larval stage provides an excellent context in which to examine not only the contributions of genetics and epigenetics to morphology, but the mechanisms responsible for plasticity and the ecological implications and potential evolutionary consequences of variation in abiotic factors to which developing fishes are naturally subjected.

C. DEVELOPMENTAL INTEGRATION

Fishes represent an excellent system in which to examine the fundamental theme of developmental integration (see Zelditch, 1996). The skull of bony fishes has long been recognized for its structural and functional complexity (DeBeer, 1985) and for its development, which requires developmental and functional coordination of the derivatives of ectoderm (including neuroectoderm, placodes, and neural crest; see Webb and Noden, 1993), mesoderm, and endoderm in both space and time.

For instance, major vertebrate sensory systems in bony fishes (e.g., the olfactory and lateral line systems) include skeletal elements that act as nonneuronal functional components of these systems (e.g., accessory nasal sacs and lateral line canals, respectively), which permit functional variation and versatility in these systems. How is the development of the peripheral nervous system and skeletal elements coordinated? How does the functional evolution of sensory system components influence the development of the skeletal elements that are essential for their function and vice versa? These are not new questions. The nature of the relationship between neuromasts and dermal bone and whether that relationship is topographical or causal are issues that have been discussed by several workers, though investigated experimentally in only three studies (Hall and Hanken, 1985). In the concluding Agenda of "The Development of the Vertebrate Skull" DeBeer (1985) poses several fundamental questions pertaining to the relationship in the patterning of neuromast distributions and dermal bones in the skull of fishes, and these questions remain unanswered. Recent descriptions of lateral line canals that are devoid of neuromasts (Wonsettler and Webb, 1997) demand that the causal relationship between neuromast distribution patterns and patterns of dermal bone ossification be readdressed. Despite the obvious process of the incorporation of lateral line canals into skeletal elements of the skull, studies of developmental osteology of the head of fishes often give little attention to the details of the pattern, order, and timing of lateral line canal morphogenesis and developmental integration into the cranial skeleton.

The development of respiratory, feeding, and locomotory mechanisms in teleost fishes requires developmental integration that extends throughout the larval, juvenile, and adult stages. During its lifetime, a fish experiences a wide spectrum of hydrodynamic conditions that put continuously changing demands on the musculoskeletal systems of the head and trunk. Environmental constraints and mechanical and physiological demands on the musculoskeletal systems of the head and trunk (e.g., feeding and respiratory mechanisms; see Hunt von Herbing and Boutilier, 1996; Hunt von Herbing et al., 1996a–c) and the sensory and motor components of the peripheral and central nervous systems (including Mauthner cell-mediated escape responses) change throughout ontogeny. These are interesting sets of dynamic, interacting systems that would be well-suited for comparative studies of hydrodynamics, biomechanics, and the ecological and evolutionary consequences of variation in locomotory behavior.

Finally, how is the development of repeated structures coordinated in space and time? Repeated (serial or metameric) skeletal structures such as vertebrae, pterygiophores, fin rays, fin spines, elements of the caudal fin (Mabee, 1993), and scales do not develop synchronously and do not develop sequentially or in one direction (Fig. 12). How is muscular development coordinated with patterns of development in serial or segmental structures of the vertebral column and median or paired fins in order to preserve functional integrity through time? Scales generally develop from the caudal peduncle and spread rostrally (e.g., Tsukamoto and Okiyama, 1993; Fig. 12), whereas the lateral line canal segments incorporated into each of the lateral line scales that compose a single lateral line canal develop in a rostral to caudal sequence. Interestingly, the anterior and posterior portions of the disjunct lateral line canal in cichlids (Webb, 1990) develop rostrocaudally and caudorostrally, respectively (Fig.

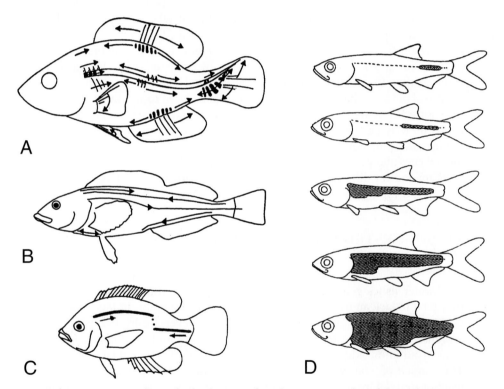

FIGURE 12 Directionality in the development of serial structures in teleost fishes. (A) Direction of development of various skeletal elements in centrarchid fishes (reprinted with permission from Mabee, 1993). (B) Direction of development of the multiple lateral line canals on the trunk of *Hexagrammos stelleri* (Wonsettler and Webb, 1997). (C) Direction of development of the anterior and posterior portions of the disjunct lateral line canal in cichlids (reprinted with permission of Cambridge University Press, from Webb, 1990). (D) Development of squamation (shaded areas) in the Pacific tarpon, *Megalops cyprinoides* (reprinted with permission from Tsukamoto and Okiyama, 1997).

12). In addition, the direction of development varies among the five lateral line canals on the trunk of hexagrammids (Wonsettler and Webb, 1997). How do these directional developmental patterns vary phylogenetically, and what are the functional consequences of such variation?

D. METAMORPHOSIS

Like insects, marine invertebrates, and amphibians, most marine fish larvae experience a distinctive metamorphic transformation. Thus, larval marine fishes provide an excellent system in which to examine the genetic mechanisms that underlie radical changes in morphology. The larval–juvenile transition may involve changes in allometric and morphometric relationships, physiological changes in muscular and internal organ systems, bone remodeling, functional changes in the nervous system (e.g., those associated with flatfish reorientation; Graf and Baker, 1990), and changes in behavior (e.g., the transition from positive to negative phototaxis; Masuda and Tsukamoto, 1996). Given the broad range of changes at metamorphosis, one is led to several fundamental questions: What are the proximate physiological and ultimate genetic mechanisms underlying metamorphosis? How does gene expression change across the metamorphic boundary? What is the nature of the interaction of environmental cues and genetic cues underlying metamorphosis? How can the genetic basis for radical differences in larval and adult forms be used as a model to understand the evolution of large-scale morphological novelty?

VI. SUGGESTED STRATEGIES FOR FUTURE INVESTIGATIONS

Freshwater fishes generally are easy to rear and manipulate, but as a group they tend to lack an overwhelming range of interesting morphological specializations of interest to evolutionary biologists. However, freshwater fishes (e.g., zebrafish, *Danio*) have been the subject of extensive and invaluable contributions to our knowledge of fundamental mechanisms of vertebrate development. Thus, much of what we know about fundamental developmental patterns in fishes is derived from species that do *not* have extensive larval specializations, radical ontogenetic changes in behavior, or a complex metamorphic threshold. However, the developmental osteology of experimental model species has been established [e.g., medaka, Langille and Hall (1987); zebrafish, Cubbage and Mabee (1996)], and such studies serve as valuable morphological baselines for future comparative studies. Only after phylogenetic hypotheses become available for model species and their relatives (e.g., Rosen and Parenti, 1981;

Parenti, 1984; Uwa and Parenti, 1988; Barman, 1991; Meyer *et al.*, 1995) will we be able to maximize the value of these fundamental data for application in developmental–evolutionary studies. Comparative developmental data are available for commercially important clupeiform, salmoniform, gadiform, and pleuronectiform fishes. If developmental and behavioral features of these species are mapped onto independently established hypotheses of phylogenetic relationships, then informative comparative syntheses of developmental morphology can be carried out. To date, there are only a few examples of comparative developmental studies on related species whose phylogenetic relationships have been established [e.g., centrarchids, Mabee (1993); Boughton *et al.*, 1991].

Marine fish larvae have many interesting morphological features that make them excellent subjects of comparative and evolutionary studies. However, their planktonic larvae generally are difficult to maintain and raise through metamorphosis and even more difficult to manipulate experimentally. As a result, only a limited number of marine species with planktonic larvae are reared routinely under controlled conditions suitable for experimentation, and these are hardly enough to allow comparative developmental studies in a formal phylogenetic context. The rearing of fishes under controlled conditions could provide many opportunities for baseline comparative developmental studies and experimental manipulations in which optimal rearing conditions can be identified through the study of the epigenetics of growth and development. Thus, both pure and applied questions can be addressed simultaneously with appropriate experimental design, an economical approach that should be of broad appeal to both investigators and funding agencies.

Finally, we need to continue to promote shipboard collaboration among biological oceanographers, plankton ecologists, fisheries scientists, systematists, and developmental and evolutionary biologists (see Collette and Vecchione, 1995) in order to maximize the archival preservation (Lavenberg *et al.*, 1984; Smith, 1989) and productive analysis of valuable and underutilized larval fish material that is obtained routinely in plankton surveys. Fish larvae collected in large plankton nets generally are crushed to some degree, thus precluding complete osteological analysis, and their delicate epithelium may be lost due to abrasion, preventing productive scanning electron microscopic and histological analyses. Continued development of more delicate collection technology, and the timely and appropriate fixation of larval fish material, would insure that it is suitable for descriptive developmental and more analytical (e.g., immunocytochemical or molecular) investigations. If collected and preserved properly, larval material can yield valuable data on developmental morphology and, for example, provide insights into the functional morphology and behavior of wild populations for comparison with laboratory-reared fishes, which could be used for more complex experimental analyses.

Thus, the study of the development and evolution of fishes through the study of the diversity of fish larvae requires an interdisciplinary approach that will require an unprecedented level of cooperation among laboratory and field scientists from both pure and applied subdisciplines. Unfortunately, workers in each of these subdisciplines use different taxa to ask fundamentally different questions, employ different methodological approaches, and, thus, generally do not speak the same language. Therefore, we must promote opportunities for discussion among developmental biologists and developmental neurobiologists working on model systems and comparative developmental fish biologists in order to address fundamental questions that address the relationship of developmental and evolutionary patterns and processes in fishes.

ACKNOWLEDGMENTS

I thank H. G. Moser and M. Fahay for critically reviewing the manuscript and for offering valuable insights and suggestions. C. Ferarris, J. J. Govoni, C. Hopkins, E. Loew, M. M. McClure, and J. G. New provided several critical references and helpful discussions. W. L. Smith assisted in the preparation of the manuscript and figures.

REFERENCES

Able, K. W., and Fahay, M. P. (1998). "The First Year in the Life of Estuarine Fishes of the Middle Atlantic Bight." Rutgers Univ. Press, Piscataway, NJ.

Able, K. W., and Kaiser, S. C. (1994). Synthesis of summer flounder habitat parameters. *NOAA Coastal Ocean Program, Decision Anal. Ser.* No. 1.

Able, K. W., Matheson, R. E., Morse W. W., Fahay, M. P., and Shepard, G. (1989). Patterns of summer flounder *Paralichthys dentatus* early life history in the mid-Atlantic bight and New Jersey estuaries. *Fish. Bull.* **88**, 1–12.

Ahlstrom, E. H., Amaoka, K., Hensley, D. A., Moser, H. G., and Sumida, B. Y. (1984). Pleuronectiformes: Development. *In* "Ontogeny and Systematics of Fishes" (H. G. Moser, W. J. Richards, D. M. Cohen, M. P. Fahay, A. W. Kendall, Jr., and S. L. Richardson, eds.) ASIH Spec. Publ. No. 1, pp. 640–670. Allen Press, Lawrence, KS.

Ahlstrom, E. H., and Ball, O. P. (1954). Description of eggs and larvae of jack mackerel (*Trachurus symmetricus*) and distribution and abundance of larvae in 1950 and 1951. *Fish. Bull. U.S.* **56**, 209–245.

Alberch, P. (1985). Problems with the interpretation of developmental sequences. *Syst. Zool.* **34**, 46–58.

Allen, J. M., Blaxter, J. H. S., and Denton, E. J. (1976). The functional anatomy and development of the swim bladder–inner ear–lateral line system in herring and sprat. *J. Mar. Biol. Assoc. U.K.* **56**, 471–486.

Allis, E. P. (1889). The anatomy and development of the lateral line system in *Amia calva*. *J. Morphol.* **2**, 463–540.

Ambrose, D. A. (1996a). Carapidae. *In* "The Early Stages of Fishes in the California Current Region" (H. G. Moser, ed.), Calif. Coop. Oceanic Fish. Invest., Atlas No. 33, pp. 533–537. Allen Press, Lawrence, Kansas.

Ambrose, D. A. (1996b). Agonidae: Poachers. In "The Early Stages of Fishes in the California Current Region" (H. G. Moser, ed.), Calif. Coop. Oceanic Fish. Invest., Atlas No. 33. pp. 844–859. Allen Press, Lawrence, Kansas.

Ancil, M. (1977). Development of biolumninescence and photophore in midshipman fish, *Porichthys notatus*. *J. Morphol.* **151**, 363–396.

Appeldoorn, R. S., Hensley, D. A., Shapiro, D. Y., Kioroglou, S., and Sanderson, B. G. (1994). Egg dispersal in a caribbean coral reef fish, *Thalassoma bifasciatum*. II. Dispersal off the reef platform. *Bull. Mar. Sci.* **54**, 271–280.

Applebaum, S., Adron, J. W., George, S. G., Mackie, A. M., and Pirie, B. J. S. (1983). On the development of the olfactory and the gustatory organs of the Dover sole, *Solea solea*, during metamorphosis. *J. Mar. Biol. Assoc. U.K.* **63**, 97–108.

Baldwin, C. C., Johnson, G. D., and Colin, P. L. (1991). Larvae of *Diploprion bifasciatum, Belonoperca chabanaudi* and *Grammistes sexlineatus* (Serranidae: Epinephelinae) with a comparison of known larvae of other epinephelines. *Bull. Mar. Sci.* **48**, 67–93.

Balon, E. K. (1981). Saltatory processes and altricial to precocial forms in the ontogeny of fishes. *Am. Zool.* **21**, 573–596.

Balon, E. K. (1990). Epigenesis of an epigeneticist: The development of some alternative concepts on the early ontogeny and evolution of fishes. *Guelph Ichthyol. Rev.* **1**, 1–42.

Barman, B. F. (1991). A taxonomic revision of the Indo Burmese species of *Danio* (Ham. Buch.) *Rec. Zool. Surv. India, Occas. Pap.* **137**, 1–91.

Batty, R. S., and Hoyt, R. D. (1995). The role of sense organs in the feeding behavior of juvenile sole and plaice. *J. Fish Biol.* **47**, 931–939.

Beacham, T. D., and Murray, C. B. (1986). Comparative developmental biology of pink salmon, *Oncorhynchus gorbuscha*, in southern British Columbia. *J. Fish Biol.* **28**, 233–246.

Beebe, W. (1934). Deep-sea fishes of the Bermuda oceanographic expeditions: Family Idiacanthidae. *Zoologica (N.Y.)* **16**, 149–241.

Berrill, N. J. (1925). The development of the skull in the sole and the plaice. *Q. J. Microsc. Sci.* **69**, 217–244.

Bisbal, G. A., and Bengtson, D. A. (1995). Development of the digestive tract in larval summer founder. *J. Fish Biol.* **47**, 277–291.

Blaxter, J. H. S. (1984). Ontogeny, systematics and fisheries. In "Ontogeny and Systematics of Fishes" (H. G. Moser, W. J. Richards, D. M. Cohen, M. P. Fahay, A. W. Kendall, Jr., and S. L. Richardson, eds.), ASIH Spec. Publ. No. 1, pp. 1–6. Allen Press, Lawrence, KS.

Blaxter, J. H. S. (1986). Development of sense organs and behaviour of teleost larvae with special reference to feeding and predator avoidance. *Trans. Am. Fish. Soc.* **115**, 98–114.

Blaxter, J. H. S. (1987). Structure and development of the lateral line. *Biol. Rev. Cambridge Philos. Soc.* **62**, 471–514.

Blaxter, J. H. S. (1988a). Pattern and variety in development. In "Fish Physiology (W. S. Hoar and D. J. Randall, eds.), Vol. 11, pp. 1–58. Academic Press, San Diego, CA.

Blaxter, J. H. S. (1988b). Sensory performance, behavior and ecology of fishes. In "Sensory Biology of Aquatic Animals" (J. Atema, R. R. Fay, A. N. Popper, R. R. Fay, and W. N. Tavolga, eds.), pp. 203–232. Springer-Verlag, New York.

Blaxter, J. H. S. (1991). Sensory systems and behaviour of larval fish. In "Marine Biology: Its Accomplishment and Future Prospect" (J. Mauchline and T. Nemoto, eds.), pp. 15–38. Hokusen-Sha, Tokyo.

Blaxter, J. H. S., and Fuiman, L. A. (1989). Function of the free neuromasts of marine teleost larvae. In "The Mechanosensory Lateral Line: Neurobiology and Evolution" (S. Coombs, P. Gorner, and H. Münz, eds.), pp. 481–500. Springer-Verlag, New York.

Blaxter, J. H. S., and Fuiman, L. A. (1990). The role of the sensory systems of herring larvae in evading predatory fishes. *J. Mar. Biol. Assoc. U.K.* **70**, 413–427.

Blaxter, J. H. S., Gray, J. A. B., and Best, A. C. G. (1983). Structure and development of the free neuromasts and lateral line system of the herring. *J. Mar. Biol. Assoc. U.K.* **63**, 247–260.

Bleckmann, H. (1993). Role of the lateral line in fish behavior. *In* "Behaviour of Teleost Fishes" (T. J. Pitcher, ed.), pp. 201–246. Chapman & Hall, London.

Boehlert, G. W. (1979). Retinal development in post larval through juvenile *Sebastes diploproa*— Adaptations to a changing photic environment. *Rev. Can. Biol.* **38**, 265–280.

Boughton, D. A., Collette, B. B., and McCune, A. R. (1991). Heterochrony in jaw morphology of needlefishes (Teleostei: Belonidae). *Syst. Zool.* **40**, 329–354.

Bowmaker, J. K. (1990). Visual pigments of fishes. *In* "The Visual System of Fish" (R. Douglas and M. Djamgoz, eds.), pp. 81–108. Chapman & Hall, London.

Brandstätter, R., and Kotrschal, K. (1989). Life history of roach, *Rutilus rutinus* (Cyprinidae, Teleostei). A qualitative and quantitative study on the development of sensory brain areas. *Brain, Behav. Evol.* **34**, 35–42.

Brewster, B. (1987). Eye migration and cranial development during flatfish metamorphosis: A reappraisal (Teleostei: Pleuronectiformes). *J. Fish Biol.* **31**, 805–833.

Brogan, M. W. (1996). Larvae of the Eastern Pacific snapper *Hoplogagrus guntheri* (Teleostei: Lutjanidae). *Bull. Mar. Sci.* **58**, 329–343.

Brönmark, C., and Miner, J. G. (1992). Predator-induced phenotypical change in body morphology in Crucian carp. *Science* **258**, 1348–1350.

Brothers, E. B. (1984). Otolith studies. *In* "Ontogeny and Systematics of Fishes" (H. G. Moser, W. J. Richards, D. M. Cohen, M. P. Fahay, A. W. Kendall, Jr., and S. L. Richardson, eds.), ASIH Spec. Publ. No. 1, pp. 50–56. Allen Press, Lawrence, KS.

Browman, H. I., and Hawryshyn, C. W. (1994). The developmental trajectory of ultraviolet photosensitivity in rainbow trout is altered by thyroxine. *Vision Res.* **34**, 1397–1406.

Bullock, T. H., and Heiligenberg, W., eds. (1986). "Electroreception." Wiley, New York.

Burke, J. S., Tanaka, M., and Seikai, T. (1995). Influence of light and salinity on behavior of larval japanese flounder (*Paralichthys olivaceus*) and implications for inshore migration. *Neth. J. Sea Res.* **34**, 59–69.

Buskey, E. J., Couler, C., and Strom, S. (1993). Locomotory patterns of microzooplankton: Potential effects on food selectivity of larval fish. *Bull. Mar. Sci.* **53**, 29–43.

Butler, J. L., Dahlin, K. A., and Moser, H. G. (1996). Growth and duration of the planktonic phase and a stage based population matrix of Dover sole, *Microstomus pacificus*. *Bull. Mar. Sci.* **58**, 29–41.

Chambers, R. C., and Trippel, E. L., eds. (1997). "Early Life History and Recruitment in Fish Populations." Chapman & Hall, New York.

Chapleau, F. (1993). Pleuronectiform relationships: A reassessment. *Bull. Mar. Sci.* **52**, 516–540.

Charter, S. R. (1996). Incertae sedis. *In* "The Early Stages of Fishes in the California Current Region" (H. G. Moser, ed.), Calif. Coop. Oceanic Fish. Invest., Atlas No. 33, pp. 141–143. Allen Press, Lawrence, Kansas.

Charter, S. R., and Moser, H. G. (1996). Mirrapinnidae: Ribbontails. *In* "The Early Stages of Fishes in the California Current Region" (H. G. Moser, ed.), Calif. Coop. Oceanic Fish. Invest., Atlas No. 33, pp. 713–716. Allen Press, Lawrence, Kansas.

Cohen, D. M. (1984). Ontogeny, systematics and phylogeny. *In* "Ontogeny and Systematics of Fishes" (H. G. Moser, W. J. Richards, D. M. Cohen, M. P. Fahay, A. W. Kendall, Jr., and S. L. Richardson, eds.), ASIH Spec. Publ. No. 1, pp. 7–10. Allen Press, Lawrence, KS.

Collette, B. B., and Vecchione, M. (1995). Interactions between fisheries and systematics. *Fisheries* **20**, 20–25.

Connaughton, V. P., Epifano, C. E., and Thomas, R. (1994). Effects of varying irradiance on feeding in larval weakfish (*Cynoscion regalis*). *J. Exp. Mar. Biol. Ecol.* **180**, 151–163.

Coombs, S., Janssen, J., and Webb, J. F. (1988). Diversity of lateral line systems: Evolutionary and functional considerations. *In* "Sensory Biology of Aquatic Animals" (J. Atema, R. R. Fay, A. N. Popper, R. R. Fay, and W. N. Tavolga, eds.), pp. 553–594. Springer-Verlag, New York.

Copp, G. H., and Kovač, V. (1996). When do fish with indirect development become juveniles? *Can J. Fish. Aquat. Sci.* **53**, 746–752.

Coughlin, D. J. (1991). Ontogeny of feeding behavior of first-feeding Atlantic salmon (*Salmo salar*). *Can. J. Fish. Aquat. Sci.* **48**, 1896–1904.

Coughlin, D. J., Strickler, J. R., and Sanderson, B. (1992). Swimming and search behaviour in clownfish, *Amphiprion perideraion* larvae. *Anim. Behav.* **44**, 427–440.

Cowan, J. H., and Houde, E. D. (1993). Relative predation potentials of schyphomedusae, cteno-phores and planktivorous fish on ichthyoplankton in Chesapeake Bay. *Mar. Ecol. Prog. Ser.* **95**, 55–65.

Cubbage, C. C., and Mabee, P. M. (1996). Development of the cranium and paired fins in the zebrafish *Danio rerio* (Ostariophysi, Cyprinidae). *J. Morphol.* **229**, 121–160.

Davis, M. W., and Olla, B. L. (1995). Formation and maintenance of aggregation in walleye pollack, *Theragra chalcogramma,* larvae under laboratory conditions—role of visual and chemi-cal stimuli. *Environ. Biol. Fishes* **44**, 385–392.

DeBeer, G. (1985). "The Development of the Vertebrate Skull." University of Chicago Press, Chicago.

Denton, E. J., and Gray, J. A. B. (1989). Some observations on the forces acting on neuromasts in fish lateral line canals. *In* "The Mechanosensory Lateral Line—Neurobiology and Evolution" (S. Coombs, P. Görner, and H. Münz, eds.), pp. 229–246. Springer Verlag, New York.

Disler, N. N. (1960). "Lateral Line Sense Organs and Their Importance in Fish Behavior." Israel Program for Scientific Translations, Jerusalem (translated from Russian, 1971).

Douglas, R. H. (1989). The spectral transmission of the lens and cornea of the brown trout (*Salmo trutta*) and goldfish (*Carassius auratus*)—Effect of age and implications for ultraviolet light. *Vision Res.* **29**, 861–869.

Dove, S., Horwitz, J., and McFall-Ngai, M. (1993). A biochemical characterization of the photophore lenses of the midshipman fish, *Porichthys notatus* Girard. *J. Comp. Physiol., A* **172**, 565–572.

Duffy, J. T., Epifanio, C. E., and Fuiman, L. A. (1997). Mortality rate imposed by three scyphozoans on red drum (*Sciaenops ocellatus*) larvae in field enclosures. *J. Exp. Mar. Biol. Ecol.* **212**, 123–131.

Dunn, J. R. (1983). The utility of developmental osteology in taxonomic and systematic studies of teleost larvae: A review. *NOAA Tech. Rep., NMFS Circ.* **450**, 1–19.

Dunn, J. R. (1984). Developmental osteology. *In* "Ontogeny and Systematics of Fishes" (H. G. Moser, W. J. Richards, D. M. Cohen, M. P. Fahay, A. W. Kendall, Jr., and S. L. Richardson, eds.), ASIH Spec. Publ. No. 1, pp. 48–49. Allen Press, Lawrence, KS.

Eaton, R. C., and DiDomenico, R. (1986). Role of the teleost escape response during development. *Trans. Am. Fish. Soc.* **115**, 128–142.

Elliott, J. K., and Leggett, W. C. (1996). The effect of temperature on predation rates of a fish (*Gasterosteus aculeatus*) and a jellyfish (*Aurelia aurita*) on larval capelin (*Mallotus villosus*). *Can. J. Fish. Aquat. Sci.* **53**, 1393–1402.

Evans, B. I., and Fernald, R. D. (1990). Metamorphosis and fish vision. *J. Neurobiol.* **21**, 1037–1052.

Fahay, M. P. (1983). Guide to the early stages of marine fishes occurring in the Western North Atlantic Ocean, Cape Hatteras to the Southern Scotian Shelf. *J. Northwest Atl. Fish. Sci.* **4**, 1–423.

Ferron, A., and Leggett, W. C. (1994). An appraisal of condition measures for marine fish larvae. *Adv. Mar. Biol.* **30**, 217–303.

Finger, T. E. (1988). Organization of chemosensory systems within the brains of bony fishes. *In* "Sensory Biology of Aquatic Animals" (J. Atema, R. R. Fay, A. N. Popper, and W. N. Tavolga, eds.), pp. 339–364. Springer-Verlag, New York.

Fink, W. L. (1981). Ontogeny and phylogeny of tooth attachment modes in actinopterygian fishes. *J. Morphol.* **167**, 167–184.

Fuiman, L. A. (1986). Burst-swimming performance of larval zebra danios and the effects of diel temperature fluctuations. *Trans. Am. Fish. Soc.* **115**, 143–148.

Fuiman, L. A. (1994). The interplay of ontogeny and scaling in the interactions of fish larvae and their predators. *J. Fish. Biol.* **45** (Suppl. A), 55–79.

Fuiman, L. A., and Batty, R. S. (1994). Susceptibility of Atlantic herring and plaice larvae to predation by juvenile cod and herring at two constant temperatures. *J. Fish Biol.* **44**, 23–34.

Fukui, A. (1997). Early ontogeny and systematics of Bothidae, Pleuronectoidei. *Bull. Mar. Sci.* **60**, 192–212.

Gatten, R. R., Sargen, J. R., and Gamble, J. C. (1983). Diet-induced changes in fatty acid composition of herring larvae reared in enclosed ecosystems. *J. Mar. Biol. Assoc. U.K.* **63**, 575–584.

Gorodilov, Y. N. (1996). Description of the early ontogeny of the Atlantic salmon, *Salmo salar,* with a novel system of interval (state) identification. *Environ. Biol. Fishes* **47**, 109–127.

Govoni, J. J., Olney, J. E., Markle, D. F., and Curtsinger, W. R. (1984). Observations on structure and evaluation of possible functions of the vexillum in larval Carapidae (Ophidiiformes). *Bull. Mar. Sci.* **34**, 60–70.

Govoni, J. J., Boehlert, G. W., and Watanabe, Y. (1986). The physiology of digestion in fish larvae. *Environ. Biol. Fishes* **16**, 1–3.

Graf, W., and Baker, R. (1990). Neuronal adaptation accompanying metamorphosis in the flatfish. *J. Neurobiol.* **21**, 1136–1152.

Hall, B. K., and Hanken, J. (1985). "The Development of the Vertebrate Skull," Foreword. University of Chicago Press, Chicago.

Hansen, A., and Zeiske, E. (1993). Development of the olfactory organ in the zebrafish, *Brachydanio rerio. J. Comp. Neurol.* **333**, 289–300.

Hara, S., Kohno, H., and Taki, Y. (1986). Spawning behavior and early life history of the rabbitfish, *Siganus guttatus* in the laboratory. *Aquaculture* **59**, 273–285.

Hara, T. J. (1992). "Fish Chemoreception," Chapman & Hall, London.

Hara, T. J. (1993). Role of olfaction in fish behaviour. *In* "Behaviour of Teleost Fishes" (T. Pitcher, ed.). pp. 171–199. Chapman & Hall, London.

Hara, T. J. (1994). The diversity of chemical stimulation in fish olfaction and gustation. *Rev. Fish Biol. Fish.* **4**, 1–35.

Hara, T. J., and Zielinski, B. (1989). Structural and functional development of the olfactory organ in teleosts. *Trans. Am. Fish. Soc.* **118**, 183–194.

Harvey, R. (1996). The olfactory epithelium in plaice (*Pleuronectes platessa*) and sole (*Solea solea*), two flatfishes with contrasting feeding behaviour. *J. Mar. Biol. Assoc. U.K.* **76**, 127–139.

Harvey, R., Blaxter, J. H. S., and Hoyt, R. D. (1992). Development superficial and lateral line neuromasts in larvae and juveniles of plaice *Pleuronectes platessa* and sole *Solea solea. J. Mar. Biol. Assoc. U.K.* **72**, 651–668.

Heath, M. R. (1992). Field investigations of the early life stages of marine fish. *Adv. Mar. Biol.* **28**, 1–133.

Helvik, J. V., and Karlsen, O. (1996). The effect of light- and dark-rearing on the development of the eyes of Atlantic halibut (*Hippoglossus hipposlossus*) yolk-sac larvae. *Mar. Freshwater Behav. Physiol.* **28**, 107–121.

Hensley, D. A., Appeldoorn, R. S., Shapiro, D. Y., Ray, M., and Turingan, R. G. (1994). Egg dispersal in a caribbean coral reef fish, *Thalassoma bifasciatum.* I. Dispersal over the reef platform. *Bull. Mar. Sci.* **54**, 256–270.

Higgs, D. M., and Fuiman, L. A. (1996a). Light intensity and schooling behavior in larval gulf menhaden. *J. Fish Biol.* **48**, 979–991.

Higgs, D. M., and Fuiman, L. A. (1996b). Ontogeny of visual and mechanosensory structure and function in Atlantic menhaden *Brevoortia tyrannus. J. Exp. Biol.* **199**, 2619–2629.

Hoff, G. R., and Fuiman, L. A. (1993). Morphometry and composition of red drum otoliths: Changes associated with temperature, somatic growth rate, and age. *Comp. Biochem. Physiol. A* **106A**, 209–219.

Holmgren, N., and Pehrson, T. (1949). Some remarks on the ontogenetical development of the sensory lines on the cheek in fishes and amphibians. *Acta Zool.* **30**, 249–314.

Hopkins, C. D. (1995). Convergent designs for electrogenesis and electroreception. *Curr. Opin. Neurobiol.* **5**, 769–777.

Hoss, D. E., and Blaxter, J. H. S. (1982). Development and function of the swim bladder–inner ear–lateral line syseem in the Atlantic menhaden, *Brevoortia tyrannus* (Latrobe). *J. Fish Biol.* **20**, 131–142.

Hulet, W. H. (1978). Structure and functional development of the eel leptocephalus *Arisoma balearicum* (De La Roche, 1809). *Philos. Trans. R. Soc. London, Ser. B* **282**, 107–138.

Hunter, J. R. (1972). Swimming and feeding behavior of larval anchovy *Engraulis mordax*. *Fish. Bull.* **70**, 821–838.

Hunter, J. R. (1981). Feeding ecology and predation of marine fish larvae. In "Marine Fish Larvae: Morphology, Ecology, and Relation to Fisheries" (R. Lasker, ed.), pp. 33–79. University of Washington Press, Seattle.

Hunter, J. R., and Thomas, G. L. (1973). Effect of prey distribution and density on the searching and feeding behaviour of larval anchovy *Engraulis mordax* Girard. In "The Early Life History of Fish" (J. H. S. Blaxter, ed.), pp. 559–574. Springer-Verlag, Berlin.

Hunt von Herbing, I., and Boutilier, R. G. (1996). Activity and metabolism of larval Atlantic cod (*Gadus morhua*) from Scotian shelf and Newfoundland source populations. *Mar. Biol.* **124**, 607–617.

Hunt von Herbing, I., Boutilier, R. G., Miyake, T., and Hall, B. K. (1996a). Effects of temperature on morphological landmarks critical to growth and survival in larval Atlantic cod (*Gadus morhua*). *Mar. Biol.* **124**, 593–606.

Hunt von Herbing, I., Miyake, T., Hall, B. K., and Boutilier, R. G. (1996b). Ontogeny of feeding and respiration in larval Atlantic cod *Gadus morhua* (Teleostei, Gadiformes): I. Morphology. *J. Morphol.* **227**, 15–35.

Hunt von Herbing, I., Miyake, T., Hall, B. K., and Boutilier, R. G. (1996c). Ontogeny of feeding and respiration in larval Atlantic cod *Gadus morhua* (Teleostei, Gadiformes): II. Function. *J. Morphol.* **227**, 37–50.

Iwai, T. (1964). Development of cupulae in free neuromasts of the Japanese Medaka *Oryzias latipes* (Temminch et Schlegel). *Bull Misaki Mar. Biol. Inst., Kyoto Univ.* **5**, 31–37.

Iwai, T. (1965). Note on the cupulae of free neuromasts in larvae of the goldfish. *Copeia*, p. 379.

Iwai, T. (1967). Structure and development of lateral line cupulae in teleost larvae. In "Lateral Line Detectors" (P. Cahn, ed.), pp. 27–44. Indiana University Press, Bloomington.

Iwai, T. (1983). Surface morphology of the naked neuromasts in anchovy larvae. *Bull. Jpn. Soc. Sci. Fish.* **49**, 1935.

Johnson, G. D. (1993). Percomorph phylogeny: Progress and problems. *Bull. Mar. Sci.* **52**, 3–28.

Johnson, G. D., and Brothers, E. B. (1993). *Schindleria*: A paedomorphic goby (Teleostei: Gobioidei). *Bull. Mar. Sci.* **52**, 441–471.

Johnson, R. K. (1984). Giganturidae: Development and relationships. In "Ontogeny and Systematics of Fishes" (H. G. Moser, W. J. Richards, D. M. Cohen, M. P. Fahay, A. W. Kendall, Jr., and S. L. Richardson, eds.), ASIH Spec. Publ. No. 1, pp. 199–201. Allen Press, Lawrence, KS.

Johnson, R. K., and Bertelsen, E. (1991). The fishes of the family Giganturidae: Systematics, development, distribution and aspects of biology. *Dana Rep.* **91**, 1–45.

Kaufman, L., Ebersole, J., Beets, J., and McIvor, C. C. (1991). A key phase in the recruitment dynamics of coral reef fishes: Post-settlement transition. *Environ. Biol. Fishes* **34**, 109–114.

Kawaguchi, K., and Moser, H. G. (1984). Stomiatoidea: Development. In "Ontogeny and Systematics of Fishes" (H. G. Moser, W. J. Richards, D. M. Cohen, M. P. Fahay, A. W. Kendall, Jr., and S. L. Richardson, eds.), ASIH Spec. Publ. No. 1, pp. 169–181. Allen Press, Lawrence, KS.

150										Jacqueline F. Webb

Kawamura, G., and Ishida, K. (1985). Changes in sense organ morphology and behaviour with growth in the flounder *Paralichthys olivaceus*. *Bull. Jpn. Soc. Sci. Fish.* **51**, 155–165.

Kawamura, G., and Munekiyo, M. (1989). Development of the sense organs of ribbonfish, *Trichiurus lepturus* larvae and juveniles. *Nippon Suisan Gakkaishi* **55**, 2075–2078.

Kawamura, G., and Washiyama, H. (1989). Ontogenetic changes in behavior and sense organ morphogenesis in largemouth bass and *Tilapia nilotica*. *Trans. Am. Fish. Soc.* **118**, 203–213.

Kawamura, G., Ysuda, R., Kumai, H., and Ohashi, S. (1984). The visual cell morphology of *Pagurus major* and its adaptive changes with shift from pelagic to benthic habitats. *Bull. Jpn. Soc. Sci. Fish.* **50**, 1975–1980.

Kawamura, G., Mori, H., and Kuwahara, A. (1989). Comparison of sense organ development in wild and reared flounder *Paralichthys olivaceus* larvae. *Bull. Jpn. Soc. Sci. Fish.* **55**, 2079–2083.

Kawamura, G., Takimoto, M., and Sobajima, N. (1990). Larval growth and age- and size-related variation in development of sense organs in the halfbeak, *Hemiramphus sajori*. *In* "The Second Asian Fisheries Forum" (R. Hirano and I. Hanyu, eds.), pp. 407–410. Asian Fisheries Society, Manila, Phillipines.

Keefe, M., and Able, K. W. (1994). Contributions of abiotic and biotic factors to settlement in summer flounder, *Paralichthys dentaus. Copeia*, pp. 458–465.

Kendall, A. W., Jr., and Matarese, A. C. (1994). Status of early life history descriptions of marine teleosts. *U.S. Fish. Bull.* **92**, 725–736.

Kendall, A. W., Jr., Ahlstrom, E. H., and Moser, H. G. (1984). Early life history stages of fishes and their characters. *In* "Ontogeny and Systematics of Fishes," (H. G. Moser, W. J. Richards, D. M. Cohen, M. P. Fahay, A. W. Kendall, Jr., and S. L. Richardson, Eds.), pp. 11–22. ASIH Special Publication No. 1, Allen Press, Lawrence, KS.

Kimmel, C. B. (1982). Development of synapses on the Mauthner neuron. *Trends Neurosci.* **5**, 47–50.

Kirschbaum, F. (1995). Reproduction and development in mormyrifomr and gymnotiform fishes. *In* "Electric Fishes—History and Behavior" (P. Moller, ed.), pp. 267–302. Chapman & Hall, London.

Kleckner, R. C., and McCleave, J. D. (1985). Spatial and temporal distribution of american eel larvae in relation to north atlantic ocean current systems. *Dana* **4**, 67–92.

Knutsen, J. A. (1992). Feeding behaviour of North Sea turbot (*Scophthalmus maximus*) and Dover sole (*Solea solea*) larvae elicited by chemical stimuli. *Mar. Biol.* **113**, 543–548.

Kohno, H., Shimizu, M., and Nose, Y. (1984). Morphological aspects of the development of swimming and feeding function in larval *Scomber japonicus. Bull. Jpn. Soc. Sci. Fish.* **50**, 1125–1137.

Kohno, H., Diani, S., and Supriatna, A. (1993). Morphological development of larval and juvenile grouper, *Epinephelus fuscoguttatus. Jpn. J. Ichthyol.* **40**, 307–316.

Konagai, M., and Rimmer, M. A. (1985). Larval ontogeny and morphology of the fire-tailed gudgeon *Hypseleotris galii* (Eleotridae). *J. Fish Biol.* **27**, 277–283.

Konishi, Y., and Okiyama, M. (1997). Morphological development of four trachichthyoid larvae (Pisces: Beryciformes), with comments on trachichthyiform relationships. *Bull. Mar. Sci.* **60**, 66–88.

Kunz, Y. W., Wildenburg, G., Goodrich, L., and Callaghan, E. (1994). The fate of ultraviolet receptors in the retina of the Atlantic salmon (*Salmo salar*). *Vision Res.* **34**, 1375–1383.

Kuronuma, K. (1984). "Rearing of Marine Fish Larvae in Japan." IDRC, Ottawa, Canada.

Kvenseth, A. M., Pittman, K., and Helvik, J. V. (1996). Eye development in Atlantic halibut (*Hippoglossus hippoglossus*): Differentiation and development of the retina from early yolk sac stages through metamorphosis. *Can. J. Fish. Aquat. Sci.* **53**, 2524–2532.

Kyle, H. M. (1921). The asymmetry, metamorphosis and origin of flat-fishes. *Philos. Trans. R. Soc. London, Ser. B* **211**, 75–129.

Langille, R., and Hall, B. K. (1987). Development of the head skeleton of the Japanese medaka, *Oryzias latipes* (Teleostei). *J. Morphol.* **193**, 135–158.

Lara, M. R. (1992). An investigation of sensory development in larvae of California halibut (*Paralichthys californicus*) using light and electron microscopy. Unpublished MS Thesis, California State University, Northridge.

Lasker, R., ed. (1981). "Marine Fish Larvae—Morphology, Ecology, and Relation to Fisheries." University of Washington Press, Seattle.

Lavenberg, R. J., McGowen, G. E., and Woodsum, R. E. (1984). Preservation and curation. *In* "Ontogeny and Systematics of Fishes." (H. G. Moser, W. J. Richards, D. M. Cohen, M. P. Fahay, A. W. Kendall, Jr., and S. L. Richardson, eds.) ASIH Special Publication No. 1, pp. 57–59. Allen Press, Lawrence, KS.

Lecomtefiniger, R. (1992). Growth history and age at recruitment of european glass eels (*Anguilla anguilla*) as revealed by otolith microstructure. *Mar. Biol.* **114**, 205–210.

Lecomtefiniger, R. (1994). Developmental stages of eel larvae (Leptecephali and elvers) of European eel (*Anguilla anguilla* L. 1758)—Migrations and metamorphosis. *Annee Biol.* **33**, 1–17.

Leis, J. M. (1989). Larval biology of butterflyfishes (Pisces, Chaetodontidae): What do we really know? *Environ. Biol. Fish.* **25**, 87–100.

Leis, J. M. (1991). The pelagic stage of reef fishes: The larval biology of coral reef fishes. *In* "The Ecology of Fishes on Coral Reefs" (P. F. Sale, ed.), pp. 183–230. Academic Press, San Diego, CA.

Leis, J. M., Olney, J. E., and Okiyama, M. (1997). Introduction to the proceedings of the symposium—Fish larvae and systematics: Ontogeny and relationships. *Bull. Mar. Sci.* **60**, 1–5.

Leis, J. M., and Rennis, D. S. (1983). "The Larvae of Indo-Pacific Coral Reef Fishes." University of Hawaii Press, Honolulu.

Leis, J. M., and Trnski, T. (1989). "The Larvae of Indo-Pacific Shorefishes." University of Hawaii Press, Honolulu.

Leis, J. M., and van der Lingen, C. D. (1997). Larval development and relationships of the perciform family Dichistiidae (= Coracinidae), the Galjoen fishes. *Bull. Mar. Sci.* **60**, 100–116.

Lekander, B. (1949). The sensory line system and the canal bones in the head of some ostariophysi. *Acta Zool.* **30**, 1–131.

Liem, K. F., and Kaufman, L. S. (1984). Intraspecific macroevolution: Functional biology of the polymorphic cichlid species *Cichlasoma minckleyi*. *In* "Evolution of Fish Species Flocks" (A. A. Echelle and I. Kornfeld, eds.), pp. 203–215. University of Maine Press, Orono.

Lindsey, C. C. (1988). Factors controlling meristic variation. *In* "Fish Physiology, Vol. XI. The Physiology of Developing Fish, Part B: Viviparity and Posthatching Juveniles." (W. S. Hoar, and D. J. Randall, eds.), pp. 197–274, Academic Press, San Diego.

Lobel, P. S. (1989). Ocean current variability and the spawning season of Hawaiian reef fishes. *Environ. Biol. Fish.* **24**, 161–171.

Lobel, P. S., and Robinson, A. R. (1986). Reef fishes at sea: Ocean currents and the advection of larvae. *In* "The Ecology of Deep and Shallow Coral Reefs," pp. 29–38. NOAA Symposium Series for Undersea Research, Vol 1, No. 1.

Loew, E. R., and McFarland, W. N. (1990). The underwater visual environment. *In* "The Visual System of Fish" (R. Douglas and M. Djamgoz, eds.), pp. 1–44. Chapman & Hall, London.

Loew, E. R., and Sillman, A. J. (1993). Age-related changes in the visual pigments of the white sturgeon (*Acipenser transmontanus*). *Can. J. Zool.* **71**, 1552–1557.

Loew, E. R., and Wahl, C. M. (1991). A short-wavelength sensitive mechanism in juvenile yellow perch, *Perca flavescens*. *Vision Res.* **31**, 353–360.

Loew, E. R., McAlary, F. A., and McFarland, W. N. (1997). Ultraviolet visual sensitivity in the larvae of two species of marine atherinid fishes. *Mar. Freshwater Behav. Physiol.* (in press).

Mabee, P. M. (1988). Supraneural and predorsal bones in fishes: Development and homologies. *Copeia*, pp. 827–838.

152

Jacqueline F. Webb

Mabee, P. M. (1989). Assumptions underlying the use of ontogenetic sequences for determining character state order. *Trans. Am. Fish. Soc.* **118**, 151–158.

Mabee, P. M. (1993). Phylogenetic interpretation of ontogenetic change: Sorting out the actual and artefactual in an empiracal case study of centrarchid fishes. *Zool. J. Linn. Soc.* **107**, 175–291.

Mabee, P. M. (1995). Evolution of pigment pattern development in centrarchid fishes. *Copeia,* pp. 586–607.

Mabee, P. M., and Trendler, T. A. (1996). Development of the cranium and paired fins in *Betta splendens* (Teleostei: Percomorpha): Intraspecific variation and interspecific comparisons. *J. Morphol.* **227**, 249–287.

Marcey, D., and Nusslein-Volhard, C. (1986). Embryology goes fishing. *Nature (London)* **321**, 380–381.

Margulies, D. (1989). Size-specific vulnerability to predation and sensory system development of white seabass, *Atractoscion nobilis,* larvae. *U. S. Fish. Bull.* **87**, 537–552.

Markle, D. F., Harris, P. M., and Toole, C. L. (1992). Metamorphosis and an overview of early-life-history stages in Dover sole *Microstomus pacificus. U. S. Fish. Bull.* **90**, 285–301.

Marliave, J. B. (1986). Lack of planktonic dispersal of rocky intertidal fish larvae. *Trans. Am. Fish. Soc.* **115**, 149–154.

Masuda, R., and Tsukamoto, K. (1996). Morphological development in relation to phototaxis and rheotaxis in the striped jack, *Pseudocaranx dentes. Mar. Freshwater Behav. Physiol.* **28**, 75–90.

Masuma, S., Tezuka, N., and Teruya, K. (1993). Embryologic and morphological development of larval and juvenile coral trout, *Plectropomus leopardus. Jpn. J. Ichthyol.* **40**, 333–342.

Matarese, A. C., Kendall, A. W., Jr., Blood, D. M., and Vinter, B. M. (1989). "Laboratory Guide to Early Life History Stages of Northeast Pacific Fishes." NOAA Tech. Rept. NMFS 80.

McCormick, M. I. (1993). Development and changes at settlement in the barbel structure of the reef fish, *Upenus tragula* (mullidae). *Environ. Biol. Fishes* **37**, 269–282.

McFarland, W. M., and Loew, E. R. (1994). Ultraviolet visual pigment in marine fishes of the family Pomacentridae. *Vision Res.* **34**, 1393–1396.

Meyer, A. (1987). Phenotypic plasticity and heterochrony in *Cichlasoma managuense* (Pisces, Cichlidae) and their implications for speciation in cichlid fishes. *Evolution (Lawrence, Kans.)* **41**, 1357–1369.

Meyer, A. (1990). Ecological and evolutionary consequences of the trophic polymorphism in *Cichlasoma citrinellum* (Pisces: Cichlidae). *Biol. J. Linn. Soc.* **39**, 279–300.

Meyer, A., Ritchie, P. A., and Witte, W. E. (1995). Predicting developmental processes from phylogenetic patterns: A molecular phylogeny of zebrafish and its relatives. *Philos. Trans. R. Soc. London, Ser. B* **349**, 103–111.

Miller, T. J., Crowder, L. B., and Rice, J. A. (1993). Ontogenetic changes in behavioural and histological measures of viual acuity in three species of fish. *Environ. Biol. Fish.* **37**, 1–8.

Milos, N., and Dingel, A. D. (1978). Dynamics of pigment pattern formation in the zebrafish, *Brachydanio rerio.* I. Establishment and regulation of the lateral line melanophore stripe during the first eight days of development. *J. Exp. Zool.* **205**, 205–216.

Mochioka, N., and Iwamizu, M. (1996). Diet of anguilloid larvae—Leptocephali feed selectively on larvacean houses and fecal pellets. *Mar. Biol.* **125**, 447–452.

Mochioka, N., Iwamizu, M., and Kanda, T. (1993). Leptocephalus eel larvae will feed in aquaria. *Environ. Biol. Fishes* **36**, 381–384.

Modin, J., Fagerholm, B., Gunnarsson, B., and Pihl, L. (1996). Changes in otolith microstructure at metamorphosis of plaice, *Pleuronectes platessa* L. *ICES J. Mar. Sci.* **53**, 745–748.

Moller, P., ed. (1995). "Electric Fishes—History and Behavior." Chapman & Hall, London.

Montgomery, J., Coombs, S., and Galstead, M. (1995). Biology of the mechanosensory lateral line in fishes. *Rev. Fish. Biol. Fisheries* **5**, 399–416.

Montgomery, J., Baker, C. F., and Carton, A. G. (1997). The lateral line can mediate rheotaxis in fish. *Nature (London)* **389**, 960–963.

Moore, J. A. (1994). What is the role of paedomorphosis in deep-sea fish evolution? *Proc.,* Indo-Pac. Fish Conf. 4th'Bangkok, Thailand, Faculty of Fisheries, Kasetsart University, pp. 448–461.

Moser, H. G. (1981). Morphological and functional aspect of marine fish larvae. In "Marine Fish Larvae—Morphology, Ecology, and Relation to Fisheries" (R. Lasker, ed.), pp. 89–131. University of Washington Press, Seattle.

Moser, H. G., ed. (1996a). "The Early Stages of Fishes in the California Current Region." Calif. Coop. Oceanic Fish. Invest., Atlas No. 33. Allen Press, Lawrence, Kansas.

Moser, H. G. (1996b). Introduction. In "The Early Stages of Fishes in the California Current Region." (H. G. Moser, ed.) California Cooperature Oceanic Fisheries Investigations, Atlas No. 33.

Moser, H. G. (1996c). Malacosteidae: Loosejaws. In "The Early Stages of Fishes in the California Current Region." (H. G. Moser, ed.), Calif. Coop. Oceanic Fish. Invest., Atlas No. 33. pp. 321–324, Allen Press, Lawrence, Kansas.

Moser, H. G. (1996d). Idiacanthidae: Blackdragons. In "The Early Stages of Fishes in the California Current Region." (H. G. Moser, ed.), Calif. Coop. Oceanic Fish. Invest., Atlas No. 33. pp. 325–327, Allen Press, Lawrence, Kansas.

Moser, H. G., and Ahlstrom, E. H. (1970). Development of lanternfishes (Family Myctophidae) in the California current. Part I. Species with narrow-eyed larvae. *Bull. L.A. Co. Mus. Nat. Hist.* 7, 1–145.

Moser, H. G., and Ahlstrom, E. H. (1996a). Bathylagidae: Blacksmelts and smoothtongues. In "The Early Stages of Fishes in the California Current Region." (H. G. Moser, ed.), Calif. Coop. Oceanic Fish. Invest., Atlas No. 33. pp. 188–207, Allen Press, Lawrence, Kansas.

Moser, H. G., and Ahlstrom, E. H. (1996b). Myctophidae: Lanternfishes. In "The Early Stages of Fishes in the California Current Region." (H. G. Moser, ed.), Calif. Coop. Oceanic Fish. Invest.. Atlas No. 33. pp. 387–475. Allen Press, Lawrence, Kansas.

Moser, H. G., and Boehlert, G. W. (1991). Ecology of pelagic larvae and juveniles of the genus *Sebastes. Environ. Biol. Fishes* 30, 203–224.

Moser, H. G., and Charter, S. R. (1996). Notacanthiformes. In "The Early Stages of Fishes in the California Current Region." (H. G. Moser, ed.), Calif. Coop. Oceanic Fish. Invest., Atlas No. 33. pp. 82–85. Allen Press, Lawrence, Kansas.

Moser, H. G., and Smith, P. E. (1993). Larval fish assemblages and oceanic boundaries (Introduction to Symposium—Advances in the Early Life History of Fishes). *Bull. Mar. Sci.* 53, 283–289.

Moser, H. G., and Sumida, B. Y. (1996). Paralichthyidae: Lefteye flounders and sanddabs. In "The Early Stages of Fishes in the California Current Region." (H. G. Moser, ed.), Calif, Coop. Oceanic Fish. Invest., Atlas No. 33. pp. 1325–1356. Allen Press, Lawrence, Kansas.

Moser, H. G., Richards, W. J., Cohen, D. M., Fahay, M. P., Kendall, A. W. Jr., and Richardson, S. L., eds. (1984). "Ontogeny and Systematics of Fishes." ASIH, Special Pub. No. 1. Allen Press, Lawrence, Kansas.

Mukai, Y., and Kobayashi, H. (1991). Morphological studies on the cupulae of free neuromasts along with the growth of larvae in cyprinid fish. *Nippon Suisan Gakkaishi* 57, 1339–1346.

Mukai, Y., and Kobayashi, H. (1992). Cupular growth rate of free neuromasts in three species of cyprinid fish. *Nippon Suisan Gakkaishi* 58, 1849–1853.

Mukai, Y., Kobayashi, H., and Yoshikawa, H. (1992). Development of free and canal neuromasts and their directions of maximum sensitivity in the larvae of ayu, *Plecoglossus altivelis. Jpn. J. Ichthyol.* 38, 411–417.

Münz, H. (1986). What influences the development of canal and superficial neuromasts? *Ann. Mus. R. Afr. Cent. Sc. Zool.* 251, 85–89.

Münz, H. (1989). Functional organization of the lateral line periphery. In "The Mechanosensory Lateral Line—Neurobiology and Evolution" (S. Coombs, P. Gorner, and H. Münz, eds.), pp. 285–298. Springer-Verlag, New York.

Nelson, J. S. (1994). "Fishes of the World," 3rd ed. Wiley, New York.

New, J. G. (1997). The evolution of vertebrate electrosensory systems. *Brain, Behav. Evol.* 50, 244–252.

New, J. G., and Tricas, T. C. (1998). Electroreceptors and magnetoreceptors: Morphology and function. In "Cell Physiology Source Book" (N. Sperelakis, ed.), 2nd ed. Academic Press, San Diego, CA (in press).

Nielsen, J. G., and Larsen, V. (1970). Remarks on the identity of the giant Dana eellarva. *Videnskab. Medde. Fra Dansk Naturhist. Foren.* 133, 149–157.

Noakes, D. L. G., and Godin, J.-G. J. (1988). Ontogeny of behavior and concurrent developmental changes in sensory systems in teleost fishes. In "Fish Physiology, Volume XI. The Physiology of Developing Fish, Part B: Viviparity and Posthatching Juveniles." (W. S. Hoar and D. J. Randall, eds.), pp. 345–395. Academic Press, San Diego.

Nobuhiro, S., Kazuhiro, O., and Naoaki, K. (1995). Organogenesis and behavioral changes during development of laboratory-reared tiger puffer, *Takifugu rubripes*. *Suisan Zoshoku (Aquiculture)* 43, 461–474.

Nobuhiro, S., Yasuhisa, K., and Tsuzumi, M. (1996). Development of free and canal neuromasts with special reference to sensory polarity in larvae of the red spotted grouper, *Epinephelus akaara*. *Suisan Zoshoku (Aquiculture)* 44, 159–168.

Northcutt, R. G. (1986). Electroreception in nonteleost bony fishes. In "Electroreception" (T. H. Bullock and W. Heiligenberg, eds.), pp. 257–286. Wiley, New York.

Northcutt, R. G. (1989). The phylogenetic distribution and innervation of craniate mechanoreceptive lateral lines. In "The Mechanosensory Lateral Line—Neurobiology and Evolution" (S. Coombs, P. Gorner, and H. Münz, eds.), pp. 17–78. Springer-Verlag, New York.

Northcutt, R. G. (1992). The phylogeny of octavolateralis ontogenies: A reaffirmation of Garstang's phylogenetic hypothesis. In "The Evolutionary Biology of Hearing" (D. Webster, A. N. Popper, and R. R. Fay, eds.), pp. 21–47. Springer-Verlag, New York.

Obara, S. (1976). Mechanisms of electroreception in ampullae of Lorenzini of the marine catfish *Plotosus*. In "Electrobiology of Nerve, Synapse, and Muscle" (J. P. Ruben, D. P. Purpura, M. V. L. Bennett, and E. R. Kandel, eds.), pp. 128–147. Raven Press, New York.

O'Connell, C. P. (1981). Development of organ systems in the Northern anchovy *Engraulis mordax*, and other teleosts. *Am. Zool.* 21, 29–446.

Olsen, K. H. (1993). Development of the olfactory organ of the Arctic charr, *Salvelinus alpinus* (L.) (Teleostei, Salmonidae). *Can. J. Zool.* 71, 1973–1984.

Otake, T., Nogami, K., and Maruyama, K. (1993). Dissolved and particulate organic matter as possible food sources for eel leptocephali. *Mar. Ecol. Prog. Ser.* 92, 27–34.

Otake, T., Ishii, T., Nakahara, M., and Nakahara, R. (1994). Drastic changes in otolith strontium–calcium ratios in leptocephali and glass eels of Japanese eel *Anguilla japonica*. *Mar. Ecol. Prog. Ser.* 112, 189–193.

Ozawa, T. (1986). "Studies on the Oceanic Ichthyoplankton in the Western North Pacific." Kyushu University Press. Fukuoka, Japan.

Pankhurst, N. W. (1984). Retinal development in larval and juvenile european eel *Anguilla anguilla*. *Can J. Zool.* 62, 335–343.

Pankhurst, N. W. (1992). Ocular morphology of the sweep *Scorpis lineolatus* and the spotty *Notolabrus celidotus* (Pisces, Teleostei) grown in low intensity light. *Brain, Behav. Evol.* 39, 116–123.

Pankhurst, N. W., and Montgomery, J. C. (1994). Uncoupling of visual and somatic growth in the rainbow trout *Oncorhynchus mykiss*. *Brain, Behav. Evol.* 44, 149–155.

Pankhurst, P. M. (1994). Age-related changes in the visual acuity of larvae of New Zealand snapper, *Pagrus auratus*. *J. Mar. Biol. Assoc. U.K.* 74, 337–349.

Pankhurst, P. M., and Butler, P. (1996). Development of the sensory organs in the greenback flounder, *Rhombosolea tapirina*. *Mar. Freshwater Behav. Physiol.* 28, 55–73.

Pankhurst, P. M., and Eagar, R. (1996). Changes in visual morphology through life-history stages of the New Zealand snapper, *Pagrus auratus*. *N. Z. J. Mar. Freshwater Res.* **30**, 79–90.

Parenti, L. R. (1984). A taxonomic revision of the andean killifish genus *Orestias* (Cyprinodontiformes, Cyprinodontidae). *Bull. Am. Mus. Nat. Hist.* **178**, 107–214.

Parichy, D. M. (1996a). Pigment patterns of larval salamanders (Ambystomatidae, Salamandridae): The role of the lateral line sensory system and the evolution of pattern-forming mechanisms. *Dev. Biol.* **175**, 265–282.

Parichy, D. M. (1996b). When neural crest and placodes collide: Interactions between melanophores and the lateral lines that generate stripes in the salamander *Ambystoma tigrinum* (Ambystomatidae). *Dev. Biol.* **175**, 283–300.

Pavlov, D. A., and Moksness, E. (1996). Sensitive stages during embryonic development of wolffish *Anarchichas lupus* L. determining the final number of rays in unpaired fins and skeletal abnormalities. *ICES J. Mar. Sci.* **53**, 731–740.

Pfeiler, E. (1986). Towards an explanation of the developmental strategy in leptocephalus larvae of marine teleost fishes. *Environ. Biol. Fish.* **15**, 3–13.

Pfeiler, E. (1989). Sensory systems and behavior of premetamorphic and metamorphic leptocephalous larvae. *Brain, Behav. Evol.* **34**, 25–34.

Pfeiler, E., and Govoni, J. J. (1993). Metabolic rates in early life history stages of elopomorph fishes. *Biol. Bull. (Woods Hole, Mass.)* **185**, 277–283.

Pfeiler, E., and Lindley, V. (1989). Chloride-type cells in the skin of the metamorphosing bonefish *Albula* sp. leptocephalus. *J. Exp. Zool.* **250**, 11–16.

Pittman, K., Skiftesvik, A. B., and Berg, L. (1990). Morphological and behavioral development of halibut, *Hippoglossus hippoglossus* (L.) larvae. *J. Fish. Biol.* **37**, 455–472.

Platt, C. (1988). Equilibrium in the vertebrates: Signals, senses and steering underwater. *In* "The Mechanosensory Lateral Line—Neurobiology and Evolution" (S. Coombs, P. Gorner, and H. Münz, eds.) pp. 783–809. Springer-Verlag, New York.

Poling, K. R., and Fuiman, L. A. (1997). Sensory development and concurrent behavioural changes in Atlantic croaker larvae. *J. Fish Biol.* **51**, 402–421.

Powers, M. K., and Raymond, P. A. (1990). Development of the visual system. *In* "The Visual System of Fish" (R. Douglas and M. Djamgoz, eds.), pp. 419–442. Chapman & Hall, London.

Radermaker, F. C., Surlemont, P., Sanna, M., Chardon, and Vandewalle, P. (1989). Ontogeny of the Weberian apparatus of *Clarias gariepinus* (Pisces, Siluriformes). *Can. J. Zool.* **67**, 2090–2097.

Röpke, A. (1993). Do larvae of mesoplelagic fishes in the Arabian Sea adjust their vertical distribution to physical and biological gradient? *Mar. Ecol.: Prog. Ser.* **101**, 223–235.

Rose, C. S., and Reiss, J. O. (1993). Metamorphosis and the vertebrate skull: Ontogenetic patterns and developmental mechanisms. *In* "The Skull" (J. Hanken and B. K. Hall, eds.), Vol. 1, pp. 289–346. University of Chicago Press, Chicago.

Rose, M. R., and Lauder, G. V. (1996). Post-Spandrel adaptationism. *In* "Adaptation" (M. R. Rose and G. V. Lauder, eds.), pp. 1–10. Academic Press, San Diego, CA.

Rosen, D. E., and Parenti, L. R. (1981). Relationships of *Oryzias*, and the groups of atherinomorph fishes. *Am. Mus. Novitiates* **2719**, 1–20.

Russell, F. S. (1976). "The Eggs and Planktonic Stages of British Marine Fishes." Academic Press, London.

Sandy, J. M., and Blaxter, J. H. S. (1980). A study of retinal development in larval herring *Clupea harengus* and sole *Solea solea*. *J. Mar. Biol. Assoc. U.K.* **60**, 59–72.

Schellart, N. A. M., and Popper, A. N. (1992). Functional aspects of the auditory system of actinopterygian fishes. *In* "Evolutionary Biology of Hearing" (D. B. Webster, R. R. Fay, and A. N. Popper, eds.), pp. 295–322. Springer-Verlag, New York.

Schmitt, E., and Kunz, Y. W. (1989). Retinal morphogenesis in the rainbow trout, *Salmo gairdneri*. *Brain, Behav. Evol.* **34**, 48–64.

Schultz, E. T., and Cowen, R. K. (1994). Recruitment of coral-reef fishes to Bermuda: Local retention or long-distance transport? *Mar. Ecol. Prog. Ser.* **109**, 15–28.

Secor, D. H., Dean, J. M., and Campana, S. E., eds. (1995). "Recent Developments in Fish Otolith Research." University of South Carolina Press, Columbia.

Senoo, S., Ang, K. J., and Kawamura, G. (1994). Study of artificial seed production of marble goby. II. Development of sense organs and mouth and feeding of reared marble goby *Oxyeleotris marmoratus* larvae. *Fish. Sci.* **60**, 361–368.

Shand, J. (1993). Changes in the spectral absorption of cone visual pigments during the settlement of the goatfish *Upeneus tragula*—the loss of red sensitivity as a benthic existence begins. *J. Comp. Physiol. A* **173**, 115–121.

Shand, J. (1994). Changes in retinal structure during development and settlement of the goatfish *Upenus tragula. Brain, Behav. Evol.* **43**, 51–60.

Smith, D. G. (1979). Guide to the Leptocephali (Elopiformes, Anguilliformes, and Notacanthiformes). NOAA Tech. Rept. NMFS Circ. **424**, 1–39.

Smith, D. G. (1989). Role of museums in studies of early life history stages of fishes. *Trans. Am. Fish. Soc.* **118**, 214–217.

Suda, Y., Shimizu, M., and Nose, Y. (1987). Morphological differences between cultivated and wild jack mackerel *Trachurus japonicus. Nippon Suisan Gakkaishi* **53**, 59–61.

Sweatman, H. (1988). Field evidence that settling coral reef fish larvae detect resident fishes using dissolved chemical cues. *J. Exp. Mar. Biol. Ecol.* **124**, 163–174.

Tanaka, M., Kawai, S., Seikai, T., and Burke, J. S. (1996). Development of the digestive organ system in Japanese flounder in relation to metamorphosis and settlement. *Mar. Freshwater Behav. Physiol.* **28**, 19–31.

Tandler, A., and Helps, S. (1985). The effects of photoperiod and water exchange rate on growth and survival of gilthead sea bream (*Sparus aurata*, Linnaeus; Sparidae) from hatching to metamorphosis in mass rearing systems. *Aquaculture* **48**, 71–82.

Tåning, A. V. (1952). Experimental study of meristic characters in fishes. *Biol. Rev. Cambridge Philos. Soc.* **27**, 69–193.

Tomoda, H., and Uematsu, K. (1996). Morphogenesis of the brain in larval and juvenile Japanese eels, *Anguilla japonica. Brain, Behav. Evol.* **47**, 33–41.

Toole, C. L., Markle, D. F., and Harris, P. M. (1993). Relationships between otolith microstructure, microchemistry, and early life history events in Dover sole, *Microstomus pacificus. U.S. Fish. Bull.* **91**, 732–753.

Tsukamoto, Y., and Okiyama, M. (1997). Growth during the early life history of the Pacific tarpon *Megalops cyprinoides. Jpn. J. Ichthyol.* **39**, 379–386.

Uwa, H., and Parenti, L. R. (1988). Morphometric and mersitic variation in ricefishes, genus *Oryzias*: A comparison with cytogenetic data. *Jpn. J. Ichthyol.* **35**, 159–166.

VanDenBossche, J., Seelye, J. G., and Zielinski, B. S. (1995). The morphology of the olfactory epithelium in larval, juvenile and upstream migrant stages of the sea lamprey, *Petromyzon marinus. Brain, Behav. Evol.* **45**, 19–24.

van der Meer, H. J. (1995). Visual resolution during growth in a cichlid fish: A morphological and behavioral case study. *Brain, Behav. Evol.* **45**, 5–33.

Vandewalle, P., Victor, D., Sanna, P., and Surlemont, C. (1989). The Weberian apparatus of a 18.5 mm fry of *Barbus barbus* (L.). *Fortsch Zool.* **35**, 363–366.

Van Netten, S. M., and Kroese, A. B. A. (1989). Dynamic behavior and micromechanical properties of the cupula. *In* "The Mechanosensory Lateral Line: Neurobiology and Evolution" (S. Coombs, P. Gorner, and H. Münz, eds.), pp. 247–264. Springer-Verlag, New York.

Verraes, W. (1976). Postembryonic development of the nasal organs, sacs and surrounding skeletal elements in *Salmo gairdneri* (Teleostei: Salmonidae), with some functional implications. *Copeia*, pp. 71–75.

Victor, B. C. (1986). Duration of the planktonic larval stage of one hundred species of Pacific and Atlantic wrasses (Family Labridae). *Mar. Biol.* **90**, 317–326.

Victor, B. C. (1991). Settlement strategies and biogeography of reef fishes. *In* "The Ecology of Fishes on Coral Reefs" (P. F. Sale, ed.), pp. 231–261. Academic Press, San Diego, CA.

Vischer, H. A. (1989). The development of lateral-line receptors in *Eigenmannia* (Teleostei, Gymnotiformes). *Brain, Behav. Evol.* **33**, 205–222.

Volk, E., Wissmar, R. C., Simenstad, C. A., and Eggers, D. M. (1984). Relationship between otolith microstructure and the growth of juvenile chum salmon (*Oncorhynchus keta*) under different prey rations. *Can. J. Fish. Aquat. Sci.* **41**, 126–133.

Watson, W. (1996a). Holocentridae: Squirrelfishes. *In* "The Early Stages of Fishes in the California Current Region." (H. G. Moser, ed.), Calif. Coop. Oceanic Fish. Invest., Atlas No. 33. pp. 686–691. Allen Press, Lawrence, Kansas.

Watson, W. (1996b). Holocentridae: Squirrelfishes. *In* "The Early Life Stages of Fishes in the California Current Region" (H. G. Moser, ed.). Calif. Coop. Oceanic Fish. Invest., Atlas No. 33, pp. 686–691. Allen Press, Lawrence, Kansas.

Watson, W. (1996c). Ostraciidae: Trunkfishes. *In* "The Early Stages of Fishes in the California Current Region." (H. G. Moser, ed.), Calif. Coop. Oceanic Fish. Invest., Atlas No. 33. pp. 1425–1427. Allen Press, Lawrence, Kansas.

Webb, J. F. (1988). Asymmetry and polymorphism in the lateral line system of the deep water flatfish, *Glyptocephalus zachirus*. *Am. Zool.* **28**, 89A.

Webb, J. F. (1989a). Developmental constraints and evolution of the lateral line system in teleost fishes. *In* "The Mechanosensory Lateral Line—Neurobiology and Evolution" (S. Coombs, P. Gorner, and H. Münz, eds.), pp. 79–98. Springer-Verlag, New York.

Webb, J. F. (1989b). Gross morphology and evolution of the mechanoreceptive lateral line system in teleost fishes. *Brain, Behav. Evol.* **33**, 34–53.

Webb, J. F. (1989c). Neuromast morphology and lateral line trunk canal ontogeny in two species of cichlids: An SEM study. *J. Morphol.* **202**, 53–68.

Webb, J. F. (1990). Ontogeny and phylogeny of the trunk lateral line system in cichlid fishes. *J. Zool.* **221**, 405–418.

Webb, J. F. (1993). Accessory nasal sacs of flatfishes: Systematic significance and functional implications. *Bull. Mar. Sci.* **52**, 541–553.

Webb, J. F. (1995). How do asymmetric cranial bones develop in the flatfish, *Glyptocephalus zachirus? Am. Zool.* **35**, 106A

Webb, J. F., and Noden, D. M. (1993). Ectodermal placodes: Contributions to the development of the vertebrate head. *Am. Zool.* **33**, 434–447.

Webb, P. W. (1981). Responses of northern anchovy, *Engraulis mordax,* larvae to predation by a biting planktivore, *Amphiprion percula. U.S. Fish. Bull.* **79**, 727–735.

Webb, P. W., and Corolla, R. T. (1981). Burst swimming performance of northern anchovy, *Engraulis mordax,* larvae. *U.S. Fish. Bull.* **79**, 143–150.

Webb, P. W., and Weihs, D. (1986). Functional locomotor morphology of early life history stages of fishes. *Trans. Am. Fish. Soc.* **115**, 115–127.

Weihs, D., and Moser, H. G. (1981). Stalked eyes as an adaptation toward more efficient foraging in marine fish larvae. *Bull. Mar. Sci.* **31**, 31–36.

Weitzman, S. H., and Vari, R. P. (1988). Miniaturization in South American freshwater fishes: An overview and discussion. *Proc. Biol. Soc. Wash.* **101**, 444–465.

Werner, R. G., and Lannoo, M. J. (1994). Development of the olfactory system of the white sucker, *Catostomus commersoni,* in relation to imprinting and homing: A comparison to the salmonid model. *Environ. Biol. Fish.* **40**, 125–140.

Williams, P. J., Brown, J. A., Gotceitas, V., and Pepin, P. (1996). Developmental changes in escape response performance of five species of marine larval fish. *Can. J. Fish. Aquat. Sci.* **53**, 1246–1253.

Williams, S. R. (1902). Changes accompanying the migration of the eye and observations on the tractus opticus and tectum opticum in *Pseudopleuronectes americanus*. *Bull. Mus. Comp. Zool.* **40**, 1–58.

Wonsettler, A. L., and Webb, J. F. (1997). Morphology and development of the multiple lateral line canals on the trunk in two species of *Hexagrammos* (Scorpaeniformes: Hexagrammidae). *J. Morphol.* **233**, 195–214.

Yamashita, K. (1982). Sensory cupulae found in pre larvae of red sea bream *Pagrus major*. *Jpn. J. Ichthyol.* **29**, 279–284.

Youson, J. H. (1988). First metamorphosis. *In* "Fish Physiology, Volume XI. The Physiology of Developing Fish, Part B: Viviparity and Posthatching Juveniles." W. S. Hoar and D. J. Randall, eds.), pp. 135–196. Academic Press, San Diego.

Zakon, H. H. (1984). Postembryonic changes in the peripheral electrosensory system of a weakly electric fish: Addition of receptor organs with age. *J. Comp. Neurol.* **228**, 557–570.

Zakon, H. H. (1986). The electroreceptive periphery. *In* "Electroreception" (T. H. Bullock and W. Heiligenberg, eds.), pp. 103–156. Wiley, New York.

Zelditch, M. L. (1996). Introduction to the symposium: Historical patterns of developmental integration. *Am. Zool.* **36**, 1–3.

Mechanisms of Larval Development and Evolution

CHAPTER 5

Heterochrony

MICHAEL W. HART* AND GREGORY A. WRAY[†]
*Section of Evolution and Ecology, University of California, Davis, California, [†]Department of
Ecology and Evolution, State University of New York, Stony Brook, New York

---

I. INTRODUCTION

Heterochrony is an ancestor–descendant difference in the relative timing of events during development. This phylogenetic *pattern* frequently has been confused with ecological and evolutionary *processes* that result in diverse larval forms that also vary in the timing of developmental events. Here we consider the role of heterochrony as a process affecting the origin and evolution of larval forms.

Discussions of heterochrony rarely distinguish patterns from processes (Raff and Wray, 1989; Hall, 1992). Reilly *et al.* (1997) define heterochronic *patterns* as **paedomorphosis**, truncation of development of a trait relative to development of the trait in an ancestor, or **peramorphosis**, extension of development relative to an ancestor. They define six different heterochronic *processes* that can produce these two patterns: **deceleration** and **acceleration**, the decrease or increase in k, the rate of development of the trait; **hypomorphosis** and **hypermorphosis**, the decrease or increase in β, the age at termination of development of the trait; and **postdisplacement** and **predisplacement**, the increase or decrease in α, the age at initiation of development of the trait. Reilly *et al.* (1997) also note that a combination of these heterochronic processes can result in a heterochronic pattern in which ancestor and descendant have the same phenotype, a pattern called **isomorphosis.**

These definitions reflect a specific interest in processes and explanations of causation acting within individual ontogenies, and this interest is characteristic of the heterochrony literature. At this lower hierarchical level, heterochronic processes are defined by the developmental parameters (k, β, α) that differ between lineages. However, for most evolutionary biologists not steeped in the heterochronic tradition, explanations of causation usually reflect processes acting among individuals and populations. This is especially true for variation in ontogenies that appears to have an adaptive basis. Acceleration, hypermorphosis, and other heterochronic phenomena can be viewed at this upper hierarchical level as patterns of development—and not evolutionary processes—that are caused by evolutionary mechanisms (mutation, recombination, drift, migration, selection) that favor one phenotype over another or constrain evolutionary changes from one phenotype to another.

For example, Garstang (1929) proposed that the origin of the gastropod veliger included a heterochronic change in differentiation of the shell. This paedomorphic pattern may have been produced in part by predisplacement (Reilly *et al.,* 1997). In addition to this heterochronic pattern, the shell serves an obvious protective function, probably favored by natural selection. The parallel advent of shelled larvae in other mollusks implies adaptation (Garstang, 1929), and some experiments have tested the defensive properties of shelled gastropod veligers (e.g., Pennington and Chia, 1985; Pennington and Hadfield, 1989). From this point of view, the origin of the veliger includes a striking pattern of heterochrony, but the process resulting in a new larval form is not a process of heterochrony but rather natural selection.

Thus, the role of heterochrony in larval evolution depends on the specific question of interest. If we ask, "What are the cellular and morphogenetic processes within organisms that result in differences in larval morphology?" then changes in the timing of events such as gene expression or tissue interactions form an important part of the answer. However, efforts to understand the origin and evolution of larval forms have focused on a more general question, "What are the ecological and evolutionary processes among organisms that cause some lineages to evolve larval forms that are different from the larvae of their ancestors and extant relatives?" Other authors in this volume review in detail the answers to this question: dispersal, population subdivision, feeding and swimming, defense against predators, recruitment to new habitats, persistence and extinction of populations, and other aspects of the demography and functional biology of embryos and larvae.

We originally intended to review cases in which heterochronic processes had an important role in larval evolution and to write a prospective account of heterochrony in this growing field of research. Heterochrony has been a potent general organizing principle in the history of evolutionary biology (Gould, 1977; Alberch *et al.*, 1979; McKinney, 1988; McKinney and McNamara, 1991; Hall, 1992; McNamara, 1995). However, it seems to us that heterochrony as a process has been relatively unimportant in both the evolution of larval forms and the revival of interest in them (Raff, 1996). Heterochronic patterns in larval development are abundant, but to our knowledge these patterns appear to follow as consequences of natural selection and other evolutionary processes acting on variation in the functional, ecological, and demographic features of a complex life cycle. Interest in heterochrony as a process in larval evolution has been restricted mainly to two historically important ideas: paedomorphosis in urodeles (Section II) and the evolution of free-living vertebrates from sessile, tunicate-like ancestors (Section III). We know of only one other example in which a heterochronic process has been implicated in the evolution of derived larval forms of marine invertebrates (Section IV).

II. URODELES

Changes in timing of larval development in salamanders are associated with two evolutionary changes in larval type that are repeated in diverse salamander lineages: "neotenic," sexually mature larvae (e.g., axolotls) and direct-developing embryos (e.g., many Plethodontidae) (Alberch and Blanco, 1996; Wake and Hanken, 1996). In both cases, larval morphogenesis is temporally decoupled from the development and maturation of gonads, reproductive behavior, and the hormonal axes that regulate their differentiation (Rosenkilde and Ussing, 1996). These examples of heterochrony in the evolution of amphibian larval forms have been important in the history of ideas about the relation between development and morphological evolution (Gould, 1977; Wake and Larson, 1987; Hall, 1992; Raff, 1996).

Such heterochronies may arise in at least two ways. (1) Changes in developmental timing may arise as indirect, pleiotropic consequences of natural selection acting on morphological or life-history traits that are related to fitness of embryos and larvae. In this case, heterochronic patterns may indicate internal, epigenetic interactions among these other traits that potentially constrain the degree or direction of response to selection (e.g., Alberch and Gale, 1985). (2) Selection could directly act on the relative timing of developmental events, such as time to metamorphosis. In this case, heterochronic patterns may indicate one of several possible responses to selection.

III. TUNICATES

Garstang (1928), Berrill (1955), and others advocated an evolutionary scheme in which the ancestral vertebrate was derived from a tunicate-like progenitor,

162

062 Michael W. Hart and Gregory A. Wray

mainly by paedomorphic loss of the sessile adult stage and accelerated development of gonads in the tadpole-like larval form. An alternative scenario proposes that the sessile adult stage of tunicates is a peramorphic addition and that the common ancestor of urochordates and vertebrates resembled a tunicate tadpole or a larvacean adult (Carter, 1957; Jollie, 1982). Gee (1996) reviews evidence suggesting that the second case is more likely than the first. In either case, a dramatic heterochrony was associated with the evolution of a new adult body plan. The functional or reproductive significance of this change is not known.

IV. TEST OF A HETEROCHRONIC HYPOTHESIS

We know of only one other hypothesis involving heterochrony as a developmental process leading directly to evolutionary changes in larval forms. Strathmann *et al.* (1992) showed that the feeding, planktonic larvae of a sea urchin respond to food limitation in laboratory culture and in the ocean by developing larger larval feeding structures while delaying the development of postlarval (juvenile) structures. This kind of phenotypic plasticity can be viewed as a kind of environment-induced heterochrony, in which abundant energy and materials are allocated to rapid juvenile development, while scarce energy and materials are husbanded for growth of a larger feeding device in the larva. Strathmann *et al.* (1992) noted that this heterochrony in the appearance of postlarval structures is mirrored in the development of sea urchin species that have nonfeeding larvae and large, yolky eggs: juvenile structures appear early in these species relative to the appearance of larval structures (including remnants of the larval feeding organs) in comparison to sea urchins with small eggs and feeding larvae. Nonfeeding larval development from large eggs is a derived condition for sea urchins: such species have evolved isomorphic heterochronies in juvenile structures via acceleration or predisplacement (or most likely both) (Reilly *et al.*, 1997).

Strathmann *et al.* (1992) suggested that if energy and materials from planktonic food are physiologically equivalent to the material sequestered by the mother in the egg, then variation in egg size could result in heterochronic changes in the development of juvenile structures merely by genetic assimilation and the action of a heterochronic mechanism that originally evolved as a phenotypically plastic response to environmental variation in planktonic food. Matsuda (1982) discusses possible laboratory evolution of paedomorphosis in axolotls by genetic assimilation of a phenotypically plastic hormonal mechanism. Under this scenario, a heterochronic process could be involved directly in the evolution of different larval forms.

A more recent experiment rejects this hypothesis for sea urchins. Bertram and Strathmann (1998) compared the growth and development of conspecific sea urchins over a naturally occurring range of egg sizes, which was produced by natural variation in food available for adults. They found consistent, minor differences in the timing of development of larval and postlarval structures between larvae from small and large eggs. However, these differences were overwhelmed by variation in the timing of development induced by small differences in the amount of food provided for larvae from both small and large eggs in laboratory culture. Thus, the material sequestered by mothers into eggs was not equivalent to the energy and materials gained from planktonic food. Heterochronic change induced by phenotypic plasticity probably cannot be genetically assimilated to produce a fixed heterochronic evolutionary change in the larval development of sea urchins merely through an increase in egg size.

Thus, the only hypothesis that explicitly invokes a developmental heterochronic process to explain contemporary evolutionary changes in larval forms has been rejected. The evolution of different feeding and nonfeeding larval forms in echinoderms involves numerous heterochronic (and other) changes in the order and relative timing of many traits (Wray and Bely, 1994), but these changes appear to follow indirectly as consequences of selection acting directly on other aspects of echinoderm life histories. We do not know of comparable hypotheses for other taxa, although they might have been proposed.

V. THE ALTERNATIVE RESEARCH PROGRAM

Larval ecology is an adaptationist research program in which many heterochronic patterns of larval development can be interpreted as consequences of natural selection acting on adaptive features of embryos and larvae and the ecological circumstances in which larvae develop (McEdward, 1995). Among the clearest examples are heterochronic differences in the development of larval feeding structures and postlarval juvenile structures between closely related species with feeding and nonfeeding planktonic larval forms. In many taxa, the evolution of eggs sufficiently large to support all of larval development from fertilization through metamorphosis permits the reduction (or loss) of redundant features of the larval feeding structures (Strathmann, 1978; Raff, 1987). Accelerated appearance of postlarval juvenile structures may become adaptive (or at least not maladaptive) as a consequence of the evolution of large eggs, resulting in heterochronic changes that were prohibited in a developmental sequence that required a functional feeding device. Such changes in developmental timing are not caused primarily by heterochronic processes like acceleration or predisplacement but by selective processes (e.g., high planktonic mortality rates that favor shorter periods from fertilization to metamorphosis).

Many examples discussed in other chapters of this book could be identified as heterochronies. Others have pointed out that almost any evolutionary change in development can formally be defined as a heterochrony (Hall, 1992; Raff, 1996). Application of this label does not provide any direct insight into the origin and evolution of larval forms. Processes operating at the level of populations, species, and perhaps higher taxa most likely are responsible for most evolutionary changes in larval development, including changes in the time of appearance or duration of development of structures. These evolutionary processes act on variations in functional and ecological features of early developmental stages, as well as emergent properties of species (such as geographic range and speciation rate) that are affected by larval dispersal and recruitment. Responses to these selective or nonselective evolutionary processes may, in many cases, be manifest as developmental heterochronies. Studies of larval ecology provide the evolutionary context for interpretation of heterochronic patterns.

VI. SUMMARY

Heterochrony in larval evolution is not an active field of research, judging by the absence of publications that explicitly invoke heterochronic processes to explain the origin and evolution of larval forms among marine invertebrates (see Raff, 1996). We doubt that this will change, mainly because the larval ecology research program provides such a useful alternative context for understanding many aspects of larval evolution. A revival of heterochrony studies on marine invertebrate larvae could follow the prescription of Alberch and Blanco (1996), although by definition it would be limited to smaller scale variations in morphological development of species with invariant order of developmental events.

REFERENCES

Alberch, P., and Blanco, M. J. (1996). Evolutionary patterns in ontogenetic transformation: From laws to regularities. *Int. J. Dev. Biol.* **40**, 845–858.
Alberch, P., and Gale, E. (1985). A developmental analysis of an evolutionary trend: Digital reduction in amphibians. *Evolution (Lawrence, Kans.)* **39**, 8–23.
Alberch, P., Gould, S. J., Oster, G. F., and Wake, D. B. (1979). Size and shape in ontogeny and phylogeny. *Paleobiology* **5**, 296–317.
Berrill, N. J. (1955). "The Origin of Vertebrates." Clarendon, Oxford.
Bertram, D. F., and Strathmann, R. R. (1998). Effects of maternal and larval nutrition on growth and form of planktotrophic larvae. *Ecology* **79**, 315–327.
Carter, G. S. (1957). Chordate phylogeny. *Syst. Zool.* **6**, 187–192.

Garstang, W. (1928). The morphology of the Tunicata, and its bearings on the phylogeny of the Chordata. *Q. J. Microsc. Sci.* **72**, 51–187.

Garstang, W. (1929). The origin and evolution of larval forms. *Rep. Br. Assoc. Adv. Sci.* pp. 77–98.

Gee, H. (1996). "Before the Backbone." Chapman & Hall, New York.

Gould, S. J. (1977). "Ontogeny and Phylogeny." Harvard University Press, Cambridge, MA.

Hall, B. K. (1992). "Evolutionary Developmental Biology." Chapman & Hall, New York.

Jollie, M. (1982). What are the "Calcichordata" and the larger question of the origin of chordates. *Zool. J. Linn. Soc.* **75**, 167–188.

Matsuda, R. (1982). The evolutionary process in talitrid amphipods and salamanders in changing environments, with a discussion of "genetic assimilation" and some other evolutionary concepts. *Can. J. Zool.* **60**, 733–749.

McEdward, L. R., ed. (1995). "Ecology of Marine Invertebrate Larvae." CRC Press, Boca Raton, FL.

McKinney, M. L., ed. (1988). "Heterochrony in Evolution." Plenum, New York

McKinney, M. L., and McNamara, K. J. (1991). "Heterochrony: The Evolution of Ontogeny." Plenum, New York.

McNamara, K. J., ed. (1995). "Evolutionary Change and Heterochrony." Wiley, New York.

Pennington, J. T., and Chia, F.-S. (1985). Gastropod torsion: A test of Garstang's hypothesis. *Biol. Bull.* (*Woods Hole, Mass.*) **169**, 391–396.

Pennington, J. T., and Hadfield, M. G. (1989). A simple nontoxic method for the decalcification of living invertebrate larvae. *J. Exp. Ma. Biol. Ecol.* **130**, 1–7.

Raff, R. A. (1987). Constraint, flexibility, and phylogenetic history in the evolution of direct development in sea urchins. *Dev. Biol.* **119**, 6–19.

Raff, R. A. (1996). "The Shape of Life." University of Chicago Press, Chicago.

Raff, R. A., and Wray, G. A. (1989). Heterochrony: Developmental mechanisms and evolutionary results. *J. Evol. Biol.* **2**, 409–434.

Reilly, S. M., Wiley, E. O., and Meinhardt, D. J. (1997). An integrative approach to heterochrony: The distinction between interspecific and intraspecific phenomena. *Biol. J. Linn. Soc.* **60**, 119–143.

Rosenkilde, P., and Ussing, A. P. (1996). What mechanisms control neoteny and regulate induced metamorphosis in urodeles? *Int. J. Dev. Biol.* **40**, 665–673.

Strathmann, R. R. (1978). The evolution and loss of feeding larval stages of marine invertebrates. *Evolution* (*Lawrence, Kans.*) **32**, 894–906.

Strathmann, R. R., Fenaux, L., and Strathmann, M. F. (1992). Heterochronic developmental plasticity in larval sea urchins and its implications for evolution of nonfeeding larvae. *Evolution* (*Lawrence, Kans.*) **46**, 972–986.

Wake, D. B., and Hanken, J. (1996). Direct development in the lungless salamanders: What are the consequences for developmental biology, evolution and phylogenesis. *Int. J. Dev. Biol.* **40**, 859–869.

Wake, D. B., and Larson, A. (1987). Multidimensional analysis of an evolving lineage. *Science* **238**, 42–48.

Wray, G. A., and Bely, A. E. (1994). The evolution of echinoderm development is driven by several distinct factors. *Development* (*Combridge, UK*), Supp., pp. 97–106.

CHAPTER 6

Hormonal Control in Larval Development and Evolution—Amphibians

CHRISTOPHER S. ROSE

Department of Biology, Dalhousie University, Halifax, Nova Scotia, Canada

I. INTRODUCTION

Among vertebrates, complex life histories producing specialized larval forms are confined to lampreys, certain teleosts, and amphibians. Of these, amphibians perhaps have best exploited the potential of a complex life history for larval diversification. Frogs, salamanders, and caecilians have each evolved a broad array of larval forms that are distinct in morphology, physiology, and ecology from both their adult counterparts and other vertebrates. Moreover, much of the phylogenetic variation present within these groups is a direct parallel of their ontogenetic variation. Differences among amphibian larval and adult morphologies commonly derive from one or two recurring patterns of ontogenetic change that affect the timing and extent of their larval development. Amphibians also are distinguished by one dominant mechanism of developmental regulation. Almost all aspects of larval development, including the culminating transformation into adult form (commonly known as metamorphosis), are under the direct and primary control of thyroid hormone (TH). This combination of a recurring tendency to exploit ontogenetic variation for evolutionary divergence and a singular reliance upon hormones for developmental regulation offers a unique opportunity to investigate the developmental basis of morphological evolution.

Before outlining the specific goals of this chapter, I will first address concerns regarding our perception of amphibian larvae as morphological entities and the conceptual relationships drawn between their morphological development and hormonal activity. Biologists commonly regard larval life-history stages as both phenotypically distinct and morphologically constant. The terms "larval state" and "larval stage" often are used to refer to an animal at any and all points throughout its larval period, meaning from the end of embryogenesis (defined here as hatching) to the start of metamorphosis (defined here as the onset of gill resorption in salamanders and tail resorption in frogs). However, whereas ecological, functional, and even some physiological characters may undergo little or no change during the larval period, the generalization of no modification does not hold for morphological characters. If internal as well as external characters are considered, it is safe to say that all amphibians undergo at least some morphological changes between embryogenesis and metamorphosis. Hence, for the purposes of this discussion, phylogenetic comparisons of developmental pattern, i.e., the sequence and relative duration of changes in morphology, offer a more informative description of evolutionary divergence than comparisons of "homologized" morphological states.

Another reason for emphasizing the primacy of pattern over states of morphological development is the dynamic relationship between morphological change and hormonal activity. Much of amphibian postembryonic development is controlled by the interaction of two parameters: hormone activity, defined here as the profile of hormone plasma concentration over larval and metamorphic stages, and tissue sensitivity, which, generally speaking, is a measure of a tissue's responsiveness to a hormone and strictly speaking is the minimum concentration required for its activation. This interaction is dynamic in the sense that hormones appear to influence morphogenesis over the course of their activity, even though the morphogenetic response itself may be confined to a single stage and/or threshold concentration of hormone. While the mechanisms involved in this interaction typically impart a high degree of developmental integration, they also permit considerable evolutionary lability. Whereas developmental patterns within species generally are invariant, those of closely related species can exhibit considerable dissociation of homologous events. Attempts to isolate the respective roles of hormone activity and tissue sensitivity in this diversification must start by comparing patterns of hormonally mediated change with patterns of hormonal activity.

What is known about the hormonally mediated development of larval amphibians? The morphological and physiological development of amphibians has received extensive attention throughout this century. All organ systems, and particularly the skin, sense organs, blood, musculoskeletal, immune, gut, and excretory systems, are known to undergo TH-mediated changes at larval and/or metamorphic stages [see chapters in Gilbert et al. (1996) for reviews]. Musculoskeletal development has received by far the most comparative analysis (albeit largely descriptive) and, thus, provides the most complete data set for exploring phylogenetic patterns in larval development and morphology.

What is known about hormone activity and tissue sensitivity in larval amphibians? The neuroendocrine control of TH production was investigated first in mammalian adults, where negative feedback loops predominate and are relatively well-understood. In contrast, the interplay of positive and negative feedback loops that coordinate the surge in TH production in larval amphibians is still the subject of considerable investigation (Denver, 1996; Kikuyama et al., 1993). Extensive research has been done on TH binding and TH-regulated protein and RNA synthesis in various aspects of tadpole remodeling, particularly tail resorption and hepatocyte and erythrocyte differentiation (Fox, 1983). The discovery of the TH receptor gene in 1986 (Sap et al., 1986; Weinberger et al., 1986) refocused research at the level of gene regulation and galvanized the search for genes that either control tissue sensitivity or are TH-regulated in specific tissue responses (Shi, 1996). Comparatively speaking, however, little is known about the roles of hormone activity and tissue sensitivity in controlling disparate patterns of larval development. TH activity has been quantified for numerous frog and salamander species, but mostly in forms exhibiting stereotypic amphibian development, i.e., oviparous forms whose free-living aquatic larvae transform in terrestrial adults. Similarly, although efforts to induce remodeling with exogenous TH have revealed profound interspecific differences in tissue sensitivity, only a few studies have addressed the cellular and molecular bases of this variation.

Because our understanding of hormone mediation in amphibian development largely derives from work on a handful of frog species, any discussion of how hormonal mediation has influenced amphibian diversification makes a relatively quick transgression from answering questions to asking them. Nonetheless, it is possible to organize our present knowledge into a framework for promoting the integration of mechanistic and comparative studies on this topic. The present chapter has three goals to this end: (1) summarize research on the pathways and levels of regulation involved in TH-mediated larval development; (2) discuss phylogenetic trends in TH-mediated development that have stimulated some mechanistic or comparative analysis of their hormonal basis; and (3) summarize what is known regarding the hormonal mediation of larval development in viviparous amphibians.

II. THE ROLE OF TH AND OTHER HORMONES IN AMPHIBIAN DEVELOPMENT

Although generally considered a monophyletic group (Trueb and Cloutier, 1991a,b), the three subgroups of Lissamphibia (frogs, salamanders, and caecilians) exhibit sufficiently distinct patterns of postembryonic development that it is difficult to specify an ancestral pattern for them beyond an aquatic-to-terrestrial transformation involving resorption of gills, tail fin, and Leydig cells, development of skin glands, closure of gill slits, and gill arch reduction (Rose and Reiss, 1993). It is still unclear whether TH mediation is ancestral to all three subgroups, as only circumstantial evidence is available for caecilians [Klumpp and Eggert, 1934, reviewed in Szarski, 1957; Welsch *et al.,* 1974]. Also, in the absence of a nonmetamorphosing lissamphibian outgroup, one must look to primitive actinopterygians, coelocanths, and lungfishes to infer the ancestral state of hormonal mediation in amphibians, i.e., the condition prior to co-option of the hypothalamus–pituitary–thyroid axis for metamorphic regulation. For these reasons, this discussion is limited to development and evolution *within* the two lissamphibian clades known to have TH mediation, salamanders and frogs.

A. The Regulation of Hormone Activity in Larval Amphibians

Almost all research on the regulation of larval hormone activity in amphibians has been done on oviparous species of three genera of frogs, the ranid *Rana,* the bufonid *Bufo,* and the pipid *Xenopus.* The majority of the work has been done on two species, *R. catesbeiana* and *X. laevis.* All of these species are characterized by distinct pre-, pro-, and climax metamorphic phases of development. Pre- and prometamorphosis refer to phases of hindlimb differentiation and growth, respectively, and climax refers to the phase of forelimb emergence, tail resorption, and remodeling of the chondrocranium, gut, and other internal organs.

The primary hormones involved (i.e., those that directly mediate tissue responses) are the following: the two forms of TH, thyroxine or tetraiodothyronine (T_4) and triiodothyronine (T_3); the major corticosteroid in larval amphibians, corticosterone (CORT); and prolactin (PRL). Though additional hormones appear to modulate the activity of these hormones, as well as regulate larval growth and metabolic activities related to metamorphosis, a summary of the evidence required to justify their inclusion here is beyond the scope of this discussion and readily available in other reviews (Denver, 1996; Hayes, 1997; Kaltenbach, 1996; Kikuyama *et al.*, 1993).

Circulating concentrations of TH, CORT, and PRL all rise during larval development. In *Rana, Bufo,* and *Xenopus,* TH (Buscaglia *et al.*, 1985; Leloup and Buscaglia, 1977; Mondou and Kaltenbach, 1979; Regard *et al.*, 1978; Suzuki and Suzuki, 1981; Weil, 1986) and CORT [Leloup-Hatey *et al.*, 1990; White and Nicoll, 1981; references in Kikuyama *et al.* (1993)] generally begin to increase in prometamorphosis and peak during climax, whereas PRL rises at late prometamorphosis or climax depending on the species [Dickhoff *et al.*, 1990; references in Denver (1996)]. These hormone profiles are the net result of numerous regulatory mechanisms acting at two levels: the neuroendocrine axes that control hormone production and the peripheral tissues that control hormone processing and hormone removal.

Despite extensive investigation, theories on the neuroendocrine control of hormone production in tadpoles remain incomplete (Fox, 1983; White and Nicoll, 1981). The present summary, which follows Kikuyama *et al.* (1993) and Denver (1996), lists only well-substantiated features and insights (Fig. 1). Over pre- and prometamorphosis, the low levels of circulating TH sequentially induce the functional maturation of neurosecretory cells in the hypothalamus and the median eminence of the neurohypophysis. Once mature, these tissues, respectively, produce and convey releasing hormones to hormone-producing cells in the pars distalis of the adenohypophysis. Interestingly, thyrotropin-releasing hormone (TRH) is produced in tadpoles, but it does not stimulate pituitary cells to release thyrotropin (TSH) until after climax in both frogs and salamanders. This function appears to be assumed in tadpoles by corticotropin-releasing factor (CRF), which also may stimulate the pituitary to secrete adrenocorticotropin (ACTH) (Denver and Licht, 1989; Denver, 1993; Gancedo *et al.*, 1992). TSH stimulates the thyroid gland to produce TH (predominantly in the form of T_4), and ACTH may stimulate the interrenal glands to produce CORT. TH in turn up-regulates PRL (Buckbinder and Brown, 1993), possibly via TRH stimulation (Norris and Dent, 1989), and also may enhance CORT production, although the neuroendocrine pathways of both interactions are still unknown (Kaltenbach, 1996). At the same time, TH appears to directly inhibit TSH and CRF production as a source of negative feedback on TH production (Denver, 1996).

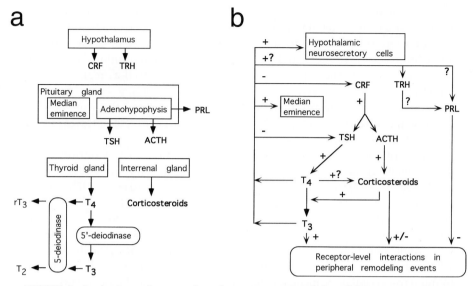

FIGURE 1 Production pathways and regulatory interactions of thyroid hormones, corticosteroids, and prolactin in tadpoles. Panel a shows the glands involved (boxes) and their respective hormone products; panel b shows the stimulatory (+) and inhibitory (−) effects of these hormones on gland maturation, hormone production and processing pathways, and tissue responses. Events occurring in peripheral tissues are enclosed in circles. These figures include only hormones that are known to directly mediate tissue responses and most regulatory interactions that are known to operate at tadpole stages; 5-deiodinase regulation has been omitted to maintain clarity. +? indicates a stimulatory effect whose level of interaction is not yet known. ? indicates one of two possible alternative routes for a +? effect (see text). Whereas most interactions occur during prometamorphosis or later, the T_4 stimulation of hypothalamic neurosecretory cells starts at premetamorphosis. Abbreviations: ACTH, adrenocorticotropin; CRF, corticotropin-releasing factor; PRL, prolactin; TRH, thyrotropin-releasing hormone; TSH, thyroid-stimulating hormone; T_2, diiodothyronine; (r)T_3, (reverse) triiodothyronine; T_4, thyroxine.

That PRL levels follow TH levels is consistent with the TH up-regulation of PRL gene expression in the pituitary (Buckbinder and Brown, 1993). This finding also refutes the traditional view that PRL functions as a "larvalizing" hormone that is high at larval stages to counter TH activity.

The relationship between TH and CORT levels is more complex and appears to involve peripheral regulation. T_4 increases the level of CORT, which in turn stimulates the conversion of T_4 to T_3 by deiodinating enzymes in peripheral tissues (Galton, 1990; Hayes and Wu, 1995a). The enzyme implicated in this finding, 5′-deiodinase (5′-D), shows low activity at premetamorphosis and high activity at climax in whole-animal assays of *Xenopus* and *Rana* (Buscaglia *et al.*, 1985; Galton, 1990). Individual tissue assays in *Rana* show that 5′-D activity and gene expression are maximal when the respective tissues are experiencing maximal remodeling, which suggests that it is the individual tissues that regulate the surges in intracellular T_3 responsible for their activation (Becker *et al.*, 1997). The net conversion of T_4 to T_3 presumably results in decreasing plasma T_4 levels, which may down-regulate CORT production, thus maintaining an autoregulating feedback loop between the two hormones (Hayes and Wu, 1995a). T_4 also appears to inhibit 5′-D activity in some tissues (Becker *et al.*, 1997), in contrast to its whole-animal effect of stimulating CORT-induced 5′-D activity. The regulation of 5′-D gene expression remains unknown.

Another kind of deiodinating enzyme, 5-deiodinase (5-D), inactivates T_4 and T_3 by converting them to rT_3 (reverse T_3) and T_2, respectively. Accounts of its activity and gene expression patterns differ among species and tissues (Becker *et al.*, 1997; Shi *et al.*, 1996; St. Germain *et al.*, 1994). 5-D exhibits increases in gene expression in *Xenopus* tail and hindlimb tissues at climax (Shi *et al.*, 1996), and its gene expression and activity are up-regulated by TH in tail at least (St. Germain *et al.*, 1994). In *Rana* tissues, 5-D activity patterns are similar to 5′-D activity patterns, suggesting that the two enzyme systems are tightly coupled to control intracellular T_3 levels (Becker *et al.*, 1997). Though 5-D activity and gene expression in *Rana* both are up-regulated by TH, they do not correlate with each other in some tissues, which suggests that additional factors are involved in regulating the activity of this enzyme (Becker *et al.*, 1995, 1997).

Other less studied mechanisms of hormone removal include the transport of TH into target cells by cellular TH binding proteins and metabolic breakdown by processes other than deiodination (Shi *et al.*, 1996).

In summary, although it is safe to say that most of the key regulatory mechanisms that govern hormone profiles have been identified, the exact expression patterns, levels of regulation, and respective contributions of many remain unresolved. Most significantly, the activity profile of TSH and the role of larval TRH are still unknown. There also is considerable evidence for additional monamine and neuropeptide involvement in the regulation of prolactin and other pituitary products (Kikuyama *et al.*, 1993; Norris and Dent, 1989). Hence, it is not yet possible to give a full account of the hormone profiles observed during larval development. Further, almost all of the preceding information has been compiled from frogs belonging to three fairly closely related, nonbasal lineages (Ford and Cannatella, 1993). Complementary studies on more primitive frogs, salamanders, and caecilians clearly are required to evaluate the general significance of these interactions for amphibian development.

B. The Hormonal Mediation of Tissue Remodeling in Larval Amphibians

1. Which Tissue Remodeling Is Hormone-Mediated?

Whether a tissue response is TH-mediated usually is determined by the classical tests of hormone supplementation, removal, and replacement. These tests determine whether a response can be (a) induced prematurely with exogenous TH, (b) prevented by blocking TH production, or (c) induced at the appropriate stage with exogenous TH when endogenous TH is blocked. Ideally, exogenous TH is applied at physiological concentrations, which are approximated most closely when the animal is first made hypothyroid by inhibiting TH production. Exogenous TH also should be applied *in vitro,* as well as *in vivo,* in order to confirm a direct effect of the hormone on target tissues. TH production can be blocked reversibly by applying thyroid inhibitors (goitrogens) or irreversibly by excising the embryonic rudiments of the thyroid gland or pituitary (thyroid-ectomy and hypophysectomy). Because some goitrogens (e.g., thiourea) may have additional side effects, results from their usage should be corroborated with other approaches. Similarly, operated animals should be checked histolog-ically to confirm that glands have been completely removed.

When the criteria of inducibility and preventability are applied to frogs with stereotypic (i.e., oviparous, indirect) development, it appears that all tissue remodeling following the primary origin and differentiation of embryonic and tadpole structures (e.g., hatching and cement glands, gills, keratinous mouthparts, lateral line organs) until the end of climax metamorphosis is TH-mediated (Fox, 1983; Gilbert *et al.,* 1996). Some tissues, however, give conflicting results. The early premetamorphic stages of hindlimb differentiation usually are considered TH-independent because they proceed normally in hypophysectomized and thyroid-blocked tadpoles [Fox (1983) for references]. However, they can be accelerated by exogenous TH in *Xenopus* at least (Knutson and Prahlad, 1971; Tata, 1968). One explanation is that gland removals and chemical inhibitors do not completely block thyroid activity. Another is that TH responsiveness develops earlier than TH dependence in some tissues. In this case, TH inducibility at early stages may indicate a facilitating, rather than essential, role of TH mediation, or it may be simply an experimental artifact.

The presumption of universal TH mediation in postembryonic tissue remod-eling is less tenable for salamanders. Despite numerous studies detailing the inducibility of external morphological changes in metamorphosing forms, rela-tively few studies have addressed internal changes, and fewer still have corrobo-rated the results with *in vitro* analysis or attempts to block thyroid activity. Gill and tail fin resorption are inducible in all metamorphosing species that have been tested. All mid-to-late larval and metamorphic changes to the cranial skeleton are TH-inducible in hemidactyliine plethodontids, with the exception of separation of the unipartite larval premaxilla into paired adult bones (Alberch *et al.*, 1985; Rose, 1995b). The skeletal remodeling that is synchronous with gill resorption in ambystomatids (Keller, 1946), as well as that occurring earlier in larval development (C. S. Rose, unpublished observation), also are inducible. That the induced larvae of other metamorphosing salamanders commonly develop normal adult-shaped heads suggests that the skeletal changes accompanying gill resorption are TH-mediated throughout these forms. Additional TH-inducible events in metamorphosing salamanders include serous and mucous gland differentiation and Leydig cell resorption (Rose, 1996), the transition from monocuspid to bicuspid teeth (Clemen, 1988; Greven and Clemen, 1990; Chibon, 1995), some of the changes in myosin and fiber type of dorsal skeletal muscle (Chanoine *et al.*, 1990; Salles-Mourlan *et al.*, 1994), and shifts in the composition of erythrocytes (i.e., hemoglobin), serum protein (Ducibella, 1974), and leukocytes (Ussing and Rosenkilde, 1995). Tests for TH involvement in the transition from undivided to divided (or pedicellate) teeth remain inconclusive (Greven and Clemen, 1990); leg formation and the postembryonic eye degeneration of cave-dwelling forms (*Proteus* and some hemidactyliine plethodontids) are TH-independent (Besharse and Brandon, 1974; Durand, 1995).

Though most studies on amphibian larvae address only T_3 mediation, both forms of TH (T_3 and T_4) probably are active inducers and they also may have different effects upon the same tissue. Whereas T_4 induces skin gland differentiation in *Bufo boreas* tadpoles, T_3 appears to promote epidermal growth at the expense of gland differentiation (Hayes and Gill, 1995). Though T_3 is a far more potent activator [references in Buscaglia *et al.* (1985)], T_4 may be compensated in this regard by its plasma concentration, which, depending on the species, may be up to 10 times higher than that of T_3 (e.g., Alberch *et al.*, 1986).

Substantial evidence exists for peripheral interactions between TH and corticosteroids in various aspects of tadpole development (Hayes, 1997). In *Bufo boreas,* TH and CORT appear to synergistically mediate tail resorption and digit development and antagonistically mediate hindlimb growth, forelimb emergence (specifically, degeneration of the opercular skin), and skin gland development (Hayes, 1995a,b; Hayes and Gill, 1995); whereas TH induces the latter three events, CORT inhibits them. Other corticosteroids also synergize with TH in some of these responses in *Rana pipiens* (Wright *et al.*, 1994), as well as in erythrocyte and skin remodeling in other frogs (Kikuyama *et al.*, 1993). Some interactions are clearly dose- and presumably stage-dependent. For example, the amount of synergism between corticosteroids and TH in *Rana* digit development increases with TH concentration (Wright *et al.*, 1994). Also, corticosteroids tend to retard limb growth at low TH levels similar to those in effect in early larval development and accelerate it at high TH levels similar to those in effect at climax metamorphosis (Wright *et al.*, 1994). However, because some corticosteroids up-regulate T_3 and possibly inhibit thyroid activity (Hayes and Wu, 1995a; Wright *et al.*, 1994), some appearances of synergism or antagonism at the hormone–tissue interface simply may reflect an augmentation or diminution of TH activity.

The possibility also exists that corticosteroids mediate some tissue responses independently of TH. When applied alone *in vivo*, CORT induces regression of the tadpole thymus, spleen, brain, and pituitary in *Bufo boreas* and the discoglossid *Bombina orientalis* (Hayes, 1995b). Again, caveats apply to any conclusion of independent hormone action on tissues that is based solely on whole-tadpole inductions. When applied alone in implants (Kaltenbach, 1958) or *in vitro* (Kikuyama *et al.*, 1983), corticosteroids usually do not induce resorption in tail tissue [but see Iwamuro and Tata (1995)]. Also, although the CORT-induced responses in spleen and thymus resemble natural metamorphic changes in these tissues, it is still unclear whether these responses actually are TH-independent. The responses induced in brain and pituitary do not parallel natural development and probably are experimental artifacts.

Considerably less is known about hormone interactions in salamanders. Evidence for non-TH mediation comes from observations that some responses that are accelerated by exogenous TH [the appearance of skin glands and the switch from ammonia to urea production in the hynobiid *Hynobius retardatus* (Wakahara and Yamaguchi, 1996) and the appearance of adult myosin in skeletal muscle of the salamandrid *Pleurodeles walt1* (Chanoine *et al.*, 1990)] still proceed, albeit at a slower rate, when endogenous TH activity is blocked. Although incomplete blocking of TH activity has not been ruled out, additional explanations are possible. One is that such tissue responses are mediated by hormones, such as corticosteroids, that normally are up-regulated by TH, but are still produced at low levels in the absence of TH; this appears to be the case for the switch from ammonia to urea production (Lamers *et al.*, 1978). Another possibility is the presence of facilitating mediation, i.e., mediators that are produced independently of TH and that normally operate in addition to or in synergy with TH but can still proceed in the absence of TH.

A more convincing case for non-TH mediation can be argued on the basis of discrepancies in the preventability of a tissue response when the thyroid axis is blocked at different levels. Whereas pituitary removal in *Hynobius retardatus* prevents the hemoglobin transition from occurring, thyroid gland removal does not (Satoh and Wakahara, 1997). Exogenous TH does not accelerate the transition in this species (Wakahara *et al.*, 1994), nor does thyroid gland inhibition retard it (Wakahara and Yamaguchi, 1996). Hence, one probably can exclude mediation by TH, TH-regulated hormones, and facilitating factors. The most likely explanation is that the hemoglobin transition in this particular species is mediated by a factor or hormone whose production and activity are under pituitary control and are completely independent of TH.

Altogether, these observations suggest that the assumptions of a developmentally predominant and phylogenetically uniform role of TH (i.e., T_3) mediation in amphibian larval development need to be reevaluated. The possibilities of differential mediation by T_4 and additional mediation by TH-dependent and -independent factors deserve more investigation with *in vitro* techniques. More comparative analysis also is required to define the morphological and phylogenetic bounds of TH-mediated development, especially in salamanders.

2. How Are TH-Mediated Responses Activated?

At first glance, activation of TH-mediated tissue responses would appear to be a relatively simple process. When sufficient TH is supplied to a responding tissue, the tissue initiates a preprogrammed morphogenetic response. At a mechanistic level, however, tissue activation poses a difficult problem to investigate. Historically, it has been conceptualized as the physical manifestation of several empirically measurable, interrelated properties: tissue competence (ability to respond to TH as a function of developmental stage), tissue sensitivity (ability to respond as a function of TH concentration), and tissue specificity (ability to effect the same response as in natural development). Such properties tell us something about when and how larval tissues can respond to TH, but not necessarily about what regulates them in normal development. Measured differences in competence and sensitivity often do not agree with observed differences in developmental timing (cf. Heady and Kollros, 1964; Kollros and McMurray, 1956; Tata, 1968; Hanken and Summers, 1988), and tissue specificity can be altered in induced development (e.g., Kollros, 1961; Rose, 1995b).

Newer research addresses tissue activation at a finer level of resolution by examining the dynamic interaction between TH activity and TH-mediated changes in cell differentiation and gene expression. Research on *Xenopus* skin differentiation indicates that adult keratin genes are activated in two steps (Miller, 1996). A low basal level of expression that is initiated in early premetamorphosis is followed by a high constitutive level at late prometamorphosis (Mathisen and Miller, 1987). Whereas the stage-specific onset of keratin gene expression is controlled by factors other than TH, the switch to high expression appears to be TH-mediated; it is preceded by both the onset of its TH responsiveness and the rise in endogenous TH. The switch also is stable as it persists through numerous cell divisions and is unaffected by the subsequent decline in TH after climax (Mathisen and Miller, 1989).

A similar multistage model is proposed for the regional differentiation of *Rana* skin (Kawai *et al.,* 1994; Yoshizato, 1996). The early premetamorphic differentiation of apical and skein cells in the epidermis is followed at later premetamorphic stages by formation of the stratum spongiosum in the dermis and the differentiation of basal cells in body (but not tail) epidermis. At climax, body epidermis undergoes a combination of adult cell differentiation and larval cell apoptosis, whereas tail epidermis exhibits wholesale apoptosis (Izutsu *et al.,* 1993). The first phase is TH-independent, the second is mediated by low TH, and the third is mediated by high TH (Kawai *et al.,* 1994).

These studies indicate that early steps in skin cell differentiation originally are TH-independent, but after they acquire TH responsiveness, the process becomes increasingly TH-dependent. In the case of skin keratinization, the "activation" of the tissue response is simply the augmentation by TH of a preexisting cell activity. In the case of skin cell differentiation, "activation" involves a series of cell differentiation events, each of which is predetermined by inductive interactions at earlier stages (see the following discussion). How such events or activities acquire TH dependence remains unclear. How they increase in TH dependence, however, can be explained by the specific properties of TH receptors.

3. TH receptors, Autoinduction, and Cross-induction

The TH receptor (TR) belongs to a superfamily of nuclear receptors that includes receptors for glucocorticoids, androgen, estrogen, and retinoic acids. Like all members of this family, TR is a ligand-dependent transcription factor. A simple description of its function is as follows. Upon binding with its ligand (TH), TR undergoes a conformational change that affects its binding to TH response elements (DNA promoter regions that are specific to TR). The result is a change (either positive or negative) in the transcription rate of genes that are regulated by those promoters. For a more complete description, the reader is referred to Shi (1996) and Shi *et al.* (1996b).

The two forms of TR to be identified and cloned in frogs, TRα and TRβ, show different patterns of expression in *Xenopus* tadpole development (Yaoita *et al.,* 1990). Both appear as mRNA and protein soon after hatching. Whereas TRα protein increases in premetamorphosis and stays high throughout prometamorphosis and climax, TRβ protein rises in parallel with TH during prometamorphosis and peaks at climax at a higher level than TRα protein (Eliceiri and Brown, 1994; Yaoita and Brown, 1990). The patterns of mRNA expression are similar to the respective protein patterns, except that TRα mRNA increases during prometamorphosis instead of remaining constant. TRβ mRNA and protein, and to a lesser extent TRα mRNA, are directly up-regulated by TH (Eliceiri and Brown, 1994; Kanamori and Brown, 1992); this phenomenon where a receptor is up-regulated by its own ligand has been termed autoinduction or autoregulation (Tata, 1994). TRα and TRβ also are differentially expressed in different body regions (Eliceiri and Brown, 1994), but quantification of tissue-specific differences remains difficult (Kawahara et al., 1991). TRβ protein shows distinct localization in some tissues, for example, the cartilage-forming cells of hindlimb buds (Fairclough and Tata, 1997).

The significance of these events to metamorphic regulation still is not entirely clear. It is clear that the up-regulation of TRβ protein by TH occurs ubiquitously and is an essential first step in activating at least some tissue responses (Baker and Tata, 1992). This process also presumably would increase the rate of removal of free TH by TR binding and allow cells the capacity to translate rapid influxes of TH into rapid increases in TH-dependent gene regulation. Further, the ontogenetic stage at which a tissue becomes TH-responsive may be defined by the onset of its capacity for TRβ autoinduction. However, TRα shows a stronger correlation than TRβ between its mRNA expression and the onset of remodeling in at least two organs in Xenopus, the hindlimb (Shi et al., 1996b) and tail (Wang and Brown, 1993). This may imply that the less TH-dependent form of TR actually plays a bigger role in the timing of tissue responses. Also, Xenopus TRβ (but not TRα) mRNA is spliced into different isoforms (Yaoita et al., 1990), which suggests that this form of TR is more likely to be involved in the specification of tissue responses. The former claim is refuted by the expression pattern of TRβ in intestinal tissues (Shi and Ishizuya-Oka, 1997). The latter claim is refuted by the finding that some Rana tissues (e.g., red blood cells) express only TRα (Schneider et al., 1993), and both claims still lack direct evidence in amphibians. In light of the apparent role of 5'-D activity in activating individual tissues (see previous discussion) it would be interesting to know which (if any) form of TR is involved in regulating the gene of this enzyme.

In vitro analyses of tadpole tail remodeling in Xenopus, Bufo, and Rana indicate that TR expression and activity in peripheral tissues are regulated additionally by interactions among TH, PRL, and corticosteroids. Whereas PRL inhibits TR autoinduction, thereby preventing TH-mediated tissue resorption [Baker and Tata, 1992; Tata et al., 1991; references in Norris and Dent (1989)], corticosteroids appear to enhance TR autoinduction (Iwamuro and Tata, 1995), thereby increasing TH binding (Kikuyama et al., 1993) and accelerating TH-mediated tissue resorption (Kaltenbach, 1996; Kikuyama et al., 1993).

One possible mechanism for these interactions is cross-induction or cross-regulation, the regulation of a receptor by a hormone other than its own (Tata, 1994). As in other vertebrates, Xenopus TR appears to regulate genes as a heterodimer (a two-protein complex) of TR and RXR, the receptor for 9-cis-retinoic acid (Wong and Shi, 1995). Though T_3 up-regulates TR, it appears to down-regulate RXR mRNA (Iwamuro and Tata, 1995). However, when corticosteroid or prolactin is combined with T_3, both appear to counteract this down-regulation, and corticosteroid further enhances RXR mRNA expression, at least temporarily (Iwamuro and Tata, 1995). Thus, corticosteroid may act to increase TH binding by increasing the amount of RXR available for making functional TR/RXR dimers. However, the precise molecular interaction(s) by which corticosteroid and PRL affect TR and RXR expression awaits further investigation.

The relationship between the receptor-level interactions of TH, corticosteroids, and PRL and their tissue- and stage-specific interactions in peripheral remodeling also has not been fully addressed. Some receptor-level interactions may be tissue-specific, as TR autoinduction is inhibited, rather than enhanced, by corticosteroids in *Rana* red blood cells (Schneider and Galton, 1995). PRL also inhibits TH-mediated resorption of the gills and tail fin in the ambystomatid salamander *Ambystoma tigrinum* (Platt and Christopher, 1977; Platt *et al.*, 1978), but it is not known whether this effect involves TR expression. How CORT–TR interactions correlate with the stage-specificity of the CORT–TH interactions observed in tail resorption and other remodeling also requires investigation. Although PRL and CORT have additional osmoregulatory roles in metamorphosis that clearly are tissue- and stage-specific (Kikuyama *et al.*, 1993; Norris and Dent, 1989), the extent to which these hormones regulate their own genes in concert with TR/RXR-regulated genes will not be known until their receptors have been identified and cloned. In view of the strong interdependence of the neuroendocrine axes that coordinate the production of TH, PRL, and CORT, it would appear that the receptor-level interactions of these hormones mainly serve to fine-tune the rate of peripheral remodeling. One possibility is that they provide a braking mechanism to compensate for irregularities in hormone production.

4. How Are TH-Mediated Responses Specified?

TH–TR/RXR binding in peripheral tissues is presumed to initiate a cascade of directly and indirectly regulated genes for transcription factors, enzymes, signaling molecules and structural proteins. Such programs, which have begun to be characterized in *Xenopus* for tail resorption (Brown *et al.*, 1996), hindlimb differentiation (Buckbinder and Brown, 1992), brain (Denver *et al.*, 1997) and intestinal remodeling (Stolow *et al.*, 1997), are only partially specific to the tissues and morphogenetic responses involved. They are further presumed to be preprogrammed, meaning that they are already programmed into the responding cells by the time TH arrives to activate them. As with all other examples of cell fate specification, there are two potential mechanisms involved: cell-autonomous processes and tissue interactions.

In vitro analyses of tadpole skin and gut remodeling offer strong evidence for tissue interaction involvement in these response specifications. Reciprocal transplants in *Rana* indicate that the disparate TH responses of body and tail epidermis that arise during the second and third phases of skin differentiation (see previous discussion) are specified by inducing signals from subepidermal mesenchyme during the first TH-independent phase [Yoshizato (1996) and references therein]. Similar studies on the tadpole small intestine in *Xenopus* indicate that the TH responses of both epidermis and the adjacent connective tissue are specified by reciprocal tissue inductions (Ishizuya-Oka and Shimozawa, 1992, 1994). There is some evidence from *in vitro* studies on *Rana* for region-specific signaling factors that may diffuse from one tissue layer into the adjacent one (Niki *et al.*, 1982, 1984) to activate or trigger the production of response-specifying transcription factors (Yoshizato, 1996). Although the players in this scenario remain unidentified, *in situ* hybridization studies on *Xenopus* intestinal remodeling suggest that the TH-regulated matrix metalloproteinase stromelysin-3 plays a role in mediating changes in the basement membrane of gut epithelium (Ishizuya-Oka *et al.*, 1996; Patterton *et al.*, 1995). Although these changes have been suggested to induce both the cell death occurring in the larval gut epithelium and the cell proliferation required to form the adult gut epithelium, it is clear that additional TH-independent specification factors must be involved.

III. THE ROLE OF TH MEDIATION IN THE EVOLUTION OF AMPHIBIAN LARVAE

For the purposes of this discussion, I have abstracted the variation in TH-mediated development into two broad, but not mutually exclusive, categories: **heterochronic variation** (or dissociation) and **repatterning.** Heterochronic variation covers differences in the timing and extent of developmental changes that are primitively TH-dependent and that are otherwise conserved among species. Repatterning covers differences in morphogenetic specification that affect the number, arrangement, or identity of primitively TH-mediated tissues. [This use of repatterning is distinct from Roth and Wake's (1985) ontogenetic repatterning, which refers to any pattern variation that has evolved in response to heterochronic changes.] In general, heterochronic variation may involve tissues that form or remodel at any point in development, whereas repatterning involves primarily stages of pattern formation, i.e., embryogenesis and conceivably metamorphosis. Although TH has diverse functions in vertebrate growth and development (Gorbman *et al.*, 1983), this hormone is not known to play any role in the patterning of embryonic events [but see Flamant and Samarut

(1998)]. Hence, this discussion pays attention to repatterning only as it relates to differences in TH-mediated larval development and morphology.

Also, as implied earlier, phases of larval and metamorphic development in amphibians are not mechanistically distinct and their demarcation usually is based on the timing of somewhat arbitrary and potentially variable morphological changes. Given the chapter's emphasis on larvae, the discussion of metamorphic remodeling is limited to developmental and evolutionary events that contribute to the diversification of larval morphologies.

A. Metamorphosing Salamanders

1. Patterns of Diversification

Oviparous metamorphosing salamanders, which are found in six of the ten salamander families (plethodontids, salamandrids, ambystomatids, hynobiids, rhyacotritonids, and dicamptodontids), exhibit a fairly conservative pattern of larval development [references in Rose (1996)]. Whereas the larval external morphology, including gills, limbs, tail fin, lateral line organs, and larval pigmentation, usually is fully developed at or soon after hatching, the internal morphology of the head undergoes a more prolonged period of development. Just prior to hatching, most cartilaginous parts of the skull, including the walls and floor of the braincase, the ear capsule, and the jaw and throat cartilages, begin to condense and chondrify. Around hatching, there is a burst of ossification, giving rise to all teeth-bearing bones and additional dermal bones in the braincase and jaw suspension. Larval development is marked by the completion of preformed parts of the chondrocranium, appearance of pedicellate teeth, addition of middle ear cartilages and skeletal capsules around the nasal sac and eye (though the latter capsule often is only a cartilage ring), formation of cartilage replacement bones in the ear capsule, middle ear, jaws, and wall of the braincase, and formation of mucous and serous glands and nasolacrimal duct in the skin. Metamorphosis, which terminates larval development, is marked by profound reorganizations of the palate (Regal, 1966) and hyobranchium (Alberch, 1987; Alberch and Gale, 1986), the transition from monocuspid to bicuspid teeth (Greven, 1989), and the resorption of gills, tail fin, and Leydig cells [reviewed in Rose (1996)].

Interestingly, metamorphosing salamanders show a pronounced discrepancy in the timing of most larval changes (Fig. 2a). On one hand, metamorphosing salamandrids, ambystomatids, hynobiids, rhyacotritonids, and dicamptodontids all exhibit a gradual and progressive elaboration of the larval skull and skin [Figs. 3a,b; references in Rose (1996)]. The roof of the nasal capsule generally appears at early posthatching stages, followed by the walls and floor

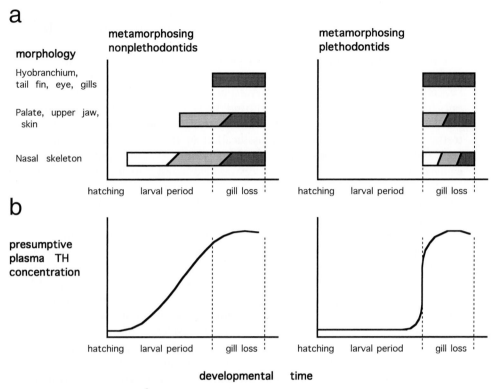

FIGURE 2   Remodeling profiles and presumptive TH profiles for different kinds of metamorphosing salamanders. Panel a shows the timing of postembryonic remodeling in different morphological regions for metamorphosing nonplethodontid and plethodontid salamanders. Whereas metamorphosing nonplethodontids exhibit progressive remodeling in some regions throughout the larval period, metamorphosing plethodontids exhibit all postembryonic remodeling in these regions, including events with high TH sensitivity, synchronously with gill loss. Stipple density indicates TH sensitivity of tissue responses as measured by induction experiments on the plethodontid *Eurycea bislineata* (Rose, 1995c): high density = low sensitivity, unstippled = high sensitivity; see Fig. 5 for a listing of individual responses and their TH sensitivities. Panel b shows the TH profiles hypothesized for metamorphosing nonplethodontid and plethodontid salamanders based on differences in the timing of their postembryonic remodeling (a) and in the dose dependency of their induced remodeling (see text).

and adjacent dermal bones (prefrontal, nasal, extension of the frontal bone, lacrimal, and septomaxilla, though the latter two do not appear until gill resorption or later in some forms). Beneath the capsule, the major bone of the upper jaw (maxilla) appears at early (ambystomatids, rhyacotritonids, dicamptodontids) or late larval (salamandrids, hynobiids) stages. In the palate, the cartilaginous pterygoid process begins to chondrify at late larval stages. Skin glands and nasolacrimal duct generally appear midway through the larval period.

On the other hand, metamorphosing plethodontids develop almost all additions to the larval skin and skull simultaneously at the stage of gill resorption, i.e., effectively at the end of the larval period (Fig. 2a; Rose, 1995a). The only features to appear consistently during the larval period in hemidactyliine plethodontids (e.g., *Hemidactylium, Eurycea,* and *Gyrinophilus*) are pedicellate teeth, the nasolacrimal duct, and ossification of the middle ear, braincase wall, and urohyal cartilages. Although desmognathine plethodontids may have evolved from a direct-developing ancestor (Titus and Larson, 1996), those species with multiyear larval periods (e.g., *Desmognathus quadramaculatus*) appear to share this distinctly abrupt pattern of postembryonic development (C. S. Rose, unpublished observation).

This discrepancy between abrupt cranial development in plethodontids and gradual cranial development in nonplethodontids has significant consequences for larval morphology (Fig. 3). In comparison with other larvae (Figs. 3a,b), plethodontid larvae retain a "stripped down," almost embryonic, skull morphology throughout their existence (Figs. 3c,d; Rose, 1995a). Even the largest plethodontid larvae *(Gyrinophilus porphyriticus)*, which surpass in size many, if not most, nonplethodontid larvae, lack a nasal skeleton, pterygoid process, maxilla, and skin glands at their ultimate state of development (Fig. 3c). Whether the absence of these structures is functionally significant or simply a consequence of shifted development remains open. The evolutionary origin of the developmental shift is obscured by the fact that the most probable sister group of plethodontids, the amphiumids (Larson and Dimmick, 1993), not only are larval reproducers but also have a highly aberrant cranial development (see the following discussion).

2. Mechanisms of Diversification

The discrepancy in TH-mediated development between metamorphosing plethodontids and nonplethodontids is consistent with two different patterns of larval TH activity, an abruptly increasing one in plethodontids and a gradually increasing one in nonplethodontids (Fig. 2b). This hypothesis can be tested directly by comparing radioimmunoassays (RIA) of larval hormone levels. Variably complete profiles of plasma TH are available for six salamander species representing four families: the plethodontid *Eurycea bislineata* (Alberch *et al.,* 1986), the salamandrid *Pleurodeles waltl* (Chanoine *et al.,* 1987), the ambystomatids *Ambystoma gracile* (Eagleson and McKeown, 1978a), *A. mexicanum* (Chanoine *et al.,* 1987; Rosenkilde *et al.,* 1982), and *A. tigrinum* (Larras-Regard *et al.,* 1981; Norman *et al.,* 1987), and the hynobiid *Hynobius nigrescens*

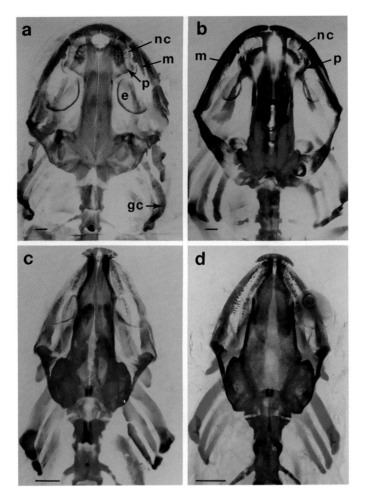

FIGURE 3   Stained and cleared whole-mounts of the larval skulls (dorsal view) of salamanders from predominantly metamorphosing families. (a) Late larval specimen of the dicamptodontid *Dicamptodon ensatus* (54 mm, snout–vent length); (b) late larval specimen of the ambystomatid *Ambystoma mexicanum* (47 mm); (c) late larval specimen of the plethodontid *Gyrinophilus porphyriticus* (48 mm); (d) adult specimen of the plethodontid *Eurycea pterophila* (34 mm). Whereas *D. ensatus* and *G. porphyriticus* larvae undergo complete metamorphic remodeling at gill loss, *A. mexicanum* and *E. pterophila* are larval reproducers and retain the late larval cranial morphology shown throughout life. The nonplethodontid larvae (*D. ensatus* and *A. mexicanum*) exhibit marked development of the upper jaw bone (maxilla), nasal capsule cartilage, and nasal roofing bone (prefrontal); both also will acquire a nasal bone at later larval stages. In contrast, the plethodontid larvae (*G. porphyriticus* and *E. pterophila*) show no trace of maxilla, nasal cartilage, or nasal roofing bone development. Abbreviations: e, eye marked by scleral cartilage (missing on left side in d); gc, gill cartilage; m, maxilla bone; nc, nasal capsule; p, prefrontal bone; blue stain = cartilage, red stain = bone. Scale bar = 1 mm. *D. ensatus*, gift of John Reiss; *G. porphyriticus* from Rose (1955a), *A. mexicanum* raised in lab by author; *E. pterophila*, Texas Natural History Museum Collection 52096 [described as *E. nana* in Rose (1995a)].

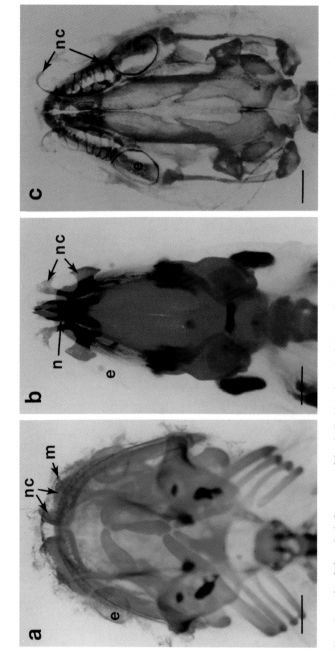

FIGURE 4 Stained and cleared whole-mounts of the skulls (dorsal view) of salamanders from exclusively larval reproducing families. (a) Early larval specimen of the cryptobranchid *Cryptobranchus alleganiensis* (22 mm, snout–vent length); (b) adult specimen of the sirenid *Pseudobranchus axanthus* (145 mm, total length); (c) late larval specimen of the proteid *Necturus maculosus* (44 mm, snout–vent length). Whereas *C. alleganiensis* larvae undergo some metamorphic remodeling at gill loss, *P. axanthus* exhibits minor traces of metamorphic remodeling and *N. maculosus* almost none. The larvae of all three taxa develop a variably modified cartilaginous nasal capsule, *C. alleganiensis* and *P. axanthus* larvae additionally develop a nasal bone (not present in *C. alleganiensis* at the stage shown), and *C. alleganiensis* larvae also develop maxilla and prefrontal bones, only the rudiments of which are detectable at the stage shown. Abbreviations: e, eye marked by scleral cartilage (missing in b); m, maxilla bone; nc, nasal bone; nc, nasal capsule; blue stain = cartilage, red stain = bone, hyobranchial skeleton removed in c. Scale bar = 1 mm. *C. alleganiensis*, gift of John Reiss; *P. axanthus*, Harvard Museum of Comparative Zoology 107677; *N. maculosus* collected by author.

(Suzuki, 1987); incomplete profiles of plasma CORT and pituitary PRL also are available for *A. tigrinum* (Carr and Norris, 1988; Norris *et al.*, 1973). The one TH profile available for plethodontids indicates that hormone levels begin to rise at a late larval stage immediately preceding the onset of gill resorption. In contrast, the TH profiles of the ambystomatid and salamandrid species all show TH increases beginning at early or midlarval stages (the profile for *H. nigrescens* does not cover larval stages). Also, the few studies of TSH expression available indicate early and midlarval stages of onset in hynobiid and ambystomatid salamanders, respectively (Eagleson and McKeown, 1978b; Yamaguchi *et al.*, 1996; Yamashita *et al.*, 1991). These observations all are generally consistent with the two profiles of TH activity proposed earlier.

However, comparisons of TH profiles are confounded by unexpectedly high intraspecific variation both between studies (cf. Rosenkilde *et al.*, 1982; Chanoine *et al.*, 1987; Larras-Regard *et al.*, 1981; Norman *et al.*, 1987) and within studies, i.e., among individuals at the same developmental stage (Alberch *et al.*, 1986; Larras-Regard *et al.*, 1981). Whereas the variation between studies can be attributed to the use of different techniques, the variation within may reflect a number of sources, including irregular hormone production and true intrastage variability (Alberch *et al.*, 1986). Though intrastage variability makes it difficult to discern, let alone compare, the developmentally significant aspects of TH profiles, comparative assays that use the same sampling and RIA techniques may help to clarify interspecific differences in their overall shape and variability.

More circumstantial evidence for a difference in TH profile between plethodontids and nonplethodontids comes from analyses of TH induction (Rose, 1996). The earliest posthatching changes in the larval ambystomatid *Ambystoma mexicanum*, appearance of the nasal capsule and maxilla, are inducible across a range of TH doses, and their induced development generally agrees with natural development. These observations, plus the presence of detectable levels of TH in larvae that are naturally undergoing these changes (Rosenkilde *et al.*, 1982), would suggest that the early larval development of this species is mediated by moderate concentrations of TH. In comparison, larvae of the plethodontid *Eurycea bislineata* exhibit distinctly different patterns of induced remodeling at low and high TH (Rose, 1995c). The remodeling induced in features that normally form midway through the 2–3 year larval period (nasolacrimal duct and endochondral bones) agrees in sequence and extent with natural remodeling only when it is induced by very low TH (0.05 nM T_4). Conversely, the remodeling induced in features that normally remodel at the stage of gill resorption agrees in sequence, extent, and rate with natural remodeling only when induced at high TH (5 nM T_4). These observations would suggest that the larval development of this species is either TH-independent or mediated by negligibly low concentrations of endogenous TH (see Fig. 2). In contrast, its metamorphic remodeling appears to be mediated by as much as a 100-fold higher concentration of TH. That some metamorphic remodeling is induced precociously at the low TH dose would further suggest that the transition from low to high TH in larvae of this species occurs quickly, just prior to the onset of gill resorption.

Until more comparative RIA data are available, however, differential TH activity remains only one of many possible mechanistic explanations for the disparate larval ontogenies of plethodontids and nonplethodontids. Unfortunately, the paucity of comparative data on hormone production, hormone interaction, and receptor regulation in metamorphosing salamanders precludes discussion of other hypotheses. Additional insights likely will come from comparisons of the onset and inducibility of hypothalamic maturation, TSH expression, and TR expression in plethodontid and nonplethodontid larvae.

B. Larval Reproducing Salamanders

1. Patterns of Diversification

Larval reproduction is a developmental mode marked by the reduction or elimination of metamorphic remodeling and the retention of larval features (usually including gills) through adulthood. The phenomenon has evolved multiple times in plethodontids, salamandrids, ambystomatids, and dicamptodontids and at least once in hynobiids. Though usually manifest as intraspecific or intrapopulational variation, it is a species-wide trait in some ambystomatids, plethodontids, and dicamptodontids. It also is the exclusive mode of development in the four other salamander families: cryptobranchids, sirenids, proteids, and amphiumids.

Plethodontid larval reproducers are distinguished by a complete absence of metamorphic remodeling (Rose, 1995a). None experience any of the changes that occur synchronously with gill resorption in metamorphosing plethodontids. Hence, larval reproduction here appears to involve a complete developmental truncation of the ancestral developmental pattern. In epigean and troglobytic forms such as *Eurycea pterophila* and *Gyrinophilus palleucus,* the midlarval morphology is simply retained in its unmodified, "stripped-down" condition into adult stages (Fig. 3d). Specialized subterranean forms such as *Typhlomolge* and *Haideotriton* show additional shape modifications to the skull and body that result from allometric growth throughout larval stages.

In contrast to plethodontid forms, salamandrid and ambystomatid larval reproducers undergo variable amounts of larval and metamorphic remodeling. Studies of adult morphology indicate that the salamandrids *Notopthalmus viridescens* and *Triturus vulgaris* complete most skull remodeling, and at least the former species completes a variable amount of hyobranchial remodeling (Marconi and Simonetta, 1988; Noble, 1929; Reilly, 1986, 1987). Among ambystomatids, *Ambystoma mexicanum* undergoes no skeletal remodeling beyond a late larval stage (Keller, 1946), but partially completes the transition from monocuspid to bicuspid teeth (Chibon, 1995), *A. talpoideum* undergoes minor additional skull remodeling (resorption of a palatal bone, variable appearance of the nasal bone; Reilly, 1987), and *A. altamirani* completes most skull and some hyobranchial remodeling (Reilly and Brandon, 1994). *A. mexicanum* also completes the shift from larval to adult hemoglobin and serum protein—at a stage comparable to gill resorption in metamorphosing congenerics (Ducibella, 1974)—but fails to complete shifts in immune cell type and function (Ussing and Rosenkilde, 1995) and in the myosin and fiber types of dorsal skeletal muscles (Chanoine *et al.*, 1987; Salles-Mourlan *et al.*, 1994). *A. tigrinum* and *T. helveticus* fail to complete the shifts in myosin (Chanoine *et al.*, 1990), hemoglobin (Cardellini *et al.*, 1978), and dentition (Chibon, 1995) that are completed in metamorphosing conspecifics. It should be noted, however, that certain metamorphosing species also fail to complete some aspects of the biochemical (e.g., Woody and Justus, 1974) and morphological remodeling (e.g., Good and Wake, 1992) exhibited by congenerics.

Because none of the cryptobranchids, sirenids, proteids, and amphiumids appear to be related as sister groups (Larson and Dimmick, 1993), larval reproduction in these forms probably evolved separately and from different ancestral metamorphic ontogenies. Despite this, cryptobranchids (the giant salamanders *Andrias* and *Cryptobranchus*), sirenids (*Siren* and *Pseudobranchus*), and proteids (the mudpuppy *Necturus* and the olm *Proteus*) share some developmental similarities with metamorphosing salamanders [Fig. 4; Rose (1996) and references therein]. The hatchlings of all genera show approximately the same stage of chondro- and osteocranial development. Posthatching development in *Necturus* is limited to larval-stage chondrification of the nasal capsule (Fig. 4c) and skin gland formation. *Siren* and *Pseudobranchus* larvae exhibit additional traces of a metamorphic pattern, including ossification of the nasal bone (Fig. 4b) and the subsequent resorption of a palatal bone. In *Siren* larvae, these events are followed by Leydig cell resorption and ossification of a much reduced maxilla, in that order. In addition to the remodeling exhibited by *Siren* larvae, *Andrias* and *Cryptobranchus* larvae show full development of prefrontal and maxilla bones (Fig. 4a) and undergo palatal remodeling, followed by gill resorption at 2 years of age. *Andrias* larvae also lose some gill arches and close all gill slits in synchrony with gill resorption. Cryptobranchid larvae (Greven and Clemen, 1980; Kerr, 1959), but not sirenid (Clemen and Greven, 1988) or proteid (Greven and Clemen, 1979) larvae, undergo a transition from monocuspid to bicuspid teeth. Hence, like other larval reproducers, cryptobranchids, sirenids, and proteids still evince, albeit to decreasing degrees, aspects of a larval developmental pattern that are shared by most metamorphosing salamanders. Amphiumids defy this characterization by forming most adult skeletal elements and undergoing gill resorption, without gill arch reduction, during or soon after embryogeny (Davidson, 1895; Hay, 1890; Kingsley, 1892; Ultsch and Arceneaux, 1988).

In summary, the general effect of larval reproduction on salamander development is to attenuate, rather than truncate, the ancestral development pattern. With the probable exception of plethodontid forms (whose nonskeletal TH-mediated remodeling has yet to be investigated), all larval reproducers retain some larval and metamorphic remodeling in common with metamorphosing relatives. However, their adult morphologies represent mosaics of larval, mid-transitional, and adult characters that are unlike any of the developmental stages found in metamorphosing relatives. The adults of certain species eventually may complete their metamorphic remodeling, but this has never been demonstrated conclusively with longitudinal data. Plethodontids offer the closest approximation to a global truncation of development. Larval reproducers in this family apparently are precluded from advancing beyond a midlarval morphology by the distinctly abrupt pattern of their ancestral metamorphic ontogenies.

2. Mechanisms of Diversification

Morphologically speaking, larval reproducing salamanders cover a broad spectrum of developmental patterns and larval forms. Mechanistically speaking, however, there appear to be only three basic categories of larval reproduction: (1) forms with fully inducible remodeling that do not naturally metamorphose; (2) inducible forms whose frequency and age of metamorphosis vary with environment (the so-called "facultative paedomorphs"); and (3) forms with largely noninducible remodeling. The first two categories occur in the six predominantly metamorphosing families. The last includes all members of the four remaining families (cryptobranchids, sirenids, proteids, and amphiumids), plus two highly specialized plethodontids, *Typhlomolge* and *Haideotriton*.

The criterion of inducibility has perpetuated the view that larval reproducers of the first category evolve by a drop in thyroid activity and those of the third by a global loss of tissue sensitivity. However, the evolutionary origin of larval reproduction cannot be determined on the basis of this criterion alone, as there is no way of resolving whether the evolutionary loss of tissue sensitivity is not preceded and possibly facilitated by a drop in thyroid activity (see Larsen, 1968). Moreover, the dichotomy of decreased thyroid activity versus diminished tissue sensitivity does not begin to explain the vast amount of dissociation that is exhibited by both metamorphosing and larval reproducing salamanders.

For larval reproducers in the first two categories, the attenuated pattern of the retained remodeling and complete inducibility of the absent remodeling are consistent with a mid-to-late larval drop in TH activity. This is substantiated by two of the three TH profiles available for these forms. Profiles for *Ambystoma mexicanum* (Rosenkilde *et al.*, 1982) and a predominantly larval reproducing population of *A. gracile* (Eagleson and McKeown, 1978a) indicate that plasma TH starts to rise at early and midlarval stages, respectively, but drops to low levels at stages equivalent to gill resorption, i.e., after hindlimb development and before sexual maturation, in metamorphosing congenerics or conspecifics.

It is difficult to explain the source of this ontogenetic drop in TH activity. A summary of the voluminous research on environmental input to the thyroid axis in larval reproducers of the second category is beyond the scope of this chapter. Also, most mechanistic research on forms of the first category has dealt with adult stages of the Mexican and American axolotls, *Ambystoma mexicanum* and *A. tigrinum*. It is debatable whether the absence of morphological change between larval and adult stages in these forms justifies the assumption of no regulatory changes between the stages. Some regulatory mechanisms clearly do change as remodeling becomes less inducible and induced animals become less viable with increasing age of treatment (Rosenkilde and Ussing, 1996; C. S. Rose, unpublished observation). This chapter accordingly focuses on studies that explicitly address TH regulation in *larval*-stage animals.

Direct application of TH to the hypothalamus induces metamorphosis in both larvae and larval reproducers of *A. tigrinum*, which suggests that hypothalamic maturation in the latter is blocked in larval development (Norris and Gern, 1976). Similar results have been obtained for *A. mexicanum* (Rosenkilde and Ussing, 1996). Whether this block is due to insufficient TH at the appropriate larval stage is unclear. *A. mexicanum* larvae at stages equivalent to gill resorption appear to have low 5'-D activity (which converts T_4 to T_3) and no 5'-D activity (which inactivates T_4 and T_3) (Galton, 1992). Also, 5'-D activity is not up-regulated by TH doses sufficient to induce complete morphological remodeling (Galton, 1992). This is contrary to tadpoles where T_4 indirectly up-regulates 5'-D activity through increasing corticosteroid levels (see previous discussion). Loss of the ability of TH to up-regulate 5'-D activity may cause a truncation of the T_3 surge required to complete hypothalamic maturation and induce metamorphic remodeling. However, the noninducibility of 5'-D activity equally could be regarded as an evolutionary consequence of a decline in TH activity effected by other mechanisms. Some aspects of muscle fiber development, which are TH-inducible in metamorphosing congenerics, are noninducible in *A. mexicanum* (Chanoine *et al.*, 1990), which suggests that this species already is starting to lose some of its tissue sensitivity in response to its diminished TH activity.

Members of the third category of larval reproducers generally are considered to have both low TH activity and widespread tissue insensitivity. However, most efforts to quantify TH activity and tissue sensitivity in these forms have dealt with adult or subadult stages [references in Baker and Stoudemayer (1951); Dundee, 1957; Gorbman, 1957; Kerkoff *et al.*, 1963; Kobayashi and Gorbman, 1962; Norris and Platt, 1973; Svob *et al.*, 1973]. To the author's knowledge, there are only five studies on inducibility in larval-stage animals. *Cryptobranchus* larvae immersed in solutions of desiccated thyroid powder underwent keratinization of the skin and resorption of Leydig cells, but no changes in the eyelids or hyobranchial apparatus (Noble and Farris, 1929). A *Siren* larva injected with T_4 exhibited the same changes in the skin, plus some regression of the gills and tailfin, but no changes in the skull or hyobranchial skeleton (Noble and Richards, 1930). *Pseudobranchus, Siren,* and *Necturus* larvae that were immersed in T_4 solutions or fed thyroid gland yielded no recognizable remodeling (Noble, 1924). *Proteus* larvae immersed in T_3 for 7 months underwent keratinization and changes in the myosin type of skeletal muscles; additional myosin and fiber type changes were induced in adults (Chanoine *et al.*, 1989). Immature specimens of *Haideotriton* immersed in T_4 exhibited regression of the tail fin, lateral line organs, and coronoid and palatal bones; partial gill resorption; and other minor skin remodeling (Dundee, 1962; Rose, 1996). However, the first three reports, which are published in abstract form only, do not indicate the morphological stage of treated specimens or controls at the start or end of treatment, and the fourth analysis does not address skeletal remodeling. Hence, it is possible that some TH-mediated changes already were complete at the start of some experiments or that some internal changes that were accelerated in some treatments had been overlooked. Because no evidence exists for excluding larval TH activity in these forms, the conclusion that all of their larval development is entirely TH-independent is premature.

Despite the prevalence of tissue insensitivity in these forms, there has been little investigation of the mechanisms involved. TR genes have been identified for both proteid genera, though the original work (Huynh *et al.*, 1996) appears to be erroneous (Safi *et al.*, 1997b). The sequenced portion of these genes shows strong similarity to the TR genes of other amphibians, and the few mutations found give no indication of deleterious effects on protein function (Safi *et al.*, 1997a). Whereas both TRα and TRβ genes are present in both proteids, only TRα has detectable expression in *Necturus* (Safi *et al.*, 1997a). This finding agrees with the earlier observation of low TR expression in *Necturus* (Galton, 1985). However, the latter study used adult specimens, whereas the former study does not specify stage(s) used.

Additional insight into the loss of tissue sensitivity in proteids comes from classical transplantation experiments. *Proteus* skin transplanted to the larvae of metamorphosing forms will undergo some remodeling, including Leydig cell resorption and keratinization (Reis, 1932). Conversely, *Necturus* skin grafted to metamorphosing forms does not undergo any changes (Noble and Richards, 1931). While the discrepancy may reflect a difference in either species or grafting technique, the *Proteus* results suggest that loss of tissue sensitivity may involve the loss or alteration of inducing signals akin to the ones presumed to convey response specification.

Given the complexity of regulatory mechanisms that have been implicated in tadpole metamorphosis, research on the cause of metamorphic failure in salamanders clearly is just beginning. Indeed, considerably more attention must be paid to preliminary questions such as the extent to which larval reproducers of the third category have abandoned TH mediation in larval development and whether neuroendocrine or peripheral regulatory mechanisms primarily are involved. Some insight into the latter question may be gleaned from a comparative examination of remodeling patterns. Despite having evolved their larval reproduction separately, cryptobranchids, sirenids, and proteids show a distinctly nested distribution of lost remodeling events (Figs. 4 and 5; Rose, 1996). Five of the six genera in question (*Proteus* is excluded for lack of data) have lost most of the metamorphic remodeling shared by metamorphosing salamanders (e.g., resorption of the coronoid bone and scleral cartilage, ossification of the septomaxilla bone, formation of tongue-supporting cartilages). *Necturus, Pseudobranchus, Siren,* and *Cryptobranchus* also have lost gill cartilage resorption; *Necturus, Pseudobranchus,* and *Siren* have lost the transition from monocuspid to bicuspid teeth, gill resorption, and the formation of a pterygoid process and prefrontal bone; *Necturus* and *Pseudobranchus* have lost Leydig cell resorption and maxilla ossification; and *Necturus* has lost formation of the nasal bone and resorption of a palatal bone. All of the events listed here are TH-inducible in metamorphosing forms. Moreover, the remodeling events that are more commonly retained in these larval reproducers (e.g., nasal capsule formation) tend to be ones that are highly sensitive to TH in metamorphosing forms; those that are more commonly lost (e.g., any involving cartilage resorption) tend to be less TH-sensitive in metamorphosing forms (Fig. 5; Rose, 1995c).

The nested distribution of lost remodeling in larval reproducers, combined with the inverse relationship between an event's frequency of loss in these forms and its TH sensitivity in metamorphosing forms, raises an intriguing evolutionary scenario (Rose, 1996). Cryptobranchid, sirenid, and proteid lineages appear to have evolved their present states of larval reproduction from similar ancestral metamorphic ontogenies. Their divergences appear to have involved a process of developmental attenuation marked by the selective loss of less sensitive remodeling events. Further, the different amounts of attenuation exhibited by different genera are consistent with two different mechanisms of

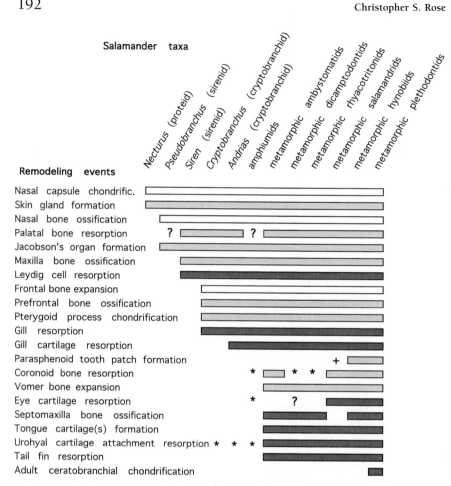

FIGURE 5 Distribution of postembryonic remodeling events across the taxa of larval reproducing and metamorphosing salamanders. Symbols: ?, insufficient data to confirm an event's occurrence; *, larval structure that is expected to undergo resorption does not form (although urohyal cartilages form in cryptobranchids, they arise separately from the rest of the hyobranchial skeleton and disappear well before gill loss); +, adult structure forms, but is not homologous with corresponding structures in other taxa. Stipple densities indicate TH sensitivity as in Fig. 2. Distribution data are based on Rose (1996) and references therein; TH sensitivity data are from Rose (1995c).

developmental change: (1) ontogenetic declines, of varying magnitudes, in TH activity (Fig. 6a) and (2) losses, of varying magnitudes, of tissue sensitivity (Fig. 6b). In the first hypothesis, less sensitive tissue responses selectively are lost because insufficient hormone is produced to activate them. In the second hypothesis, the less sensitive responses selectively are lost not because of a change in TH activity, but because these responses somehow are predisposed to lose their tissue sensitivity more readily than more sensitive responses.

The first hypothesis clearly is plausible based on what is known about first-category larval reproducers. Indeed, one implication of the first hypothesis is that first- and third-category larval reproducers evolve by similar mechanisms and probably represent different stages of the same evolutionary transition. Evaluation of the second hypothesis requires a deeper understanding of the mechanisms controlling tissue activation. For example, no mechanistic basis exists for identifying developmental factors or properties that may bias the likelihood of a loss in tissue sensitivity. How the requirement of a high TH concentration for tissue activation might translate into a predisposition to lose tissue sensitivity without an accompanying change in hormone levels is a mystery. Because cryptobranchid, sirenid, and proteid lineages would appear to have undergone only partial losses in tissue sensitivity, one might deduce that the mechanisms involved did not include any that are essential to tissue activation. On this basis, one may exclude any changes that would cause a functional inactivation of one or both TR proteins. This deduction is supported by the findings of little sequence divergence in proteid TR genes and recovered inducibility of *Proteus* tissues when transplanted to metamorphosing species (see previous discussion).

Direct tests of the two hypotheses are obfuscated by the strong probability that the original mechanistic changes responsible for the evolution of larval reproduction in cryptobranchids, sirenids, and proteids have been obscured by subsequent changes in TH activity and/or tissue sensitivity (Larsen, 1968; Martin and Gordon, 1995). However, cryptobranchids still retain a discrete larval period capped by gill resorption and some palatal and hyobranchial remodeling, which makes them prime candidates for searching for vestigial or even functional aspects of metamorphic regulation.

In addition to larval reproduction, salamanders exhibit direct development, which involves the reduction of larval morphology, coupled with an earlier onset (or predisplacement) of metamorphic remodeling, to produce a hatchling with an almost fully adult morphology. Though direct development (which shall be limited in this discussion to oviparous forms laying terrestrial eggs) has evolved only between two and five times in one family of salamanders (plethodontids), it is exhibited by over 50% of all salamander species and is tied to substantial repatterning of cranial morphology (Duellman, 1993; Wake and Marks, 1993; Wake and Roth, 1989). However, as little attention has been paid to the role of TH in direct development in salamanders, discussion of this phenomenon will be restricted to its occurrence in frogs.

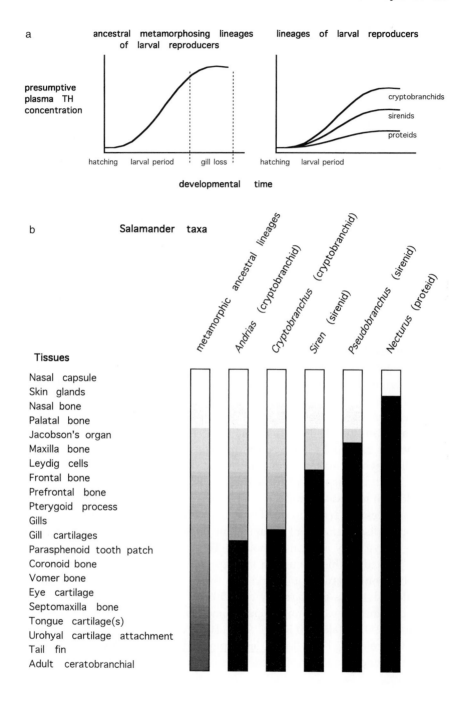

C. INDIRECT- AND DIRECT-DEVELOPING FROGS

1. Patterns of Diversification

Some members of almost all frog families still exhibit the primitive pattern of indirect development, producing a free-living tadpole that progresses through distinct pre-, pro-, and climax metamorphic phases of remodeling. Limb development in these forms generally is perceived as a conserved component of TH-mediated development and thus used as a metric for identifying heterochronic variation in other tissues. Although interspecific comparisons reveal frequent discrepancies in the timing of individual ossifications (e.g., Hanken and Hall, 1984) and aspects of chondrocranial remodeling (e.g., Rose and Reiss, 1993) and skin differentiation (e.g., Kawai et al., 1994), more systematic analysis clearly is needed to resolve the phylogenetic expression of this level of variation and its evolutionary history.

Frogs also have evolved some major deviations from their stereotypic tadpole-to-froglet development, the most dramatic being direct development. Though direct development has evolved at least 12 times in frogs and occurs in almost 20% of all species (Duellman, 1993; Duellman and Trueb, 1986; Ford and Cannatella, 1993), detailed studies of internal development are limited to a handful of genera: *Breviceps* (Swanepoel, 1970), *Leiopelma* (Stephenson, 1951, 1955), *Arthroleptella* (de Villiers, 1929), and *Eleutherodactylus* (Hanken et al., 1992; Lynn, 1942; Schlosser and Roth, 1997). Modern mechanistic

FIGURE 6 Developmental hypotheses for the evolution of larval reproduction in cryptobranchid, sirenid, and proteid salamanders. Panels a and b (hypotheses 1 and 2) present alternative explanations for how *Necturus, Pseudobranchus, Siren, Cryptobranchus,* and *Andrias* lineages may have undergone losses of TH-mediated remodeling events to acquire their present nested distribution of retained events. Both explanations are consistent with the inverse relationship observed between a remodeling event's range of occurrence in these taxa and its TH sensitivity in metamorphosing forms (Fig. 5). Panel a shows the disparate attenuations of TH profile that may be hypothesized for cryptobranchid, sirenid, and proteid lineages (right) in their evolutionary divergence from ancestral metamorphosing lineages (left). In this case, the loss of remodeling events in each lineage is determined by the extent of the drop in TH activity. Although sirenids and cryptobranchids are given only one profile each, minor differences in the amount of remodeling exhibited by the two genera in each group would suggest slightly different levels of attenuation. Panel b shows the disparate losses of TH sensitivity that may be alternately hypothesized for cryptobranchid, sirenid, and proteid lineages in their evolutionary divergence from ancestral metamorphosing lineages (left column). In this case, the loss of remodeling events in each lineage is predisposed by the TH sensitivity of individual tissues; less sensitive events are more readily lost than more sensitive ones. The gray scale indicates the presumed distribution of TH sensitivity among tissues: white, high sensitivity; gray, low sensitivity; black, loss of sensitivity.

investigations are limited to one of the more than 500 species belonging to the latter genus: *E. coqui* (Hanken *et al.*, 1997).

The effect of direct development upon larval development is best characterized for the leptodactylid *E. coqui* (Elinson, 1990; Hanken *et al.*, 1992; Schlosser and Roth, 1997). Cranial ossification, limb development, and tail resorption in this frog show clear evidence of predisplacement. Most cranial bones appear before hatching, as opposed to between mid-premetamorphosis and the end of climax as in indirect-developing frogs (Gaudin, 1978; Hanken and Hall, 1988; Kemp and Hoyt, 1969; Trueb, 1985; Trueb and Hanken, 1992; Wiens, 1989). Also, the first bones to appear in *E. coqui* are in the lower jaw and jaw suspension, rather than the braincase as in other frogs. Limb development and tail resorption also have been shifted to early (Elinson, 1990) and late (Townsend and Stewart, 1985) embryogeny, respectively. In addition to being predisplaced, chondrocranial development in *E. coqui* shows evidence of repatterning (Hanken *et al.*, 1992). Certain cartilages that typically are present in tadpoles and disappear at climax, e.g., the suprarostral cartilage and trabecular horns of the upper jaw, never appear in *E. coqui* embryos. Other tadpole cartilages, such as those of the lower jaw, hyobranchium, and jaw suspension, appear, but their embryonic condensations directly assume a midmetamorphic configuration; they subsequently are remodeled into adult shapes before hatching. Muscle and nerve development in *E. coqui* shows a complementary array of predisplaced adult features and repatterned larval features (Schlosser and Roth, 1997). Although it would appear that the loss of larval muscles in *E. coqui* has kept pace with the reduction of larval skeleton, there is at least one indication that nerve evolution has lagged behind these changes.

In summary, changes to the primitively TH-mediated component of *E. coqui* development include the unequal or mosaic predisplacement of adult features and the repatterning (i.e., loss or reconfiguration) of larval structures that primitively underwent resorption or remodeling at metamorphosis. Although unraveling of the cellular basis of this repatterning awaits a comparative analysis of cell lineages, two kinds of change in cell specification would appear to be involved. In the example of larval cartilages that originally were destined for resorption, the neural crest precursor cells that normally would migrate and condense to form them either do not arise in *E. coqui* embryos or have been respecified to contribute to other structures. In the example of larval cartilages originally destined for remodeling, precursor cells that normally would condense into cartilages of larval shape and progress through a series of TH-mediated shape changes have been respecified in *E. coqui* to assume a midmetamorphic shape directly. In this case, the information required to make larval morphology, rather than the larval morphology itself, seems to have been excised from the developmental program. These changes are in addition to embryonic modifications of TH-independent features, including loss of the cement and hatching glands, operculum, lateral line organs, and keratinous mouthparts, reduced gill development, specialization of the tail into a respiratory organ, and development of a keratinous egg tooth to facilitate hatching (Elinson, 1990; Townsend and Stewart, 1985).

An indirect-developing leptodactylid frog, the ceratophryine *Lepidobatrachus laevis,* presents a contrasting example of how repatterning can affect TH-mediated development (Hanken, 1993; Ruibal and Thomas, 1988). The tadpoles of this species are distinguished from other tadpoles (including other ceratophryines) by an enlarged head with hypertrophied jaw cartilages. The lower jaw cartilages are much longer than in other tadpoles, the jaw suspension is oriented at a significantly larger angle to the floor of the braincase, and the median cartilage of the lower jaw is unpaired, rather than paired (Ruibal and Thomas, 1988). These configurations, which develop directly from embryonic condensations (Jennings *et al.,* 1991), resemble the midmetamorphic configurations of other frogs. Hence, as in *E. coqui,* some chondrocranial features in *Lepidobatrachus* appear to have been repatterned to assume midmetamorphic configurations directly. Unlike *E. coqui,* where such configurations are attained only transiently *intra ovum,* the midmetamorphic jaw configuration of *Lepidobatrachus* is retained throughout the tadpole period and presumably completes its remodeling at climax metamorphosis. Various buccopharyngeal features also appear to assume midmetamorphic states directly (Wassersug and Heyer, 1988), although some of these result from the reduction or absence of embryonic derivatives, e.g., keratinous mouthparts and gill filter plates. Other examples of repatterning in *Lepidobatrachus* tadpoles result in altogether new configurations that bear no resemblance to midmetamorphic development in other frogs. The ceratohyal cartilage is displaced laterally, the gill arch cartilages are more robust and less ornate, and the upper jaw cartilage is grossly elongated to match the width of the lower jaw (this cartilage is resorbed, rather than lengthened, in metamorphosing tadpoles).

2. Mechanisms of Diversification

The early timing of most primitively TH-mediated remodeling (e.g., cranial ossification, limb development, and tail resorption) in *E. coqui* relative to other frogs suggests that the evolution of direct development in eleutherodactylid frogs involved a predisplacement and acceleration of TH activity. This hypothesis is supported by the precocious timing of several aspects of the thyroid axis in *E. coqui*. The development of the thyroid gland and median eminence of the hypothalamus, the onset of TSH production in the pituitary, and the onset of TRβ protein expression in peripheral tissues all occur in embryogenesis, as opposed to during posthatching stages as in indirect-developing frogs (Hanken *et al.,* 1997; Jennings and Hanken, 1998). Acceleration of thyroid axis activity does not explain the unequal predisplacement of features (e.g., the jaw bones relative to other bones), nor their apparent loss of TH sensitivity (see the following discussion). These events would require additional mechanistic changes in individual tissues.

Several larger questions also are raised by this hypothesis. First, which mechanisms may lead to precocious development of the thyroid axis, and is there any connection between this shift and the presence of maternal TH and/or TR in the eggs? Like indirect-developing frogs, *E. coqui* provisions its oviposited eggs with TH, both T_3 and T_4 (Hanken *et al.*, 1997); TRα protein also may be present, as in other frog species (Eliceiri and Brown, 1994; Kawahara *et al.*, 1991). Although the function of TH and TR in amphibian eggs is completely unstudied, several lines of evidence suggest a developmental role for maternally derived, yolk-stored TH in teleost fish embryos and larvae [summarized by Brown and Bern (1989)]. Also, recent evidence for T_3-TR involvement in chick neurulation (Flamant and Samarut, 1998) indicates that maternal TH may play a role in the patterning of early brain development. The maternal provisioning of amphibian eggs with this particular hormone–receptor pair thus provides a potential mechanism for introducing shifts in the timing of TH-mediated development, starting with hypothalamic maturation. More investigation is required to determine whether maternal TH and TR are provisioned or function any differently in *E. coqui* than in other frogs. The issue becomes more intriguing when one considers viviparous amphibians (see the following discussion), which by retaining eggs in the oviduct or internal pouches have the potential to maintain a continuous supply of maternal TH to developing embryos (Fitzgerald *et al.*, 1979).

Second, to what extent is TH mediation still functional in the developmental regulation of direct-developing amphibians, and what does this tell us about the evolutionary history of their TH mediation? The only features of *E. coqui* development found to be TH-inducible, tail resorption (Elinson, 1994) and a urea cycle enzyme (Callery and Elinson, 1996), require activating concentrations that are far above those normally required to induce remodeling in tadpoles. Because the equivalent endogenous concentrations probably are not attained by *E. coqui* embryos, it would appear that this residual inducibility is vestigial, i.e., the last detectable trace of TH sensitivity in development that is otherwise TH-independent. In this case, tissues that primitively were under TH control may now be mediated by other systemic factors, as evidenced by *in vitro* and transplant analyses of hindlimb development (Elinson, 1994). However, the question of whether any tissues have fully abandoned their TH mediation can only be fully resolved by examining the extent to which their remodeling is still dependent on TH-regulated gene expression.

If, in fact, *E. coqui* development is TH-independent, the requisite changes in developmental timing and mediation that must have evolved in the eleutherodactylid lineage may have happened in two sequences (Lynn, 1961). The most likely scenario is that thyroid axis development was predisplaced first, and this event was accompanied by the predisplacement of TH-dependent remodeling and followed by a breakdown in TH mediation (or switch to non-TH mediation). In this case, the embryonic onsets of TH activity and TR expression in *E. coqui* may be interpreted as relics of the evolutionary stage at which *E. coqui* ancestors abandoned their reliance on TH mediation. Alternatively, TH mediation may have been lost first, and this loss was followed by predisplacements of both thyroid axis development and peripheral remodeling. However, there is no precedence for metamorphosing amphibians abandoning their TH mediation without experiencing contingent changes in developmental pattern. Also, the second sequence obviously is less parsimonious in its requirement for numerous predisplacements of presumably independent developmental events (although such predisplacements conceivably may arise as a result of having deleted a major part of the larval developmental program; see the following discussion). Despite these shortcomings, the second sequence can only be ruled out by the discovery of at least some eleutherodactylid species that still utilize TH mediation in their direct development.

The third question raised by the hypothesis of accelerated TH activity is how to account for the underlying changes in embryonic pattern. Some progenitor cells of cranial skeleton in metamorphosing vertebrates presumably are programmed with the information for making a larval shape, followed by a series of TH-mediated intermediate shapes and a terminal adult shape. The embryos of *E. coqui* and *Lepidobatrachus* demonstrate that it is possible for such cells to "skip ahead" in the program, i.e., to excise the larval and early TH-mediated shapes. *Lepidobatrachus* further demonstrates that one TH-mediated shape can be dissociated from subsequent ones by shifting it to embryonic development. Such deviations would suggest that mechanisms involved in embryonic and metamorphic patterning are not inherently different and that TH mediation is easily decoupled from them.

As alluded to previously, there is the alternative explanation that direct development in *E. coqui* evolved via the excision of a major part of the larval developmental program (Schlosser and Roth, 1997). Possible causes include the loss of regulatory (presumably genetic) mechanisms that (1) normally inhibit adult development in larval frogs and/or (2) normally promote the development and maintenance of tadpole-specific structures (Schlosser and Roth, 1997). In this case, precocious thyroid axis development in *E. coqui* could be regarded more as a product than a cause of the global predisplacement of adult features. This hypothesis is supported by the widespread loss of tadpole-specific structures and tissue shapes and arrangements in *E. coqui*. It rests on the untested assumption that the genetic controls of larval and adult developmental programs in frogs are sufficiently independent to allow the decoupling of the two programs on an organism-wide scale.

This hypothesis also has been argued on the basis that TH mediation actually may function to relieve the inhibition of adult development in larval frogs (Elinson, 1990). Thus, one can envision that a breakdown of the mechanisms that normally inhibit adult development at larval stages would lead to development becoming both more direct and less dependent on TH (Elinson, 1990). The view of TH as a repressor remover in frogs runs contrary to the finding that most TH-regulated genes in different tissues appear to be up-regulated, rather than down-regulated, by TH. Also, the few down-regulated genes that have been identified so far do not code for transcription factors, but instead code for extracellular glyoproteins in tadpole-specific tissues (Furlow et al., 1997). However, as the assays used to identify TH-regulated genes do not isolate all of them, it certainly is possible that there are still unidentified down-regulated genes with widespread repressor functions.

The hypothesis of TH as repressor remover may be examined in a new light when one considers that TR in the absence of TH functions as a transcriptional repressor, rather than activator, of genes that are controlled by TH response elements (Shi et al., 1996a,b). In this scenario, tadpole-specific structures and shapes would be maintained by TR regulation, and adult-specific structures would be activated by the onset of TH–TR regulation. However, because both the termination of tadpole structures and the onset of adult structures now are tied to the same regulatory switch (appearance of TH), it becomes difficult to explain a shift in the timing of the switch without invoking a simultaneous shift in TH activity. More research obviously is required to resolve the relative contributions of genetic and hormonal mechanisms to the evolution of direct development.

IV. THE ROLE OF HORMONE MEDIATION IN THE EVOLUTION OF AMPHIBIAN VIVIPARITY

Though a full discussion of amphibian viviparity is beyond the scope of this chapter, the phenomenon merits some mention here for its numerous occurrences and diverse manifestations in the three lissamphibian subgroups and the pervasive role of hormonal mediators in its evolution. In its broadest definition (which is the only one that can fully describe the phenomenon in amphibians), viviparity refers to live-bearing or the retention of embryos by parents (Wake, 1992). The embryos/larvae of viviparous amphibians are carried by a parent either externally, on the legs (the discoglossid frog *Alytes*) or back (dendrobatid frogs, most hemiphractine hylid frogs including some *Gastrotheca* species, the sooglossid frog *Sooglossus seychellensis*), or internally, in dorsal skin pouches (the pipid frog *Pipa pipa,* other *Gastrotheca* species), inguinal pouches (the myobatrachid frog *Assa darlingtoni*), oviduct (the bufonid frogs *Nectophrynoides occidentalis, N. liberiensis, N. tornieri,* and *N. viviparus; Eleutherodactylus jasperi;* the salamandrids *Salamandra atra, S. salamandra,* and *Mertensiella luschiani;* some caeciliad and scolecomorphid caecilians; all typhlonectid caecilians), vocal sac (the rhinodermatid frog *Rhinoderma darwinii*), or stomach (the myobatrachid frog *Rheobatrachus silus*) [Wake (1982, 1989) and Zug (1993) and references therein]. Embryos/larvae that are maintained inside the body are either ovoviviparous, i.e., entirely yolk-dependent, or dependent on additional nutrition from lysed eggs *(S. atra, M. l. anatalyana)* and/or secretions from oviduct membranes (older embryos of the previous two taxa, *N. occidentalis, N. liberiensis,* all viviparous caecilians except typhlonectids), pseudoplacentae (typhlonectids), and possibly other epithelial membranes *(R. darwinii)*. Larvae are released to the environment at various stages depending on the species and, in certain cases, the environment.

The effects of viviparity on larval development are variable. In some species that facultatively release animals at postmetamorphic stages (e.g., *S. salamandra*), embryos complete most, if not all, of the larval and metamorphic development exhibited by oviparous conspecifics and do so in a similar sequence (Alberch and Blanco, 1996). Other species that consistently release after metamorphosis exhibit modifications to larval development that are similar to ones evolved by direct-developing forms (e.g., the vascularized tail of *Pipa pipa,* expanded "bell gills" of *Gastrotheca,* and loss of tadpole mouthparts in *Nectophrynoides*) (Wake, 1982, 1989). What is known about skull development in the hemiphractine hylid *Cryptobatrachus* and the bufonid *N. tornieri* (Orton, 1949) suggests changes similar to those described for direct-developing frogs. Similarly, the viviparous caecilians *Dermophis mexicanus* and *Typhlonectes compressicaudus* show changes in ossification sequence relative to oviparous forms that are similar to that observed in *E. coqui* (Wake *et al.,* 1985; Wake and Hanken, 1982). Some species that release after metamorphosis also develop new specializations for intraoviduct feeding, such as oral fimbriae *(N. occidentalis)* and fetal dentition (all viviparous caecilians and *S. atra*) (Wake, 1982, 1989).

Evolving viviparity requires at least three fundamental mechanistic changes, two of which have a hormonal basis (Callard *et al.,* 1992; Packard *et al.,* 1989). These are (1) the evolution of internal fertilization, (2) an increase in the duration of embryo retention, and (3) specializations of the egg and maternal membranes to permit gas exchange and other forms of mother–embryo transfer or signaling. Internal fertilization, which is already present in most salamanders and all caecilians and only required for oviductal viviparity, probably arises for reasons unrelated to viviparity (Packard *et al.,* 1989). Steps (2) and (3) involve functional changes in the corpus luteum. The corpus luteum, an ephemeral endocrine gland derived from postovulatory follicle cells, produces steroid hormones (primarily progesterone) that collectively are responsible for several aspects of oviparous development. These include activating oocyte nuclear maturation and ovulation; stimulating the production of egg casings, oviduct vascularity, and oviduct secretions; and inhibiting oviduct contractility and vitellogenesis for immature oocytes (Browning, 1973). In viviparous vertebrates, the corpus luteum generally is believed to prolong its progesterone production following ovulation in order to increase oviduct (or skin) vascularity and inhibit oviduct contractility, vitellogenesis, and ovulation during the time of embryo retention (Callard et al., 1992; Packard et al., 1989). There is also some indication that prolonged progesterone production may provide a local immunosuppressive effect to prevent embryo rejection by the mother [references in Wake (1982)]. Other enhanced functions of the corpus luteum in viviparous forms may include promoting oocyte atresia (cell death) to decrease clutch size and provisioning eggs during vitellogenesis with progesterone and estrogen (Browning, 1973). Oocyte atresia, which indirectly may lead to increased egg size, is most important in ovoviviparous forms (and direct developers), which are reliant solely upon yolk reserves to complete their gestation. The steroid loading of eggs is conjectured to allow embryos the capacity to regulate their own gestation period by releasing hormones that regulate maternal reproductive functions (Browning, 1973).

The empirical evidence for these effects in viviparous amphibians comes mainly from studies on *Salamandra salamandra, S. atra, Nectophrynoides occidentalis,* and *Gastrotheca riobambae.* The data include observations on the developmental timing and immunohistochemistry of the corpus luteum and the effects of steroid induction and embryo or ovary removal [reviewed by Browning (1973); Guillette, (1989); Xavier, (1977)]. The corpora lutea of these animals produce significant amounts of progesterone, which, along with estrogen, appear to mediate maternal specializations arising around the time of ovulation/fertilization and delay the development of new oocytes. However, the corpora lutea tend not to remain active for the full gestation period, which leaves open the possibility that the timing of birth is under embryonic control. There is some evidence that embryos release factors that modulate maternal functions during gestation (e.g., prolong luteal life in *N. occidentalis,* inhibit gastric function in stomach-brooding frogs) [reviewed by Guillette (1989)]. Less attention has been paid to the possibility of maternal factors that modulate embryonic and larval development. Progesterone appears to slow larval development in *N. occidentalis* in order to postpone birth until after estivation (Xavier, 1977). The abrupt disappearance of fetal dentition in viviparous caecilians, which correlates with birth, rather than embryo size, points to a maternal, rather than embryonic, hormone signal (Wake, 1982). It is possible that most modifications to the TH-mediated larval development of viviparous forms involve the same embryonic control mechanisms as in direct-developing forms. However, viviparity offers the additional opportunity for maternal hormones to influence TH-mediated events in the embryo/larva throughout its gestation. There has been no attempt to date to investigate this phenomenon or to ascertain the level of TH mediation still in effect during the larval development of viviparous forms.

V. CONCLUDING REMARKS AND DIRECTIONS FOR FUTURE RESEARCH

Integration of the current explosion in mechanistic research on amphibian (i.e., frog) metamorphosis with the century-long accumulation of comparative descriptions of frog and salamander development is truly a schizophrenic task. The two research activities are distinguished by vastly different investigative styles and objectives. Whereas comparative morphologists have been drawn to certain organ systems by their functional significance and evolutionary diversity (e.g., musculoskeletal remodeling), developmental biologists are drawn to others by the ease of tissue isolation and manipulation in culture and by the uniform, easily quantified and relatively conservative nature of their hormone responses (e.g., tail, liver, and skin remodeling). Despite its dramatic impact upon morphological diversity, musculoskeletal remodeling has received perhaps the least mechanistic investigation of all hormonally mediated systems; the skeleton is all but omitted in overviews of this research (Gilbert *et al.*, 1996). Similarly, despite their profoundly variable larval development, salamanders have received relatively little mechanistic investigation, presumably because researchers prefer the more spectacular, albeit stereotypic, phenomenon of frog metamorphosis. Caecilians, which are mostly viviparous but include some direct- and indirect-developing oviparous forms (Wake, 1989; Zug, 1993), thus far have received only descriptive analysis of their development.

Aside from interspecific comparisons, little mechanistic research on amphibian development has been done in a phylogenetic context. Explanations of evolutionary transitions in development usually are based on identifying a key difference(s) in proximate developmental mechanisms. There are two basic limitations to this approach. The first concerns the task of resolving primary from secondary causation. Any mutation in a gene for one of the many hormones, enzymes, receptors, signaling and transcription factors, and hormone binding proteins involved in TH mediation has some potential to cause an alteration or breakdown of regulating mechanisms and consequent transition in developmental pattern. Once such a change becomes genetically fixed in a particular lineage, there is the potential for additional mutations to accumulate in "TH mediation genes," be passed on to descendent lineages, and thus obscure the primary mechanistic change(s) responsible for the original transition in developmental pattern (Martin and Gordon, 1995). For the evolutionary transitions described here (metamorphic ontogeny → larval reproduction and metamorphic ontogeny → direct development), certain mechanistic changes involving rate of TH activity (attenuation and predisplacement/acceleration of TH production) have been inferred to be primary on the basis of comparative developmental evidence. Large-scale changes in tissue sensitivity, when present, may be deemed secondary only because they offer a less parsimonious primary mechanism than a single change in TH activity. However, the matter of distinguishing primary from secondary changes ultimately is a question of resolving their evolutionary sequence. Hence, confirming these inferences and unraveling the exact mechanisms involved in individual transitions will require a synthesis of mechanistic and comparative approaches.

The second limitation of a strictly mechanistic approach is its inability to explain much of the underlying diversity in developmental patterns. For example, what is currently known about the regulatory mechanisms controlling TH-mediated development does not offer any *a priori* explanations for why most larval reproducers lose certain remodeling events before others or why some larval reproducers (plethodontids) tend to lose all remodeling, whereas others (cryptobranchids, sirenids, and proteids) show a nested distribution of remodeling. It appears that the morphological diversity of larval reproducers reflects variation not only in the control of tissue sensitivity and TH activity in individual lineages, but in the settings of tissue sensitivity and TH activity in their ancestral metamorphosing lineages. Again, a comparative approach is required to unravel the evolutionary history of these parameters and their impact upon particular evolutionary transitions.

In closing, a number of additional intriguing developmental and evolutionary phenomena remain in amphibian larvae that probably involve hormonal mediation, but are relatively unstudied in this regard. These include caecilian development in general, the development of specialized cannibal morphs in salamander and frog larvae (Pfennig, 1992; Walls *et al.,* 1993), the potential revolution of metamorphosis by direct-developing (Titus and Larson, 1996; Wassersug and Duellman, 1984) and larval reproducing amphibians (Shaffer and McKnight, 1996), and the effects of disparate reproductive modes [reviewed by Duellman and Trueb (1986)] on hormonally mediated larval development. The possibilities of maternal hormone transfer to embryos and the exchange of larval hormones among conspecifics (see Hayes and Wu, 1995b) raise new questions regarding the role of developmental setting in the evolution of larval development and morphology. Finally, not enough attention has been paid to the old question of why salamanders frequently evolve larval reproduction and frogs and caecilians never do. Research on frogs has revealed significant hormonal interplay between metamorphosis and sexual maturation. For at least two events at sexual maturation, the tissues involved require prior exposure to TH in order to become responsive to their respective steroid activators (Robertson and Kelley, 1996; Tata, 1994). This observation begs the question of whether salamanders exhibit a similar dependency, and, if so, how they evolve the capacity to become sexually mature without going through metamorphosis. Though frogs can teach us much about the hormonal mediation of larval development, salamanders and caecilians clearly have a few stories of their own to tell.

Note added in proof: Recent research on *Xenopus* suggests that amphibians have capitalized upon the dual repression and activation functions of TR to orchestrate biphasic patterns of gene expression. Genes for the matrix metalloproteinase, stromelysin-3, and the signaling molecule, hedgehog, are expressed in embryonic development and later in climax metamorphosis when they are directly up-regulated by TH to participate in tissue-specific remodeling (Patterton *et al.*, 1995; Stolow and Shi, 1995). The intervening gap or reduction in expression coincides with the expression of Trα which, in the absence of TH, is expected to repress these genes and TRβ (Puzianowska-Kuznicka *et al.*, 1997; Ranjan *et al.*, 1994). This gap is thought to be terminated by the arrival of TH which up-regulates TRβ to activate these genes. Additional data on other TH-responsive genes with embryonic expression suggest that this pattern of gene regulation may apply to most, if not all, the genes that are up-regulated in *Xenopus* metamorphosis (Puzianowska-Kuznicka *et al.*, 1997). Developmentally, the designation of Trα and Trβ as asynchronous repressors and activators of gene expression would prevent the precocious activation of TH-regulated genes that may interfere with larval growth and development. Evolutionarily, this designation may have played a key role in establishing temporally distinct genetic controls for the development of larval and adult features.

ACKNOWLEDGMENTS

Thanks to the editors for helpful suggestions and to Liz Callery, Rick Elinson, and Hung Fang for enjoyable discussions on the topic. This work was supported by a Canadian MRC postdoctoral fellowship to C.S.R. and NSERC funding to B. K. Hall.

REFERENCES

Alberch, P. (1987). Evolution of a developmental process: Irreversibility and redundancy in amphibian metamorphosis. *In* "Development as an Evolutionary Process" (R. A. Raff and E. C. Raff, eds.), pp. 23–46. Liss, New York.

Alberch, P., and Blanco, M. J. (1996). Evolutionary patterns in ontogenetic transformation: From laws to regularities. *Int. J. Dev. Biol.* **40**, 845–858.

Alberch, P., and Gale, E. A. (1986). Pathways of cytodifferentiation during the metamorphosis of the epibranchial cartilage in the salamander, *Eurycea bislineata. Dev. Biol.* **117**, 233–244.

Alberch, P., Lewbart, G. A., and Gale, E. A. (1985). The fate of larval chondrocytes during the metamorphosis of the epibranchial in the salamander, *Eurycea bislineata. J. Embryol. Exp. Morphol.* **88**, 71–83.

Alberch, P., Gale, E. A., and Larsen, P. R. (1986). Plasma T$_4$ and T$_3$ levels in naturally metamorphosing *Eurycea bislineata* (Amphibia; Plethodontidae). *Gen. Comp. Endocrinol.* **61**, 153–163.

Baker, B. S., and Tata, J. R. (1992). Prolactin prevents the autoinduction of thyroid hormone receptor mRNAs during amphibian metamorphosis. *Dev. Biol.* **149**, 463–467.

Baker, C. L., and Stoudemayer, M. B. (1951). The influence of thyroxine, epinephrine, and X-rays on metamorphosis of some neotenous urodeles. *J. Tenn. Acad. Sci.* **26**, 32–41.

Becker, K. B., Schneider, M. J., Davey, J. C., and Galton, V. A. (1995). The type III 5-deiodinase in *Rana catesbeiana* is encoded by a thyroid-hormone-responsive gene. *Endocrinology (Baltimore)* **136**, 4424–4431.

Becker, K. B., Stephens, K. C., Davey, J. C., Schneider, M. J., and Galton, V. A. (1997). The type 2 and type 3 iodothyronine deiodinases play important roles in coordinating development in *Rana catesbeiana* tadpoles. *Endocrinology (Baltimore)* **138**, 2989–2997.

Besharse, J. C., and Brandon, R. A. (1974). Postembryonic eye degeneration in the troglobitic salamander *Typhlotriton spelaeus. J. Morphol.* **144**, 381–406.

Brown, C. L., and Bern, H. A. (1989). Thyroid hormones in early development, with special reference to teleost fishes. *In* "Development, Maturation, and Senescence of Neuroendocrine Systems" (M. P. Schreibman and C. G. Scanes, eds.), pp. 289–306. Academic Press, San Diego, CA.

Brown, D. D., Wang, Z., Furlow, J. D., Kanamori, A., Schwartzman, R. A., Remo, B. F., and Pinder, A. (1996). The thyroid hormone-induced tail resorption program during *Xenopus laevis* metamorphosis. *Proc. Natl. Acad. Sci. U.S.A.* **93**, 1924–1929.

Browning, H. C. (1973). The evolutionary history of the corpus luteum. *Biol. Reprod.* **8**, 128–157.

Buckbinder, L., and Brown, D. D. (1992). Thyroid hormone-induced gene expression changes in the developing frog limb. *J. Biol. Chem.* **267**, 25786–25791.

Buckbinder, L., and Brown, D. D. (1993). Expression of the *Xenopus laevis* prolactin and thyrotropin genes during metamorphosis. *Proc. Natl. Acad. Sci. U.S.A.* **90**, 3820–3824.

Buscaglia, M., Leloup, J., and de Luze, A. (1985). The role and regulation of monodeiodination of thyroxine to 3,5,3'-triiodothyronine during amphibian metamorphosis. *In* "Metamorphosis" (M. Balls and M. Bownes, eds.), pp. 273–293. Clarendon Press, Oxford.

Callard, I. P., Fileti, L. A., Perez, L. E., Sorbera, L. A., Giannoukos, G., Klosterman, L. L., Tsang, P., and McCracken, J. A. (1992). Role of the corpus luteum and progesterone in the evolution of vertebrate viviparity. *Am. Zool.* **32**, 264–275.

Callery, E. M., and Elinson, R. P. (1996). Developmental regulation of the urea-cycle enzyme arginase in the direct developing frog *Eleutherodactylus coqui. J. Exp. Zool.* **275**, 61–66.

Cardellini, P., Gabrion, J., and Sala, M. (1978). Electrophoretic patterns of larval, neotenic and adult haemoglobin of *Triturus helveticus* Raz. *Acta Embryol. Exp.* **2**, 151–161.

Carr, J. A., and Norris, D. O. (1988). Interrenal activity during metamorphosis of the tiger salamander, *Ambystoma tigrinum. Gen. Comp. Endocrinol.* **71**, 63–69.

Chanoine, C., d'Albis, A., Lenfant-Guyot, M., Janmot, C., and Gallien, C. L. (1987). Regulation by thyroid hormone of terminal muscle differentiation in the skeletal dorsal muscle. II. Urodelan amphibians. *Dev. Biol.* **123**, 33–42.

Chanoine, C., Guyot-Lenfant, M., d'Albis, A., Durand, J., Perasso, F., Salles-Mourlan, A.-M., Janmot, C., and Gallien, C. L. (1989). Thyroidal status and myosin isoenzymatic pattern in the skeletal dorsal muscle of urodelan amphibians—the perennibranchiate *Proteus anguinus*. *Cell Differ. Dev.* 28, 135–144.

Chanoine, C., Guyot-Lenfant, M., Saadi, A., Perasso, F., Salles-Mourlan, A.-M., and Gallien, C. L. (1990). Myosin structure and thyroidian control of myosin synthesis in urodelan amphibian skeletal muscle. *Int. J. Dev. Biol.* 34, 163–170.

Chibon, P. (1995). La région buccale et les dents. *In* "Traité de Zoologie: Anatomie, Systématique, Biologie" (M. Delsol, ed.), Vol. 14, Part IA, pp. 941–952. Masson, Paris.

Clemen, G. (1988). Experimental analysis of the capacity of dental laminae in *Ambystoma mexicanum* Shaw. *Arch. Biol.* 99, 111–132.

Clemen, G., and Greven, H. (1988). Morphological studies on the mouth cavity of Urodela. IX. Teeth of the palate and the splenials in *Siren* and *Pseudobranchus* (Sirenidae: Amphibia). *Z. Zool. Syst. Evolutionsforsch.* 26, 135–143.

Davidson, A. (1895). A contribution to the anatomy and phylogeny of *Amphiuma means* (Gardner). *J. Morphol.* 11, 375–410.

Denver, R. J. (1993). Acceleration of anuran amphibian metamorphosis by corticotropin-releasing hormone-like peptides. *Gen. Comp. Endocrinol.* 91, 38–51.

Denver, R. J. (1996). Neuroendocrine control of amphibian metamorphosis. *In* "Metamorphosis: Postembryonic Reprogramming of Gene Expression in Amphibian and Insect Cells" (L. I. Gilbert, J. R. Tata, and B. G. Atkinson, eds.), pp. 433–464. Academic Press, San Diego, CA.

Denver, R. J., and Licht, P. (1989). Neuropeptide stimulation of thyrotropin secretion in the larval bullfrog: Evidence for a common neuroregulator of thyroid and interrenal activity in metamorphosis. *J. Exp. Zool.* 252, 101–104.

Denver, R. J., Pavgi, S., and Shi, Y.-B. (1997). Thyroid hormone-dependent gene expression program for *Xenopus* neural development. *J. Biol. Chem.* 272, 8179–8188.

de Villiers, C. G. S. (1929). The development of a species of *Arthroleptella* from Jonkershoek, Stellenbosch. *S. Afr. J. Sci.* 26, 481–510.

Dickhoff, W. W., Brown, C. L., Sullivan, C. V., and Bern, H. A. (1990). Fish and amphibian models for developmental endocrinology. *J. Exp. Zool., Suppl.* 4, 90–97.

Ducibella, T. (1974). The occurrence of biochemical metamorphic events without anatomical metamorphosis in the axolotl. *Dev. Biol.* 38, 175–186.

Duellman, W. E. (1993). "Amphibian Species of the World: Additions and Corrections." Mus. Nat. Hist., University of Kansas, Spec. Publ. No. 21. Lawrence, Kansas.

Duellman, W. E., and Trueb, L. (1986). "Biology of Amphibians." McGraw-Hill, New York.

Dundee, H. A. (1957). Partial metamorphosis induced in *Typhlomolge rathbuni. Copeia,* pp. 52–53.

Dundee, H. A. (1962). Response of the neotenic salamander *Haideotriton wallacei* to a metamorphic agent. *Science* 135, 1060–1061.

Durand, J. P. (1995). La régression de l'oeil chez les amphibiens cavernicoles. *In* "Traité de Zoologie: Anatomie, Systématique, Biologie" (M. Delsol, ed.), Vol. 14, Part IA, pp. 631–637. Masson, Paris.

Eagleson, G. W., and McKeown, B. A. (1978a). Changes in thyroid activity of *Ambystoma gracile* (Baird) during different larval, transforming, and postmetamorphic phases. *Can. J. Zool.* 56, 1377–1381.

Eagleson, G. W., and McKeown, B. A. (1978b). Localization of the pituitary lactotropes and thyrotropes within *Ambystoma gracile* by histochemical and immunochemical methods. A developmental study of two populations. *Cell Tissue Res.* 189, 53–66.

Eliceiri, B. P., and Brown, D. D. (1994). Quantitation of endogenous thyroid hormone receptors a and b during embryogenesis and metamorphosis in *Xenopus laevis. J. Biol. Chem.* 269, 24459–24465.

Elinson, R. P. (1990). Direct development in frogs: Wiping the recapitulationist slate clean. *Semin. Dev. Biol.* 1, 263–270.

Elinson, R. P. (1994). Leg development in a frog without a tadpole (*Eleutherodactylus coqui*). *J. Exp. Zool.* **270**, 202–210.

Fairclough, L., and Tata, J. (1997). An immunocytochemical analysis of the expression of thyroid hormone receptor alpha and beta proteins during natural and thyroid hormone-induced metamorphosis in *Xenopus*. *Dev. Growth Differ.* **39**, 273–283.

Fitzgerald, K. T., Guillette, L. J., Jr., and Duvall, D. (1979). Notes on birth, development and care of *Gastrotheca riobambae* tadpoles in the laboratory (Amphibia, Anura, Hylidae). *J. Herpetol.* **13**, 457–460.

Flamant, F., and Samarut, J. (1998). Involvement of thyroid hormone and its α receptor in avian neurulation. *Dev. Biol.* **197**, 1–11.

Ford, L. S., and Cannatella, D. C. (1993). The major clades of frogs. *Herpetol. Monogr.* **7**, 94–117.

Fox, H. (1983). "Amphibian Morphogenesis." Humana Press, Clifton, NJ.

Furlow, J. D., Berry, D. L., Wang, Z., and Brown, D. D. (1997). A set of novel tadpole specific genes expressed only in the epidermis are down-regulated by thyroid hormone during *Xenopus laevis* metamorphosis. *Dev. Biol.* **182**, 284–298.

Galton, V. A. (1985). 3,5,3'-Triiodothyronine receptors and thyroxine 5'-monodeiodonating activity in thyroid hormone-insensitive amphibia. *Gen. Comp. Endocrinol.* **57**, 465–471.

Galton, V. A. (1990). Mechanisms underlying the acceleration of thyroid hormone-induced tadpole metamorphosis by corticosterone. *Endocrinology (Baltimore)* **127**, 2997–3002.

Galton, V. A. (1992). Thyroid hormone receptors and iodothyronine deiodinases in the developing Mexican axolotl, *Ambystoma mexicanum*. *Gen. Comp. Endocrinol.* **85**, 62–70.

Gancedo, B., Corpas, I., Alonso-Gomez, A. L., Delgado, M. J., Morreale de Escobar, G., and Alonso-Bedate, M. (1992). Corticotropin-releasing factor stimulates metamorphosis and increases thyroid hormone concentration in prometamorphic *Rana perezi* larvae. *Gen. Comp. Endocrinol.* **87**, 6–13.

Gaudin, A. J. (1978). The sequence of cranial ossification in the California toad, *Bufo bufo* (Amphibia, Anura, Bufonidae). *J. Herpetol.* **12**, 309–318.

Gilbert, L. I., Tata, J. R., and Atkinson, B. G., eds. (1996). "Metamorphosis: Postembryonic Reprogramming of Gene Expression in Amphibian and Insect Cells." Academic Press, San Diego, CA.

Good, D. A., and Wake, D. B. (1992). Geographic variation and speciation in the torrent salamanders of the genus *Rhyacotriton* (Caudata: Rhyacotritonidae). *Univ. Calif., Berkeley, Publ. Zool.* **126**, 1–91.

Gorbman, A. (1957). The thyroid gland of *Typhlomolge rathbuni*. *Copeia*, pp. 41–43.

Gorbman, A., Dickhoff, W. W., Vigna, S. R., Clark, N. B., and Ralph, C. L. (1983). "Comparative Endocrinology." Wiley, New York.

Greven, H. (1989). Teeth of extant amphibia: Morphology and some implications. *In* "Trends in Vertebrate Morphology" (H. Splechtna and H. Hilgers, eds.), Vol. 35, pp. 451–455. Fischer, Stuttgart.

Greven, H., and Clemen, G. (1979). Morphological studies on the mouth cavity of urodeles. IV. The teeth of the upper jaw and the palate in *Necturus maculosus* (Rafinesque) (Proteidae: Amphibia). *Arch. Histol. Jpn.* **42**, 445–457.

Greven, H., and Clemen, G. (1980). Morphological studies on the mouth cavity of urodeles. VI. The teeth of the upper jaw and the palate in *Andrias davidianus* (Blanchard) and *A. japonicus* (Temminck) (Cryptobranchidae: Amphibia). *Arch. Histol. Jpn.* **1**, 49–59.

Greven, H., and Clemen, G. (1990). Effect of hypophysectomy on the structure of normal and ectopically transplanted teeth in larval and adult urodeles. *Acta Embryol. Morphol. Exp.* **11**, 33–43.

Guillette, L. J., Jr. (1989). The evolution of vertebrate viviparity: Morphological modifications and endocrine control. *In* "Complex Organismal Functions: Integration and Evolution in

Vertebrates, Dahlem Konferenzen" (D. B. Wake and G. Roth, eds.), pp. 219–233. Wiley, New York.

Hanken, J. (1993). Model systems versus outgroups: Alternative approaches to the study of head development and evolution. *Am. Zool.* **33**, 448–456.

Hanken, J., and Hall, B. K. (1984). Variation and timing of the cranial ossification sequence of the oriental fire-bellied toad, *Bombina orientalis* (Amphibia, Discoglossidae). *J. Morphol.* **182**, 245–255.

Hanken, J., and Hall, B. K. (1988). Skull development during anuran metamorphosis: I. Early development of the first three bones to form—the exoccipital, the parasphenoid and the frontoparietal. *J. Morphol.* **195**, 247–256.

Hanken, J., and Summers, C. H. (1988). Skull development during anuran metamorphosis: III. Role of thyroid hormone in chondrogenesis. *J. Exp. Zool.* **246**, 156–170.

Hanken, J., Klymkowsky, M. W., Summers, C. H., Seufert, D. W., and Ingebrigtsen, N. (1992). Cranial ontogeny in the direct-developing frog, *Eleutherodactylus coqui* (Anura: Leptodactylidae), analyzed using whole-mount immunohistochemistry. *J. Morphol.* **211**, 95–118.

Hanken, J., Jennings, D. H., and Olsson, L. (1997). Mechanistic basis of life-history evolution in anuran amphibians: Direct development. *Am. Zool.* **37**, 160–171.

Hay, O. P. (1890). The skeletal anatomy of *Amphiuma* during its earlier stages. *J. Morphol.* **4**, 11–34.

Hayes, T. B. (1995a). Interdependence of corticosterone and thyroid hormones in larval toads (*Bufo boreas*). I. Thyroid hormone-dependent and independent effects of corticosterone on growth and development. *J. Exp. Zool.* **271**, 95–102.

Hayes, T. B. (1995b). Histological examination of the effects of corticosterone in larvae of the western toad, *Bufo boreas* (Anura: Bufonidae), and the Oriental fire-bellied toad, *Bombina orientalis* (Anura: Discoglossidae). *J. Morphol.* **226**, 297–307.

Hayes, T. B. (1997). Steroids as potential modulators of thyroid activity in anuran metamorphosis. *Am. Zool.* **37**, 185–194.

Hayes, T. B., and Gill, T. N. (1995). Hormonal regulation of skin gland differentiation in the toad (*Bufo boreas*): The role of the thyroid hormones and corticosterone. *Gen. Comp. Endocrinol.* **99**, 161–168.

Hayes, T. B., and Wu, T. H. (1995a). Interdependence of corticosterone and thyroid hormones in toad larvae (*Bufo boreas*). II. Regulation of corticosterone and thyroid hormones. *J. Exp. Zool.* **271**, 103–111.

Hayes, T. B., and Wu, T. H. (1995b). Role of corticosterone in anuran metamorphosis and potential role in stress-induced metamorphosis. *Neth. J. Zool.* **45**, 107–109.

Heady, J. E., and Kollros, J. J. (1964). Hormonal modification of the development of plical skin glands. *Gen. Comp. Endocrinol.* **4**, 124–131.

Huynh, T. D. H., Gallien, C. L., Durand, J. P., and Chanoine, C. (1996). Cloning and expression of a thryoid hormone receptor alphaI in the perennibranchiate amphibian *Proteus anguinus*. *Int. J. Dev. Biol.* **40**, 537–543.

Ishizuya-Oka, A., and Shimozawa, A. (1992). Connective tissue is involved in adult epithelial development of the small intestine during anuran metamorphosis in vitro. *Roux's Arch. Dev. Biol.* **201**, 322–329.

Ishizuya-Oka, A., and Shimozawa, A. (1994). Inductive action of epithelium on differentiation of intestinal connective tissue of *Xenopus laevis* tadpoles during metamorphosis in vitro. *Cell Tissue Res.* **277**, 427–436.

Ishizuya-Oka, A., Ueda, S., and Shi, Y.-B. (1996). Transient expression of stromelysin-3 mRNA in the amphibian small intestine during metamorphosis. *Cell Tissue Res.* **283**, 325–329.

Iwamuro, S., and Tata, J. R. (1995). Contrasting patterns of expression of thyroid hormone and retinoid X receptor genes during hormonal manipulation of *Xenopus* tadpole tail regression in culture. *Mol. Cell. Endocrinol.* **113**, 235–243.

Izutsu, Y., Kaiho, M., and Yoshizato, K. (1993). Different distribution of epidermal basal cells in the anuran larval skin correlates with the skin's region-specific fate at metamorphosis. *J. Exp. Zool.* **267**, 605–614.

Jennings, D. H., and Hanken, J. (1998). Mechanistic basis of life history evolution in anuran amphibians: thyroid gland development in the direct-developing frog. *Eleutherodactylus coqui. Gen. Comp. Endocrinol.* **111**, 225–232.

Jennings, D., Hanken, J., and Ruibal, R. (1991). Developmental basis of trophic specialization in larval anurans. *Am. Zool.* **31**, 45A.

Kaltenbach, J. C. (1958). Direct steroid enhancement of induced metamorphosis in peripheral tissues. *Anat. Rec.* **131**, 569–570.

Kaltenbach, J. C. (1996). Endocrinology of amphibian metamorphosis. *In* "Metamorphosis: Postembryonic Reprogramming of Gene Expression in Amphibian and Insect Cells" (L. I. Gilbert, J. R. Tata, and B. G. Atkinson, eds.), pp. 403–431. Academic Press, San Diego, CA.

Kanamori, A., and Brown, D. D. (1992). The regulation of thyroid hormone receptor b genes by thyroid hormone in *Xenopus laevis. J. Biol. Chem.* **267**, 739–745.

Kawahara, A., Baker, B. S., and Tata, J. R. (1991). Developmental and regional expression of thyroid hormone receptor genes during *Xenopus* metamorphosis. *Development (Cambridge, UK)* **112**, 933–943.

Kawai, A., Ikeya, J., Kinoshita, T., and Yoshizato, K. (1994). A three-step mechanism of action of thyroid hormone and mesenchyme in metamorphic changes in anuran larval skin. *Dev. Biol.* **166**, 477–488.

Keller, R. (1946). Morphogenetische Untersuchungen am Skelett von *Siredon mexicanus* Shaw mit besonderer Berücksichtigung des Ossifikationsmodus beim neotenen Axolotl. *Rev. Suisse Zool.* **53**, 329–426.

Kemp, N. E., and Hoyt, J. A. (1969). Sequence of ossification in the skeleton of growing and metamorphosing tadpoles of *Rana pipiens. J. Morphol.* **129**, 415–444.

Kerkoff, P. R., Tong, W., and Chaikoff, I. L. (1963). I131 Utilization by salamanders: *Taricha, Amphiuma* and *Necturus. Endocrinology (Baltimore)* **73**, 185–192.

Kerr, T. (1959). Development and structure of some actinopterygian and urodele teeth. *Proc. Zool. Soc. London* **133**, 401–422.

Kikuyama, S., Niki, K., Mayumi, M., Shibayama, R., Nishikawa, M., and Shintake, N. (1983). Studies on corticoid action on the tadpole tail in vitro. *Gen. Comp. Endocrinol.* **52**, 395–399.

Kikuyama, S., Kawamura, K., Tanaka, S., and Yamamoto, K. (1993). Aspects of amphibian metamorphosis: Hormonal control. *Int. Rev. Cytol.* **145**, 105–148.

Kingsley, J. S. (1892). The head of an embryo *Amphiuma. Am. Nat.* **26**, 671–680.

Klumpp, W., and Eggert, B. (1934). Die Schilddrüse und die branchiogenen Organe von *Ichthyophis glutinosus* L. *Z. Wiss. Zool.* **146**, 329–381.

Knutson, T. L., and Prahlad, K. V. (1971). 3,5,3′-Triiodo-L-thyronine uptake and development of metamorphic competence by *Xenopus laevis* embryonic tissues. *J. Exp. Zool.* **178**, 45–58.

Kobayashi, H., and Gorbman, A. (1962). Thyroid function in *Amphiuma. Gen. Comp. Endocrinol.* **2**, 279–282.

Kollros, J. J. (1961). Mechanisms of amphibian metamorphosis: Hormones. *Am. Zool.* **1**, 107–114.

Kollros, J. J., and McMurray, V. M. (1956). The mesencephalic V nucleus in anurans. II. The influence of thyroid hormone on cell size and cell number. *J. Exp. Zool.* **131**, 1–26.

Lamers, W. H., Vink, C., and Charles, R. (1978). Role of thyroid hormones in the normal and glucocorticosteroid hormone-induced evolution of carbamoyl-phosphate synthase (ammonia) activity in axolotl liver. *Comp. Biochem. Physiol. B* **59B**, 103–110.

Larras-Regard, E., Taurog, A., and Dorris, M. (1981). Plasma T_4 and T_3 levels in *Ambystoma tigrinum* at various stages of metamorphosis. *Gen. Comp. Endocrinol.* **43**, 443–450.

Larsen, J. H., Jr. (1968). Ultrastructure of thyroid follicle cells of three salamanders (*Ambystoma, Amphiuma,* and *Necturus*) exhibiting varying degrees of neoteny. *J. Ultrastruct. Res.* **24,** 190–209.

Larson, A., and Dimmick, W. W. (1993). Phylogenetic relationships of the salamander families: An analysis of congruence among morphological and molecular characters. *Herpetol. Monogr.* **7,** 77–93.

Leloup, J., and Buscaglia, M. (1977). La triidothyronine, hormone de la métamorphose des amphibiens. *C. R. Séances Acad. Sci., Sér. D* **284,** 2261–2263.

Leloup-Hatey, J., Buscaglia, M., Jolivet-Jaudet, G., and Leloup, J. (1990). Interrenal function during the metamorphosis in anuran amphibia. *Fortschr. Zool.* **38,** 139–154.

Lynn, W. G. (1942). The embryology of *Eleutherodactylus nubicola,* an anuran which has no tadpole stage. *Carnegie Inst. Washington Contrib. Embryol.* **541,** 27–62.

Lynn, W. G. (1961). Types of amphibian metamorphosis. *Am. Zool.* **1,** 151–161.

Marconi, M., and Simonetta, A. M. (1988). The morphology of the skull in neotenic and normal *Triturus vulgaris meridonalis* (Boulenger) (Amphibia Caudata Salamandridae). *Monit. Zool. Ital.* [N.S.] **22,** 365–396.

Martin, C. C., and Gordon, R. (1995). Differentiation trees, a junk DNA molecular clock, and the evolution of neoteny in salamanders. *J. Evol. Biol.* **8,** 339–354.

Mathisen, P. M., and Miller, L. (1987). Thyroid hormone induction of keratin genes: A two-step activation of gene expression during development. *Genes Dev.* **1,** 1107–1117.

Mathisen, P. M., and Miller, L. (1989). Thyroid hormone induces constitutive keratin gene expression during *Xenopus laevis* development. *Mol. Cell. Biol.* **9,** 1823–1831.

Miller, L. (1996). Hormone-induced changes in keratin gene expression during amphibian skin metamorphosis. *In* "Metamorphosis: Postembryonic Reprogramming of Gene Expression in Amphibian and Insect Cells" (L. I. Gilbert, J. R. Tata, and B. G. Atkinson, eds.), pp. 599–624. Academic Press, San Diego, CA.

Mondou, P. M., and Kaltenbach, J. C. (1979). Thyroxine concentrations in blood serum and pericardial fluid of metamorphosing tadpoles and of adult frogs. *Gen. Comp. Endocrinol.* **39,** 343–349.

Niki, K., Namiki, H., Kikuyama, S., and Yoshizato, K. (1982). Epidermal tissue requirement for tadpole tail regression induced by thyroid hormone. *Dev. Biol.* **94,** 116–120.

Niki, K., Yoshizato, K., Namiki, H., and Kikuyama, S. (1984). In vitro regression of tadpole tail by thyroid hormone. *Dev., Growth Differ.* **26,** 329–338.

Noble, G. K. (1924). The "retrograde metamorphosis" of the Sirenidae; Experiments on the functional activity of the thyroid of the perennibranchs. *Anat. Rec.* **29,** 100.

Noble, G. K. (1929). Further observations on the life-history of the newt, *Triturus viridescens. Am. Mus. Novit.* **348,** 1–22.

Noble, G. K., and Farris, E. J. (1929). A metamorphic change produced in *Cryptobranchus* by thyroid solutions. *Anat. Rec.* **42,** 59.

Noble, G. K., and Richards, L. B. (1930). A metamorphic change produced in *Siren* by thyroxin injections. *Anat. Rec.* **45,** 275.

Noble, G. K., and Richards, L. B. (1931). The criteria of metamorphosis in urodeles. *Anat. Rec.* **48,** 58.

Norman, M. F., Carr, J. A., and Norris, D. O. (1987). Adenohypophysial-thyroid activity of the tiger salamander, *Ambystoma tigrinum,* as a function of metamorphosis and captivity. *J. Exp. Zool.* **242,** 55–66.

Norris, D. O., and Dent, J. N. (1989). Neuroendocrine aspects of amphibian metamorphosis. *In* "Development, Maturation, and Senescence of Neuroendocrine Systems" (M. P. Schreibman and C. G. Scanes, eds.), pp. 63–90. Academic Press, San Diego, CA.

Norris, D. O., and Gern, W. A. (1976). Thyroxine-induced activation of hypothalamo-hypophysial axis in neotenic salamander larvae. *Science* **194,** 525–527.

Norris, D. O., and Platt, J. E. (1973). Effects of pituitary hormones, melatonin, and thyroidal inhibitors on radioiodide uptake by the thyroid glands of larval and adult tiger salamanders, *Ambystoma tigrinum* (Amphibia: Caudata). *Gen. Comp. Endocrinol.* 21, 368–376.

Norris, D. O., Jones, R. E., and Criley, B. B. (1973). Pituitary prolactin levels in larval, neotenic and metamorphosed salamanders (*Ambystoma tigrinum*). Gen. Comp. Endocrinol. 20, 437–442.

Orton, G. L. (1949). Larval development of *Nectophrynoides tornieri* (Roux), with comments on direct development in frogs. *Ann. Carnegie Mus.* 31, 257–271.

Packard, G. C., Elison, R. P., Gavaud, J., Guillette, L. J., Jr., Lombardi, J., Schindler, J., Shine, R., Tyndale-Biscoe, H. C., Wake, M. H., Xavier, F. D. J., Yaron, Z. (1989). Group Report: How are reproductive systems integrated and how has viviparity evolved. *In* "Complex Organismal Functions: Integration and Evolution in Vertebrates, Dahlem Konferenzen" (D. B. Wake and G. Roth, eds.), pp. 281–293. Wiley, New York.

Patterton, D., Hayes, W. P., and Shi, Y.-B. (1995). Transcriptional activation of the matrix metalloproteinase gene *stromelysin*-3 coincides with thyroid hormone-induced cell death during frog metamorphosis. *Dev. Biol.* 167, 252–262.

Pfennig, D. W. (1992). Proximate and functional causes of polyphenism in an anuran tadpole. *Funct. Ecol.* 6, 167–174.

Platt, J. E., and Christopher, M. A. (1977). Effects of prolactin on the water and sodium content of larval tissues from neotenic and metamorphosing *Ambystoma tigrinum. Gen. Comp. Endocrinol.* 31, 243–248.

Platt, J. E., Christopher, M. A., and Sullivan, C. A. (1978). The role of prolactin in blocking thyroxine-induced differentiation of tail tissue in larval and neotenic metamorphosing *Ambystoma tigrinum. Gen. Comp. Endocrinol.* 35, 402–408.

Puzianowska-Kuznicka, M., Damjanovski, S., and Shi, Y. (1997). Both thyroid hormone and 9-cis retinoic acid receptors are required to efficiently mediate the effects of thyroid hormone on embryonic development and specific gene regulation in *Xenopus laevis. Mol. Cell. Biol.* 17, 4738–4749.

Ranjan, M., Wong, J., and Shi, Y. (1994). Transcriptional repression of *Xenopus* TRβ gene is mediated by a thyroid hormone response element located near the start site. *J. Biol. Chem.* 269, 24699–24705.

Regal, P. J. (1966). Feeding specializations and the classification of terrestrial salamanders. *Evolution (Lawrence, Kans.)* 20, 392–407.

Regard, E., Taurog, A., and Nakashima, T. (1978). Plasma thyroxine and triiodothyronine levels in spontaneously metamorphosing *Rana catesbeiana* tadpoles and in adult anuran amphibia. *Endocrinology (Baltimore)* 102, 674–684.

Reilly, S. M. (1986). Ontogeny of cranial ossification in the eastern newt, *Notophthalmus viridescens* (Caudata: Salamandridae), and its relationship to metamorphosis and neoteny. *J. Morphol.* 188, 315–326.

Reilly, S. M. (1987). Ontogeny of the hyobranchial apparatus in the salamanders *Ambystoma talpoideum* (Ambystomatidae) and *Notophthalmus viridescens* (Salamandridae): The ecological morphology of two neotenic strategies. *J. Morphol.* 191, 205–214.

Reilly, S. M., and Brandon, R. A. (1994). Partial metamorphosis in the Mexican stream ambystomatids and the taxonomic significance of the genus *Rhyacosiredon* Dunn. *Copeia,* pp. 656–662.

Reis, K. (1932). La métamorphose des greffes hétéroplastiques de la peau des amphibiens néoténiques (*Proteus anguineus*). *C. R. Séances Soc. Biol. Ses. Fil.* 109, 1015–1017.

Robertson, J. C., and Kelley, D. B. (1996). Thyroid hormone controls the onset of androgen sensitivity in the developing larynx of *Xenopus laevis. Dev. Biol.* 176, 108–123.

Rose, C. S. (1995a). Skeletal morphogenesis in the urodele skull: I. Postembryonic development in the Hemidactyliini (Amphibia: Plethodontidae). *J. Morphol.* 223, 125–148.

Rose, C. S. (1995b). Skeletal morphogenesis in the urodele skull: II. Effect of developmental stage in TH-induced remodeling. *J. Morphol.* **223**, 149–166.

Rose, C. S. (1995c). Skeletal morphogenesis in the urodele skull: III. Effect of hormone dosage in TH-induced remodeling. *J. Morphol.* **223**, 243–261.

Rose, C. S. (1996). An endocrine-based model for developmental and morphogenetic diversification in metamorphic and paedomorphic urodeles. *J. Zool.* **239**, 253–284.

Rose, C. S., and Reiss, J. O. (1993). Metamorphosis and the vertebrate skull: Ontogenetic patterns and developmental mechanisms. *In* "The Skull" (J. Hanken and B. K. Hall, eds.), Vol. 1, pp. 289–346. University of Chicago Press, Chicago.

Rosenkilde, P., and Ussing, A. P. (1996). What mechanisms control neoteny and regulate induced metamorphosis in urodeles. *Int. J. Dev. Biol.* **40**, 665–673.

Rosenkilde, P., Mogensen, E., Centervall, G., and Jørgensen, O. S. (1982). Peaks of neuronal membrane antigen and thyroxine in larval development of the Mexican axolotl. *Gen. Comp. Endocrinol.* **48**, 504–514.

Roth, G., and Wake, D. B. (1985). Trends in the functional morphology and sensorimotor control of feeding behavior in salamanders: An example of the role of internal dynamics in evolution. *Acta Biotheor.* **34**, 175–192.

Ruibal, R., and Thomas, E. (1988). The obligate carnivorous larvae of the frog, *Lepidobatrachus laevis* (Leptodactylidae). *Copeia,* pp. 591–604.

Safi, R., Begue, A., Hänni, C., Stehelin, D., Tata, J. R., and Laudet, V. (1997a). Thyroid hormone receptor genes of neotenic amphibians. *J. Mol. Biol.* **44**, 595–604.

Safi, R., Deprez, L., and Laudet, V. (1997b). Thyroid hormone receptor genes in perennibranchiate amphibians. *Int. J. Dev. Biol.* **41**, 533–535.

Salles-Mourlan, A.-M., Guyot-Lenfant, M., Chanoine, C., and Gallien, C. L. (1994). Pituitary-thyroid axis controls the final differentiation of the dorsal skeletal muscle in urodelan amphibians. *Int. J. Dev. Biol.* **38**, 99–106.

Sap, J., Muñoz, A., Damm, K., Goldberg, Y., Ghysdael, J., Leutz, A., Beug, H., and Vennström, B. (1986). The *c-erb-A* protein is a high-affinity receptor for thyroid hormone. *Nature (London)* **324**, 635–640.

Satoh, S. J., and Wakahara, M. (1997). Hemoglobin transition from larval to adult types in a salamander (*Hynobius retardatus*) depends on activity of the pituitary gland, but not that of the thyroid gland. *J. Exp. Zool.* **278**, 87–92.

Schlosser, G., and Roth, G. (1997). Evolution of nerve development in frogs. II. Modified development of the peripheral nervous system in the direct-developing frog *Eleutherodactylus coqui* (Leptodactylidae). *Brain, Behav. Evol.* **50**, 94–128.

Schneider, M. J., and Galton, V. A. (1995). Effect of glucocorticoids on thyroid hormone action in cultured red blood cells from *Rana catesbeiana* tadpoles. *Endocrinology (Baltimore)* **136**, 1435–1440.

Schneider, M. J., Davey, J. C., and Galton, V. A. (1993). *Rana catesbeiana* tadpole red blood cells express an α, but not a β, *c-erbA* gene. *Endocrinology (Baltimore)* **133**, 2488–2495.

Shaffer, H. B., and McKnight, M. L. (1996). The polytypic species revisited: genetic differentiation and molecular phylogenies of the tiger salamander *Ambystoma tigrinum* (Amphibia: Caudata) complex. *Evolution* **50**, 417–433.

Shi, Y.-B. (1996). Thyroid hormone-regulated early and late genes during amphibian metamorphosis. *In* "Metamorphosis: Postembryonic Reprogramming of Gene Expression in Amphibian and Insect Cells" (L. I. Gilbert, J. R. Tata, and B. G. Atkinson, eds.), pp. 505–538. Academic Press, San Diego, CA.

Shi, Y.-B., Wong, J., and Puzianowska-Kuznicka, M. (1996a). Thyroid hormone receptors: mechanisms of transcriptional regulation and roles during frog development. *J. Biomed. Sci.* **3**, 307–318.

Shi, Y.-B., Wong, J., Puzianowska-Kuznicka, M., and Stolow, M. A. (1996b). Tadpole competence and tissue-specific temporal regulation of amphibian metamorphosis: Roles of thyroid hormone and its receptors. *BioEssays* **18**, 391–399.

Shi, Y.-B., and Ishizuya-Oka, A. (1997). Autoactivation of *Xenopus* thyroid hormone receptor β genes correlates with larval epithelial apoptosis and adult cell proliferation. *J. Biomed. Sci.* **4**, 9–18.

Stephenson, N. G. (1951). On the development of the chondrocranium and visceral arches of *Leiopelma archeyi*. *Trans. Zool. Soc. London* **27**, 203–253.

Stephenson, N. G. (1955). On the development of the frog, *Leiopelma hochstetteri* Fitzinger. *Proc. Zool. Soc. London* **124**, 785–795.

St. Germain, D. L., Schwartzman, R. A., Croteau, W., Kanamori, A., Wang, Z., Brown, D. D., and Galton, V. A. (1994). A thyroid hormone-regulated gene in *Xenopus laevis* encodes a type III iodothyronine 5-deiodinase. *Proc. Natl. Acad. Sci. U.S.A.* **91**, 7767–7771.

Stolow, M. A., and Shi, Y. (1995). *Xenopus* sonic hedgehog as a potential morphogen during embryogenesis and thyroid hormone-dependent metamorphosis. *Nucl. Acids Res.* **23**, 2555–2562.

Stolow, M. A., Ishizuya-Oka, A., Su, Y., and Shi, Y.-B. (1997). Gene regulation by thyroid hormone during amphibian metamorphosis: Implications on the role of cell-cell and cell-extracellular matrix interactions. *Am. Zool.* **37**, 195–207.

Suzuki, S. (1987). Plasma thyroid hormone levels in metamorphosing larvae and adults of a salamander, *Hynobius nigrescens*. *Zool. Sci.* **4**, 849–854.

Suzuki, S., and Suzuki, M. (1981). Changes in thyroidal and plasma iodine compounds during and after metamorphosis of the bullfrog, *Rana catesbeiana*. *Gen. Comp. Endocrinol.* **45**, 74–81.

Svob, M., Musafija, A., Frank, F., Durovic, N., Svob, T., Cuckovic, S., and Hlaca, D. (1973). Response of tail fin of *Proteus anguinus* to thyroxine. *J. Exp. Zool.* **184**, 341–344.

Swanepoel, J. H. (1970). The ontogenesis of the chondrocranium and of the nasal sac of the microhylid frog *Breviceps adspersus pentheri* Werner. *Ann.—Univ. Stellenbosch Ser. A* **45**, 1–119.

Szarski, H. (1957). The origin of the larva and metamorphosis in Amphibia. *Am. Nat.* **91**, 283–301.

Tata, J. R. (1968). Early metamorphic competence of *Xenopus* larvae. *Dev. Biol.* **18**, 415–440.

Tata, J. R. (1994). Autoregulation and crossregulation of nuclear receptor genes. *Trends Endocrinol. Metab.* **5**, 283–290.

Tata, J. R., Kawahara, A., and Baker, B. S. (1991). Prolactin inhibits both thyroid hormone-induced morphogenesis and cell death in cultured amphibian larval tissues. *Dev. Biol.* **146**, 72–80.

Titus, T. A., and Larson, A. (1996). Molecular phylogenetics of desmognathine salamanders (Caudata: Plethodontidae): A reevaluation of evolution in ecology, life history and morphology. *Syst. Biol.* **45**, 451–472.

Townsend, D. S., and Stewart, M. M. (1985). Direct development in *Eleutherodactylus coqui* (Anura: Leptodactylidae): A staging table. *Copeia,* pp. 423–436.

Trueb, L. (1985). A summary of osteocranial development in anurans with notes on the sequence of cranial ossification in *Rhinophrynus dorsalis* (Anura: Pipoidea: Rhinophrynidae). *S. Afr. J. Sci.* **81**, 181–185.

Trueb, L., and Cloutier, R. (1991a). Towards an understanding of the amphibians: Two centuries of systematic history. *In* "Origins of the Higher Groups of Tetrapods. Controversy and Consensus" (H.-P. Schultze and L. Trueb, eds.), pp. 175–193. Cornell University Press (Comstock), Ithaca, NY.

Trueb, L., and Cloutier, R. (1991b). A phylogenetic investigation of the inter- and intrarelationships of the Lissamphibia (Amphibia: Temnospondyli). *In* "Origins of the Higher Groups of Tetrapods. Controversy and Consensus" (H.-P. Schultze and L. Trueb, eds.), pp. 223–313. Cornell University Press (Comstock), Ithaca, NY.

Trueb, L., and Hanken, J. (1992). Skeletal development in *Xenopus laevis* (Anura: Pipidae). *J. Morphol.* **214**, 1–41.

Ultsch, G. R., and Arceneaux, S. J. (1988). Gill loss in larval *Amphiuma tridactylum. J. Herpetol.* **22**, 347–348.

Ussing, A. P., and Rosenkilde, P. (1995). Effect of induced metamorphosis on the immune system of the axolotl, *Ambystoma mexicanum. Gen. Comp. Endocrinol.* **97**, 308–319.

Wakahara, M., and Yamaguchi, M. (1996). Heterochronic expression of several adult phenotypes in normally metamorphosing and metamorphosis-arrested larvae of a salamander *Hynobius retardatus. Zool. Sci.* **13**, 483–488.

Wakahara, M., Miyashita, N., Sakamoto, A., and Arai, T. (1994). Several biochemical alterations from larval to adult types are independent on morphological metamorphosis in a salamander, *Hynobius retardatus. Zool. Sci.* **11**, 583–588.

Wake, D. B., and Marks, S. B. (1993). Development and evolution of plethodontid salamanders: A review of prior studies and a prospectus for future research. *Herpetologica* **49**, 194–203.

Wake, D. B., and Roth, G. (1989). The linkage between ontogeny and phylogeny in the evolution of complex systems. *In* "Complex Organismal Functions: Integration and Evolution in Vertebrates, Dahlem Konferenzen" (D. B. Wake and G. Roth, eds.), pp. 361–377. Wiley, New York.

Wake, M. H. (1982). Diversity within a framework of constraints. Amphibian reproductive modes. *In* "Environmental Adaptation and Evolution" (D. Mossakowski and G. Roth, eds.), pp. 87–106. Fischer, Stuttgart.

Wake, M. H. (1989). Phylogenesis of direct development and viviparity in vertebrates. *In* "Complex Organismal Functions: Integration and Evolution in Vertebrates, Dahlem Konferenzen" (D. B. Wake and G. Roth, eds.), pp. 235–250. Wiley, New York.

Wake, M. H. (1992). Evolutionary scenarios, homology and convergence of structural specializations for vertebrate viviparity. *Am. Zool.* **32**, 256–263.

Wake, M. H., and Hanken, J. (1982). Development of the skull of *Dermophis mexicanus* (Amphibia: Gymnophiona), with comments on skull kinesis and amphibian relationships. *J. Morphol.* **173**, 203–223.

Wake, M. H., Exbrayat, J.-M., and Delsol, M. (1985). The development of the chondrocranium of *Typhlonectes compressicaudus* (Gymnophiona), with comparison to other species. *J. Herpetol.* **19**, 68–77.

Walls, S. C., Bélanger, S. S., and Blaustein, A. R. (1993). Morphological variation in a larval salamander: Dietary induction of plasticity in head shape. *Oecologia* **96**, 162–168.

Wang, Z., and Brown, D. D. (1993). Thyroid hormone-induced gene expression program for amphibian tail resorption. *J. Biol. Chem.* **268**, 16270–16278.

Wassersug, R. J., and Duellman, W. E. (1984). Oral structures and their development in egg-brooding hylid frog embryos and larvae: Evolutionary and ecological implications. *J. Morphol.* **182**, 1–37.

Wassersug, R. J., and Heyer, W. R. (1988). A urvey of internal oral features of leptodacyloid larvae (Amphibia: Anura). *Smithson. Contrib. Zool.* **457**, 1–99.

Weil, M. R. (1986). Changes in plasma thyroxine levels during and after spontaneous metamorphosis in a natural population of the green frog, *Rana clamitans. Gen. Comp. Endocrinol.* **62**, 8–12.

Weinberger, C., Thompson, C. C., Ong, E. S., Lebo, R., Gruol, D. J., and Evans, R. M. (1986). The *c-erb-A* gene encodes a thyroid hormone receptor. *Nature (London)* **324**, 641–646.

Welsch, U., Schubert, C., and Storch, V. (1974). Investigations on the thyroid gland of embryonic, larval and adult *Ichthyophis glutinosus* and *Ichthyophis kohtaoensis* (Gymnophiona, Amphibia). *Cell Tissue Res.* **155**, 245–268.

White, B. A., and Nicoll, C. S. (1981). Hormonal control of amphibian metamorphosis. *In* "Metamorphosis: A Problem in Developmental Biology" (L. I. Gilbert and E. Frieden, eds.), pp. 363–396. Plenum, New York.

Wiens, J. J. (1989). Ontogeny of the skeleton of *Spea bombifrons* (Anura: Pelobatidae). *J. Morphol.* **202**, 29–51.

Wong, J., and Shi, Y.-B. (1995). Coordinated regulation of and transcriptional activation by *Xenopus* thyroid hormone and retinoic X receptors. *J. Biol. Chem.* **270**, 18479–18483.

Woody, A.-Y., and Justus, J. T. (1974). Comparative studies of hemoglobins from the clouded tiger salamander before and after metamorphosis. *J. Exp. Zool.* **188**, 215–224.

Wright, M. L., Cykowski, L. J., Lundrigan, L., Hemond, K. L., Kochan, D. M., Faszewski, E. E., and Anuszewski, C. M. (1994). Anterior pituitary and adrenal cortical hormones accelerate or inhibit tadpole hindlimb growth and development depending on stage of spontaneous development or thyroxine concentration in induced metamorphosis. *J. Exp. Zool.* **270**, 175–188.

Xavier, F. (1977). An exceptional reproductive strategy in Anura: *Nectophyrnoides occidentalis* Angel (Bufonidae), an example of adaptation to terrestrial life by viviparity. *In* "Major Patterns in Vertebrate Evolution" (M. K. Hecht, P. C. Goody, and B. M. Hecht, eds.), pp. 545–552. Plenum, New York.

Yamaguchi, M., Tanaka, S., and Wakahara, M. (1996). Immunohisto- and immunocytochemical studies on the dynamics of TSH and GTH cells in normally metamorphosing, metamorphosed, and metamorphosis-arrested *Hynobius retardatus*. *Gen. Comp. Endocrinol.* **104**, 273–283.

Yamashita, K., Iwasawa, H., and Watanabe, Y. G. (1991). Immunocytochemical study on the dynamics of TSH cells before, during, and after metamorphosis in the salamander, *Hynobius nigrescens*. *Zool. Sci.* **8**, 609–612.

Yaoita, Y., and Brown, D. D. (1990). A correlation of thyroid hormone receptor gene expression with amphibian metamorphosis. *Genes Dev.* **4**, 1917–1924.

Yaoita, Y., Shi, Y.-B., and Brown, D. D. (1990). *Xenopus laevis* α and β thyroid hormone receptors. *Dev. Biol.* **87**, 7090–7094.

Yoshizato, K. (1996). Cell death and histolysis in amphibian tail during metamorphosis. *In* "Metamorphosis: Postembryonic Reprogramming of Gene Expression in Amphibian and Insect Cells" (L. I. Gilbert, J. R. Tata, and B. G. Atkinson, eds.), pp. 647–671. Academic Press, San Diego, CA.

Zug, G. R. (1993). "Herpetology, An Introductory Biology of Amphibians and Reptiles." Academic Press, San Diego, CA.

CHAPTER 7

Hormonal Control in Larval Development and Evolution—Insects

H. Frederik Nijhout

Department of Zoology, Duke University, Durham, North Carolina

I. THE DEVELOPMENTAL HORMONES OF INSECTS

Virtually every aspect of the postembryonic development of insects is controlled by hormones. The timing of many developmental processes, as well as the choice of alternative developmental pathways, often is under hormonal rather than genetic control. The hormonally controlled developmental events in insects are enormously diverse. They range from the molting cycle, to the complex transformations of metamorphosis, to the diversity of alternative phenotypic forms a given individual can develop into, such as the castes of ants, bees, and termites and the distinctive seasonal forms of many species. The control points for these diverse developmental events occur almost entirely during the larval life of insects.

The endocrine system that manages this diversity of developmental transformations is, by contrast, very simple indeed. It involves not more than a handful of hormones, of which two, the ecdysteroids and juvenile hormones (JH), are responsible for virtually the entire panoply of events. Exactly how so few hormones can have such a broad diversity of actions, and at the same time such great specificity of regulatory effects on development, is the overarching subject of this chapter. As I consider how this system works, I will address questions about how it evolved and its implications for the morphological evolution of insects.

Figure 1 is a summary diagram of the principal endocrine players. The central nervous system controls all aspects of endocrine secretion via a set of regulatory neurosecretory hormones, a system analogous to the pituitary–hypothalamic axis of vertebrates. The two endocrine glands of primary concern for the regulation of development are the prothoracic glands, which secrete some of the ecdysteroids, and the corpora allata, which secrete the juvenile hormones. (I use the plural form for these hormones because they come in several molecular forms, but the JHs have the same functions, differing only in that some molecular variants are taxon-specific. Ecdysteroids also have similar functions, except that some are prohormones and relatively inactive until chemically modified.) Secretory activity of the prothoracic glands is controlled by the prothoracicotropic hormones (PTTH), produced by neurosecretory cells in the brain and released from a pair of neurohemal organs called

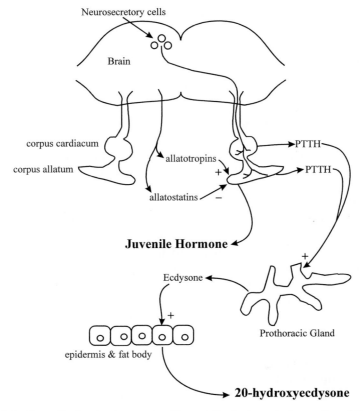

FIGURE 1   Generalized endocrine system of insects. The brain controls secretion of JH by the corpora allata via stimulatory and inhibitory neurohormones. The brain also controls ecdysone secretion by the prothoracic glands via the neurosecretory hormone PTTH. Ecdysone is a prohormone and is transformed into the active form, 20-hydroxyecdysone, by peripheral tissues like the epidermis and fat body. The diverse developmental decisions larvae make during the course of their lives are controlled largely by JH and 20-hydroxyecdysone, as outlined in the text.

the corpora cardiaca (except in the Lepidoptera, where the corpora allata also act as the neurohemal site for brain neurosecretions; Nijhout, 1975a; Agui *et al.*, 1980). In many species the prothoracic glands only secrete a prohormone, ecdysone, which is converted into the active hormone, 20-hydroxyecdysone, in the epidermis and fat body. In Lepidoptera such as *Manduca sexta,* the prothoracic glands actually secrete a pro-prohormone, 3-dehydroecdysone, that is converted to ecdysone in the hemolymph before being converted to the active hormone in the peripheral tissues (Warren *et al.*, 1988; Kiriishi *et al.*, 1990). Secretion of JH by the corpora allata is controlled by the brain via the secretion of activating and inhibiting hormones, the allatotropins and allatostatins (Nijhout, 1994). In addition, the corpora allata and prothoracic glands may be under direct nervous control in some species, and some of the neurosecretory regulators may be delivered directly to the gland via neurosecretory axons.

II. WHY HORMONES?

Why should hormones be involved in the control of development at all? All evidence from developmental genetics over the past two decades suggests that embryonic development is controlled by a largely self-organizing cascade of gene activation, in which one round of gene expression sets up the appropriate environment that stimulates the next round of gene expression (e.g., Lawrence, 1992). Gradients of maternal gene products in the egg stimulate (or inhibit) zygotic gene expression in a concentration-dependent manner. The products of these genes in turn diffuse and stimulate (or inhibit) expression of another set of genes, whose products in turn regulate the expression of genes yet farther downstream. Nowhere is there any evidence of a role for hormones, nor is there reason to suspect that these cascades of gene expression depend on any external control whatsoever.

The regulation of embryonic development is, therefore, very different from the regulation of postembryonic development, which, wherever it has been studied, is dominated by hormonal control mechanisms. One of the reasons why hormones are insignificant early in development whereas they come to play a dominant role later in development has to do with the dimensions of morphogenetic fields (Gilbert et al., 1996). Morphogenetic fields can be operationally defined as the regions (i) within which cells are in communication with one another, (ii) that develop as recognizable entities, and (iii) within which development is self-organizing. The dimensions of such fields have been estimated and typically are on the order of several hundred micrometers; they seldom, if ever, are greater than a millimeter (Wolpert, 1969). These dimensions exceed those of most eggs and embryos, and this implies that all cells of an embryo potentially are in communication with each other and that the whole embryo can act as a single field, as indeed it does, at least during the early stages of embryogenesis.

Later in development, after some growth has occurred, morphogenetic fields have become small relative to the size of the organism, and developmental control essentially becomes a local process. Now if it becomes necessary to coordinate development in distant parts, or to make different things happen simultaneously in different parts of the developing animal (as would be necessary during metamorphosis), some kind of centralized control and long-range signaling system will be required. Hormones provide that long-range signaling mechanism. Central control resides in the central nervous system, which regulates the secretion of developmental hormones through nervous input to the endocrine glands or, more commonly, via neurosecretory hormones that act either as developmental hormones in their own right or as activators or inhibitors of endocrine glands elsewhere in the body.

III. THE FUNCTION OF HORMONES IN INSECT LARVAL DEVELOPMENT

The role of hormones in insect development is very different from their role in metabolic and physiological regulation. Rather than acting in a quantitative manner in which the magnitude of their effect is proportional to their concentration (as is the case, for example, with the diuretic hormone of insects and insulin and the antidiuretic hormone of vertebrates), the developmental hormones of insects act in an all-or-none fashion. Rather than acting on a system that is always prepared to respond, insect developmental hormones act during discrete windows of tissue sensitivity. The timing and pattern of tissue sensitivity and the timing of hormone secretion are regulated independently and form an interactive system of great flexibility. Table I lists the various general properties of hormones in insect developmental systems that

TABLE I   Characteristics of Insect Endocrine Systems in the Regulation of Development

Empirical findings		Possible mechanisms
Type	Description	
1. Hormone-sensitve periods	Tissues only are responsive to hormone during discrete temporal windows	Response is contingent on another developmental event Response is contingent on activation of receptor mechanism
2. Tissue specificity	Not all tissues are responsive at a given time	Tissue-specific expression or activation of receptors
3. Spatial heterogeneity	Different tissues have different responses to the same hormone	Tissue-specific differences in response pathways downstream of receptor binding
4. Temporal heterogeneity	Same tissue responds differently to the same hormone at different times in development	Developmental changes in response pathways downstream of receptor binding

have been found empirically in the course of the past three decades or so, as well as the various mechanisms that underlie each of those properties. I will briefly discuss each of them in turn.

A. HORMONE-SENSITIVE PERIODS

The fact that insect tissues are not responsive to developmental hormones at all times is now well-established. The imaginal disks of *Drosophila* acquire competence to respond to ecdysteroids sometime during the first half of the second (penultimate) larval instar (Riddiford, 1993; Riddiford and Ashburner, 1991). In larvae of the ant *Pheidole bicarinata,* a JH-sensitive period for the induction of soldier development occurs during an approximately day-long period in the middle of the last larval instar (Wheeler and Nijhout, 1983). The prothoracic glands of larval *Manduca* become responsive to stimulation by PTTH only during the last portion of the instar (Watson and Bollenbacher, 1988). In *Manduca,* responsiveness to the eclosion hormone begins only a few hours prior to the actual release of the hormone (Reynolds *et al.,* 1979). In honey bee larvae, a JH-sensitive period for queen induction occurs during the fourth day of larval development, at the time of the molt to the final larval instar (Rachinsky and Hartfelder, 1990). In the bloodsucking bug *Rhodnius prolixus,* the decision to metamorphose, that is, the decision to either molt to an adult or to make an additional larval instar, is made during a well-defined JH-sensitive period (Nijhout, 1983). In *Manduca* there is a brief JH-sensitive period 12–24 hr prior to the molt to the final larval instar, during which the color of the cuticle of the final instar is determined (Truman *et al.,* 1973). The seasonal form of the Buckeye butterfly *Precis coenia* is determined during a 24-hr-long ecdysteroid-sensitive period that begins 28 hr after pupation (Rountree and Nijhout, 1996). Additional examples are shown in Fig. 5 and discussed by Nijhout and Wheeler (1982), Wheeler (1986), and Nijhout (1994).

In each of these cases, exposure to the hormone outside the sensitive period has no effect on development. Hormone-sensitive periods can be very brief (as in the cases of *Rhodnius* and *Precis* noted earlier) or quite long. The timing and duration of a hormone-sensitive period depends on the species, the tissue, and the function of the hormone during the developmental period in question. Certain ecdysteroid-sensitive periods are well-defined. For instance, the sensitive period for seasonal morph induction in the European map butterfly *Araschnia levana* lasts from 3 to 10 days after pupation (Koch and Bückmann, 1987), whereas that of *Precis coenia* lasts from 28 to 48 hr after pupation (Rountree and Nijhout, 1996). Juvenile hormone-sensitive periods range in duration from a few hours to several days during a given instar (many instances are discussed later). It seems reasonable to assume that the timing and duration of hormone-sensitive periods have evolved to optimize their specific functions, so that the developmental events triggered by the hormone can be timed to occur at suitable points within the animal's life cycle or within a specific developmental program. As yet no one has addressed the issues of adaptation or optimization experimentally.

The mechanisms underlying hormone-sensitive periods probably are diverse. The simplest reason for a hormone-sensitive period is that hormonal responsiveness is contingent on some progressive developmental event. For instance, some JH-sensitive periods are dependent on the ecdysteroid-induced molting cycle and occur at fixed times relative to the initiation of ecdysteroid secretion. The ecdysteroid signal initiates a series of cellular events that culminate in the secretion of a new cuticle, and, at some point, the developmental process reaches a stage that can be influenced by JH. At a later time, development has progressed so far that it can no longer be altered by external signals like JH. In traditional terms, JH controls a determination switch. As development progresses, tissues become competent to respond to a JH signal; once tissues are committed, JH can no longer change their state of determination. The simple form of this model assumes that hormone receptors are always present.

Another mechanism for controlling a hormone-sensitive period is through temporal regulation of cellular receptors for the hormone. Hormone-sensitive periods that do not have a clear external synchronizing signal, such as the ecdysteroid-sensitive periods for seasonal morph determination in *Precis coenia* and *Araschnia levana* (Koch and Bückmann, 1987; Rountree and Nijhout, 1996), may be controlled in this fashion. The dynamics of receptor expression or activation could be readily tied to specific developmental stages or events, so that hormone-sensitive periods are timed to occur at appropriate times in development.

B. TISSUE SPECIFICITY OF THE RESPONSE

Not all tissues respond to a hormonal signal. This is, of course, a feature shared by homeostatic and regulatory hormones as well. The cause of tissue specificity probably is that tissues differ in the time (or whether or not) they express or activate hormone receptors. Tissue- and time-specific patterns of hormone receptor expression have been documented in several cases (Talbot *et al.*, 1993; Riddiford and Truman, 1994; Cherbas and Cherbas, 1996; Truman, 1996). In the trivial extreme case, some tissues never respond to a hormone because the function of the tissue is irrelevant to that of the hormone (for instance, epidermal cells do not respond to diuretic hormone). Such tissues fail to respond because they never express receptors for the hormone.

Sometimes tissue specificity is due to the fact that the hormone-sensitive periods for different tissues do not coincide, and this can have a complex developmental consequence. If two sets of tissues have nonoverlapping hormone-sensitive periods, then secretion of the hormone during the first period will produce a different response than secretion of the hormone during the second. Then, by regulating whether the hormone is secreted during one, both, or neither sensitive period, the animal can control an array of alternative developmental options. This is a particularly important regulatory mechanism when an animal needs to choose from more than two alternative pathways. For instance, at metamorphosis a *Pheidole* ant larva has to decide whether to pupate or molt to an additional larval instar (regulated by JH) and, if it chooses to pupate, whether to form a worker or a soldier pupa (also regulated by JH). This triple choice is managed through two binary choices spaced well apart in larval life. During about the middle of the last larval instar there is a JH-sensitive period that manages the switch between the soldier versus worker developmental pathway and nothing else. Several days later, there is another JH-sensitive period during which the commitment to metamorphose is made, but that has no effect on whether the animal will become a worker or a soldier (Wheeler and Nijhout, 1983; Wheeler, 1991). Likewise in honey bees, the JH-sensitive period for queen determination occurs several days before the JH-sensitive period for metamorphosis (Wirtz, 1973; Rachinsky and Hartfelder, 1990). In lower termites there appear to be three successive JH-sensitive periods in a pseudergate instar, during which either soldier characters, replacement reproductive characters, or alate imaginal characters are determined (Nijhout and Wheeler, 1982).

C. Spatial Heterogeneity of the Response

Not all tissues respond in the same way to the same hormone. In response to ecdysteroids at metamorphosis, for instance, most epidermal cells of caterpillars undergo cell division and subsequently secrete a new cuticle, but the epidermal cells in the prolegs undergo programmed cell death (Fain and Riddiford, 1977; Riddiford, 1985). In higher Diptera, ecdysteroids provoke a complex evagination of the imaginal disks, requiring changes in cell shape and specific morphogenetic movements (Oberlander, 1985), while at the same time they induce mitosis in some epidermal cells and cell death in others (Lockshin, 1985).

In order to obtain different responses to the same hormonal signal, either the cells in different tissues must use different receptors or the receptors must activate different response elements. Some tissues differ in the types of ecdysteroid receptor heterodimers and isoforms they express. This may be related to the differential responses of those tissues, but the connection between receptor isoform and response type has yet to be demonstrated critically (Talbot *et al.*, 1993; Truman, 1996; Cherbas and Cherbas, 1996).

D. TEMPORAL HETEROGENEITY OF THE RESPONSE

At different times in development, a given tissue can respond in different ways to the same hormone. The most common case of this is in holometabolous metamorphosis, where a general epidermal cell can secrete either a larval, pupal, or adult cuticle in response to ecdysteroids, depending entirely on its prior developmental history. In *Drosophila* an imaginal disk initially is unresponsive to ecdysteroids; then, after the second larval instar, it evaginates in response to ecdysteroids, and only after evagination does it secrete a cuticle in response to the same hormone (Oberlander, 1985; Riddiford, 1993).

As in the previous case, the reason cells respond differently at different times must be due either to the expression of different receptors or to the fact that, as development progresses, the receptors are linked to different (developmental stage- and location-specific) response pathways, so that different genes are transcribed or different enzymes activated.

IV. CONTROL OF MOLTING

Insects, like all arthropods, have a nonliving external skeleton, or cuticle, that cannot expand (there are exceptions, which I will return to later). In order for the animal to grow, this inextensible cuticle must be shed periodically so that a new, larger cuticle can be made. The process of shedding an old cuticle and manufacturing a new one is called molting or, because it recurs periodically, the molting cycle. Insects do not grow as adults, so molting is strictly a larval activity. When insects are maintained under ideal conditions, larvae of many species feed and grow continuously and therefore must molt regularly.

The causal factors, or stimuli, for the molting process have been analyzed at two levels. The majority of studies have dealt with the immediate or proximal causes, namely, the hormonal stimulus for molting and its effect on the physiology and biochemistry of the epidermis. Fewer studies have dealth with the deeper causal factors by which the timing of the molts within the life cycle of the larva are regulated. I deal with each of these two levels of causation in turn.

It is well-established that secretion of the molting hormones, the ecdysteroids, is controlled by the brain via the secretion of the prothoracicotropic hormone (PTTH). In *Manduca,* the secretion of PTTH at each larval molt is tightly controlled by a photoperiodic clock. PTTH is secreted only during well-defined photoperiodic gates, and whether a larva secretes PTTH during any given gate depends on its prior growth history (Truman, 1972). PTTH directly stimulates ecdysteroid synthesis in the prothoracic glands. As noted previously, the prothoracic glands actually secrete a prohormone, ecdysone, which is converted into the active hormone, 20-hydroxyecdysone, in the epidermis and fat body. The 20-hydroxyecdysone acts directly on the cells of the epidermis and stimulates the complex sequence of events that causes these cells to separate from the cuticle (a process known as apolysis), undergo a round of cell division (by which the integument grows in surface area), partially digest the old cuticle, and synthesize the new cuticle of the next and larger larval instar. There is evidence that the rising and falling phases of the pulse of ecdysone that stimulates the larval–pupal molt stimulate the different stages of this complex sequence of events (Riddiford, 1985; Hiruma et al., 1991). In the case of a larval–larval molt, however, it is not clear how the different stages in the sequence of events of the molting cycle are controlled or, indeed, whether they are controlled independently.

The hormonal control of molting is the same whether it is a larva → larva molt, a larva → pupa molt, a pupa → adult molt, or, as in the case of Hemimetabola, a larva → adult molt (Bollenbacher et al., 1981; Nijhout, 1994). The characteristics of the stage the animal molts to are controlled separately, by juvenile hormone, as discussed later. In fact, it appears that the hormonal control of molting by ecdysteroids is the same throughout the arthropods. Evidently there has been little evolution in the endocrine control mechanism of molting. By contrast, there has been enormous evolutionary divergence in the response mechanism to the molting hormone. At the beginning of the response chain there is a modest divergence in ecdysteroid receptor isoforms, as suggested by the fact that tissues with different responses to ecdysteroids often express different isoforms (Talbot et al., 1993; Cherbas and Cherbas, 1996; Truman, 1996). Presumably different receptors are at the head of different signaling pathways that regulate different cellular responses. Ecdysteroid receptors have some sequence similarity to thyroid hormone receptors, which may be a sign of functional convergence, though it is more likely to be a shared primitive character. In either event, receptor diversity probably contains little information about the evolution of the diversity of ecdysteroid responses.

Downstream of the receptor there has evolved such a great diversity of responses that almost every region of integument, in every species, differs in some way in the characteristics of the cuticle it synthesizes in response to the ecdysteroid signal. Even if we ignore for the moment the interstage differences in the types of cuticle synthesized in response to ecdysteroids, within a single individual we find vastly different cuticles in the joints, in the middle of a sclerite, around sensillae, over the compound eye, on the mandibles and tarsal claws, etc. An enormously complex cell biology is necessary to secrete and mold a butterfly scale, one of the most elaborate and intricate unicellular structures in the animal world (Ghiradella, 1985). Yet neighboring epidermal cells on a butterfly wing can manufacture either such a scale or a bland, simple, featureless cuticle in response to the same ecdysteroid signal. Neither the hormone nor the receptor contains the necessary information to account for the details of structure produced, nor for differences among the many kinds of cuticle. It is conventional to assume that the suite of proteins synthesized or activated in response to the hormonal signal must be different in different kinds of cells and that ultimately this accounts for the differences in structure and texture of the various types of cuticle. Indeed, Willis (1996) and others have shown that the cuticles from different regions of the body differ in at least a subset of proteins they contain. But the mechanisms by which such protein differences interact with other elements within the cells to produce the variety of structural and mechanical types of cuticles remain largely unexplored (but see Overton, 1966; Ghiradella, 1989).

The second view of the control of molting does not ask how hormones control the molting cycle, but why those hormones are secreted at some particular time and not another. In larvae that grow under ideal conditions, molts occur on a predictably regular schedule, but it has proven very difficult to ascertain exactly what the first stimulus is that initiates the process each time. The ultimate stimulus for molting is known only in the Hemiptera. In bloodsucking bugs such as *Rhodnius prolixus* and members of the genera *Triatoma* and *Dipetalogaster,* the actual taking of a large blood meal stimulates stretch receptors in abdominal nerves, which in turn stimulate the brain to secrete PTTH (Wigglesworth, 1934; Nijhout, 1984). These animals molt only after they take a blood meal. If *Rhodnius* larvae are not fed, or are fed only small meals that do not achieve the requisite degree of stretch, they may live for a year or more, but will never molt. In other Hemiptera, such as the milkweed bug *Oncopeltus fasciatus,* abdominal stretch also is the ultimate stimulus for molting, but in this case a single large meal is not necessary. Simple gradual growth of abdominal volume by normal small meals over a period of days or weeks eventually induces a molt when a critical degree of stretch is achieved. Interestingly, a simple injection of saline into the abdomen can produce sufficient stretch to stimulate *Oncopeltus* larvae to molt (Nijhout, 1979). It has proven difficult to demonstrate whether or not stretch reception also stimulates molting in other insects, even in cases where the initiation of a molt is tightly correlated with the achievement of a critical weight (Nijhout, 1994). Not all insects need to grow in order to molt. Beetles (Beck, 1971) and certain caterpillars (Yin and Chippendale, 1974, 1976) continue to molt at irregular intervals even when starved or in diapause. Such larvae actually become smaller at each molt.

V. THE HORMONAL CONTROL
OF METAMORPHOSIS

The various properties of developmental endocrine control systems outlined previously are nowhere better illustrated than in the various roles JH plays in the control of metamorphosis. Next I will briefly outline old and current models of JH action. Our current understanding of the role of JH in metamorphosis has revealed the mechanics of an extraordinarily flexible developmental system that has been exploited by insects in myriad ways to control not only their metamorphosis but also the development of many specialized features of the larval and adult stages.

A. The High–Low–No Model of JH Action
in Metamorphosis

The conventional model for the control of metamorphosis suggests that the progression from larval to pupal and then to adult morphology is controlled by the concentration of JH. Under this model, levels of circulating JH are assumed to be high during early larval life but decline during the last larval instar. When a larva molts in the presence of high JH levels, it manufactures another larval cuticle and thus molts to another, now larger, larval instar. When a molt occurs in the presence of a diminished level of JH, the larva pupates because low levels of JH regulate the expression of pupa-specific genes. During the pupal stage, JH disappears entirely, and when the pupa now molts in the absence of JH, adult-specific genes become expressed. This model was first proposed by Schneiderman and Gilbert (1964) and given a suitable molecular genetic interpretation by Williams and Kafatos (1971). At the time, this was a reasonable model, consistent with some of the earliest experimental data, and is so charmingly simple and direct that it continues to be widely adopted in textbooks of biology [e.g., most recently by Kalthoff (1996)].

B. The Status Quo Model of JH Action

However, not everyone accepted such a simple role for JH in metamorphosis. Early on, Willis (1974) pointed out that JH appeared to have its effect only during tissue-specific critical periods in development, and Williams (1961) and Riddiford (1972, 1996) came to emphasize the "status quo" action of JH rather than its directing effect during metamorphosis.

The "status quo" model for JH action differs from the high–low–no model in that it assumes that JH has only *one* function, namely, the preservation of the current state of gene expression. In the presence of JH, then, a larva remains a larva because it continues to express a larval genetic and developmental "program," and therefore it simply forms a larger version of itself at each molt. In the absence of JH the larva progresses to the next developmental stage when it molts, but JH does not control the characteristics of that next stage. As far as is known, a holometabolous insect larva molts to a pupa, and a pupa molts to an adult, because of an intrinsic developmental program within each tissue. Progression to the next stage simply requires the absence of JH. Accordingly, if a pupa is given a high dose of JH, it maintains the status quo and molts to an additional pupal stage (Williams, 1961).

C. JUVENILE HORMONE-SENSITIVE PERIODS

Experiments over the past two decades have failed to find critical evidence for the high–low–no model. The first major prediction of this model is that larval epidermal cells should have two thresholds of JH sensitivity. When JH is above both thresholds epidermal cells secrete a larval cuticle, when between the two they secrete pupal cuticle, and when below both they secrete adult cuticle. However, in the complete absence of JH, larval epidermal cells actually secrete pupal cuticle (Fukuda, 1944; Riddiford, 1985). Imaginal disks of *Hyalophora* and *Manduca,* by contrast, can bypass the pupal stage and go from a larval state directly to the adult. When a larva is allatectomized (i.e., has its corpora allata, the glands that produce JH, removed), metamorphosis will occur in the complete absence of JH, and many of the imaginal disks of such larvae produce an adult cuticle without having first made a pupal cuticle. Typically, the transformation to the adult is incomplete and the animal is a nonviable mosaic of pupal and adultoid characters. This premature adult expression can be inhibited by a small dose of JH given just prior to the pupal molt. The imaginal disks of allatectomized animals "rescued" by such a small dose of JH secrete a normal pupal cuticle (Kiguchi and Riddiford, 1978). A small amount of JH evidently is necessary for the imaginal disk to develop normal pupal characteristics. This finding, on first sight, appears to provide strong, albeit circumstantial, evidence for the high–low–no model.

As it turns out, it is not the dosage but the precise timing of this small dose of JH that controls the development fate of the imaginal disks. In normal larval–pupal development of *Manduca* there actually is a brief peak of JH secretion that coincides with an equally brief JH-sensitive period, just prior to the larval–pupal molt (Fig. 2). It is this coordinated event that is critical for the development fate of the imaginal disks, as we will see in the following.

When, in the mid-1970s, it became possible to measure JH titers accurately, little evidence was found to support the second major prediction of the high–

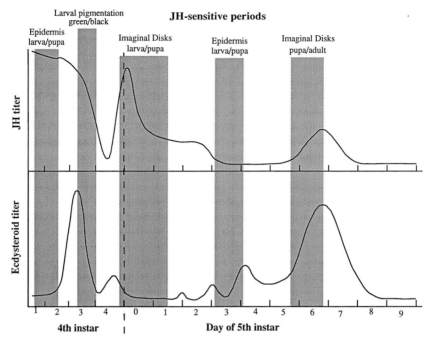

FIGURE 2 Juvenile hormone-sensitive periods during the last larval instar of *Manduca sexta*.
Top panel shows the profile of JH titers; bottom panel shows the profile of ecdysteroid titers
during the last larval instar. The functions of the various sensitive periods are explained in the
text. Hormone titer data are taken from Baker *et al.* (1987), Bollenbacher *et al.* (1981), Fain and
Riddiford (1975), Nijhout and Williams (1974b), and Wolfgang and Riddiford (1986).

low–no model: a stepwise decline in JH titers during metamorphosis. Rather
than being high during most of larval life, declining to an intermediate level
during the last larval instar and pupal molt, and vanishing during the pupal
stage and adult molt, the JH titer profile proved to be rather erratic, with sharp
peaks and valleys. The titers are best known for *Manduca sexta* (Fig. 2), where
the level of JH in the blood typically drops and rises again just prior to each
larval molt (Baker *et al.*, 1987). In the last larval instar, the JH titer rises
sharply during the molt and falls again equally sharply during the first day of
the instar. It then remains at a low plateau until the larva reaches a well-
defined critical weight (Nijhout and Williams, 1974b), after which JH vanishes
entirely, only to rise again for a day-long period immediately after the larva
enters the prepupal stage (Nijhout and Williams, 1974b; Fain and Riddiford,
1975; Baker *et al.*, 1987). In other species, including the Hemimetabola, the
JH titers are equally erratic [for instance, see Lanzrein *et al.*, (1985)].

It turns out that the erratic pattern of JH secretion has a function. Each of the peaks and valleys shown in Fig. 2 is meaningful in orchestrating the orderly transformation of larva to pupa: different tissues have different windows of sensitivity to JH, so that most do not experience this fluctuation but only monitor the presence or absence of JH during their time window. How this works can best be seen by overlaying the various juvenile hormone-sensitive periods that have been found so far on top of the JH titer curve (Fig. 2). We will begin in the middle with the JH-sensitive period that has received the most attention, namely, the one that controls larval pupal commitment of the epidermis in the middle of the fifth instar (Fig. 2; epidermis, larva–pupa). Riddiford (1976, 1978, 1985) has shown that this period is cued by the first secretion of ecdysteroids that occurs in the absence of JH. If JH continues to be absent during this period, the epidermis switches to pupal commitment (though it will not molt and synthesize the new pupal cuticle until a second ecdysteroid secretion takes place), but if JH is supplied exogenously during the mid or late final larval instar, it is able to inhibit pupal commitment of the epidermis only during the brief period indicated. A JH-sensitive period that may be homologous to this one occurs late during the penultimate larval instar (Fig. 2; epidermis, larva–pupa), because if JH is removed (by allatectomy or by a ligature behind the head) prior to this period, the penultimate larva undergoes premature pupation (Truman, 1972).

Pupal commitment of the imaginal disks occurs separately from that of the epidermis, during the first 24–48 hr of the last larval instar, while JH titers are still high enough to inhibit pupal commitment of the epidermis. Pupal commitment of the imaginal disks can be prevented, however, by application of exogenous JH during this early sensitive period (Kremen and Nijhout, 1989). These findings indicate that the imaginal disk may have a higher threshold of sensitivity to JH, so that it is able to commit to pupal development in the presence of JH titers that are high enough to maintain the status quo of the epidermis, but are readily maintained in their larval state by raising the JH titer through exogenous application. This early JH-sensitive period (Fig. 2; imaginal disks, larva–pupa) is known to be complex, because different imaginal disks switch to pupal commitment at different times during the 48-hr period indicated (Kremen and Nijhout, 1989, 1998). For any one disk the actual JH-sensitive period is a subset of the 48-hr period shown in Fig. 2.

There is a second JH-sensitive period for the imaginal disk near the end of the last larval instar (Fig. 2; imaginal disks, pupa–adult), just after entry into the prepupal stage and coincident with a brief peak of endogenous JH secretion. If JH is absent during this period, the already pupal-committed imaginal disk now becomes committed to adult development and will express adult features when the animal molts. In the presence of JH (the normal condition), the status quo of pupal commitment is maintained. This JH-sensitive period (and its accompanying peak of JH secretion) is the one that led to the initial confusion about whether low levels of JH must be present for normal pupation (Nijhout, 1994).

An unusual JH-sensitive period occurs just prior to the molt to the fifth larval instar (Fig. 2; larval pigmentation). The presence or absence of JH during this period controls the pigmentation of the integument of the final larval instar (Truman et al., 1973). In the presence of JH the final instar will develop a typical green integument, whereas in the absence of JH melanin is deposited in the integument and the final instar caterpillar is black. This JH-sensitive period has no known function in Manduca sexta, but in the related species, Manduca quinquemaculata, a normal color polyphenism in the last larval instar is believed to be controlled by this mechanism. When reared under short day conditions, final instar larvae of this species are dark brown. If reared under long day conditions they are green (Hudson, 1966). It is assumed that this polyphenism is controlled by environmentally sensitive variation in the JH titer during this JH-sensitive period. In M. sexta there is a genetic black larva mutant whose phenotype is due to a diminished JH titer during the JH-sensitive period and is the basis of a very sensitive bioassay for JH.

D. Hemimetabolous Metamorphosis

At metamorphosis, hemimetabolous insects molt directly from larva to adult without the intervention of a pupal stage. Typically a greater progression of morphology is seen during larval life than we find in the Holometabola, such as external wing pads that grow visibly during the last few larval instars and pigment patterns and overall body form that gradually approach those of the adult. There are many exceptions, of course, as in those orders with aquatic larvae (such as the Odonata), where larval–adult divergence can be almost as great as that found in insects with complete metamorphosis.

The hormonal control of metamorphosis in Hemimetabola is somewhat simpler than that in Holometabola, though it has its own complexities. Here metamorphosis also depends on a decline in JH. In most cases studied so far, the JH titer drops to undetectable levels during the first few days of the last larval instar, and the molt to the adult occurs in the absence of JH. Juvenile hormone titers earlier in larval life are every bit as variable as those in Manduca. The JH titer profile of the roaches Diploptera punctata (Tobe et al., 1985; Kikukawa and Tobe, 1986) and Nauphoeta cinerea (Lanzrein et al., 1985) exhibits a profile of apparently erratic peaks and valleys, reminiscent of those in Fig. 2, that begs the question of function. In Diploptera there are JH-sensitive periods for metamorphosis during the first one-third of the final (fourth) and penultimate (third) nymphal instars (Kikukawa and Tobe, 1986). If JH is absent during either of these periods, the next molt will be a metamorphic molt to the adult stage. Interestingly, if JH is removed by allatectomy during earlier larval instars, a premature metamorphosis does not follow. Evidently, the system of very young larvae is unable to undergo metamorphosis, and the presence or absence of JH is irrelevant to the maintenance of the larval state during the early larval instars (Kikukawa and Tobe, 1986).

E. Definition of a JH-Sensitive Period

The timing of JH-sensitive periods has been determined only approximately in most systems and typically is known only to the nearest 12–24 hr or so. The lack of accuracy in finding the precise beginning and end of a JH-sensitive period is due largely to the inability to synchronize the molting cycle of large groups of animals sufficiently closely; different individuals in an experimental cohort are not of the same physiological age. In the bloodsucking bug *Rhodnius prolixus,* the timing of the JH-sensitive period in the last larval instar can be determined with great accuracy because the timing of the molting cycle can be controlled accurately by giving the animal a blood meal (see previous discussion). In this hemimetabolous insect, the JH-sensitive period is complex and has some individual variability. The JH-sensitive period lasts from day 3 to day 9 after a blood meal, but any individual animal only requires the presence of JH during a 2–4-day period somewhere within that 6-day window. JH is equally effective late or early in this window, but individuals differ in the length of time they require JH. In some, metamorphosis is prevented by a 2-day exposure; others require as much as 4 days. Therefore, if JH sensitivity is tested by single applications of JH, animals that require a long exposure to JH would appear to be less sensitive and require larger doses of the hormone than animals that require only a brief exposure (Nijhout, 1983). Some confusion in the literature as to the dosage requirement for JH undoubtedly is due to this effect.

VI. ALL-OR-NONE AND MOSAIC EFFECTS OF JH?

In discussing the status quo model for JH action, I noted that JH appears to have an all-or-none effect on metamorphosis. This statement obviously needs to be refined because we have seen that many tissues differ in their JH-sensitive period and in their threshold of sensitivity to JH. So the presence of JH at any given brief period of time, or at a given low concentration (so that it is below the threshold of some responses), has a mosaic effect: some parts of a tissue or organ will retain the status quo whereas others will progress in their differentiation. This mosaicism of the response to JH extends to the cellular and subcellular levels. When the last instar larvae of a variety of insects are exposed to JH sometime during the middle of their JH-sensitive periods, they form mosaic cuticles with areas of pure larval and pure adult (or pupal) cuticles, but in certain places of their bodies they also form cuticles that express *both* larval and adult (or pupal) characters simultaneously (Lawrence, 1969; Willis *et al.*, 1982; Willis, 1996). Apparently, an epidermal cell can secrete a cuticle with mixed larval and adult features (such as larval pigmentation, but adult surface sculpturing). Different features of the cuticle secreted by a single epidermal cell switch their developmental fate independent of each other, which implies that different aspects of an epidermal cell's physiology and behavior can be switched independently.

VII. A MODULAR MECHANISM OF DEVELOPMENTAL CONTROL

The mechanism I described earlier for the control of differentiation during metamorphosis clearly is more complex than that proposed by the simple high–low–no model. But it is also far more interesting biologically, because it demonstrates the operation of an essentially modular control mechanism of great flexibility. Control over development is modular in both time and space. It is compartmentalized in time because different developmental "decisions" are made at different points throughout the larval instar, and it is compartmentalized in space because at any one time only one or a few tissues have their prospective developmental fates altered by a hormone. The decisions in each case are binary: "switch to a new stage or remain as you are," as proposed by the status quo model. There is no evidence that development is modulated by varying concentrations of the hormone. Instead, different portions of a tissue, and even different aspects of a cell's physiology and behavior, respond in an all-or-none fashion to a particular threshold value of hormone. Not all tissues have the same threshold of sensitivity. The imaginal disks of *Precis* and *Manduca*, for instance, are considerably less sensitive to JH than the abdominal epidermis (that different tissues have different sensitivity thresholds to JH has led to much confusion and occasionally has been misinterpreted as evidence for the high–low–no model of JH action).

The timing of tissue sensitivity, the threshold of that sensitivity, the characteristics of the actual response, and the temporal pattern of JH secretion and metabolism are controlled independently. This modular structure of insect development makes it simple to program multiple alternative decisions and to alter the development of one set of features without affecting that of others.

Developmental modularity has been a critical component in the evolution of complex life cycles and alternative phenotypes in insects, as discussed in the section on polyphenisms.

VIII. SIZE AT METAMORPHOSIS

We have seen that the control of the molting cycle can be studied at two levels, depending on whether one is interested in proximate (hormonal and biochemical) or ultimate (physiological and environmental) causes (see Section IV). The control of metamorphosis also can be analyzed at those two levels. I have already dealt with proximal hormonal control at some length; I now step farther back and ask what kind of information a larva uses to decide when to begin metamorphosis. More precisely, on the basis of what information does a larva decide whether it is currently in the last larval instar, so that the next time it molts it will not form a larger larva but will molt to a pupa or adult?

The number of larval instars is not simply a genetically fixed character. Although many insects have a characteristic number of larval instars when they grow under ideal conditions, that number can vary when larvae experience suboptimal conditions such as food deprivation or parasitization. If larval growth is slowed down the number of molts often increases, so that the adult stage is reached in a larger number of smaller incremental steps in larvae that grow slowly than in those that grow more rapidly. Because the study of control mechanisms requires variation, the control mechanisms over instar number and size at molting are studied most readily in larvae growing under suboptimal conditions.

A. THRESHOLD SIZE

The mechanism by which a larva "decides" whether it is in the last larval instar is best understood in the tobacco hornworm *Manduca sexta*. In this species, the larva monitors its own size and begins metamorphosis only after a well-defined threshold size is reached. In order to understand how this works, it is necessary to recognize that there are actually two measures of size for a larva: (1) the size of the hard parts of its exoskeleton, which do not change during an instar, and (2) its biomass, which continually increases as it feeds and assimilates nutrients. Both of these measures of size play a role in the control of the timing of metamorphosis. Both measures also are functionally related. The size of the exoskeleton of a given instar is determined largely by the size the previous instar attained before it molted. Thus, if two larvae of identical initial size are given different amounts of food, so that one molts at a lower weight than the other, then, in the next instar, the size of the exoskeleton of the first larva will be smaller than that of the second. Growth in biomass therefore can affect the size of the exoskeleton. Size of the exoskeleton, in turn, establishes the size of the mouthparts, which affects the rate at which a larva can grow and accumulate biomass during a given instar (Nijhout and Williams, 1974a; Nijhout, 1975b, 1981).

Although the two measures of size are coupled functionally, they play a very different roles in the control of the timing of metamorphosis. Just as there are two measures of size, there are also two ways in which the time of metamorphosis can be expressed: one is the specification of the number of the instar after which the metamorphic molt will occur, and the other is the time *within* that instar at which the process of metamorphosis begins. Interestingly, the latter has the greatest effect on the final size of the adult. This is so because insects increase in size exponentially. Consequently, most of the growth actually occurs during the final larval instar. *Manduca* larvae, for instance, begin the final instar at a weight of about 1 g and grow to between 7 and 9 g during the instar. More than 80% of an individual's growth occurs during the last larval instar. It is inherent in an exponential growth trajectory that, no matter how many instars there are, most of the growth will occur during the last one.

Larvae of *Manduca* are able to ascertain both size of their exoskeleton (or rather, something associated with the absolute size of the instar they are in) as well as their body mass within the instar. Whether a larva will behave developmentally and physiologically as if it is in its last instar can be predicted accurately by measuring the width of its head capsule. Larvae with head capsules wider than 5 mm are in the final larval instar and those with smaller head capsules are not, regardless of their prior growth history (Nijhout, 1975b). Exactly what the larva itself is measuring is not known, only that it is highly correlated with head capsule width and appears to emerge as a measure of *absolute* size. The 5-mm head capsule size thus constitutes a *threshold size for metamorphosis*, in that all larvae will begin metamorphosis in the instar whose head capsule size exceeds 5 mm. *Manduca* typically has five larval instars, but if a larva grows poorly during an early instar it may not reach the threshold size when it molts to the fifth instar; thus it subsequently will molt to a sixth larval instar before pupating (Fig. 3).

B. Critical Weight

Once a *Manduca* larva is in the final instar several aspects of its physiology change. We have already seen that the imaginal disks become committed to pupal development during the first few days of the final instar, even though

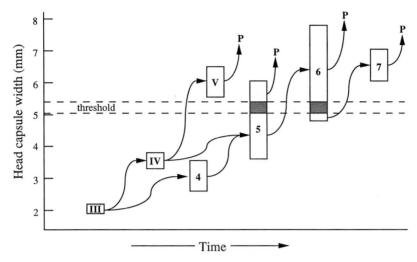

FIGURE 3   Threshold size for metamorphosis in *Manduca sexta*. Vertical bars indicate the range of head capsule sizes measured in cohorts of normal animals (Roman numerals) and animals that were starved or underfed in early instars (Arabic numerals). Normally growing animals surpass the critical body size for metamorphosis (threshold) for the first time in the fifth larval instar. When animals are underfed early in larval life the size range in each instar is increased, so that some fifth instar larvae are still below the threshold size, and these continue to form a supernumerary (sixth) instar. If some of those are still below threshold size they will form a seventh instar. Redrawn from Nijhout (1975b).

JH titers in the blood are still high. But for our present purposes, the most interesting change is that the brain's PTTH-secreting mechanism suddenly becomes sensitive to inhibition by JH. While JH is circulating, the brain is inhibited from secreting the PTTH that would start the next molt (Nijhout and Williams, 1974b; Rountree and Bollenbacher, 1986). At the same time, JH also inhibits the ability of the prothoracic glands to respond to PTTH (Gruetzmacher *et al.*, 1984; Bollenbacher, 1988; Sakurai and Gilbert, 1990). We believe this inhibition initially evolved as a safety mechanism to prevent the onset of a metamorphic molt while there is still some JH in the blood. Low remaining levels of JH might prevent pupal commitment of the most JH-sensitive tissues, so if a molt occurred before JH had been fully cleared the animal would molt to a nonviable larval–pupal mosaic.

It turns out that the inhibition of PTTH and ecdysteroid secretion by JH also is a critical part of a surprising regulatory mechanism of JH secretion. The last larval instar of the silkworm *Bombyx mori* has an unusual pattern of ecdysteroid secretion. In contrast to the earlier instars (where the level of ecdysteroids in the blood is elevated slightly during the first few days of the instar), the ecdysteroid titer is very low during the first few days of the final larval instar (Gu and Chow, 1993, 1996). By manipulating ecdysteroid levels during the last two instars, Gu and Chow (1996) showed that when the ecdysteroid titer is elevated during the early portion of any instar the corpora allata continue to secrete JH throughout the entire instar and an additional larval molt ensues. The low levels of ecdysteroids at the beginning of the last larval instar are crucial for normal cessation of JH secretion in the middle of the instar. Only if ecdysteroid levels are low during the first few days of the instar do the corpora allata cease JH secretion in the middle of the instar. Earlier I noted that, in the last larval instar, PTTH and ecdysteroid secretion come under inhibition by JH. It appears that this inhibition leads to lower than normal ecdysteroid titers. These, in turn, are essential for the cessation of JH secretion later in the instar.

The overall effect of this mutual inhibitory mechanism is that PTTH and ecdy-steroid secretion will not begin until all JH secretion has stopped and the remain-ing JH has been cleared. The timing of the metamorphic molt therefore is deter-mined in large part by the timing of JH clearance. If larvae of *Manduca* are allatectomized during the early days of the final larval instar, they enter the pre-pupal stage 24–36 hr later. By artificially varying the growth rate of caterpillars, it can be shown that larvae always enter the prepupal stage 24–36 hr after they reach a weight of 5 g. Measurements of JH titers show that JH begins to decline rapidly when an animal reaches the 5-g weight, presumably because the corpora allata cease secreting JH at that time. It takes JH approximately 24 hr to disappear. PTTH secretion then occurs during the next available photoperiodic gate (PTTH secretion can only occur during a brief window of the day, determined by a photoperiod-driven circadian clock; Truman, 1972). If larvae are treated with JH soon *after* they have reached the 5-g critical weight, PTTH secretion and the pupal molt are delayed in proportion to the dose received. Thus, the timing of metamorphosis within the last larval instar is determined by the time when JH is cleared from the blood, which, in turn, depends on when the larva reaches its critical weight. During the 24–36-hr time lag required to clear JH and meet the next photoperiodic gate, the larva continues to feed and grow and can gain an additional 2–4 g before entering the prepupal stage. The critical weight at which secretion of JH stops is not an absolute value but is proportional to the size of the larval exoskeleton (which, in turn, depends on the weight at the previous molt). The 5-g value mentioned earlier holds only for animals growing under ideal laboratory conditions; the actual relation between body size (measured as head capsule width) and critical weight is shown in Fig. 4.

C. SIZE ASSESSMENT AND CONVERGENT GROWTH

The experimental data suggest that a larva of *Manduca* can assess both its threshold size and its critical weight with great accuracy. Unfortunately, we

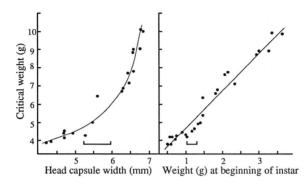

FIGURE 4 Relationship between head capsule size and critical weight for metamorphosis in final instar larvae of *Manduca sexta*. Individuals with large exoskeletons have a larger critical size than individuals with small exoskeletons. The size of a larva is measured by the inextensible parts of its exoskeleton and usually is indexed as the size of its head capsule or the initial weight of the instar. The range of these measures in final instar larvae growing under optimal conditions is indicated by the horizontal square brackets. For such animals the critical weight is about 5 g. Redrawn from Nijhout (1981).

are still completely in the dark as to exactly how it accomplishes this. Stretch receptors, such as those found in the Hemiptera, do not appear to play a role in measuring the critical weight in the last larval instar (Nijhout, 1994). It also is not clear what the larva uses to assess the threshold size for metamorphosis so that it emerges as a measure of absolute size.

That most insect larvae can ascertain their absolute size with considerable accuracy is evident from the fact that the sizes of adult insects vary relatively little around their species-specific means. The problem of how developing animals in general assess their body size is one of the oldest still unanswered questions in developmental biology. The problem actually extends to animal parts as well, because it appears that the size regulation in organs and append-ages is largely an autonomous feature of the structure itself (Bryant and Simpson, 1984). However, whereas the sizes of organs and appendages may be controlled autonomously and locally rather than centrally, the control of body size, at least in insects, must have a pathway that involves the central nervous system. This is because the timing of molting and metamorphosis is controlled via neuroendocrine secretions from the brain (which control the secretion of ecdysteroids and JH; Fig. 1). The brain therefore is indirectly in control of body size.

The integration of growth and the size-monitoring mechanisms can be far more complex than the pattern I have just outlined for *Manduca*. In the cockroach *Blatella germanica,* for instance, the coefficient of variation in body size increases during the early larval instars, due to individual variation in growth rate. But in later instars individuals that are relatively small tend to grow more than those that are relatively large, and there is an overall *decrease* in variation in body size (Tanaka, 1981). Similarly, in the waterstrider *Limnoporus canaliculatus,* body sizes can diverge considerable during early larval growth. But Klingenberg (1996) has shown that in the early instars a weak negative correlation also exists between the relative size of the animal (relative to the mean for the population) and the increment it will grow at the next molt. The effect of this negative correlation is that relatively small individuals molt with a larger increment in the size of their exoskeleton than do relatively large individuals. Exactly how these size-regulating mechanisms work physiologically is not known, but they have the overall effect of keeping individuals with different growth rates from drifting too far away from each other and produce an essentially convergent growth trajectory to a common species-characteristic size.

IX. LARVAL DIAPAUSE

Most insects have the ability to undergo a period of developmental and physiological quiescence as an adaptation for surviving seasonally recurring adverse conditions (Tauber *et al.,* 1986). In temperate regions this phenomenon is referred to as diapause and typically is a strategy for overwintering. In regions with prolonged dry periods, the developmental standstill during the dry season is called aestivation. Across the insects, diapause can occur in any of the developmental stages (egg, larva, pupa, adult), but a given species diapauses at a single species-specific point in its life cycle. For insects with only one generation per year, diapause is a genetically programmed stage in the life cycle. For insects with multiple generations per year, diapause is facultative and is stimulated by environmental conditions such as short day photoperiods. All instances of facultative diapause that have been studied have been found to be under hormonal control (Denlinger, 1985; Nijhout, 1994).

   The endocrine control of larval diapause has been best studied in the final instar larvae of certain Lepidoptera. In diapausing final instar larvae of the rice stem borer *Chilo suppressalis* and the Southwestern corn borer *Diatraea grandiosella,* failure to continue development and molt to the pupal stage it is due to the suppression of PTTH and ecdysteroid secretion. This suppression comes about through the fact that diapausing larvae do not cease JH secretion in the middle of the instar (as they normally would), so that the continued elevated titer of JH inhibits the onset of the pupal molt, as outlined previously. This phenomenon is demonstrated by allatectomizing a diapausing larva. Allatectomy breaks diapause and causes larvae to pupate normally (Yagi and Fukaya, 1974; Yin and Chippendale, 1976, 1979). Diapause continues as long as the JH titer remains elevated. Both species occasionally molt during their diapause period, but these molts are to an additional larval instar of the same size but with a pigmentation pattern very different from that of a nondiapausing larva (Chippendale and Yin, 1976; Yin and Chippendale, 1974, 1976). In larvae of the European corn borer *Ostrinia nubilalis,* the onset of diapause also is due to an elevated titer of JH, but JH does not appear to remain high during the entire diapause period. It is not clear what maintains developmental arrest after JH declines (Bean and Beck, 1980).

Larval diapause thus appears to have evolved by capturing an endocrine control mechanism originally designed to prevent premature metamorphosis in the presence of JH. Larvae diapause simply by keeping their levels of JH high. Of course, many additional physiological changes occur in diapausing larvae that allow them to survive the rigors of winter. Such adaptations presumably evolved as further responses to the elevated JH level.

X. POLYPHENIC DEVELOPMENT AND ITS HORMONAL REGULATION

Polyphenism refers to the ability of animals with the same genotype to develop into two or more discrete alternative phenotypes. Insect polyphenisms can involve simple changes in pigmentation or more extensive quantitative changes in body proportions. Some polyphenisms are sufficiently dramatic that the alternative morphs originally were described as different species (Nijhout, 1991, 1994). Polyphenisms generally are believed to be evolved adaptations to a variable environment. In particular, they are adaptations to predictable environmental changes such as seasonal variations in temperature and food availability (Shapiro, 1976; Moran, 1992).

Polyphenisms can be classified in a variety of ways, depending on one's purpose. One way is to subdivide them into two categories: sequential and alternative (Table II). All of these polyphenisms, though phenomenologically very diverse, share two regulatory features that suggest that they belong to a

TABLE II Classification of Insect Polyphenisms

Sequential	Alternative
Metamorphosis	Castes in termites and social Hymenoptera
Hypermetamorphosis	Phases in aphids and migratory locusts
Heteromorphosis	Seasonal forms in most insect orders
	Wing length in Orthoptera, Hemiptera, and Homoptera
	Dispersal strategies in most insect orders
	Horn length and mating tactics in scarab beetles

common class of developmental phenomena. The first and perhaps most important commonality is that in all of these polyphenisms the developmental switch that determines which alternative adult phenotype will develop is *controlled by the same hormones that control metamorphosis,* namely, JH and ecdysteroids. The second thing they have in common is that in all of them the developmental switch is *cued by a token stimulus from the environment.* What this means is that the environmental change that induces the developmental switch is not the same as the environmental change to which the alternative phenotype is an adaptation (Moran, 1992). For instance, many seasonal polyphenisms are adaptations to the harsh temperatures or reduced food availability of winter, but the environmental cue that actually causes the developmental switch to the winter form is the decreasing day length of autumn. Short day lengths themselves have no effect on fitness, but act as a token stimulus to predict the onset of an environmental change that *does.* So, unlike a norm of reaction, which is a direct developmental response to an environmental change such as a decrease in temperature, a polyphenism is an evolved morphological and physiological adaptation to a predictably recurring alternative environment. The evolution and ecology of polyphenism are discussed by Greene in Chapter 11 of this volume. Here we will be concerned primarily with the mechanisms by which the developmental switch is accomplished.

A. CASTES IN THE SOCIAL HYMENOPTERA

Ants, and the social wasps and bees, have evolved various degrees of caste differentiation. Colonies consist of one or more reproductive females, the queens, and their sterile daughters, the workers. In some ants the sterile female caste is subdivided further into minor and major workers, also called workers and soldiers, respectively. At hatching, larvae are pluripotent and can develop into any of the castes, depending entirely on stimuli received during their development. The environmental cues are diverse, but seem to act through a common physiological mechanism as suggested by the fact that, wherever it has been studied, a larva can be induced to switch to one or another developmental pathway by the application of JH at a critical period in its development (Wheeler, 1986). A summary diagram of the timing of these critical periods for various social Hymenoptera is shown in Fig. 5. The JH-sensitive periods can occur at any point during larval life. Those that occur early in larval life typically have a more profound effect on the morphology of the adult than those that occur late in larval life or in the pupal stage. For instance, in species where determination occurs late in development, such as in paper wasps *(Polistes),* caste determination is largely behavioral and is established through a dominance hierarchy among females; the JH level biases the outcome of

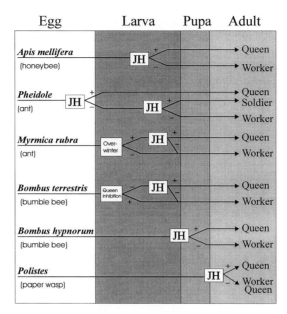

FIGURE 5   Control of caste polyphenism in social Hymenoptera. Switch points in development are controlled by JH. Plus signs indicate the pathway taken when JH is above threshold during the critical period indicated. Based on Wheeler (1986).

competition for dominance (Röseler *et al.*, 1984). In bumble bees *(Bombus)* the queen differs from the workers primarily in size and behavior. In species where determination occurs earlier in larval life, such as in honey bees and ants, the queen differs from workers in many morphological details. The necessity of this relationship between timing of determination and adult morphological divergence is not obvious, however, because in holometabolous insects the adult is built up from imaginal disks and rearrangements of larval parts all occur during the prepupal and pupal stages, no matter when determination occurred.

The JH-sensitive periods, shown in Fig. 5 as "black boxes," have a more or less complex physiology associated with the way in which they control the outcome of development, as illustrated by the following two case histories of an ant and the honey bee. In the ant *Pheidole bicarinata*, a larva will develop into a worker unless it experiences a high level of JH during a brief critical period in the third larval instar, in which case it will develop into a soldier (Wheeler and Nijhout, 1981, 1983). The level of JH is believed to be affected by the quality of nutrition a larva receives. In colonies fed mainly on carbohydrates few larvae develop into soldiers, whereas in colonies fed on highly proteinaceous food many larvae develop into soldiers, presumably because in such larvae the JH titer is elevated during the critical period. Two developmental changes occur if JH is present during the critical period. First, the size at which the larva will cease feeding and begin metamorphosis is delayed, so that a larger adult will develop. Second, the growth pattern of some of the imaginal disks is reprogrammed so that in the prepupal stage they grow faster and larger than they normally would have, resulting in the development of large-headed soldiers (Wheeler and Nijhout, 1981, 1983; Nijhout and Wheeler, 1996). Perhaps the most interesting regulatory feature of this system comes into play in a well-nourished colony, once a sufficient number of soldiers is present and the development of additional soldiers needs to be curbed. Adult soldiers produce a soldier-inhibiting pheromone that raises the threshold of JH sensitivity of the larvae. In the presence of soldiers it is still possible to induce larvae to develop into soldiers, but a much higher dose of JH is required than in the absence of soldiers (Wheeler and Nijhout, 1984). Hence the proportion of soldiers in a colony is regulated with some precision. A dynamic balance is established through the interaction of nutrition, which raises the JH titer and induces soldier formation, and the soldier-inhibiting pheromone, which raises the threshold of sensitivity to JH and reduces the production of soldiers.

Larvae of the honey bee, *Apis mellifera,* also are pluripotent and can develop into either queens or workers. Those that are fed a particularly nourishing secretion, royal jelly, produced by the workers' salivary glands, will develop into queens (Wilson, 1971). This diet leads to an elevated JH level during a JH-sensitive period in the early part of the last larval instar (Nijhout and Wheeler, 1982; Rachinsky and Hartfelder, 1990). Queen development also can be induced in a larva fed a normal diet, which would develop into a normal worker, by the application of JH during this critical period (Wirtz, 1973). The enriched diet of presumptive queen larvae causes them to grow slightly faster, so that they pupate at a somewhat larger size than worker larvae (Dogra *et al.,* 1977). The JH signal during the critical period alters various growth processes in the prepupa and pupa, so that queens develop a distinctive morphology and the queen pupal stage is only half the length of that of the worker.

B. CASTES IN TERMITES

Caste determination has been best studied in the lower termites. There is no terminal worker caste, but the function of workers in the colony is performed by larvae that are permanently arrested in the larval stage, called pseudergates. Pseudergates continue to molt without growing, and most eventually will die without further differentiation. A few will, at some point, metamorphose into either soldiers, called imagoes, which are winged adults that will disperse and found new colonies, or replacement reproductives, which are wingless and, as their name implies, develop if the queen of the colony should die. In a normal colony, queens and soldiers produce pheromones that inhibit pseudergates from developing into replacement reproductives and soldiers, respectively (Lüscher, 1972; Miller, 1969). Hence, pseudergates develop into replacement reproductives or soldiers only if the inhibitory pheromones disappear, by death of the queen, for instance, or are diluted by colony growth.

Pseudergates differ in their sensitivity to the inhibitory pheromones according to the stage in their stationary molting cycle. Soldier inhibition has a window of sensitivity during the last portion of each instar, whereas induction of supplementary reproductives has a sensitivity window in the early portion of each pseudergate instar (Lüscher, 1953; Springhetti, 1972). These windows of pheromone sensitivity correspond to periods of JH sensitivity. If JH is applied to a pseudergate late in the instar, the animal metamorphoses to a soldier, after first molting to a presoldier stage (Lüscher, 1974). Metamorphosis to a replacement reproductive appears to require the absence of JH during the early portion of the instar, as suggested by the small size of the corpora allata of

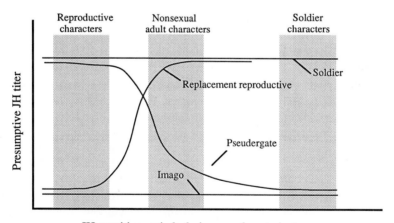

JH-sensitive periods during pseudergate instar

FIGURE 6   Model for the control of caste polyphenism in lower termites. The pseudergate is a terminal larval form that undergoes stationary molts. During each molting cycle, it has the option to transform into any of the three adult castes or to remain a pseudergate. These developmental decisions are controlled by JH during at least three JH-sensitive periods (gray bars). The JH titer curves shown are hypothetical. If the JH titer is high during the appropriate sensitive period, reproductive characters and nonsexual adult characters (such as wings) are suppressed and soldier characters are induced. Pseudergates induced to tansform into replacement reproductives have a foreshortened instar and molt before the soldier-sensitive period begins. Redrawn from Nijhout and Wheeler (1982).

pseudergates preparing to metamorphose to this form (Stuart, 1979). Nijhout and Wheeler (1982) proposed a general model for the way in which variation in JH titers interacting with a succession of JH-sensitive periods can control the alternative developmental options of pseudergates (Fig. 6).

C. Seasonal Color Polyphenisms in Butterflies

Many species of butterflies have distinctive seasonal forms. One of the forms usually is an adaptation to the harsh conditions of winter (in the temperate zones) or the dry season (in the tropics). Winter and autumn forms typically have darker and duller wing patterns than do summer forms. These duskier colors allow them to blend with the dried foliage in their environment. In some species of the genus *Precis,* the dry season forms have evolved both color patterns and wing shapes that make them excellent leaf mimics (e.g., Nijhout, 1991). In *Precis coenia* the autumn morph also has a modified behavior in that it is migratory. In some cases the two seasonal forms have diverged so much in color pattern (and sometimes in wing morphology) that they initially were described as different species (Nijhout, 1991). Larvae can develop into either seasonal morph, depending on the temperature and photoperiod that they experience during critical periods in their development. These environmental variables alter patterns of endocrine secretion, and these in turn affect development.

The control over this polyphenic switch was described first by Koch and Bückmann (1987) for the European map butterfly *Araschnia levana.* Under short day lengths, this species enters pupal diapause. The adults that emerge in the spring have an orange and brown pigmentation pattern. Under long day conditions there is direct development without diapause, and the adults have a black and white color pattern. Koch and Bückmann (1987) showed that the polyphenic switch is controlled entirely by the timing of ecdysteroid secretion. An ecdysteroid-sensitive period extends from 3 to 10 days after pupation. If ecdysteroids are present during this window, the animal will develop the summer phenotype. Otherwise, the spring phenotype develops. Direct-developing animals begin to secrete ecdysteroids soon after pupation in order to initiate adult development (Nijhout, 1994) and therefore develop the summer phenotype. In animals that enter diapause, ecdysteroid secretion is delayed until the end of winter, so that the ecdysteroid-sensitive period passes in the absence of the hormone and the animal develops the spring phenotype. If animals that have been reared under short day lengths (and would have diapaused as pupae) are injected with ecdysteroids within the critical period, they develop a normal summer phenotype.

The seasonal polyphenism of *Precis coenia* also is controlled by ecdysteroids during a critical period. The main difference between this and the previously discussed species is that *Precis* does not diapause as a pupa. Its ecdysteroid-sensitive period is much briefer, extending from 28 to 48 hr after pupation (Rountree and Nijhout, 1996). In long day animals ecdysteroid secretion begins about 24 hr after pupation, so that the ecdysteroid titers are high during the sensitive period and the summer phenotype develops. In short day animals ecdysteroid secretion is delayed by about 24 hr, so that ecdysteroids are low during the sensitive period and the animal develops the autumn phenotype.

In the Asian comma butterfly, *Polygonia c-aureum*, the polyphenic switch between summer and autumn phenotypes is controlled by a novel neurosecretory hormone from the brain, the summer morph producing hormone (SMPH; Endo *et al.*, 1988). In addition, ecdysteroid injections during the early pupal stage can induce development of the summer morph (Masaki *et al.*, 1989), which suggests that SMPH might act via stimulation of ecdysteroid secretion. However, SMPH is dissimilar to any of the known forms of PTTH (and the analogous molecule known as bombyxin), and it is not clear whether its action is ecdysteroidogenic or whether it acts independently on the development of the summer phenotype (Tanaka *et al.*, 1997).

D. DISPERSAL POLYPHENISMS

Many insect polyphenisms are associated with differences in dispersal strategies at different times of the year (Zera and Denno, 1997). Wing length polyphenisms are among these. Numerous hemimetabolous insects such as grasshoppers, crickets, plant hoppers, and waterstriders have at least two distinct dispersal forms. One is long-winged and typically has small ovaries and reduced reproductive capacity, whereas the other is short-winged and typically has much enlarged ovaries and a greater reproductive potential. Dispersal polyphenisms also occur among the Holometabola, although they are, on the whole, much rarer in this taxon. Again, experiences during larval life bias development in the direction of one form or the other. Environmental variables such as crowding, food quality, photoperiod, and temperature experienced during larval development appear to be used as cues by different species (Zera and Denno, 1997).

In many cases, there is an important genetic component to variation in wing length in that the sensitivity to the environmental cue can be under polygenic or monogenic control. Both types of cues (strictly environmental as well as genetically modulated) are believed to be mediated by changes in JH levels during critical periods in larval development (Harrison, 1980; Roff,

1986), though this has been demonstrated critically in only a single species of cricket (Pener, 1991; Zera and Denno, 1997).

XI. LARVAL HETEROMORPHOSIS

One of the most puzzling polyphenisms is one exhibited by the larvae of many insects that go through a series of distinctive larval morphologies in the course of their development. This process, known as heteromorphosis, is best known in certain species with parasitic larvae, such as blister beetles (Meloidae) and mantispids (Neuroptera). In these species the early larval instars are long-legged, highly active, and search out a suitable host to parasitize. Once inside the host the larva "metamorphoses" to a grublike form, grows, and eventually goes through a normal holometabolous metamorphosis. Less dramatic but more common are the successive changes in pigmentation patterns or body proportions that characterize the various larval instars of many caterpillars and other insect larvae. This polyphenism is puzzling because we are still completely in the dark as to its control. Juvenile hormone does not appear to play a role in the progression of heteromorphic characters in saturniid caterpillars (Staal, 1971; Willis, 1974), and it may be that we are dealing with a fixed preprogrammed developmental sequence.

However, there is some evidence that, like metamorphosis, heteromorphosis is cued by the size of the larva. In *Manduca,* for instance, the head cuticle of the fourth instar larva has a surface sculpturing that is dictinctly different from that of the fifth instar larva. But when the growth of early instars is disrupted so that the larva is forced to go through supernumerary larval molts, the switch in cuticle sculpturing does not occur at the fourth to fifth larval molt, but rather at the molt when the head capsule of the new instar exceeds a particular critical value, regardless of the number of instars that have preceded it (Nijhout, 1975b). The switch in sculpturing also is not simply a characteristic of the final larval instar, because it happens at a head capsule size significantly smaller than that that characterizes the final larval instar; individuals with a typical fifth instar cuticle can still be in the penultimate larval instar (Fig. 7).

A far more dramatic case of plasticity in larval morphology is that of facultative mimicry in the caterpillars of the geometrid moth *Nemoria arizonaria.* In the spring, when the oaks are flowering, these caterpillars resemble the catkins of the oaks on which they feed, whereas in the summer they resemble oak twigs (Greene, 1989). The switch in morphology is cued by chemical changes in the host plant. Exactly how this food signal is translated into a developmental switch, and whether hormones are involved, is discussed by Greene in Chapter 11 of this volume. The existence of developmentally plastic larval morphologies whose expression can be triggered by an environmental signal and others that

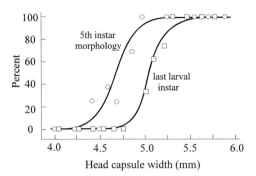

FIGURE 7   Larval heteromorphosis in *Manduca sexta*. Left curve shows the percentage of larvae
of different sizes that have fifth instar type sculpturing on the larval cuticle (the remainder have
fourth instar type sculpturing). Right curve shows the percentage of larvae of different body sizes
that actually are in the final larval instar and will pupate at the next molt (the remainder will
undergo a supernumerary larval molt). Fifth instar larvae of small body size can have a fourth
instar type cuticle, and larvae with a fifth instar type cuticle are not necessarily in the last larval
instar. Redrawn from Nijhout (1975b).

are contingent on the achievement of a particular body size provides us with
systems that can be manipulated experimentally. It is in such systems that the
physiological control pathway of heteromorphosis can best be studied.

REFERENCES

Agui, N., Bollenbacher, W. E., Granger, N. A., and Gilbert, L. I. (1980). Corpus allatum is release
    site for insect prothoracicotropic hormone. *Nature (London)* **285**, 669–670.
Baker, F. C., Tsai, L. W., Reuter, C. C., and Schooley, D. A. (1987). In vivo fluctuation of JH,
    JH acid, ecdysteroid titer, and JH esterase activity during development of 5th stadium *Manduca
    sexta*. *Insect Biochem.* **17**, 989–996.
Bean, D. W., and Beck, S. D. (1980). The role of juvenile hormone in the larval diapause of the
    European corn borer, *Ostrinia nubilalis*. *J. Insect Physiol.* **26**, 579–584.
Beck, S. D. (1971). Growth and retrogression in larvae of *Trogoderma glabrum* (Coleoptera:
    Dermestidae). 1. Characteristics under feeding and starvation conditions. *Ann. Entomol. Soc.
    Am.* **64**, 149–155.
Bollenbacher, W. E. (1988). The interendocrine regulation of larval-pupal development in the
    tobacco hornworm, *Manduca sexta*: A model. *J. Insect Physiol.* **34**, 941–947.
Bollenbacher, W. E., Smith, S. L., Goodman, W., and Gilbert, L. I. (1981). Ecdysteroid titer
    during larval–pupal–adult development of the tobacco hornworm, *Manduca sexta*. *Gen. Comp.
    Endocrinol.* **44**, 302–306.
Bryant, P. J., and Simpson, P. (1984). Intrinsic and extrinsic control of growth in developing
    organs. *Q. Rev. Biol.* **59**, 387–415.
Cherbas, P., and Cherbas, L. (1996). Molecular aspects of ecdysteroid hormone action. In "Meta-
    morphosis" (L. I. Gilbert, J. R. Tata, and B. G. Atkinson, eds.), pp. 175–221. Academic Press,
    San Diego, CA.

Chippendale, G. M., and Yin, C.-M. (1976). Endocrine interactions controlling the larval diapause of the southwestern corn borer, *Diatraea grandiosella*. *J. Insect Physiol.* **22**, 989–995.

Denlinger, D. (1985). Diapause. *In* "Comprehensive Insect Physiology, Biochemistry and Pharmacology" (G. A. Kerkut and L. I. Gilbert, eds.), Vol. 8, pp. 353–412. Pergamon, New York.

Dogra, G. S., Ulrich, G. M., and Rembold, H. (1977). A comparative study of the endocrine system of the honeybee larvae under normal and experimental conditions. *Z. Naturforsch.* **32**, 637–642.

Endo, K., Masaki, T., and Kumagai, K. (1988). Neuroendocrine regulation of the development of seasonal morphs in the Asian comma butterfly, *Polygonia c-aureum* L.: Difference in activity of the summer-morph-producing hormone from brain-extracts of the long-day and short-day pupae. *Zool. Sci.* **5**, 145–152.

Fain, M. J., and Riddiford, L. M. (1975). Juvenile hormone titers in the hemolymph during later larval development of the tobacco hornworm, *Manduca sexta* (L.). *Biol. Bull. (Woods Hole, Mass.)* **149**, 506–521.

Fain, M. J., and Riddiford, L. M. (1977). Requirements for molting of the crochet epidermis of the tobacco hornworm larva in vivo and in vitro. *Wilhelm Roux's Arch. Dev. Biol.* **181**, 285–307.

Fukuda, S. (1944). The hormonal mechanism of larval molting and metamorphosis in the silkworm. *J. Fac. Sci., Imp. Univ. Tokyo* **4**, 477–532.

Ghiradella, H. (1985). Structure and development of iridescent lepidopteran scales: The Papilionidae as a showcase family. *Ann. Entomol. Soc. Am.* **78**, 252–264.

Ghiradella, H. (1989). Structure and development of iridescent butterfly scales: Lattices and laminae. *J. Morphol.* **202**, 69–88.

Gilbert, S. F., Opitz, J. M., and Raff, R. A. (1996). Resynthesizing evolutionary and developmental biology. *Dev. Biol.* **173**, 357–372.

Greene, E. (1989). A diet-induced developmental polymorphism in a caterpillar. *Science* **243**, 643–646.

Gruetzmacher, M. C., Gilbert, L. I., Granger, N. A., Goodman, W., and Bollenbacher, W. E. (1984). The effect of juvenile hormone on prothoracic gland function during the larval-pupal development of *Manduca sexta*: An *in situ* and *in vitro* analysis. *J. Insect Physiol.* **30**, 331–340.

Gu, S.-H., and Chow, Y.-S. (1993). Role of low ecdysteroid levels in the early last larval instar of *Bombyx mori*. *Experientia* **49**, 806–809.

Gu, S.-H., and Chow, Y.-S. (1996). Regulation of juvenile hormone biosynthesis by ecdysteroid levels during the early stages of the last two larval instars of *Bombyx mori*. *J. Insect Physiol.* **42**, 625–632.

Harrison, R. G. (1980). Dispersal polymorphisms in insects. *Annu. Rev. Ecol. Syst.* **11**, 95–118.

Hiruma, K., Hardie, J., and Riddiford, L. M. (1991). Hormonal regulation of epidermal metamorphosis *in vitro*: Control of expression of a larval-specific cuticle gene. *Dev. Biol.* **144**, 369–378.

Hudson, A. (1966). Proteins in the haemolymph and other tissues of the developing tomato hornworm, *Protoparce quinquemaculata* Haworth. *Can. J. Zool.* **44**, 541–555.

Kalthoff, K. (1996). "Analysis of Biological Development." McGraw-Hill, New York.

Kiguchi, K., and Riddiford, L. M. (1978). A role of juvenile hormone in pupal development of the tobacco hornworm, *Manduca sexta*. *J. Insect Physiol.* **24**, 673–680.

Kikukawa, S., and Tobe, S. S. (1986). Critical periods for juvenile hormone sensitivity during larval life of female *Diploptera punctata*. *J. Insect Physiol.* **32**, 1035–1042.

Kiriishi, S., Rountree, D. B., Sakurai, S., and Gilbert, L. I. (1990). Prothoracic gland synthesis of 3-dehydoecdysone and its hemolymph 3β-reductase mediated conversion to ecdysone in representative insects. *Experientia* **46**, 716–721.

Klingenberg, C. P. (1996). Individual variation of ontogenies: A longitudinal study of growth and timing. *Evolution (Lawrence, Kans.)* **50**, 2412–2428.

Koch, P. B., and Bückmann, D. (1987). Hormonal control of seasonal morphs by the timing of ecdysteroid release in *Araschnia levana* (Nymphalidae: Lepidoptera). *J. Insect Physiol.* **33**, 823–829.

Kremen, C., and Nijhout, H. F. (1989). Juvenile hormone controls the onset of pupal commitment in the imaginal disks and epidermis of *Precis coenia* (Lepidoptera: Nymphalidae). *J. Insect Physiol.* **35**, 603–612.

Kremen, C., and Nijhout, H. F. (1998). Control of pupal commitment in the imaginal disks of *Precis coenia* (Lepidoptera: Nymphalidae). *J. Insect Physiol.* **44**, 287–296.

Lanzrein, B., Gentinetta, V., Abegglen, H., Baker, F. C., Miller, C. A., and Schooley, D. A. (1985). Titers of ecdysone, 20-hydroxyecdysone and juvenile hormone III throughout the life cycle of a hemimetabolous insect, the ovoviviparous cockroach *Nauphoeta cinerea. Experientia* **41**, 913–917.

Lawrence, P. A. (1969). Cellular differentiation and pattern formation during metamorphosis of the milkweed bug *Oncopeltus. Dev. Biol.* **19**, 12–40.

Lawrence, P. A. (1992). "The Making of a Fly." Blackwell, Oxford.

Lockshin, R. A. (1985). Programmed cell death. In "Comprehensive Insect Physiology, Biochemistry and Pharmacology" (G. A. Kerkut and L. I. Gilbert, eds.). Vol. 2, pp. 301–317. Pergamon, New York.

Lüscher, M. (1953). Kann die Determination durch eine monomolekulare Reaktion ausgelöst werden? *Rev. Suisse Zool.* **60**, 524–528.

Lüscher, M. (1972). Environmental control of juvenile hormone (JH) secretion and caste differentiation in termites. *Comp. Endocrinol., Suppl.* **3**, 509–514.

Lüscher, M. (1974). Die Kompetenz zur Soldatenbildung bei Larven (Pseudergaten) der Termite *Zootermopsis angusticollis. Rev. Suisse Zool.* **81**, 710–714.

Masaki, T., Endo, K., and Kumagai, K. (1989). Neuroendocrine regulation of the development of seasonal morphs in the Asian comma butterfly, *Polygonia c-aureum* L.: Stage-dependent changes in the activity of summer-morph-producing hormone of the brain-extracts. *Zoological Science (Tokyo)* **6**, 113–119.

Miller, E. M. (1969). Caste differentiation in lower termites. In "Biology of Termites" (K. Krishna and F. M. Weesner, eds.), pp. 283–307. Academic Press, New York.

Moran, N. (1992). The evolutionary maintenance of alternative phenotypes. *Am. Nat.* **139**, 971–989.

Nijhout, H. F. (1975a). The brain-retrocerebral neuroendocrine complex of *Manduca sexta* (L.) (Lepidoptera: Sphingidae). *Int. J. Insect Morphol. Embryol.* **4**, 529–538.

Nijhout, H. F. (1975b). A threshold size for metamorphosis in the tobacco hornworm, *Manduca sexta. Biol. Bull. (Woods Hole, Mass.)* **149**, 214–225.

Nijhout, H. F. (1979). Stetch-induced moulting in *Oncopeltus fasciatus. J. Insect. Physiol.* **25**, 277–281.

Nijhout, H. F. (1981). Physiological control of molting in insects. *Am. Zool.* **21**, 631–640.

Nijhout, H. F. (1983). Definition of a juvenile hormone-sensitive period in *Rhodnius prolixus. J. Insect Physiol.* **29**, 669–677.

Nijhout, H. F. (1984). Abdominal stretch reception in *Dipetalogaster maximus* (Hemiptera: Reduviidae). *J. Insect Physiol.* **30**, 629–633.

Nijhout, H. F. (1991). "The Development and Evolution of Butterfly Wing Patterns." Smithsonian Institution Press, Washington, DC.

Nijhout, H. F. (1994). "Insect Hormones." Princeton University Press, Princeton, NJ.

Nijhout, H. F., and Wheeler, D. E. (1982). Juvenile hormone and the physiological basis of insect polymorphisms. *Q. Rev. Biol.* **57**, 109–133.

Nijhout, H. F., and Wheeler, D. E. (1996). Growth models of complex allometries in holometabolous insects. *Am. Nat.* **148**, 40–56.

Nijhout, H. F., and Williams, C. M. (1974a). Control of moulting and metamorphosis in the tobacco hornworm, *Manduca sexta* (L.): Growth of the last instar larva and the decision to pupate. *J. Exp. Biol.* **61**, 481–491.

Nijhout, H. F., and Williams, C. M. (1974b). Control of moulting and metamorphosis in the tobacco hornworm, *Manduca sexta* (L.): Cessation of juvenile hormone secretion as a trigger for pupation. *J. Exp. Biol.* **61**, 493–501.

Oberlander, H. (1985). The imaginal disks. *In* "Comprehensive Insect Physiology, Biochemistry, and Pharamacology" (G. A. Kerkut and L. I. Gilbert, eds.), Vol. 2, pp. 151–182. Pergamon, New York.

Overton, J. (1966). Microtubules and microfibrils in morphogenesis of the scale cells of *Ephestia kühniella. J. Cell Biol.* **29**, 293–305.

Pener, M. P. (1991). Locust phase polymorphism and its endocrine relations. *Adv. Insect Physiol.* **23**, 1–79.

Rachinsky, A., and Hartfelder, K. (1990). Corpora allata activity, a prime regulating element for caste-specific juvenile hormone titre in honey bee larvae (*Apis mellifera carnica*). J. Insect Physiol. **36**, 189–194.

Reynolds, S. E., Taghert, P. E., and Truman, J. W. (1979). Eclosion hormones and bursicon titres and the onset of hormonal responsiveness during the last day of adult development in *Manduca sexta. J. Exp. Biol.* **78**, 77–86.

Riddiford, L. M. (1972). Juvenile hormone in relation to the larval–pupal transformation of the Cecropia silkworm. *Biol. Bull.* (*Woods Hole, Mass.*) **142**, 310–325.

Riddiford, L. M. (1976). Hormonal control of insect epidermal cell commitment *in vitro. Nature* (*London*) **259**, 115–117.

Riddiford, L. M. (1978). Ecdysone-induced change in cellular commitment of the epidermis of the tobacco hornworm, *Manduca sexta,* at the initiation of metamorphosis. *Gen. Comp. Endocrinol.* **34**, 438–446.

Riddiford, L. M. (1985). Hormone action at the cellular level. *In* "Comprehensive Insect Physiology, Biochemistry and Pharmacology" (G. A. Kerkut and L. I. Gilbert, eds.), Vol. 8, pp. 37–84. Pergamon, New York.

Riddiford, L. M. (1993). Hormones and *Drosophila* development. *In* "The Development of *Drosophila*" (M. Bate and A. Martinez Arias, eds.), pp. 899–939. Cold Spring Harbor Lab., Cold Spring Harbor, NY.

Riddiford, L. M. (1996). Juvenile hormone: The status of its "status quo" action. *Arch. Insect Biochem.* **32**, 271–286.

Riddiford, L. M., and Ashburner, M. (1991). Role of juvenile hormone in larval development and metamorphosis of *Drosophila melanogaster. Gen. Comp. Endocrinol.* **82**, 172–183.

Riddiford, L. M., and Truman, J. W. (1994). Hormone receptors and the orchestration of development during insect metamorphosis. *In* "Perspectives in Comparative Endocrinology" (K. G. Davey, R. E. Peter, and S. S. Tobe, eds.), pp. 389–394. National Research Council of Canada, Ottawa.

Roff, D. A. (1986). The evolution of wing dimorphism in insects. *Evolution* (*Lawrence, Kans.*) **40**, 1009–1020.

Röseler, P.-F., Röseler, I., Strambi, A., and Augier, R. (1984). Influence of insect hormones on the establishment of dominance hierarchies among foundresses of the paper wasp, *Polistes gallicus. Behav. Ecol. Sociobiol.* **15**, 133–142.

Rountree, D. B., and Bollenbacher, W. E. (1986). The release of the prothoracocotropic hormone in the tobacco hornworm, *Manduca sexta,* is controlled intrinsically by juvenile hormone. *J. Exp. Biol.* **120**, 41–58.

Rountree, D. B., and Nijhout H. F. (1996). Hormonal control of a seasonal polyphenism in *Precis coenia* (Lepidoptera: Nymphalidae). *J. Insect Physiol.* **41**, 987–992.

Sakurai, S., and Gilbert, L. I. (1990). Biosynthesis and secretion of ecdysteroids by the prothoracic glands. *In* "Molting and Metamorphosis" (E. Ohnishi and H. Ishizaki, eds.), pp. 83–106. Springer-Verlag, Berlin.

Schneiderman, H. E., and Gilbert, L. I. (1964). Control of growth in insects. *Science* 143, 325–333.

Shapiro, A. M. (1976). Seasonal polyphenism. *Evol. Biol.* 9, 259–333.

Springhetti, A. (1972). The competence of *Kalotermes flavicollis* Fabr. (Isoptera) pseudergates to differentiate into soldiers. *Monit. Zool. Ital.* [N.S.] 6, 97–111.

Staal, G. (1971). The role of juvenile hormone in the morphogenetic development of larval instars in lepidoptera. *Endocrinol. Exp.* 5, 35–38.

Stuart, A. M. (1979). The determination and regulation of the neotenic reproductive caste in the lower termites: With special reference to the genus *Zootermopsis*. *Sociobiology* 4, 223–237.

Talbot, W. S., Swyryd, E. A., and Hogness, D. S. (1993). *Drosophila* tissues with different metamorphic responses to ecdysone express different ecdysone receptor isoforms. *Cell (Cambridge, Mass.)* 73, 1323–1337.

Tanaka, A. (1981). Regulation of body size during larval development in the German cockroach, *Blatella germanica*. *J. Insect Physiol.* 27, 587–592.

Tanaka, D., Sakurama, T., Mitsumasu, K., Yamakana, A., and Endo, K. (1997). Separation of bombyxin from a neuropeptide of *Bombyx mori* showing summer-morph-producing hormone (SMPH) activity in the Asian comma butterfly, *Polygonia c-aureum* L. *J. Insect Physiol.* 43, 197–201.

Tauber, M. J., Tauber, C. A., and Masaki, S. (1986). "Seasonal Adaptations of Insects." Oxford University Press, New York.

Tobe, S. S., Ruegg, R. P., Stay, B., Baker, F. C., Miller, C. A., and Schooley, D. A. (1985). Juvenile hormone titre and regulation in the cockroach *Diploptera punctata*. *Experientia* 41, 1029–1034.

Truman, J. W. (1972). Physiology of insect rhythms: I. Circadian organization of the endocrine events underlying the moulting cycle of larval tobacco hornworms. *J. Exp. Biol.* 57, 805–820.

Truman, J. W. (1996). Metamorphosis of the insect nervous system. In "Metamorphosis" (L. I. Gilbert, J. R. Tata, and B. G. Atkinson, eds.), pp. 283–320. Academic Press, San Diego, CA.

Truman, J. W., Riddiford, L. M., and Safranek, L. (1973). Hormonal control of cuticle coloration in the tobacco hornworm, *Manduca sexta*: Basis for an ultrasensitive bioassay for juvenile hormone. *J. Insect Physiol.* 19, 195–203.

Warren, J. T., Sakurai, S., Rountree, D. B., and Gilbert, L. I. (1988). Synthesis and secretion of ecdysteroids by the prothoracic glands of *Manduca sexta*. *J. Insect Physiol.* 34, 561–576.

Watson, R. D., and Bollenbacher, W. E. (1988). Juvenile hormone regulates the steroidogenic competence of *Manduca sexta* prothoracic glands. *Mol. Cell. Endocrinol.* 57, 251–259.

Wheeler, D. E. (1986). Developmental and physiological determinants of caste in social Hymenoptera: Evolutionary implication. *Am. Nat.* 128, 13–34.

Wheeler, D. E. (1991). The developmental basis of worker caste polymorphism in ants. *Am. Nat.* 138, 1218–1238.

Wheeler, D. E., and Nijhout, H. F. (1981). Soldier determination in ants: New role for juvenile hormone. *Science* 213, 361–363.

Wheeler, D. E., and Nijhout, H. F. (1983). Soldier determination in *Pheidole bicarinata*: Effect of methoprene on caste and size within castes. *J. Insect Physiol.* 29, 847–854.

Wheeler, D. E., and Nijhout, H. F. (1984). Soldier determination in the ant *Pheidole bicarinata*: Inhibition by adult soldiers. *J. Insect Physiol.* 30, 127–135.

Wigglesworth, V. B. (1934). The physiology of ecdysis in *Rhodnius prolixus* (Hemiptera). II. Factors controlling moulting and "metamorphosis." *Q. J. Microsc. Sci.* 77, 191–222.

Williams, C. M. (1961). The juvenile hormone. II. Its role in the endocrine control of molting, pupation, and adult development in the cecropia silkworm. *Biol. Bull. (Woods Hole, Mass.)* 121, 572–585.

Williams, C. M., and Kafatos, F. C. (1971). Theoretical aspects of the action of juvenile hormone. *Mitt. Schweiz. Entomol. Ges.* 44, 151–162.

Willis, J. H. (1974). Morphogenetic action of insect hormones. *Annu. Rev. Entomol.* 19, 97–115.

Willis, J. H. (1996). Metamorphosis of the cuticle, its proteins, and their genes. In "Metamorphosis" (L. I. Gilbert, J. R. Tata, and B. G. Atkinson, eds.), pp. 253–282. Academic Press, San Diego, CA.

Willis, J. H., Rezaur, R., and Sehnal, F. (1982). Juvenoids cause some insects to form composite cuticles. J. Embryol. Exp. Morphol. 71, 25–40.

Wilson, E. O. (1971). "The Insect Societies." Harvard University Press, Cambridge, MA.

Wirtz, P. (1973). Differentiation in the honeybee larva. Meded. Landbouw hogesch. Wageningen 73–75, 1–66.

Wolfgang, W. J., and Riddiford, L. M. (1986). Larval cuticular morphogenesis in the tobacco hornworm, Manduca sexta, and its hormonal regulation. Dev. Biol. 113, 305–316.

Wolpert, L. (1969). Positional information and the spatial pattern of cellular differentiation. J. Theor. Biol. 25, 1–47.

Yagi, S., and Fukaya, M. (1974). Juvenile hormone as a key factor in regulating larval diapause in the rice stem borer Chilo suppressalis (Lepidoptera: Pyralidae). Appl. Entomol. Zool. 9, 247–255.

Yin, C.-M., and Chippendale, G. M. (1974). Juvenile hormone and the induction of larval polymorphism and diapause of the Southwestern corn borer, Diatraea grandiosella. J. Insect Physiol. 20, 1833–1847.

Yin, C.-M., and Chippendale, G. M. (1976). Hormonal control of larval diapause and metamorphosis in the Southwestern corn borer, Diatraea grandiosella. J. Exp. Biol. 64, 303–310.

Yin, C.-M., and Chippendale, G. M. (1979). Diapause of the southwestern corn borer, Diatraea grandiosella: Further evidence showing juvenile hormone to be the regulation. J. Insect Physiol. 25, 513–523.

Zera, A. J., and Denno, R. F. (1997). Physiology and ecology of dispersal polymorphism in insects. Annu. Rev. Entomol. 42, 207–231.

CHAPTER 8

Cell Lineages in Larval Development and Evolution of Echinoderms

RUDOLF A. RAFF

Department of Biology and Indiana Molecular Biology Institute, Indiana University, Bloomington, Indiana

I. EMBRYOS AND CELL LINEAGES

The embryonic cell lineages of echinoderms contribute to understanding complementary functional and evolutionary aspects of their larval biology. Most marine animals have complex life histories in which the embryo gives rise to a feeding larva distinct from the adult. The sea urchin pluteus larva is a prime example. The early cell lineages of indirect-developing species generate the cells that produce these larval forms. These larvae feed and grow; thus, the eggs that produce them are relatively small. Only a few cell cleavage cycles are needed to produce the requisite cells and initiate larval morphogenesis. The origins of metazoan development have been debated extensively [reviewed by Rieger (1994)]. There are essentially two hypotheses as to the ancestral form of development. The first is that the earliest metazoans, apparently small animals, had a simple life history in which the adult developed directly (Conway Morris, 1993; Grell and Ruthmann, 1991; Rieger et al., 1991). Complex life histories in this scenario arose later. The second hypothesis holds that a complex life history was basal to animals (Davidson et al., 1995; Jägersten, 1972; Rieger, 1994).

Planktonic larvae suffer great attrition, requiring species that reproduce this way to release large numbers of eggs. As only large metazoans can afford to produce large numbers of planktotrophic larvae (Strathmann, 1985; Vance, 1973), I favor a hypothesis that the first metazoans were small and were direct-developers with indeterminate growth. As a consequence they produced a small number of large eggs. These early metazoans invented the major aspects of axial organization and the genetic regulatory systems that control their development (Slack et al., 1993). Thus, the basic bilaterian metazoan adult body plan would have had an early origin. The late Precambrian metazoan radiation, whose results make up the Cambrian "explosion," resulted in larger body sizes and opened new ecological niches. Larger size made possible the evolution of complex life histories, involving the production of large numbers of small eggs and feeding larvae. Planktotrophic larvae thus would have originated after sufficiently large adult body sizes were achieved. The exact timing of the metazoan radiation is in dispute (Wray et al., 1996). Large metazoans do not appear in the fossil record until shortly before and during the Cambrian "explosion" [reviewed in Raff (1996)], but they may have been present as poorly fossilized soft-bodied forms much earlier.

Species that have plantotrophic larvae do so by investing less in each egg than do direct-developers; instead they produce large numbers of small eggs, each provisioned with only sufficient yolk to build a feeding larval stage. They must develop rapidly to the point of generating feeding structures. Their embryonic cell lineages thus are adaptations to rapidly cleaving the mass of the egg into embryonic cells; no growth is possible until larval cellular differentiation and morphogenesis can assemble feeding structures. Development of the adult rudiment takes place in a feeding larva at a more leisurely pace. Generally cleavage is limited to about 10 cycles (Davidson et al., 1995). In indirect development, this small number of precise cleavages economically lays out the larval body plan. The adult arises from a subset of cells that resume division in the feeding larva (Davidson et al., 1995). Thus, there are evolutionarily conserved cell lineages in echinoderms that arose with the invention of complex life histories, whose function is the production of specialized larval cells. Echinoderms appeared in the fossil record of the Cambrian, at least 535 million years ago. The living echinoderm classes appeared within 50 million years of that time (Campbell and Marshall, 1987; Smith, 1988). Indirect development and the basic features of echinoderm development and larvae thus have one-half billion year histories, and possibly longer. As will be shown herein, in some circumstances these highly conserved cell lineages have, unexpectedly, been highly modified. In species that only secondarily evolved direct development, cell lineages have been radically reorganized and now directly give rise to adult cells and body structures.

II. FEATURES OF ECHINODERM CELL LINEAGES

A. CONSERVED FEATURES

It generally has been believed that early development is highly conserved, both as an extrapolation of Von Baer's law and on theoretical grounds (Arthur, 1988). Shared larval forms indeed have been widely, and apparently successfully, used to phylogenetically unite otherwise disparate phyla. The expectation thus is that basic features of early development, including cell lineage, should be conserved. Ancient features of echinoderm cell lineages are the following: (1) maternal determination of the animal–vegetal (A–V) axis; (2) a radial pattern of cleavage; (3) equal cell divisions throughout cleavage; (4) determination of the larval dorsal–ventral (D–V) axis during early cleavage; and (5) segregation of determinants into final positions after the third cleavage (Wray, 1994). Wray (1994) noted that most of these features are present in other deuterostomes as well and, thus, predate the origin of echinoderms.

Some of these ancient features are highly conserved among echinoderms. For example, there is no known counterexample to a maternally determined A–V axis. An interesting question is whether ancient axial and cell lineage features are conserved because they are developmentally constrained, or whether such features are relatively unconstrained but limited for other reasons. Vermeij (1987) has suggested that selection can be conflated with constraint. Thus, "If many phenotypes are forbidden under normally prevailing ecological conditions, theoretically independent traits often become highly correlated, and variation in these traits is constrained within well-defined limits. The resulting pattern of covariation may be so pervasive that it is often interpreted as a manifestation of developmental constraints" (Vermeij, 1987). One difficulty in assessing the full range of properties of echinoderm cell lineages and their evolution is that so few cell lineages have been determined; one starfish species and two sea urchins have been characterized. As the vast preponderance of data on cell lineage function and evolution in echinoderm development comes from echinoids, I focus on them.

B. AXIS DETERMINATION

Figure 1 shows the conservation of A–V axis determination with respect to variation in the first four cleavages in various echinoderm classes. All exhibit

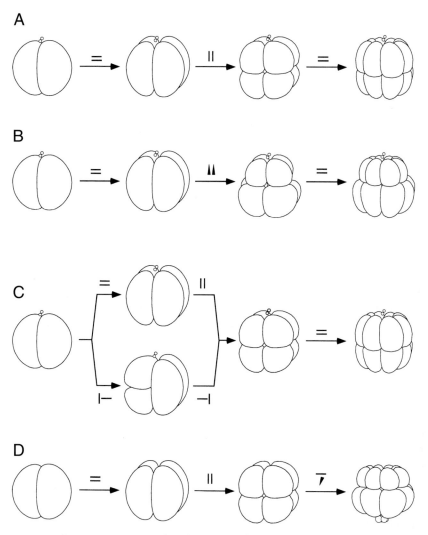

FIGURE 1 Cleavage geometry in echinoderms. First cleavage is parallel to the A–V axis in all. (A) Most species of crinoids, asteroids, ophiuroids, and holothurians exhibit radial cleavage to the fourth cleavage and produce two-tiered, 16-cell embryos with equal-size cells. A few echinoids have revolved this pattern. (B) Some species of crinoids and ophiuroids display a modification of pattern A, in which there is an unequal third cleavage. (C) Variation in second cleavage occurs within broods in some asteroids and ophiuroids. This pattern is compensated for in the third cleavage. (D) The characteristic unequal third cleavage pattern of most echinoids, which results in embryos having three cell sizes at the 16-cell stage. Figure courtesy of G. A. Wray (1991), with permission.

radial cleavage. First and second cleavage planes are parallel to the animal–vegetal axis, and the third cleavage plane lies perpendicular to it. The second primary axis, the dorsoventral (D–V) (or aboral–oral) axis, lies perpendicular to the A–V axis. It has a constant relationship to the first two cleavage planes in some species, but this specification is conditional and is not determined until after the second cleavage. The relationship of the planes of the first two cleavages to the D–V axis has been demonstrated by cell labeling studies with a few sea urchin species (Hörstadius and Wolsky, 1936; Kominami, 1988; Cameron et al., 1989; Wray and Raff, 1990). The boundary between the labeled and unlabeled halves of embryos with one blastomere labeled at the 2-cell stage reveals the dorsoventral axis. The relationship between cleavage plane and axis is constant within some species. However, that relationship varies between species, such that the D–V axis can lie parallel to first cleavage plane (Lytechinus variegatus, Heliocidaris tuberculata), or offset from it by 45° (Strongylocentrotus purpuratus). These constant relationships indicate that the D–V axis is specified before the first cleavage. However, experiments in which separated blastomeres from 2-cell or 4-cell embryos form normal pluteii show that the D–V axis is labile and that it does not become determined until later in cleavage [reviewed by Henry and Raff (1992)]. In a few species (Hemicentrotus pulcherimus, Paracentrotus lividus), the D–V axis lies randomly among embryos in any of these three relationships, indicating a lack of specification.

The third primary larval axis is the left–right (L–R) axis. This axis is of particular interest in echinoderms because, in the indirect-developing species, the bilateral symmetry of the larva has to be broken to produce the adult rudiment. This process is well-known in echinoids. In the pluteus, left and right coeloms initially form by enterocoelous budding from the wall of the archenteron. A portion of the left coelom is elaborated to produce the hydrocoel, in which pentameral adult symmetry is generated. The hydrocoel interacts with an ectodermal invagination, the vestibule, to form the oral (ventral) structures of the developing adult rudiment [reviewed by Okazaki (1975)]. The vestibular invagination also forms on the left side and does so autonomously from the hydrocoel (Czihak, 1960, 1965; Henry and Raff, 1990). Although the molecular mechanism establishing the left side of the sea urchin embryo is not yet known, the vertebrate L–R axis has been shown to be determined by the action of a maternally expressed member of the TGF-β family, Vg1 (Hyatt et al., 1996; Robertson, 1997). The autonomous development of left-side structures in sea urchin rudiment development is consistent with a role for maternal determinants, but the L–R axis is not rigidly specified in the zygote. McCain and McClay (1994) investigated the establishment of the left–right axis by following left–right axis reversal by marked and bisected embryos of the sea urchin Lytechinus variegatus. They showed that left–right symmetry does not become fixed until the late blastula stage. Determination of the left–right axis occurs coincidentally with determination of the dorsoventral axis, a result strikingly like that observed in early development of the nematode, *Caenorhabditis elegans,* by Wood (1991).

C. Cell Lineages and Gene Expression Territories

Davidson (1989, 1990) placed sea urchin cell lineages into a context of gene expression. He suggested that the invariant cell lineage pattern of sea urchins like *S. purpuratus* is coupled to cell specification processes, such that cell fates within embryo territories arise from the effects of cell–cell interactions acting in combination with invariant cell cleavages. These processes operate during cleavage stages and generate a small number of discrete territories that each contain the products of several particular cell lineages. Each territory is characterized by the expression of a discrete set of histospecific genes. These territories then undergo defined local morphogenetic programs. Regulation of expression of the histospecific genes that define territories is a consequence of these genes having a complex regulatory architecture that integrates a large number of inputs (including positive- and negative-acting trans-acting factors) that limit expression to a particular cell lineage or territory (Kirchhamer and Davidson, 1996; Kirchhamer *et al.,* 1996). Experimental manipulation of these regulatory elements can be shown to change the site of expression and other properties. Commitment to particular cell fates is tied to the expression of transcription factors. Thus, experimental overexpression of the *otx* gene causes all larval ectoderm to transform to aboral ectoderm (Mao *et al.,* 1996). The strength of this system of control is that gene expression territories can be blocked out without the action of higher level regulators that govern body patterning later in development.

The Davidson (1989, 1990) hypothesis puts forward a coherent explanation for rapid larval differentiation as a direct consequence of gene expression patterns established through the cell lineage program of the embryo. Each species program is tied to a conserved pattern. However, there is striking experimental and evolutionary flexibility in the operation of the mechanisms that establish cell territories in sea urchin embryos. Experimental results have demonstrated a great regulative capacity in sea urchin embryos. For example, embryos that arise from separated blastomeres may establish normal territories, but through synthetic cell lineages not seen in the intact embryo (Raff, 1996).

There are other phenomena that bear upon the integration of territories in an embryo, the most crucial being signaling among cell lineages (Ettensohn and McClay, 1988; Henry *et al.,* 1989; Ransick and Davidson, 1993, 1995;

Wikramanayake and Klein, 1997; Wikramanayake *et al.*, 1995). Signaling events induce or prevent neighboring cells from assuming particular fates and provide important embryo-wide aspects of patterning. From studies of morphological differentiation of isolated mesomere caps, or pairs of mesomeres, Henry *et al.* (1989) concluded that signals both from the vegetal tier of the embryo as well as among mesomeres limit the fate of these cells in the intact embryo. By using cell separation in conjunction with territory-specific molecular markers, Wikramanayake *et al.* (1995) compared the establishment of aboral–oral (D–V) differentiation in the ectoderm of two species of sea urchins. They found that in *S. purpuratus* mesomeres, much of axial establishment was autonomous by the 8-cell stage, with both oral and aboral ectodermal marker gene expression appearing without evident requirement for a signal from vegetal blastomeres; full differentiation, including some morphological features and correct restriction of the expression of the oral marker, did not occur without such a signal(s). In contrast, in *L. pictus* mesomeres the oral marker, but not the aboral marker, was expressed, indicating a requirement for a signal from vegetal blastomeres. It is significant that D–V axial differentiation differs between these two closely related sea urchins that form similar pluteus larvae. It is most likely that the difference lies in a heterochronic shift in vegetal signal timing. In a more detailed examination of *L. pictus* mesomeres, Wikramanayake and Klein (1997) concluded that the two ectodermal territories, oral ectoderm and aboral ectoderm, which arise from distinct cell lineages and exhibit distinct patterns of gene expression, differ in their establishment. Expression of the oral gene marker is autonomous, but expression of the aboral marker requires a positive signal from vegetal blastomeres that specifies aboral ectoderm and a negative signal that inactivates expression of oral territory genes in the aboral territory.

The importance of signaling along the A–V axis also is indicated by treatment of sea urchin embryos with Li^+ ions, which induces more vegetal fates and patterns of gene expression in animal poleward blastomeres of the embryo (Livingston and Wilt, 1989, 1990; Nocente-McGrath *et al.*, 1991). Morphological and molecular markers show that Li^+ changes cell fates and results in more cells assuming mesenchyme and endodermal fates at the expense of ectodermal fates. Li^+ interacts with the inositol phosphate–protein kinase C second messenger pathway (Berridge *et al.*, 1989). Livingston and Wilt (1992) provided evidence of this link in sea urchins by demonstrating that phorbol esters (which are activators of protein kinase C) produce the same effects as Li^+ treatment of sea urchin embryos. There is contradictory evidence, however, and the mechanism by which Li^+ affects embryos is not clear. Klein and Melton (1996) showed that dorsalization of frog (*Xenopus laevis*) embryos by Li^+ is mediated by inhibition of glycogen synthase kinase-3β rather than on IP_3-

mediated signal pathways. Li$^+$ has been observed to inhibit the GSK-3 family in mammal and *Drosophila* cells as well (Stambolik *et al.*, 1996).

Analogously, treatment of sea urchin embryos with Ni^{2+} ions radializes embryos and results in the initiation of multiple skeletal foci around the blastocoel. Reciprocal cell transplant experiments show that this effect is due to effects of Ni^{2+} ions on the ectoderm, not on the primary mesenchyme cells themselves (Armstrong *et al.*, 1993). Other cell transplantation experiments in which extra primary mesenchyme cells are injected into normal embryos yield normal numbers of skeletal foci, but with more cells in each (Ettensohn, 1990). The ectoderm thus regulates aspects of patterning of larval architecture by influencing primary mesenchyme cells inside the embryo.

These experiments illustrate the operations of global patterning involving embryo cell lineages, and they show an underlying dissociability of processes operating in embryos that makes the evolution of early development possible (Raff, 1996). Evolutionary change is discussed in the following sections.

III. EVOLUTION OF AXIS AND CELL LINEAGE DETERMINATION

A. Evolutionary Changes in Axis Determination

Generally, the majority of experimental work on cell determination and embryonic axes is carried out on echinoid species that develop indirectly through a feeding pluteus larva. These species have an A–V axis that is rigidly fixed maternally, whereas D–V and L–R axes are determined in the embryo. Studies on axis formation in a direct-developing sea urchin, *Heliocidaris erythrogramma*, show that although patterns of axial determination may be highly conserved in evolution, this does not necessarily mean that mechanistic constraints are operating. Although maternal determination of the A–V axis is conserved in this species, most of the other features of axial determination have been modified, some radically. A number of features modified in the evolution of direct development in *H. erythrogramma* are either basal to most echinoderms or basic to echinoids (Fig. 2). These changes include a switch to maternal L–R (equal to adult D–V axis) determination, a reversion to equal fourth and fifth cleavages (no micromeres are formed), a reversion to commitment of skeletogenic cells after the fourth cleavage, and no commitment of coelomic precursors from fifth cleavage micromeres. Other features such as maternal specification of the A–V axis and radial cleavage are conserved.

Figure 2B shows that evolutionary modifications of generally conserved features have taken place within the lineage ancestral to the direct-developing

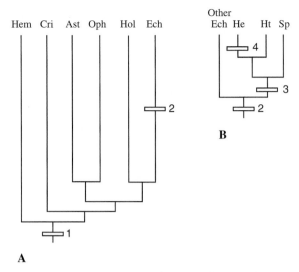

FIGURE 2   Cell lineage events in echinoderm and sea urchin evolution. (A) An echinoderm class phylogeny (hem, hemichordates; cri, crinoids; ast, asteroids; oph, ophiuroids; ech, echinoids). (B) A simplified echinoid phylogeny (He, *H. erythrogramma;* Ht, *H. tuberculata;* Sp, *S. purpuratus*). (1) The primitive feartures of echinoderm cell lineage discussed in the text. (2) Cell lineage features of living sea urchins: fourth cleavage unequal, specification of skeletogenic lineage at fourth cleavage, commitment of skeletogenic lineage at the fourth cleavage. (3) D–V axis specified, but not committed, before first cleavage, with specification of some coelomic precursors as fifth cleavage small micromeres. (4) D–V and L–R axes committed before first cleavage, reversion to equal fourth cleavage, reversion to commitment of skeletogenic cells after the fourth cleavage, and no commitment of coelomic precursors from fifth cleavage micromeres. Cell lineage features from Wray and Raff (1990) and Wray (1994). Phylogeny based on Wray (1994) and Smith *et al.* (1993).

echinoid species *H. erythrogramma.* The features modified recently in *H. erythrogramma* evolution date at least to the Permo–Triassic boundary, 250 MYA, when the living echinoid lineages originated. Its lineage diverged from *H. tuberculata,* 8–12 MYA, placing an upper limit on the time of evolution of direct development in *H. erythrogramma* and the evolutionary modification of these features (McMillan *et al.,* 1992; Smith *et al.,* 1990).

The most striking modification of axial determination is the shift in determination of the D–V axis from embryonic to maternal. In the indirect-developing sea urchin *S. purpuratus,* isolated 2-cell blastomeres not only form normal larvae but metamorphose and develop into sexually mature adults, showing that the first two blastomeres are equivalent (Cameron *et al.,* 1996). The first two blastomeres of *H. erythrogramma* do not develop equivalently, indicating the existence of a maternally determined D–V axis (Henry and Raff, 1990). It should be noted that even so, the development of axial structures in *H.*

erythrogramma also involves inductive interactions in the embryo (Henry and Raff, 1994). In addition, Henry *et al.* (1990) fertilized *H. erythrogramma* eggs that had been elongated by being drawn into silicone tubes. This constrained both the site of fertilization and the first cleavage plane. Vital dye marking of these embryos demonstrated that establishment the D–V axis depended upon neither the site of fertilization nor the plane of first cleavage. *H. erythrogramma* has reduced its larval axes. The larval L–R axis corresponds to the adult D–V axis. There is evidence that in *H. erythrogramma* a remnant of the primitive pluteus D–V axis is retained in the development of some relict features, including larval serotonergic neurons, cilia, and larval spicules (Bisgrove and Raff, 1989; Emlet, 1995).

B. EVOLUTIONARY REORGANIZATION OF CELL LINEAGES

Comparison of 32-cell fate maps of the starfish *Asterina pectinifera*, the indirect-developing sea urchins *S. purpuratus* and *P. lividus,* and the direct-developing sea urchin *H. erythrogramma* shows that echinoderm fate maps exhibit both conserved and variable features (Fig. 3). Fate maps for members of other echinoderm classes are not yet available and will be necessary to make any meaningful generalizations about conservation and change. However, certain patterns are evident.

The derived fate map for *H. erythrogramma* shown in Fig. 3 results from profound modifications in cell lineages associated with the evolution of direct development from a large egg (Wray and Raff, 1990; Wray, 1994; Raff, 1996). Modifications include (1) a transformation of the most dorsal vegetal cells to an ectodermal fate, with concomitant loss of dorsal contribution to internal cell fates; (2) large-scale shifts in allocations of ectodermal cells; (3) a change in sites of origin of the ciliary band; and (4) loss of a larval mouth. These changes presumably have arisen as a consequence of the fact that *H. erythrogramma* larvae do not feed. As a result of having to generate adult structures without growth, allocation of cells must be done from the existing embryonic blastomeres. One way developmental modifications can be achieved without greatly disrupting development is through heterochrony (the dissociation of the relative timing of events in development between ancestral and descendant ontogenies). Some of these changes observed in *H. erythrogramma* are heterochronic, but many others are not and instead involve changes in maternal determinants, cell fate, cleavage patterns, and cell lineage behavior. The D–V polarity in internal cell fates in *H. erythrogramma* is in stark contrast to the nearly radial symmetry of *S. purpuratus*. However, this dramatic asymmetry has a precursor in the more modest asymmetry in contribution of cell precursors to

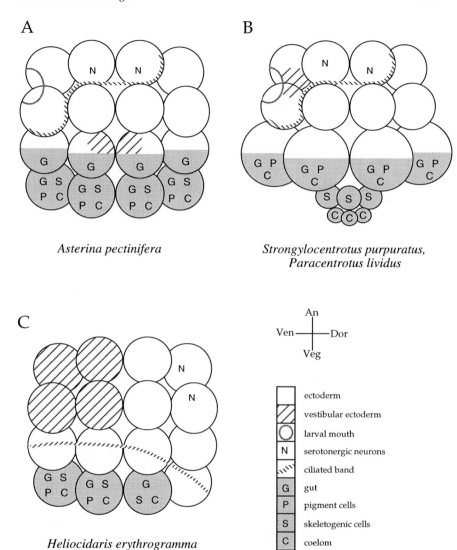

FIGURE 3 Fate maps and cleavage patterns of three echinoderms, showing conserved as well as highly variable features. The 32-cell fate maps are known for only three echinoderms. (A) The asteroid *Asterina pectinifera* (Kominami, 1984; Kuraishi and Osanai, 1992). (B) The indirect-developing echinoid *Strongylocentrotus purpuratus* (Cameron *et al.*, 1987, 1989, 1990, 1991). (C) The derived direct-developing echinoid *Heliocidaris erythrogramma* (Wray and Raff, 1989, 1990). Figure courtesy of G. A. Wray (1994), with permission.

coelomic fates that has been documented in *S. purpuratus* by Cameron *et al.* (1991). Although asymmetric in origin in *H. erythrogramma,* gut and pigment cells arise from radially symmetric precursors in *S. purpuratus.*

Figure 4 shows some of the modifications of cell lineage that occurred during the evolution of direct development. Equivalent blastomeres of an 8-cell indirect-developing and a *H. erythrogramma* embryo are compared. Cleavage patterns are different, as are patterns of blastomere commitment. Most notably, *H. erythrogramma,* unlike the indirect-developing echinoid pattern, has no unequal fourth and fifth cleavages and produces no micromeres. *H. erythrogramma* has reverted to the primitive equal pattern of echinoderm cleavage. In addition, the fourth division micromeres of indirect developers give rise to

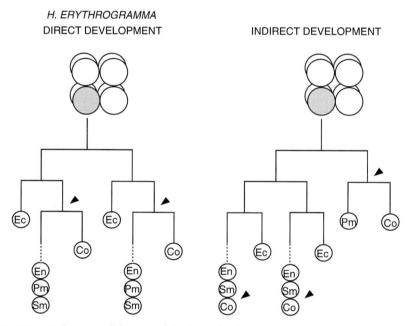

FIGURE 4   Embryonic cell lineages of the direct-developing sea urchin *H. erythrogramma* contrasted with those of an indirect-developing sea urchin. The cell lineage of a single vegetal pole cell of an 8-cell embryo is traced for both. The four vegetal pole blastomeres of the indirect developer all produce the same set of lineages. The ventral blastomeres of *H. erythrogramma* differ from the dorsal vegetal blastomere lineages (not shown). The arrowheads indicate coelomic founder cells. In indirect development, the micromeres produced at the fourth cleavage divide to produce a primary mesenchyme cell founder (Pm) and a coelomic founder (Co). In *H. erythrogramma,* primary mesenchyme cells arise later. Abbreviations: Co, coelomic founder cell; Ec, ectodermal; En, endodermal; Pm, primary mesenchyme; Sm, secondary mesenchyme. Reproduced with permission from Raff (1996). © 1996 University of Chicago Press.

two founder cells, for coelom and primary mesenchyme, at the fifth cleavage. This does not occur in *H. erythrogramma*, although both coelomic and primary mesenchyme cell types are produced in this species. Heterochronic changes include an early commitment of ectoderm founders and a later commitment of both coelomic and primary mesenchyme founder cells.

The evolutionary modifications in cell lineages exhibited by *H. erythro-gramma* are important because they underlie much of the reorganization of the embryo and also because they show that rapid and radical changes are possible in otherwise long-conserved cell lineage features in echinoderms. One might propose that such long persistence requires developmental constraints. The changes that have occurred in less than 10 MYR during the evolution of *H. erythrogramma* show that this is not the case. Given the conservation of cell lineage patterns in indirect-developing echinoids, it is notable that cell deletion in indirect-developing species demonstrates a remarkable degree of developmental regulation, that is, the ability of partial embryos to restore missing aspects of the embryo. For example, if the micromeres are removed, the embryo nonetheless produces primary mesenchyme cells and skeleton. This cell fate is generated synthetically by conversion of secondary mesenchyme cells derived from the macromeres (Ettensohn and McClay, 1988). The evolutionary changes in cell lineage observed in *H. erythrogramma* may have been made possible by the flexible developmental regulatory properties of indirect-developing echinoids (Raff, 1996).

C. Evolution of Gene Expression Patterns within Cell Lineages

Although the cell lineages of *H. erythrogramma* have been modified substantially many of the major larval cell types of indirect-developing sea urchin embryos are present. This implies that much of the same molecular machinery is being assembled. Has cell lineage modification been accompanied by substantial changes in gene expression? The answer appears to be that the overall catalogue of genes expressed is similar, but that striking differences in pattern of expression have evolved. We have found fewer than 0.1% differences in transcript catalogues between *H. tuberculata* and *H. erythrogramma* in the more prevalent half of mRNAs expressed in late gastrulae–early larvae (Haag and Raff, 1998). Even so, profound differences in gene expression have accompanied the evolution of direct development. I mention two examples briefly.

H. erythrogramma produces over 1000 primary mesenchyme cells, as opposed to the usual 32 primary mesenchyme cells for indirect-developing sea urchins. The much larger number of cells in the *H. erythrogramma* gastrula appears to result from a higher nuclear–cytoplasmic ratio in the blastomeres of this large-egged species and, consequently, more rounds of cleavage (Parks *et al.*, 1988). The behavior of the primary mesenchyme cell lineages differs in *H. erythrogramma* in another way as well. Skeleton secretion is delayed. Instead of beginning with the start of gastrulation, it is initiated several hours later in the early larva. Only a tiny remnant of the larval skeleton is produced (Emlet, 1995), and secretion of adult skeleton begins in earnest. This heterochrony in skeleton secretion by primary mesenchyme cells is accompanied by a heterochrony in expression of the *msp 130* gene by these cells (Parks *et al.*, 1988). This gene expression heterochrony is controlled by a delay in timing of the action of trans-acting factors that regulate transcription of the *msp 130* gene in the primary mesenchyme cells (Klueg *et al.*, 1997).

In the second example, we (Hahn *et al.*, 1995; Kissinger *et al.*, 1997; Kissinger and Raff, 1998) defined the actin gene family and its expression in *Heliocidaris* species. We chose these genes because actin genes are highly conserved in mammals and highly regulated in expression in cell lineage territories in *S. purpuratus*. We expected them to serve as conserved markers of cell lineage homology in *H. erythrogramma*. However, the sea urchin actin family has been evolving rapidly both in number of family members and in expression pattern. Thus, *H. tuberculata*, with the same mode of development, is very similar to *S. purpuratus* in the expression of CyIII actin in its aboral ectoderm territory, has a modified pattern of CyI expression, but does not express CyII actin at all. *H. erythrogramma* expresses CyI actin similarly to *H. tuberculata*, expresses CyII, but does not express CyIII, which is represented by only a pseudogene in this species. The loss of CyIII actin reflects the loss of the characteristic aboral ectoderm of the pluteus in *H. erythrogramma*. The pattern of expression of CyI, as well as some other ectoderm-expressed genes (Haag and Raff, 1998), indicates that new ectoderm gene expression territories have evolved. Some gene expression territories (primary mesenchyme) may be relatively conserved in histospecific genes expressed, but others (ectodermal territories) are not. Thus, neither cell lineage patterns nor even histospecific patterns of gene expression necessarily are conserved in evolution, despite their tight regulation within a species.

IV. SUMMARY AND CONCLUSIONS

Although too few echinoderm classes have had their cell lineages determined, the existing sea urchin cell lineage data show that cell lineage information is important to both functional studies of developmental mechanisms as well as to studies of the evolution of development. Many cell lineage features are highly conserved. Most notably, maternal determination of the A–V axis and its relationship to the first two cleavage planes may be a good candidate for a deep constraint. Some other features that have been long conserved nonetheless have evolved dramatically in the few million years separating *H. erythrogramma* from its conserved indirect-developing relatives. This observation implies that many conserved features are retained not because development is constrained, but because selection may be maintaining a suite of features involved in producing a feeding larva. When selection is relaxed, perhaps as a result of evolution of a large egg and loss of a need for larval feeding, rapid evolutionary changes in axial and cell lineage features can occur.

Echinoderms offer exceptionally good experimental possibilities for the study of cell lineage and developmental evolution. *H. erythrogramma* has been studied in the most detail, but other direct-developing echinoderms now being investigated offer interesting possibilities. Examples include *Clypeaster rosaceus* and other sea urchin species such as *Brisaster latifrons* that produce large eggs and facultatively feeding plutei (Emlet, 1986; Hart, 1996), *Peronella japonica,* which forms a nonfeeding abbreviated pluteus (Okazaki and Dan, 1954; Amemiya, 1996; Amemiya and Arakawa, 1996), *Holopneustes purpurescens,* a large-egged Australian direct-developing sea urchin with asymmetric pigmentation of the egg (Morris, 1995), *Asthenosoma ijimai,* a shallow water echinothurioid with some pluteus remnant structures (Amemiya and Emlet, 1992), *Abatus cordatus,* an Antarctic brooding heart urchin with a very large egg and no remnant of larval structures [Schatt, 1985; see also Poulin and Féral (1996)], and the varied modes of development (planktonic, lecithotrophic, and brooding) all found within the single starfish genus *Patiriella* (Byrne, 1995; Byrne and Cerra, 1996). No cell lineage information is available for any of these species yet, but all offer exciting possibilities to ask, for example, if their cell lineages have been modified convergently with those of *H. erythrogramma.* This wealth of comparative material offers exceptional prospects for understanding the processes of evolution of development and evolutionary transitions among larval forms.

ACKNOWLEDGMENTS

I thank Greg Wray for the use of Figures 1 and 3 and Eric Haag for critically reading the manuscript.

REFERENCES

Amemiya, S. (1996). Complete regulation of development throughout metamorphosis of sea urchin embryos devoid of macromeres. *Dev., Growth Differ.* 38, 465–476.

Amemiya, S., and Arakawa, E. (1996). Variation of cleavage pattern permitting normal development in a sand dollar, *Peronella japonica*: Comparison with other sand dollars. *Dev. Genes Evol.* **206**, 125–135.

Amemiya, S., and Emlet, R. B. (1992). The development and larval form of an echinothurioid echinoid, *Asthenosoma ijimai*, revisited. *Biol. Bull.* (*Woods Hole, Mass.*) **182**, 15–30.

Armstrong, N., Hardin, J., and McClay, D. R. (1993). Cell-cell interactions regulate skeleton formation in the sea urchin embryo. *Development* (*Cambridge, UK*) **119**, 833–840.

Arthur, W. (1988). "A Theory of the Evolution of Development." Wiley, Chichester.

Berridge, M. J., Downes, C. P., and Hanley, M. R. (1989). Neural and developmental actions of lithium: A unifying hypothesis. *Cell* (*Cambridge, Mass.*) **59**, 411–419.

Bisgrove, B. W., and Raff, R. A. (1989). Evolutionary conservation of the larval serotonergic nervous system in a direct developing sea urchin. *Dev., Growth Differ.* **31**, 363–370.

Byrne, M. (1995). Changes in larval morphology in the evolution of benthic development by *Patiriella exigua* (Asteroidea: Asterinidae), a comparison with the larvae of *Patiriella* species with planktonic development. *Biol. Bull.* (*Woods Hole, Mass.*) **188**, 293–305.

Byrne, M., and Cerra, A. (1996). Evolution of intragonadal development in the diminutive asterinid sea stars *Patiriella vivipara* and *P. parvivipara* with an overview of development in the Asterinidae. *Biol. Bull.* (*Woods Hole, Mass.*) **191**, 17–26.

Cameron, R. A., Hough-Evans, B., Britten, R. J., and Davidson, E. H. (1987). Lineage and fate of each blastomere of the eight-cell sea urchin embryo. *Genes Dev.* **1**, 75–84.

Cameron, R. A., Fraser, S. E., Britten, R. J., and Davidson, E. H. (1989). The oral-aboral axis of a sea urchin embryo is specified by first cleavage. *Development* (*Cambridge, UK*) **106**, 641–647.

Cameron, R. A., Fraser, S. E., Britten, R. J., and Davidson, E. H. (1990). Segregation of oral from aboral ectoderm precursors is completed at fifth cleavage in the embryogenesis of *Strongylocentrotus purpuratus*. *Dev. Biol.* **137**, 77–85.

Cameron, R. A., Fraser, S. E., Britten, R. J., and Davidson, E. H. (1991). Macromere cell fates during sea urchin development. *Development* (*Cambridge, UK*) **113**, 1085–1091.

Cameron, R. A., Leahy, P. S., and Davidson, E. H. (1996). Twins raised from separated blastomeres develop into sexually mature *Strongylocentrotus purpuratus*. *Dev. Biol.* **178**, 514–519.

Campbell, K. S. W., and Marshall, C. R. (1987). Rates of evolution among Paleozoic echinoderms. *In* "Rates of Evolution" (K. S. W. Campbell and M. F. Day, eds.), pp. 61–100. Allen & Unwin, London.

Conway Morris, S. (1993). The fossil record and the early evolution of the Metazoa. *Nature* (*London*) **361**, 219–225.

Czihak, G. (1960). Untersuchungen über die Coelomanlagen und die Metamorphose des Pluteus von *Psammechinus miliaris* (Gmelin). *Zool. Jahrb., Abt. Anat. Ontog. Tierre* **78**, 235–256.

Czihak, G. (1965). Entwicklungsphysiologische Untersuchungen an Echiniden. Experimentelle Analyse der Coelomentwicklung. *Wilhelm Roux' Arch. Entwicklungsmech. Org.* **155**, 709–729.

Davidson, E. H. (1989). Lineage-specific gene expression and the regulative capacities of the sea urchin embryo: a proposed mechanism. *Development* **105**, 421–445.

Davidson, E. H. (1990). How embryos work: a comparative view of diverse modes of cell fate specification. *Development* **108**, 365–389.

Davidson, E. H., Peterson, K. J., and Cameron, R. A. (1995). Origin of bilaterian body plans: Evolution of developmental regulatory mechanisms. *Science* **270**, 1319–1325.

Emlet, R. B. (1986). Facultative planktotrophy in the tropical echinoid *Clypeaster rosaceus* (Linnaeus) and a comparison with obligate planktotrophy in *Clypeaster subdepressus* (Gray) (Clypeasteroida: Echinoidea). *J. Exp. Mar. Biol. Ecol.* **95**, 183–202.

Emlet, R. B. (1995). Larval spicules, cilia, and symmetry as remnants of indirect development in the direct developing sea urchin *Heliocidaris erythrogramma*. *Dev. Biol.* **167**, 405–415.

Ettensohn, C. A. (1990). The regulation of primary mesenchyme cell patterning. *Dev. Biol.* **150**, 261–271.

Ettensohn, C. A., and McClay, D. R. (1988). Cell lineage conversion in the sea urchin embryo. *Dev. Biol.* **125**, 396–409.

Grell, K. G., and Ruthmann, A. (1991). Placozoa. In "Microscopic Anatomy of the Invertebrates" (F. W. Harrison and J. A. Westfall, eds.), Vol. 2, pp. 13–27. Wiley-Liss, New York.

Haag, E. H., and Raff, R. A. (1998). Isolation and characterization of three mRNAs enriched in embryos of the direct-developing sea urchin *Heliocidaris erythrogramma*: evolution of larval ectoderm. *Dev. Genes Evol.* **208**, 188–204.

Hahn, J.-H., Kissinger, J. C., and Raff, R. A. (1995). Structure and evolution of CyI cytoplasmic actin-encoding genes in the indirect- and direct-developing sea urchins *Heliocidaris tuberculata* and *Heliocidaris erythrogramma*. *Gene* **153**, 219–224.

Hart, M. W. (1996). Evolutionary loss of larval feeding: Development, form and function in a facultatively feeding larva, *Brisaster latifrons*. *Evolution (Lawrence, Kans.)* **50**, 174–187.

Henry, J. J., and Raff, R. A. (1990). Evolutionary changes in the process of dorsoventral axis determination in the direct developing sea urchin, *Heliocidaris erythrogramma*. *Dev. Biol.* **141**, 55–69.

Henry, J. J., and Raff, R. A. (1992). Development and evolution of embryonic axial systems and cell determination in sea urchins. *Semin. Dev. Biol.* **3**, 35–42.

Henry, J. J., and Raff, R. A. (1994). Progressive determination of cell fates along the dorsoventral axis in the sea urchin *Heliocidaris erythrogramma*. *Roux's Arch. Dev. Biol.* **204**, 62–69.

Henry, J. J., Amemiya, S., Wray, G. A., and Raff, R. A. (1989). Early inductive interactions are involved in restricting cell fates of mesomeres in sea urchin embryos. *Dev. Biol.* **136**, 140–153.

Henry, J. J., Wray, G. A., and Raff, R. A. (1990). The dorsoventral axis is specified prior to the first cleavage in the direct developing sea urchin *Heliocidaris erythrogramma*. *Development (Cambridge, UK)* **110**, 875–884.

Hörstadius, S., and Wolsky, A. (1936). Studien über die Determination der Bilateral-symmetrie des jungen Seeigel Keimes. *Wilhelm Roux' Arch. Entwicklungsmech. Org.* **135**, 69–113.

Hyatt, B. A., Lohr, J. L., and Yost, H. J. (1996). Initiation of vertebrate left-right axis formation by maternal Vg1. *Nature (London)* **384**, 62–65.

Jägersten, G. (1972). "Evolution of the Metazoan Life Cycle. A Comprehensive Theory." Academic Press, New York.

Kirchhamer, C. V., and Davidson, E. H. (1996). Spatial and temporal information processing in the sea urchin embryo: Modular and intramodular organization of the *CyIIIa* gene cis-regulatory system. *Development (Cambridge, UK)* **122**, 333–348.

Kirchhamer, C. V., Yuh, C.-H., and Davidson, E. H. (1996). Modular cis-regulatory organization of developmentally expressed genes: Two genes transcribed territorially in the sea urchin embryo, and additional examples. *Proc. Natl. Acad. Sci. U.S.A.* **93**, 9322–9328.

Kissinger, J. C., Hahn, J.-H., and Raff, R. A. (1997). Rapid evolution in a conserved gene family. Evolution of the actin gene family in the sea urchin genus *Heliocidaris* and related genera. *Mol. Biol. Evol.* **14**, 654–665.

Kissinger, J. C., and Raff, R. A. (1998). Evolutionary changes in sites and timing of expression of actin genes in embryos of the direct- and indirect-developing sea urchins *Heliocidaris erythrogramma* and *H. tuberculata*. *Dev. Genes Evol.* **208**, 82–93.

Klein, P. S., and Melton, D. A. (1996). A molecular mechanism for the effect of lithium on development. *Proc. Natl. Acad. Sci. U.S.A.* **93**, 8455–8459.

Klueg, K. M., Harkey, M. A., and Raff, R. A. (1997). Mechanisms of evolutionary changes in timing, spatial expression, and mRNA processing in the *msp130* gene in a direct-developing sea urchin, *Heliocidaris erythrogramma*. *Dev. Biol.* **182**, 121–133.

Kominami, T. (1984). Allocation of mesodermal cells during early embryogenesis in the starfish, *Asterina pectinifera. J. Embryol. Exp. Morphol.* 84, 177–190.

Kominami, T. (1988). Determination of the dorsoventral axis in early embryos of the sea urchin, *Hemicentrotus pulcherrimus. Dev. Biol.* 127, 187–196.

Kuraishi, R., and Osanai, K. (1992). Cell movements during gastrulation of starfish larvae. *Biol. Bull.* (*Woods Hole, Mass.*) 183, 258–268.

Livingston, B. T., and Wilt, F. H. (1989). Lithium evokes expression of vegetal-specific molecules in the animal blastomeres of sea urchin embryos. *Proc. Natl. Acad. Sci. U.S.A.* 86, 3669–3673.

Livingston, B. T., and Wilt, F. H. (1990). Range and stability of cell fate determination in isolated sea urchin blastomeres. *Development* (*Cambridge, UK*) 108, 403–410.

Livingston, B. T., and Wilt, F. H. (1992). Phorbol esters alter cell fate during development of sea urchin embryos. *J. Cell Biol.* 119, 1641–1648.

Mao, C.-A., Wikramanayake, A. H., Gan, L., Chuang, C.-K., Summers, R. G., and Klein, W. H. (1996). Altering cell fates in sea urchin embryos by overexpressing SpOtx, an orthodenticle-related protein. *Development* (*Cambridge, UK*) 122, 1489–1498.

McCain, E. R., and McClay, D. R. (1994). The establishment of bilateral asymmetry in sea urchin embryos. *Development* (*Cambridge, UK*) 120, 395–404.

McMillan, W. O., Raff, R. A., and Palumbi, S. R. (1992). Population genetic consequences of developmental evolution in sea urchins (Genus *Heliocidaris*). *Evolution* (*Lawrence, Kans.*) 46, 1299–1312.

Morris, V. B. (1995). Apluteal development of the sea urchin *Holopneustes purpurescens* Agassiz (Echinodermata: Echinoidea: Euechinoidea). *Zool. J. Linn. Soc.* 114, 349–364.

Nocente-McGrath, C., McIsaac, R., and Ernst, S. G. (1991). Altered cell fate in LiCl-treated sea urchin embryos. *Dev. Biol.* 147, 445–450.

Okazaki, K. (1975). Normal development to metamorphosis. *In* "The Sea Urchin Embryo" (G. Czihak, ed.), pp. 177–232. Springer-Verlag, Berlin.

Okazaki, K., and Dan, K. (1954). The metamorphosis of partial larvae of *Peronella japonica* Mortensen, a sand dollar. *Biol. Bull.* (*Woods Hole, Mass.*) 106, 83–99.

Parks, A. L., Parr, B. A., Chin, J.-E., Leaf, D. S., and Raff, R. A. (1988). Molecular analysis of heterochronic changes in the evolution of direct developing sea urchins. *J. Evol. Biol.* 1, 27–44.

Poulin, E., and Féral, J. P. (1996). Why are there so many species of brooding Antarctic echinoids? *Evol.* 50, 820–830.

Raff, R. A. (1996). "The Shape of Life. Genes, Development, and the Evolution of Animal Form." University of Chicago Press, Chicago.

Ransick, A., and Davidson, E. H. (1993). A complete second gut induced by transplanted micromeres in the sea urchin embryo. *Science* 259, 1134–1138.

Ransick, A., and Davidson, E. H. (1995). Micromeres are required for normal vegetal plate specification in sea urchin embryos. *Development* (*Cambridge, UK*) 121, 3215–3222.

Rieger, R. M. (1994). The biphasic life cycle—A central theme of metazoan evolution. *Am. Zool.* 34, 484–491.

Rieger, R. M., Haszprunar, G., and Schuchert, P. (1991). On the origin of the Bilateria: Traditional views and recent alternative concepts. *In* "The Early Evolution of Metazoa and the Significance of Problematic Taxa" (A. M. Simonetta and S. Conway Morris, eds.), pp. 107–113. Clarendon Press, Oxford.

Robertson, E. J. (1997). Left-right asymmetry. *Science* 275, 1280.

Schatt, P. (1985). Development of *Abatus*. Ph.D. Dissertation, Université Pierre et Marie Curie, Paris.

Slack, J. M. W., Holland, P. W. H., and Graham, C. F. (1993). The zootype and the phylotypic stage. *Nature* (*London*) 361, 490–492.

Smith, A. B. (1988). Phylogenetic relationship, divergence times, and rates of molecular evolution for camarodont sea urchins. *Mol. Biol. Evol.* 5, 345–365.

Smith, M. J., Boom, J. D. G., and Raff, R. A. (1990). Single copy DNA distances between two congeneric sea urchin species exhibiting radically different modes of development. *Mol. Biol. Evol.* 7, 315–326.

Smith, M. J., Arndt, A., Gorski, S., and Fajber, E. (1993). The phylogeny of echinoderm classes based on mitochondrial gene arrangements. *J. Mol. Evol.* 36, 545–554.

Stambolik, V., Ruel, L., and Woodgett, J. R. (1996). Lithium inhibits glycogen synthase kinase-3 activity and mimics *Wingless* signalling in intact cells. *Curr. Biol.* 6, 1664–1668.

Strathmann, R. R. (1985). Feeding and nonfeeding larval development and life-history evolution in marine invertebrates. *Annu. Rev. Ecol. Syst.* 16, 339–361.

Vance, R. R. (1973). On reproductive strategies in marine benthic invertebrates. *Am. Nat.* 107, 339–352.

Vermeij, G. J. (1987). "Evolution and Escalation." Princeton University Press, Princeton, NJ.

Wikramanayake, A. H., Brandhorst, B. P., and Klein, W. H. (1995). Autonomous and non-autonomous differentiation of ectoderm in different sea urchin species. *Development* 121, 1497–1505.

Wikramanayake, A. H., and Klein, W. H. (1997). Multiple signaling events specify ectoderm and pattern the oral-aboral axis in the sea urchin embryo. *Development (Cambridge, UK)* 124, 13–20.

Wood, W. B. (1991). Evidence from reversal of handedness in *C. elegans* embryos for early cell interactions determining cell fates. *Nature (London)* 349, 536–538.

Wray, G. A. (1994). The evolution of cell lineage in echinoderms. *Am. Zool.* 34, 353–363.

Wray, G. A., and Raff, R. A. (1989). Evolutionary modification of cell lineage in the direct-developing sea urchin *Heliocidaris erythrogramma*. *Dev. Biol.* 132, 458–470.

Wray, G. A., and Raff, R. A. (1990). Novel origins of lineage founder cells in the direct-developing sea urchin *Heliocidaris erythrogramma*. *Dev. Biol.* 141, 41–54.

Wray, G. A., Levinton, J. S., and Shapiro, L. H. (1996). Molecular evidence for deep Precambrian divergences among metazoan phyla. *Science* 274, 568–573.

CHAPTER 9

Cell Lineages in Larval Development and Evolution of Holometabolous Insects

LISA M. NAGY* AND MIODRAG GRBIĆ†

*Department of Molecular and Cellular Biology, University of Arizona, Tucson, Arizona, †Department of Zoology, University of Western Ontario, London, Ontario, Canada

The Origin and Evolution of Larval Forms

I. INTRODUCTION

In his 1929 presidential address, "The Origin and Evolution of Larval Forms," presented to the zoology section of the British Association for the Advancement of Science, Garstang argued against the then popular Haeckelian approach to evolution. Under this theory, all modifications to life history and larval morphology occurred as terminal additions, so that an organism's life history represented a linked chain of adult ancestors. Garstang argued instead for viewing the adult as a final complex resulting from a number of differentiating cell lineages, some of which can be pruned off at no consequence and others of which become solely adaptive for the larva. Using gastropod torsion as an example, he struggled with logical explanations for the origin of phenotypic jumps in the evolutionary record or saltatory mutations. Although we still are at a loss to explain how saltatory mutations become integrated into either developmental systems or populations, we now know that developmental mechanisms can evolve rapidly and by multiple means and certainly are not restricted to terminal additions (see Raff, 1996).

II. CELL LINEAGE AS A TOOL TO STUDY THE EVOLUTION OF DEVELOPMENT

The study of cell lineage within and between closely related species has been a useful tool for understanding how individuals develop and ontogenies evolve. Cell lineage studies began in the 1870s in the context of the controversy over Haeckel's "biogenetic" law. In the most extreme interpretation of this theory, early development of any metazoan recapitulated the nondifferentiated tissues of a spongelike ancestor. The cells of an early cleavage embryo were thought to have no more relation to the adult than "snowflakes to an avalanche" (Stent, 1985, pp. 1–2). This theory easily was repudiated by Whitman (1878, 1887) and many others since, who have shown with careful cell lineage studies that individual cells in early embryos have extremely reproducible fates and species-specific characteristics. For example, the nematode *Caenorhabditis elegans* was found to have essentially an invariant cell lineage; the 1059 cells in the adult hermaphrodite form through nearly identical cleavage patterns in each individual (Sulston and Horvitz, 1977). The embryos of most species in other invertebrate phyla studied, including echinoderms, mollusks, annelids, and ascidians, have stereotypical cleavage patterns that separate cells early in development into lineage groups with specific developmental fates.

These early lineage studies, coupled with the results from early experimental perturbations of embryos, led to a view of development as a repeated division of the egg into discrete lineages, whose fate was controlled by the differential distribution of cytoplasmic determinants. Cell lineage was believed to be an indicator of cell-autonomous mechanisms of determination. It is now apparent, however, that an overt pattern of regular cleavage should not be confused with a primarily lineage-based, or cell-autonomous, patterning mechanism. Even in those species whose cleavage patterns are extremely regular and reproducible, patterning involves an interplay between lineage and intercellular communication [reviewed in Schnabel and Priess (1997)]. Only rarely are developmental fates determined by entirely cell-autonomous means.

When comparing the early development of different species, changes in cell lineage often can serve as clear indicators of ontogenetic change. However, invariant cell lineage sometimes can be a red herring. Developmental mechanisms can evolve while the cell lineage pattern remains invariant (Sommer, 1997). Nevertheless, lineage should not be abandoned as a means of studying comparative development, but its limitations as a mechanistic explanation should be understood.

Insect embryos represent an extreme example of the dissociation of cell lineage from cell patterning. Other than late patterning in the nervous system, tissues within the insect embryo are not generated with cell lineages reproducible from individual to individual. Early insect development originates with a series of nuclear divisions that occur in the absence of cell membranes. These divisions create a bag of as yet undifferentiated nuclei, known as a syncytium. Groups of nuclei obtain patterning information via a gradient mechanism according to their positions along the anteroposterior and dorsoventral egg axes. These gradients generate domains of gene expression that are maintained as the embryo cellularizes and provide the genetic information necessary for specifying the developmental fates of the cells in the early embryo [reviewed in Akam (1987); Ingham, (1988)]. This process varies among different insect species, but in all insects examined to date at least some portion of the embryo derives from the patterning that takes place in the syncytium and appears to be lineage-independent (Sander, 1976; Nagy, 1994).

In *Drosophila,* there are a few examples where the domains of gene expression established in the early embryos correspond with lineage compartments. In these cases, a group of cells clustered together in the early embryo form what is called a "polyclone." Polyclones begin as a small group of founder cells, which at their origin do not share a common lineage, but which subsequently behave as lineage compartments. These compartments are maintained by the stable inheritance of discrete patterns of gene expression, in particular through the activity of "selector" genes. The combination of particular selector gene activities provide the compartments with information concerning their specific developmental fates, e.g., which part of the body to specify (Garcia-Bellido *et al.*, 1973; Crick and Lawrence, 1975). Compartment boundaries then function as an important landmark in the generation of subsequent cell interactions important for refining the more global patterning information down to the level of an individual cell [reviewed in Lawrence and Struhl (1996)]. Polyclones thereby link gene action with cell lineage and anatomy.

III. GENE NETWORKS AS A FUNCTIONAL ANALOGUE OF CELL LINEAGE

In spite of the appeal of exchanging classical cell lineage of individual cells with an analysis of polyclones and the selector genes that maintain them as a tool for comparative insect development, identified selector genes that generate strict lineage compartments are rare [reviewed in Blair (1995)]. However, as the direct linkage of domains of gene expression and developmental fate now has been established in many model developmental systems (Davidson, 1989, 1991; Gilbert *et al.*, 1996), if we abandon the need to understand the commonality of the cell lineage of a particular character, domains of gene expression become a functional analogue of the classical concept of cell lineage. If we substitute "groups of cells under similar genetic control" for "differentiating cell lineages" in the statement from Garstang quoted earlier, much of our current view of how development evolves is consistent with Garstang's statement.

From a phylogenetic perspective, a comparative examination of groups of cells under the control of similar gene networks can provide a new measure for understanding the origin and evolution of insect larvae. Instead of an analysis of the evolution of development by comparing groups of cells that have a common genealogy, clusters of cells that share common gene expression patterns can become markers for change. The strategy here is first to identify genes or gene networks that act as important markers of patterning and, secondly, to identify patterns in how gene networks operate. What, for example, is the degree of connectivity between elements within a particular gene network? How are separate gene networks linked together? By what means do they become dissociated? Then, by comparing the activity of these gene networks in different species, one can ask how cell fates are orchestrated in related cell lineages and how the operation of these gene networks evolves.

A. MECHANISMS OF DISSOCIATION

Gene networks link cells with a common developmental fate in different ways. One well-studied class of gene networks relevant to development and evolution is that controlled through the previously mentioned selector genes. The role of selector genes is to provide the genetic address or information that maintains positional orientation and to orchestrate cell fates within a given position (Garcia-Bellido, 1975). When expressed at the wrong time or place via mutation or in transgenic overexpression experiments, they can direct cells to take on novel identities. In some cases, selector genes control specific anatomical features. For example, the homeotic or HOX genes are a family of transcription factors that control the positional identity of segments and appendages, presumably by activating a cascade of downstream genes (Lewis, 1978). These genes are famous for their bizarre mutant phenotypes, e.g., gain-of-function mutations where a walking leg develops where the mouthparts or antennae should be. *Pax-6*, another transcription factor, when misexpressed in *Drosophila*, can elicit the formation of extra eyes (Halder *et al.*, 1995). Note, however, that ectopic eyes only form on specific parts of wings, legs, and antennae, i.e., from imaginal disk tissues. Selector genes cannot direct *any* cell to take on a new fate: the responding tissue must be competent to respond to the signal. In other cases, selector genes control more generic positional information. The *Drosophila engrailed* gene functions to define "posterior" in all the segment primordia throughout the life of the insect [reviewed in Blair (1995)]. Thus, selector genes have in common the ability to function as triggers for activating specific gene networks.

Modifications in the timing and expression of selector genes have been postulated to provide a mechanism to generate dissociable or heterochronic events in evolution (Goldschmidt, 1938). It has long been realized that development consists of a series of nested gene networks, some of which experimentally are dissociable from one another (Needham, 1933; Bonner, 1978). By changing the timing of expression of a selector gene in a descendant relative to its ancestor, a structure can be turned on in a novel time or place. Particularly relevant to this type of hypothesis are those selector genes that have been dubbed "master regulatory genes" (Callaerts *et al.*, 1997). This name has been applied to those genes that, either through dominant gain-of-function mutation or ectopic expression experiments, can elicit the *complete* development of a structure such as an eye or a wing.

Master regulatory genes trigger gene networks, but the concept also triggers alarm in the minds of many developmental biologists. Does a gene deserve this name if it is one of many genes that can elicit the same structure? Ectopic expression of two other genes in the *Drosophila* eye development pathway elicits ectopic eyes (Bonini *et al.*, 1997; Chen *et al.*, 1997; Pignoni *et al.*, 1997). In this case is *Pax-6* really a "master" or only an "executive committee" member (see Desplan, 1997)? How *complete* does the invoked structure have to be to merit this title? Is the generation of wing bristles in novel locations sufficient to tag a gene a master regulator of the wing?

Among the many problems associated with the concept of master regulatory genes are three common misconceptions concerning these genes that are relevant to the issue of the evolution of cell lineage and morphology. The first misconception is that these masters represent a unique class of genes with the potential to create saltatory change in evolution. Although these genes create saltatory phenotypes in *Drosophila* mutants, there is still no evidence that these phenotypes mimic evolution. The second misconception is that these genes operate by controlling a downstream hierarchy. Whereas they could function through downstream hierarchies, very few gene networks known in development function in a truly hierarchical manner. The final misconception is that these genes are in sufficient "control" that a change in expression could result in a fully integrated change in development. Whereas "master regulatory genes" provide the mechanism to dissociate structures, any dissociation in embryogenesis must be balanced by a mechanism of integration. Selector genes can, in some circumstances, regulate the capacity of cells in which they are activated to induce the surrounding tissue to respond appropriately, e.g., the ectopic eyes that result from *Pax-6* misexpression form primitive neural connections. To form a fully functioning novel organism, however, the new structure must be integrated fully into the new location.

B. Mechanisms of Integration

An understanding of mechanisms of integration may emerge from studying the regulatory mechanisms that maintain homeostasis or stability. Within physiological and intracellular regulatory systems, these two states often are achieved by a combination of negative feedback and/or checkpoints, which can hold a process poised for activation until all of the independent conditions necessary to proceed are met. Both of these states heavily rely on the activity of repressors in gene-regulatory or biochemical pathways. A similar reliance on repressors to create pattern in time and space is used during early development in *Drosophila*. Many differentiation pathways begin by a general activation of gene expression (see Lawrence, 1992). Pattern then is established by localized repression of gene activity to create such features as stripes in the embryo or patterned neuroblast determination in the ventral ectoderm (Carroll, 1990; Artavanis-Tsakonas *et al.*, 1995). Second order repression, or repressors of a repressor, resulting in an apparent activation, also is quite common. Because negative feedback and checkpoints are such a prominent feature of developmental, physiological, biochemical, and gene pathways (see Weintraub, 1993), they also are likely targets for evolutionary modification. Examples of how morphological changes can be triggered by flipping negative switches in preexisting gene networks will follow in Section V.D.

IV. PATTERNS IN THE EVOLUTION OF HOLOMETABOLOUS INSECT LARVAE

It has become possible to explore the genetic mechanisms underlying the origins of morphological diversity and the changes in life history by using the approach described earlier. Insect larvae are morphologically diverse, and insect life history is understood in considerable detail. Insects therefore serve as a good system for this type of study. We begin by describing the morphological features and patterns of life histories seen within extant and fossil insects, with an emphasis on discerning the most obvious patterns of change observed over time. We then discuss possible developmental mechanisms that could be responsible for these morphological modifications. To do this, we review a few examples of how "gene network analogues" of cell lineage vary between species and outline some examples of preliminary evidence of how gene networks are modified during insect larval evolution.

A. WHAT TYPES OF LIFE HISTORIES ARE FOUND WITHIN THE INSECTS?

In winged (or pterygote) insects, larvae possess the full complement of body segments, but do not exhibit the complete adult morphology. Winged insects can be grouped into two classes according to the type of metamorphosis they undergo. The first class is the hemimetabolous or exoptyergote insects (external wings), where immatures emerge with the basic adult body plan but have incompletely developed wings and genitalia. These insects are called "hemimetabolous" due to the subjective consideration that the difference between the last larval instar and the adult is greater than the changes between the earlier larval instars. The second class of insects is the holometabolous or endoptyergote (internal wings) insects, where larvae are distinctly different from adults and the transformation into the adult includes a pupal stage [see Sehnal et al., (1996) for a more detailed description of these two categories and additional life-history variants that exist]. Primitive wingless insects typically are called "ametabolous" (without metamorphosis). However, their development is quite similar to that of hemimetabolous insects. Larvae possess the complete adult body plan, albeit incompletely developed genitalia. They differ from hemimetabolous insects in that they never develop wings and the degree of change between each of the molts is gradual.

Phylogenetically, holometabolous development is restricted to advanced groups of insects. The larval stages of holometabolous insects differ radically from the adult. They frequently are adapted for life in ephemeral habitats. The mouthparts are modified, the compound eyes are absent or nonfunctional, and in many, but not all, species larval legs reappear on the abdomen. Wings, genitalia, and sometimes legs and mouthparts are not visible externally during the holometabolous larval stages. Although it is commonly assumed that these appendages have been internalized as imaginal discs in all holometabolous insects, imaginal discs are not found in all holometabolous larvae (Svacha, 1992). Many orders have no invaginated disc structures, whereas others develop them only in the final larval instar. In the advanced Holometabola, different combinations of the adult appendages can be internalized as imaginal discs during varying stages of the life history (Svacha, 1992). An adult dipteran develops nearly entirely from imaginal discs (Auerbach, 1936), Lepidoptera develop only wings and genitalia from imaginal discs (Kim, 1959; Eassa, 1953), and some Coleoptera have imaginal discs that only partially invaginate (Tower, 1903). All holometabolous larvae nonetheless transform into adults through a complex metamorphosis involving a nonfeeding pupal stage, which is lethargic-to-immobile and frequently hidden. In nearly all species, whether or not they have imaginal discs, wings make their first external appearance in the pupal instar.

B. Life-History and Morphological Trends Apparent from the Insect Fossil Record

How did this distribution of metamorphic types evolve? Data on fossil insect larvae are scarce, although we can piece together a plausible scenario for the evolution of insect life histories from the existing data. The oldest described insects are wingless Collembolans (springtails), reported from the early Devonian (Greenslade and Whalley, 1988). Devonian springtails are remarkably similar to modern Collembolans. It is not until the Carboniferous that representatives of most of the modern winged insect orders appear and life histories characteristic of most hemimetabolous and holometabolous insects are represented. Aquatic nymphs of the basal hemimetabolous orders, Protoodonata (dragonfly) and Ephemeroptera (mayfly), are abundant in Carboniferous fossil beds and already are morphologically distinct from the adults, which in some cases were gigantic, with wingspans as large as 71 cm (Kukalova-Peck, 1991).

Hemimetabolous Paleozoic species underwent many larval instars. Wings grew progressively larger in each instar, but did not become completely outstretched and functional until the older larval stages (Kukalova-Peck, 1991). There is a trend in the fossil record toward a reduction in the number of instars with nonfunctional wing buds and development in concealed habitats, presumably hidden from predators. Interestingly, the presumed selection for fewer molts did not result in any cases of direct development, as seen in many other phyla. For example, in the crustaceans, the postulated sister group to the insects, direct development in peracarids is thought to have evolved by condensing the ancestral larval stages into the embryo.

Major climatic changes took place during the Permian, leading to dramatic changes in floral and fauna. During this period, a major extinction of primitive pterygote insect orders lacking a metamorphic stage took place and holometabolous insects underwent a major radiation. Two major advantages of holometabolous development are thought to be responsible for this radiation: (1) wing formation was delayed to a single penultimate instar, allowing larvae without external wing buds to hide from predators in microhabitats; (2) larvae were functionally different from adults and could exploit different ecological niches (Kukalova-Peck, 1991). Over time, holometabolous development increasingly evolved toward greater functional and morphological disparity between larva and adults, presumably by selection for increasing niche separation. Interestingly, all primarily wingless insects never evolved the ability to undergo metamorphosis, consistent with the hypothesis that selective pressure to get rid of nonfunctional wings may have driven the evolution of holometaboly.

C. Origin Theories

It is uniformly agreed that holometaboly evolved from some form of hemimetabolous development; however, the exact nature of the changes required for the origin of holometaboly is not resolved. Two different suggestions, with multiple variations, have been made. In one version, the pupal stage is proposed to be a terminal addition into an otherwise conservative life history. In this case, larval stages between hemi- and holometabolous insects are presumed equivalent. The only difference is that the latter have internalized their wings. The holometabolous pupal stage is simply a modification of the terminal larval stage typical of the hemimetabolous life history (Poyaroff, 1914; Hinton, 1948; Sehnal et al., 1996). This theory was justified by the reasoning that a transitional form was required to mold the adult from a larval stage that had grown increasingly divergent in form. This is, however, inconsistent, with known constraints on metamorphosis: radical transformations in morphology are known to occur in some hemimetabolous species without a pupal instar (Sehnal et al., 1996).

The second model, encapsulated in the Berlese–Imms theory (see Imms, 1937, 1948), proposes that to make the transition from a hemimetabolous to a holometabolous life cycle, three changes are necessary: (1) a precocious emergence of the hatchling from the egg; (2) an early insertion of multiple molts into the late embryonic stage of the ancestral life history; and (3) a condensation of all of the hemimetabolous larval stages into a single pupal instar. In this theory, holometabolous larvae essentially are free-living, feeding embryos with a prolonged life. Berlese and Imms support their theory with a claim that a decrease in maternal yolk contributions would lead to an earlier emergence of the embryo. However, Sehnal et al. (1996) argue that if any trend in maternal yolk contributions to the egg are apparent within the insects, it is in the opposite direction.

Although this theory has been out of vogue for some time, Truman et al. (1999) provide novel evidence in support of the Berlese–Imms theory based on new information from embryonic JH titers in hemimetabolous insects. The pupal stage of holometabolous insects is characterized by a unique pattern of hormone levels. Truman and colleagues found that hormone titers during late embryogenesis in ametabolous and hemimetabolous insects have profiles similar to those of the pupal hormone titers in holometabolous insects. If these hormone profiles are the conservative feature in evolution, then to maintain their unique characteristics, the evolution of holometaboly would have evolved by inserting multiple instars into late embryonic development prior to this unique molt.

In the Berlese–Imms (see Imms, 1937) and Truman et al. (1999) models, embryos hatch earlier without wings, creating a feeding larva that could take advantage of resources unavailable to the winged adult. Resource partitioning then could rapidly lead to an increasing morphological divergence between larval and adult stages. Under this scenario, in the primitive holometabolous life history, wings would not develop until the pupal stage. Holometabolous larvae are not simply hemimetabolous larva with inpocketed wings. Instead, precocious development of imaginal discs as invaginated structures occurred as a secondary evolutionary event. This is consistent with the distribution of imaginal discs within holometabolous insects described earlier.

D. Summary of the Changes Observed in Insect Life Histories

Several features thus are apparent concerning the origin and evolution of holometabolous larvae. First, as seen in the fossil record, there was a decrease in the number and timing of molts and perhaps the insertion of new stages into embryonic development. Second, the evolution of holometabolous larval stages corresponded with a dissociation of the growth of wings and genitalia from larval growth. This dissociation could be considered a classic example of heterochrony, in which the appearance of a structure during development is delayed relative to its ancestor. As larval stages became even more specialized, additional morphological features became dissociated from larval growth, including the compound eyes, mouthparts, and thoracic legs. Third, a secondary change in the opposite direction occurred, seen as an acceleration of the growth of these same structures, facilitated by the evolution of a mechanism (or mechanisms) to internalize the developing adult structures as imaginal discs. Finally, the appearance of novel structures like larval ocelli and the reappearance of abdominal legs in the larva also are seen.

V. WHAT GENETIC ALTERATIONS WERE NECESSARY FOR THE ORIGIN AND SUBSEQUENT EVOLUTION OF LARVAE?

What developmental changes are required for the changes in life history and morphology described earlier? On the basis of known principles of the operation of developmental systems, we can speculate about the mechanistic triggers that underlie the observed patterns. Implicit in our analysis is that the ultimate causality of larval change arises from natural selection. Abbreviation in the number of molts likely resulted from selection for a more rapid life cycle, delay in the production of wings from selection against juveniles with nonfunctional wings, subsequent divergence in form between larva and adult for resource partitioning, and loss of larval legs by selection for wiggling around in mucky, near-liquid food sources. However, our concern is to understand the internal mechanisms that modulate the regulation of form.

A. POSSIBLE MECHANISMS TO INSERT NOVEL STAGES INTO LIFE HISTORIES

One of the problems with accepting the Berlese–Imms theory over a theory of the pupal stage evolving as a terminal addition to the life history is that there are many more examples of terminal additions in ontogeny, and it seems intuitively easier for development to evolve via terminal additions. We are, however, familiar with a qualitatively similar modification to early embryology from our studies on wasp development (Grbić et al., 1996), which serves to illustrate how insertions into early ontogeny within insect life histories are indeed feasible. In polyembryonic wasp development, as described later, a period of massive proliferation is inserted into the life history prior to gastrulation, compared to the insertion of multiple repetitions of a late embryonic stage in the life history of a hemimetabolous insect, as proposed in the Berlese–Imms theory.

Drastic departures from the general plan of insect development are seen in taxa that develop in very different environmental circumstances and that have extremely modified life-history strategies. One such group is the wasps (Hymenoptera) that develop as parasites of other arthropods. Parasitic wasps exploit two different strategies: ectoparasitism, where the wasp oviposits her egg on the surface of the host, and endoparasitism, where the wasp oviposits her egg within the host body cavity. In ectoparasitism, embryos develop as most insects do, protected from adverse environmental influences by a tough chorion and supplied with sufficient yolk to support development. In contrast, eggs of endoparasitic species are deposited within the host cavity. These embryos emerge from the chorion early in development and continue development in the highly nutritive environment of the host hemolymph (Grbić *et al.*, 1996; 1998; Grbić and Strand, 1998). The most extreme case of endoparasitism is seen in the evolution of polyembryony, where multiple embryos develop from a single egg. For example, embryos of the polyembryonic wasp *Copidosoma floridanum* proliferate to form up to 2000 clonal embryos from the single egg [reviewed in Grbić and Strand (1998)].

Endoparasitism has arisen many times in the Hymenoptera. Polyembryony also has arisen at least five times independently, always from an endoparastic ancestor (Strand and Grbić, 1997). Endoparasitic wasp development is associated with drastic changes in cellular and developmental aspects of embryogenesis. Embryos of these monoembryonic endoparasitic species cellularize early in development, there is a reduction in egg chorion and egg size, and modifications in the extraembryonic membrane facilitate the uptake of nutrients from the host tissue. All of these features also are characteristic of polyembryonic species. It is likely that the developmental changes associated with endoparasitism facilitate the evolution of polyembryony. For example, an extraembryonic membrane for nutrient absorption appears to be an important preadaptation for the evolution of polyembryony.

What types of mechanistic changes underlie the evolution of polyembryony? *Drosophila* embryos undergo syncytial cleavage, whereby nuclear division generates several thousand nuclei residing in a common cytoplasm before cellularization, resulting in a syncytial blastoderm. Maternally encoded transcription factors, distributed within the syncytium, initiate a cascade of downstream genes that regulate segmentation (St. Johnston and Nusslein-Vollhard, 1992). When we examined whether markers of the gradient-based patterning system that functions in *Drosophila* were conserved in the polyembryonic wasp *Copidosoma,* we found that they were conserved, but their expression had been shifted *in toto* to later in development, once the 2000 progeny were initiating individual development (Grbić *et al.*, 1996). This implies that the trigger or triggers for polyembryonic development include the capacity to delay the onset of segmentation and to initiate the program much later in time. This also implies that the segmentation gene network itself can, in this case, function as a dissociable unit.

Drosophila genetics have not yet provided any clues about potential mechanisms underlying this particular type of delay in development. Perhaps this is not surprising, as this type of modification results from multiple, sequential changes. Such changes cannot be uncovered in a mutagenesis screen that makes random, and frequently single, hits in the genome. However, by comparing early patterning mechanisms between wasp species with different life histories, we may be able to uncover the mechanistic basis of the evolution of polyembryony and, consequently, learn more about how insertions into life histories can occur.

B. Dissociation of Morphological Structures

Patterns in the evolution of insect morphology suggest that dissociation of developmental features has been one of the major ways that holometabolous insects evolve. In the Berlese–Imms (see Imms, 1937) or Truman *et al.* (1999) scenarios described earlier, wings, genitalia, and compound eyes are delayed in their development in holometabolous insects relative to hemimetabolous insects. Truman *et al.* (1998) hypothesize that the founder cells for these adult structures may have become differentially sensitive to hormone levels, for example, via a change in the activity of an esterase that degrades hormones. Changes in selector genes also could easily play critical roles in dissociation. For example, modification of the timing of expression of the selector gene *Pax-6* could underlie the changes in timing of the development of the compound eye. As mentioned earlier, ectopic expression of the *Pax-6* gene results in the formation of ectopic eyes (Halder *et al.*, 1995). *Pax-6* orthologues have been found throughout metazoans and are likely to be conserved in all insects. One can imagine an adaptive scenario whereby, with increasing niche separation, functional adaptations for vision during flight might not be advantageous for larval vision in rotting fruit. Naturally occurring polymorphisms in how the *cis*-acting regulatory regions of the *Pax-6* gene respond to hormone levels (or some downstream target of hormone regulation) could, under proper conditions, be selected for delayed onset. There also is evidence to suggest that regulatory regions of selector genes are polymorphic in natural populations (Gibson and Hogness, 1996). One problem with this hand-waving scenario is that, without an eye, a developing larva clearly would be at a disadvantage. Interestingly, the larval ocelli are the only known example of a light-sensing structure not regulated by *Pax-6* (Halder *et al.*, 1995). Perhaps the delay in the development of the compound eyes of the adult corresponded with the independent evolution of larval ocelli.

Other strongly conserved selector genes, like the homeotic genes, also could be involved in dissociation of the development of an appendage from the rest of the larva. Because the homeotic genes can change one appendage into another—antenna or mouthpart into leg—the homeotic selector genes long were thought to specify given appendages. However, when the role of a given homeotic gene is compared between *Drosophila* and a crustacean, or *Drosophila* and a butterfly, it is clear that these genes do not specify appendage identity, but rather specify axial positional information. The same axial information reads out into the production of very different appendages in each species. Nonetheless, as with *Pax-6*, a change in the timing of the expression of a homeotic gene easily could delay the onset of differentiation of an appendage (see Section V.D).

It is worth noting that not all selector genes seem involved in dissociative trends. For example, in all insects examined, the selector gene *engrailed* is expressed in the posterior compartment of each segment. Its expression pattern or assumed role does not vary. Similarly, not all of the proposed "master regulatory" genes from *Drosophila* may be important for evolutionary changes. Although the *vestigial* gene has been proposed to control wing development (Kim *et al.*, 1996), it has not yet been isolated outside the Diptera. Interestingly, if wings have dissociated from the rest of development several times independently, there may not be one conserved "master regulator" for wing development in insects. Different groups of insects might employ different molecules to dissociate wing development from larval growth.

C. ACCELERATION OF ADULT STRUCTURES

Imaginal discs likely evolved several times independently (Svacha, 1992) and separately from the evolution of a holometabolous life history. It also is likely that not all appendages evolved the capacity to invaginate as imaginal discs simultaneously. Another class of genes that likely was critical for the evolution of imaginal discs is the zinc finger class transcription factors, *snail* and *escargot*. Unlike the selector genes discussed earlier, together these genes are expressed in all of the imaginal disks. *snail* functions in multiple cell types that undergo invaginations (Alberga *et al.*, 1991), which would be a requirement for internalized disk development. *escargot* functions to maintain diploidy in disk cells while all of the larval cells around them become polyploid (Fuse *et al.*, 1994; 1996). Polyploidization of the larval cells also is a derived feature unique to higher holometabolous insects (Svacha, 1992). Interestingly, *escargot* appears to have arisen via a gene duplication event and also may be unique to higher holometabolous insects. The presence of these two genes in the disc primordia may have been crucial for the evolution of imaginal discs.

D. Appearance of Larval Legs: Atavisms or Novelties?

The development and evolution of holometabolous larval abdominal append-ages serve as a useful illustration of what we have learned concerning the types of genetic changes that underlie morphological changes in evolution. Abdominal legs, or prolegs, are present on the larva of some members of all holometabolous insects. Larval prolegs are present in very different arrange-ments, in terms of both number and segmental distribution (Fig. 1). Figure 2 illustrates the pattern of larval proleg evolution within the Diptera. Flies have radiated in many ecological niches, ranging from soil, rotting plant mate-rial, living plant and animal tissue, water, and even extreme habitats such as hot springs, geysers, and petroleum pools (some Ephedridae and Stratyomi-dae). They exhibit a plethora of life-history strategies, including saprophagy, predation, parasitism, and symbiosis. By mapping the presence and absence of prolegs onto the established dipteran phylogeny (McAlpine and Wood, 1989), we see unambiguous rampant convergence of this trait, with apparent repeated losses and gains of this trait throughout the history of the Diptera.

Regardless of phylogenetic position, the presence of larval prolegs correlates with aquatic habitat, and in many cases with a predatory lifestyle, where prolegs likely allowed active foraging for food. In terrestrial habitats, dipteran larvae develop on the nutritive substrate chosen by the ovipositing female. Because these larvae do not have to forage for food, mutations that acted on suppression of the developmental program responsible for the proleg development might not be deleterious. Although we have not done as thorough an analysis in other holometabolous orders, we expect to see a similar degree of convergence and correlation of the presence or absence of larval legs with larval habitat.

What is the origin of these abdominal legs? Are they atavisms or novelties? Can a comparative analysis of limb development shed light on the mechanisms of larval leg evolution? The homeotic mutants are famous for transforming the character of one segment, and its characteristic appendage, into another. In spite of this dramatic effect on segment character, the homeotic genes do not function to activate appendage formation early in development. From genetic studies it is known that the ground state of each *Drosophila* trunk segment appears to be "express limb." By ground state, we mean how the segment would differentiate if it lacked all homeotic gene input. This ground state to "express limb" is defined by using the same positional information that sets up the embryonic axes during the earliest stages of development. Limb primordia arise at the intersection of anteroposterior and dorsoventral axes [reviewed in Cohen (1993); Williams and Nagy (1995)]. Primordia are

Diptera

Lepidoptera

Hymenoptera

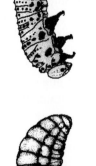

FIGURE 1 Diagram of representative patterns of larval legs and prolegs in holometabolous insects: orders Lepidoptera, Hymenoptera, and Diptera.

marked by a cluster of cells expressing *Distalless* (*Dll*), a gene that encodes a homeodomain protein required for proximodistal (P–D) limb outgrowth. This promotion of limb primordia at the boundary between the A–P and D–V axes occurs in *all* of the trunk segments of *Drosophila*.

To repress the development of limbs in the abdomen requires the activity of the abdominal homeotic genes. The *Drosophila* homeotic genes *Ultrabithorax* (*Ubx*) and *abdominal-A* (*abd-A*) regulate the regional presence of appendages by *repressing* limb formation in the abdomen [Lewis (1978); Vachon *et al.* (1992); reviewed in Carroll (1995)]. In the absence of the abdominal *HOX* genes, limb primordia develop in every segment (Lewis, 1978; Vachon *et al.*, 1992). Early in development, homeotic genes simply modify the response of each segment to limb-inducing information present in *every* trunk segment [Lewis (1978); Carroll *et al.* (1995); reviewed in Carroll (1995)]. Later in development, the homeotic genes function to determine the individual characteristics of an appendage. The only exception to this rule may be the homeotic gene *AbdB*, which is thought to promote genital development (Averof and Akam, 1995).

Both the homeotic genes *Ubx* and *abdA* are present in other arthropods, including crustaceans, chelicerates, myriapods, and onychophorans (Averof and Akam, 1995; Cartwright *et al.*, 1993; Grenier *et al.*, 1997). However, rather than repressing limbs, they are co-expressed in the same cells as the limb patterning gene *Distalless* (*Dll*; Averof and Akam, 1995; Panganiban *et al.*, 1995). Their expression may be neutral, or even serve an activating role for limb development. This led to the hypothesis that *Ubx* and *abdA* repressing limbs was a synapomorphy for pterygote insects. The repression of hexapod abdominal limbs also may, however, have evolved independently several times. Kukalova-Peck (1998) reports the presence of segmented abdominal leglets on primitive hemimetabolous and holometabolous insects. Palopoli and Patel (1998) also report variation in the degree of co-expression of Ubx–abdA and Dll protein expression in the third thoracic and first abdominal appendages in collembolans, grasshoppers, and beetles.

Boudreaux (1979) predicted that larval abdominal legs result from a derepression of "leg" genes in the abdomen and suggested that this derepression event may be a ground plan apomorphy for all of the Endopterygota, except the Coleoptera. Using a combination of genetic analyses in *Drosophila* and comparative gene expression, molecular evidence was found supporting Boudreaux's predicted derepression event (Warren *et al.*, 1994). Several aspects of the gene network that patterns *Drosophila* limbs were shown to be conserved during the development of the larva of the butterfly *Precis coenia*. Particularly interesting was the observed localized and late-acting regulation of the abdominal HOX genes, which resulted in the derepression of limb development, as marked by the appearance of small bilateral rings of *Dll* expression in the

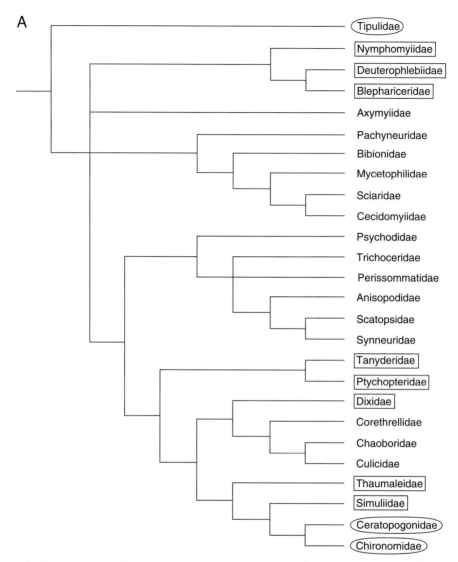

A

- Tipulidae
- Nymphomyiidae
- Deuterophlebiidae
- Blephariceridae
- Axymyiidae
- Pachyneuridae
- Bibionidae
- Mycetophilidae
- Sciaridae
- Cecidomyiidae
- Psychodidae
- Trichoceridae
- Perissommatidae
- Anisopodidae
- Scatopsidae
- Synneuridae
- Tanyderidae
- Ptychopteridae
- Dixidae
- Corethrellidae
- Chaoboridae
- Culicidae
- Thaumaleidae
- Simuliidae
- Ceratopogonidae
- Chironomidae

FIGURE 2 Convergence in the presence–absence of prolegs within families of dipteran insects. Diptera represent a monophyletic order of 120,000 described species. Traditionally, this order is subdivided into suborders Nematocera and Brachicera. Nematoceran flies include primitive dipteran families, whereas the Brachicera represent an assemblage of all other derived families. Families not marked: all species with no prolegs. Boxes: all species with larval prolegs. Ovals: most species without prolegs or very few with prolegs (invariably associated with aquatic habitat and predatory lifestyle). (A) The distribution of prolegs in Nematocera suggests that the most primitive species had well-developed abdominal prolegs. In higher nematocerans, which radiated to terrestrial habitats, prolegs were lost in many families. In some higher species, abdominal prolegs were lost from only a portion of the abdomen (only 2–3 pairs retained in Ptychopteridae and Dixidae). Invariably all basal groups containing species with prolegs are aquatic, predatory,

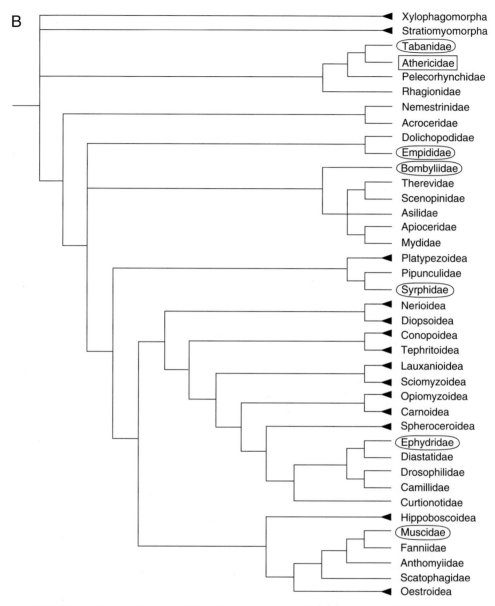

B

Xylophagomorpha
Stratiomyomorpha
(Tabanidae)
[Athericidae]
Pelecorhynchidae
Rhagionidae
Nemestrinidae
Acroceridae
Dolichopodidae
(Empididae)
(Bombyliidae)
Therevidae
Scenopinidae
Asilidae
Apioceridae
Mydidae
Platypezoidea
Pipunculidae
(Syrphidae)
Nerioidea
Diopsoidea
Conopoidea
Tephritoidea
Lauxanioidea
Sciomyzoidea
Opiomyzoidea
Carnoidea
Spheroceroidea
(Ephydridae)
Diastatidae
Drosophilidae
Camillidae
Curtionotidae
Hippoboscoidea
(Muscidae)
Fanniidae
Anthomyiidae
Scatophagidae
Oestroidea

FIGURE 2 (*Continued*) and inhabit fast streams (Nymphomidae, Deuterophlebiidae, and Blephariceridae). In Tipulidae, terrestrial species do not have prolegs, whereas aquatic species have well-developed prolegs. In these families, 5–8 pairs of prolegs are arranged on abdominal segments and are lacking on the thorax. Prolegs are absent from terrestrial nematocerans, as well as from aquatic nematocerans that are free swimmers (like the mosquito larva Culicidae). However, aquatic Ptychopteridae and Dixidae have 2–3 pairs of abdominal prolegs. In more advanced families (Thaumatelidae, Simulidae, Ceratopogonidae, and Chironomidae), prolegs again are

abdominal segments. Abdominal limbs in the butterfly larva develop as a late-acting repression of the early repression of limbs in the abdomen.

The mechanism by which this derepression works remains unclear. Some factor upstream of the abdominal HOX genes is acting in a very discrete spatial and temporal pattern to repress the HOX genes. In *Precis,* this factor must be restricted to the middle abdominal segments and a small dorsolateral region within these segments. It also acts much later in development than the initial events that establish thoracic limbs. Thoracic legs already are well-developed when abdominal leg development is initiated.

There is a preoccupation in the insect literature with determining whether prolegs are homologous with thoracic legs or derived secondarily (Snodgrass, 1954; Hinton, 1955; Birket-Smith, 1984). There is an additional feature of

---

present and associated with aquatic habitat. However, these species display a different proleg arrangement than the more primitive Nematoceran species. Prolegs are present only on the first thoracic and last abdominal segments in Thaumatelidae, Simulidae, Ceratopogonidae, and Chironomidae. (B) In contrast to the primitive nematoceran families, it appears that terrestrial habitat and lack of larval prolegs is the ancestral state of Brachicera. Some brachiceran groups independently evolved prolegs at least seven times. In each of these cases, the evolution of prolegs is associated with a switch from the terrestrial habitat to an aquatic lifestyle. Fossil evidence suggests that Ragaonidae appear to be the most primitive Brachicerans. Extant Ragaonidae are terrestrial, developing on decaying organic matter, and do not have prolegs. However, more distally within the Tabanomorpha cluster, Athericidae larvae, which are aquatic predators, have eight pairs of abdominal prolegs. Their sister group, Tabanidae, contains prolegless terrestrial species and aquatic species that have well-developed prolegs on seven abdominal segments. Empididae, in general, is a terrestrial group; however, two aquatic genera have larvae with prolegs on eight abdominal segments. Species in the superfamily Asiloidea are terrestrial predators generally with no abdominal prolegs. All members of the superfamily Platypezoidea are terrestrial species, developing on decaying organic matter. Their larvae are completely without prolegs. The superfamily Syrphideae, representative of the family Pipinculidae, is parasitic, without prolegs. Family Syrphidae contains terrestrial species with no prolegs and some aquatic species that have well-developed abdominal prolegs. Superfamilies Neiroidae, Diopsoidea, Conopoidea, Tephritoidea, Lauxanoidea, Sciomyzoidea, Opiomyzoidea, Carnoidea, and Spheroceroidea are terrestrial and their larvae develop on decaying organic matter, plants, as parasites, and as predators of other insects. Their larvae do not have prolegs. Superfamily Ephydroidea contains several terrestrial families whose immatures develop on decaying organic matter, including Drosophilidae. Although in family Ephedrydae some genera are terrestrial, with developing immatures feeding on plants and carrion, several aquatic species have well-developed prolegs on eight abdominal segments. Finally, the superfamily Muscoideae contains terrestrial families whose larvae develop by feeding on plants or decaying organic matter and lack abdominal prolegs. A few rare muscid species that develop in aquatic habitats (for example, *Phaonia exoleta* larvae develop in water holes in trees and are predators of mosquito larvae) have well-developed abdominal legs on 6–7 abdominal segments. Phylogeny and larval leg distributions were modified from McAlpine and Wood (1989); Tsacas and Disney, (1974); and Smith (1989); for alternative phylogeny of Nematocera, see Ooesterbroek and Courtney (1995).

interest in Fig. 2 relevant to this question. Primitive dipteran larvae possess only abdominal prolegs, with no visible external thoracic larval legs. Their thoracic legs presumably are internalized as imaginal discs. Several families secondarily have gained a pair of thoracic prolegs on the first thoracic segments and lost all but the terminal abdominal prolegs. Does this larva no longer have a first thoracic leg disc or does it have both a larval leg and an imaginal disc? This also is the only example in which the larval legs do not appear as a continuous series, making it difficult to imagine a single mechanism that could regulate the appearance of these two legs.

How have prolegs evolved independently so many times? Are these repeated losses and gains achieved by the mechanism inferred from the butterfly studies? Do individual species evolve specific changes in developmental circuitry that allow for the development of prolegs? The repeated secondary appearance of abdominal legs in the larva of endopterygote insects could be explained through changes anywhere in the gene regulatory network described earlier. Either the particular switch that has been implicated in the development of butterfly larval legs is particularly easy to trigger or multiple independent regulatory events, through novel modifications of the gene network, may be occurring repeatedly. We suggest that prolegs indeed are leg homologues, but the distinct mechanism that triggers their growth may have unique features from species to species. Prolegs thus are neither atavisms nor novelties. Minor features of the gene network are novel, but the structure is atavistic.

The study of the evolution of abdominal limbs in insects suggests that structures can appear and disappear by hooking into preexisting, fully functioning networks. HOX genes repress the limb gene network in the abdomen early in insect evolution. Later in evolution, a new repressor comes along and represses this repressor, activating the original switch and causing a secondary reappearance of legs. In both cases, integrated changes are triggered by throwing new repressors into the circuitry. In this scenario, multiple separate mutations affecting the same developmental process are not needed: modifications are linked directly into the preexisting network.

VI. SUMMARY AND CONCLUSIONS

Patterns in the evolution of insect morphology suggest that insertions of novel stages into life histories, dissociation of developmental features, and shuffling of old gene circuits to make new morphologies have been common in the evolution of holometabolous insect larvae. We have discussed various mechanisms of how these features might evolve that emerge from using the concept of "groups of cells that express the same gene networks" as an analogue for the classical concept of cell lineage.

It appears that some organs, like the eye, may be regulated by selector genes or, perhaps more accurately, by "master organ genes." Ectopic activation of such a gene theoretically can result in the appearance of structures in novel times and places. How then does this new structure achieve integration into the whole system at its new time and place? We suggest that the use of selector genes as modulators of repressible gene networks and second order repression as a means of activation provides a mechanism for integrated change. This mechanism requires that the gene network to be modulated has the capacity to be repressed or held in check. How a network becomes capable of dissociation is a separate question. The simplest networks likely evolved negative feedback very early, as this property seems crucial to the survival of any organism in a changing environment. Genes capable of turning on and off networks that regulate morphological structures might be a late event in the evolution of gene network organization. Once present however, such a gene could modulate the timing of activation of a gene network. Although some of the selector genes known from *Drosophila* may have evolved early in the history of metazoans, some may be unique to *Drosophila*. Additional comparative molecular analyses within the insects may reveal multiple selector genes regulating the same structure in different species, i.e., the wing. Interestingly, this mechanism of modulating repressible circuits does not allow for the generation of novelty, only the shuffling of parts in the descendant relative to the ancestor.

It also is apparent that there is strong convergence of morphological traits within holometabolous insects. All of the features we have discussed—polyembryony, imaginal discs, abdominal prolegs—likely evolved multiple, independent times within the insects. There has been insufficient comparative molecular analysis to determine whether there is also a multiplicity of underlying mechanisms. It is possible that a comparative mechanistic analysis will inform us of the underlying organizational principles that regulate the repeated occurrence of morphological structures.

ACKNOWLEDGMENTS

The authors are grateful for the thoughtful comments received from T. Williams, G. Ballovine, D. Lambert, and E. Jockusch.

REFERENCES

Akam, M. (1987). The molecular basis of for metameric pattern in the *Drosophila* embryo. *Development (Cambridge, UK)* 101, 1–22.

Alberga, A., Boulay, J. L., Kempe, E., Dennefeld, C., and Haenlin, M. (1991). The *snail* gene required for mesoderm formation in *Drosophila* is expressed dynamically in derivatives of all three germ layers. *Development (Cambridge, UK)* 111, 983–992.

Artavanis-Tsakonas, S., Matsuno, K., and Fortini, M. (1995). Notch signaling. *Science* 268, 225–232.

Auerbach, C. (1936). The development of legs, wings, and halteres in wild type and some mutant strains of *Drosophila melanogaster*. *Trans.—R. Soc. Edinburgh* 57, 787–816.

Averof, M., and Akam, M. (1995). Hox genes and the diversification of insect and crustacean body plans. *Nature (London)* 376, 420–423.

Birket-Smith, S. J. R. (1984). "Prolegs, Legs and Wings of Insects." Scandinavian Science Press Ltd., Copenhagen.

Blair, S. S. (1995). Compartments and appendage development in *Drosophila*. *BioEssays* 17, 299–309.

Bonini, N., Bui, Q., Gray-Board, G., and Warrick, J. (1997). The *Drosophila eyes absent* gene directs ectopic eye formation in a pathway conserved between flies and vertebrates. *Development (Cambridge, UK)* 124, 4819–4826.

Bonner, J. T. (1978). "The Evolution of Complexity." Princeton University Press, Princeton, Nd.

Boudreaux, H. B. (1979). "Arthropod Phylogeny, with Special Reference to Insects." Krieger Publ. Co., Melbourne, FL.

Callaerts, P., Halder, G., and Gehring, W. J. (1997). PAX-6 in development and evolution. *Annu. Rev. Neurosci.* 20, 483–532.

Carroll, S. B. (1990). Zebra stripes in fly embryos: Activation of stripes or repression of interstripes? *Cell (Cambridge, Mass.)* 60, 9–16.

Carroll, S. B. (1995). Homeotic genes and the evolution of chordates. *Nature (London)* 376, 479–485.

Carroll, S. B., Weatherbee, S. D., Langland, J. A. (1995). Homeotic genes and the regulation and evolution of insect wing number. *Nature (London)* 375, 58–61.

Cartwright, P., Dick, M., and Buss, L. W. (1993). Hom/Hox type Homeoboxes in the Chelicerate *Limulus polyphemus*. *Mol. Phyl. Evol.* 2(3), 185–192.

Chen, R., Amoui, M., Zhang, Z., and Mardon, G. (1997). *Dachshund* and *eyes absent* proteins form a complex and function synergistically to induce ectopic eye development in *Drosophila*. *Cell (Cambridge, Mass.)* 91, 893–903.

Cohen, S. M. (1993). Imaginal disc development. *In* "The Development of *Drosophila melanogaster*" (M. Bate and A. Martinez-Arias, eds.). pp. 747–842. Cold Spring Harbor Lab. Press, Cold Spring Harbor, NY.

Crick, F. H. C., and Lawrence, P. A. (1975). Compartments and polyclones in insect development. *Science* 189, 340–347.

Davidson, E. (1989). How embryos work: A comparative view of diverse modes of cell fate specification. *Development (Cambridge, UK)* 108, 365–390.

Davidson, E. (1991). Spatial mechanisms of gene regulation in metazoan embryos. *Development (Cambridge, UK)* 113, 1–26.

Desplan, C. (1997). Eye development: Governed by a dictator or a junta? *Cell (Cambridge, Mass.)* 91, 862–864.

Eassa, Y. E. (1953). The development of imaginal buds in the head of *Pieris brassicae* Linn. (Lepidoptera). *Trans. R. Entomol. Soc. London* 104, 39–50.

Fuse, N., Hirose, S., and Hayashi, S. (1994). Diploidy of *Drosophila* imaginal cells is maintained by a transcriptional repressor encoded by *escargot*. *Genes Dev.* 8, 2270–2281.

Fuse, N., Hirose, S., and Hayashi, S. (1996). Determination of wing cell fate by the *escargot* and *snail* genes in *Drosophila*. *Development (Cambridge, UK)* 122, 1059–1067.

Garcia-Bellido, A. (1975). Genetic control of wing disc development in Drosophila. Ciba Found. Symp. 29, 161–183.

Garcia-Bellido, A., Ripoll, P., and Morata, G. (1973). Developmental compartmentalization of the wing disc of Drosophila. Nature (London) 245, 251–253.

Garstang, W. (1929). The origin and evolution of larval forms. Rep. Br. Assoc. Adv. Sci. Sect. D, pp. 77–98.

Gibson, G., and Hogness, D. S. (1996). Effect of polymorphism in the Drosophila regulatory gene Ultrabithorax on homeotic stability. Science 271, 200–203.

Gilbert, S. F., Opitz, J. M., and Raff, R. A. (1996). Resynthesizing evolutionary and developmental biology. Dev. Biol. 173, 357–372.

Goldschmidt, R. (1938). "Physiological Genetics." McGraw-Hill, New York.

Grbić, M., Nagy, L. M., and Strand, M. (1996). Polyembryonic development: Insect pattern formation in a cellularized environment. Development (Cambridge, UK) 122, 795–804.

Grbić, M., Nagy, L. M., and Strand, M. R. (1998). Development of polyembryonic insects: a major departure from typical insect embryogenesis. Dev., Genes Evol. 208, 69–81.

Greenslade, P., and Whalley, P.E.S. (1988). The systematic position of Rhyniella precursor Hirst and Maulik (Collembola), the earliest known hexapod. In "Second International Seminoar on Apterygota" (R. Dallai, ed.), pp. 319–23. University of Siena, Siena, Italy.

Grenier, J. K., Garber, T. L., Warren, R., Whitington, P. M., and Carroll S. (1997). Evidence for a Precambrian origin of arthropod Hox genes. Curr. Biol. 7, 547–543.

Halder, G., Callaerts, P., and Gehring, W. J. (1995). Induction of ectopic eyes by targeted expression of the eyeless gene in Drosophila. Science 267, 1788–1792.

Hinton, H. E. (1948). The origin and function of the pupal stage. Trans. R. Entomol. Soc. London 99, 395–409.

Hinton, H. E. (1955). On the structure, function, and distribution of the prolegs of the panorpoidea, with a criticism of the Berlese-Imms theory. Trans. R. Entomol. Soc. London 106, 455–540.

Imms, A. D. (1937). "Recent Advances in Entomology." Blakiston, Philadelphia.

Imms, A. D. (1948). "A General Textbook of Entomology." Dutton, New York.

Ingham, P. (1988). The molecular genetics of embryonic pattern formation in Drosophila. Nature (London) 336, 25–34.

Kim, C. W. (1959). The differentiation centre inducing development from larval to adult leg in Pieris brassicae (Lepidoptera). J. Embryol. Exp. Morphol. 7, 572–582.

Kim, J., Sebring A., Esch, J. J., Kraus, M. E., Vorwerk, K., Magee, J., and Carroll, S. B. (1996). Integration of positional signals and regulation of wing formation and identity by Drosophila vestigial gene. Nature (London) 382, 133–138.

Kukalova-Peck, J. (1991). Fossil history and evolution of hexapod structures. In "Insects of Australia" (I. E. Naumann and CSIRO, eds.) pp. 141–180. Melbourne University Press, Melbourne.

Kukalova-Peck, J. (1998). Arthropod phylogeny and 'basal' morphological structures. In "Arthropod Relationships" (R. A. Fortey and R. H. Thomas, eds.), Syst. Asso. Spec. Vol. Ser. 55, pp. 249–268. Chapman & Hall, London.

Lawrence, P. (1992). "The Making of a Fly: The Genetics of Animal Design." Blackwell, Oxford.

Lawrence, P., and Struhl, G. (1996). Morphogens, compartments, and pattern: Lessons· from Drosophila. Cell (Cambridge, Mass.) 85, 951–961.

Lewis, E. B. (1978). A gene complex controlling segmentation in Drosophila. Nature (London) 276, 565–570.

McAlpine, J. F., and Wood, D. M., ed. (1989). "Manual of Neartic Diptera." Agriculture, Ottawa, Canada.

Nagy, L. M. (1994). A glance posterior. Curr. Biol. 9, 811–814.

Needham, J. (1933). On the dissociability of the fundamental processes in ontogenesis. *Biol. Rev. Cambridge Philos. Soc.* **8**, 180–223.

Ooesterbroek, F. L. S., and Courtney, G. (1995). Phylogeny of the nematocerous families of Diptera (Insecta). *Zool. J. Linn. Soc.* **115**, 267–311.

Palopoli, M., and Patel, N. (1998). Evolution of the interaction between *HOX* genes and a downstream target. *Curr. Biol.* **8**, 587–590.

Panganiban, G., Sebring, A., Nagy, L., and Carroll, S. (1995). The development of crustacean limbs and the evolution of arthropods. *Science* **270**, 1363–1365.

Pignoni, F., Hu, B., Zavitz, K. H., Xiao, J., Garrity, P. A., and Zipursky, S. L. (1997). The eye-specification proteins *So* and *Eya* form a complex and regulate multiple steps in *Drosophila* eye development. *Cell (Cambridge, Mass.)* **91**, 881–891.

Poyaroff, E. (1914). Essai d'une théorie de la nymphe des Insectes Holometaboles. *Arch. Zool. Exp. Gen.* **54**, 221–265.

Raff, R. A. (1996). "The Shape of Life." University of Chicago Press, Chicago.

Sander, K. (1976). Specification of the basic body pattern in insect embryogenesis. *Adv. Insect Physiol.* **12**, 125–239.

Schnabel, R., and Priess, J. R. (1997). Specification of cell fates in the early embryo. In "The Nematode *C. elegans*" (T. B. D. Riddle, B. Meyer, and J. R. Priess, eds.). Cold Spring Harbor Lab. Press, Cold Spring Harbor, NY.

Sehnal, F., Svacha, P., and Zrzavy, J. (1996). 'Evolution of insect metamorphosis.' In "Metamorphosis: Postembryonic Reprogramming of Gene expression in Amphibian and Insect Cells" (L. I. Gilbert, J. R. Tata, and B. G. Atkinson, eds.), pp. 3–58. Academic Press, San Diego, CA.

Smith, K. G. V. (1989). "An Introduction to the Immature Stages of British Flies." Royal Entomological Society of London.

Snodgrass, R. E. (1954). Insect metamorphosis. *Smithson. Misc. Collect.* **122**, (9), 1–124.

Sommer, R. J. (1997). Evolutionary changes of developmental mechanisms in the absence of cell lineage alterations during vulva formation in the Diplogastridae (Nematoda). *Development (Cambridge, UK)* **124**, 243–251.

Stent, G. (1985). The role of cell lineage in development. *Philos. Trans. R. Soc. London* **15**, 3–19.

St. Johnston, D., and Nusslein-Volhard, C. (1992). The origin of pattern and polarity in the *Drosophila* embryo. *Cell (Cambridge, Mass.)* **68**, 201–219.

Strand, M. R., and Grbić, M. (1997). The development and evolution of polyembryonic insects. *Curr. Top. Dev. Biol.* **35**, 121–160.

Sulston, J. E., and Horvitz, H. R. (1977). Postembryonic cell lineage of the nematode *Caenorhabditis elegans*. *Dev. Biol.* **56**, 110–156.

Svacha, P. (1992). What are and what are not imaginal discs: Reevaluation of some basic concepts (Insecta, Holometabola). *Dev. Biol.* **154**, 101–117.

Tower, W. L. (1903). The origin and development of the wings of Coleoptera. *Zool. Jahrb., Abt. Anat. Ontog. Tiere* **17**, 517–572.

Truman, J. W., Riddiford, L. M., and Ball, E. E. (1999). The pronymphal stage and the evolution of insect metamorphosis: Evidence from the endocrine and nervous systems. Submitted for publication.

Tsacas, L, and Disney, R. H. L. (1974). Two new African species of *Drosophila* (Diptera: Drosophilidae) whose larvae feed on Simulium larvae (Diptera: Simulidae). *Tropenmed. Parasitol.* **25**, 360–377.

Vachon, G., Cohen, B., Pfeifle, C., McGuffin, M. E., Botas, J., and Cohen, S. (1992). Homeotic genes of the bithorax complex repress limb development in the abdomen of the *Drosophila* embryo through the target gene *Distalless*. *Cell. (Cambridge, Mass.)* **71**, 437–450.

Warren, R., Nagy, L., Seleque, J., and Carroll, S. (1994). Evolution of homeotic gene regulation and function in flies and butterflies. *Nature (London)* **372**, 458–461.

Lisa M. Nagy and M. Grbić

Weintraub, H. (1993). The MyoD family and myogenesis: Redundancy, networks, and thresholds. *Cell (Cambridge, Mass.)* **75**, 1241–1244.

Whitman, C. O. (1878). The embryology of *Clepsine*. *Q. J. Microsc. Sci.* **18**, 215–315.

Whitman, C. O. (1887). A contribution to the history of germ layers in *Clepsine*. *J. Morphol.* **1**, 105–182.

Williams, T. W., and Nagy, L. M. (1995). Brine shrimp add salt to the sea. *Curr. Biol.* **5**, 1330–1333.

PART III

Larval Functional Morphology, Physiology, and Ecology

CHAPTER 10

Development and Evolution of Aquatic Larval Feeding Mechanisms

S. LAURIE SANDERSON* AND SARAH J. KUPFERBERG†

*Department of Biology, College of William and Mary, Williamsburg, Virginia, †Department of Animal Ecology, Umeå University, Umeå, Sweden and Department of Integrative Biology, University of California, Berkeley, California

301

I. INTRODUCTION

How larval vertebrates feed is of intense interest to developmental and evolutionary biologists, ecologists, and functional morphologists for a number of
reasons. (1) The feeding mechanisms of larval protovertebrates are hypothesized to have been of substantial importance in vertebrate evolution (Jollie,
1982; Mallatt, 1985; Northcutt and Gans, 1983). (2) Relatively minor alterations in developmental pathways can result in extensive morphological and
functional changes [see Lauder et al. (1989), ontogenetic repatterning in Wake
and Roth (1989)]. (3) Many of the selective advantages attributed to the larval
stage of vertebrates, such as niche separation from adults, storage of energy,
and growth, relate to larval feeding. (4) Larval fish feeding is thought to be
an important factor in stock recruitment, directly through the avoidance of
mortality from starvation during the critical period at first feeding (e.g., Houde
and Schekter, 1980; Hunter, 1980) and indirectly through larval growth rates
that affect vulnerability to predation (e.g., Houde, 1987; Pepin, 1991).
(5) Although a great deal is known about the functional morphology of feeding
in adult fishes, there are limited data on the functional morphology of feeding
in the larval precursors of these adults, which also must function as organisms
(see Lauder et al., 1989). (6) Larval feeding is of primary importance to the
population biology, ecology, and evolution of amphibians. (7) Amphibian
larval feeding can have profound effects on the distribution and abundance
of their prey organisms in aquatic ecosystems.

   Although the literature on vertebrate larval feeding is vast, there have been
no reviews of the subject. There has been little interdisciplinary synthesis to link
developmental, functional, and ecological aspects of vertebrate larval feeding
mechanisms in an evolutionary framework. Our goals are to summarize current
knowledge on vertebrate larval feeding and suggest directions for future research. In the pursuit of both of these goals, we emphasize the advantages of
an interdisciplinary approach to the analysis of vertebrate larval feeding.

   We include references to aquatic invertebrate larvae in order to compare
the variation in structures and evolutionary processes that exists among organisms that vary greatly in design, yet must solve similar foraging and life-history

problems. For reviews of invertebrate larval feeding, see Strathmann (1987) for marine organisms and Resh and Rosenberg (1984) for aquatic insects.

The intersection of development, functional morphology, and trophic ecology has the potential to be a productive area of research in larval fish biology. Although Osse noted in 1990 that there had been limited investigation of the ecomorphology of fish larvae, particularly the relationship between developmental processes and changing food selection, this area of research still remains largely unexplored. On the basis of the limited available literature, Hunt von Herbing *et al.* (1996b) suggested that the differential fitness and survival of fish larvae may be determined to a large extent by inter- and intraspecific differences in the timing of the development of feeding systems (e.g., A in Fig. 1). Houde and Schekter (1980) emphasized that feeding strategy, specifically the relationship between prey consumption rate and prey concentration and the change in this relationship with growth, may be important in regulating year class success (e.g., B in Fig. 1). Mark *et al.* (1989) urged investigation of

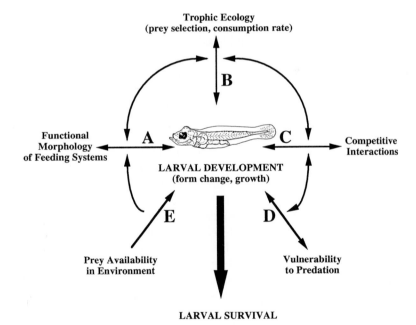

FIGURE 1 Larval development affects and is affected by feeding mechanisms (A), trophic ecology (B), and competition and predation (C and D). Prey availability in the environment also affects larval development (E). By influencing larval development, these factors determine larval survival. Arrows indicate direction of influence.

the role of developmental processes in determining larval fish food selection and species interactions (e.g., B and C in Fig. 1). Data summarized by Miller *et al.* (1988) indicate that vulnerability to predation is a function of larval fish size relative to predator size [e.g., D in Fig. 1, but see Bertram and Leggett (1994)]. Finally, as demonstrated experimentally by Gotceitas *et al.* (1996), temporal variation in prey abundance during larval development can affect larval fish growth and survival (e.g., E in Fig. 1).

The growth made possible by larval feeding is a key feature in the ability of amphibians to recruit new individuals to their populations (Wilbur, 1980). Time to and size at metamorphosis are correlates of fitness for both frogs and salamanders (Smith, 1987; Semlitsch *et al.,* 1988). The ability to feed and grow quickly is especially important for species breeding in ephemeral bodies of water. Successful recruitment and the age structure of desert anuran populations are determined by the ability of larvae to metamorphose before desiccation of the habitat (Tinsley and Tocque, 1995). Additionally, the size of an individual determines its risk of predation and competition. Feeding thus determines how quickly an individual can outgrow these risks.

The effects of amphibian larvae as consumers vary with the feeding mode and trophic level (Dickman, 1968; Seale, 1980; Morin *et al.,* 1988; Brönmark *et al.,* 1991; Liebold and Wilbur, 1992). In North American temperate ponds, microphagous feeding by ranid and hylid tadpoles on suspended particles can significantly reduce the standing crop of phytoplankton, shift the state of nitrogen from mostly particulate to mostly dissolved, and reduce the total rate of primary production. Although all microphagous tadpoles are suspension feeders, some tadpoles first scrape attached algae to create a suspension, which is then filtered. For tadpoles that graze periphyton (attached algae), larval feeding also leads to significant reductions in standing crop biomass. An exception to this pattern is that tadpole grazing on epiphytic algae can change the outcome of competition between the macroalgal host and microalgal epiphytes, thereby increasing overall algal standing crop biomass (Kupferberg, 1997a). Obligate macrophagous tadpoles that eat insect larvae and/or tadpoles have significant effects on the species composition and secondary production of the communities that assemble in plant-held waters of tropical rainforests (Pimm and Kitching, 1987; Jenkins and Kitching, 1990; Caldwell, 1993). Even tadpoles that consume other anuran eggs and larvae only facultatively can exclude their prey species from breeding sites (Petranka *et al.,* 1994).

Larval salamanders, as predators, also significantly influence the aquatic food webs in which they are embedded (Dodson, 1970; Sprules, 1972; Giguere, 1979; Morin, 1983, 1987; Gustafson, 1993; Parker, 1993a, 1994). Pond-dwelling larval salamanders that eat zooplankton strongly influence the species composition and size distribution of their prey. As keystone predators in temporary ponds, larval salamanders can reverse the outcome of competition among tadpoles by selectively consuming the competitively dominant species. Benthivorous stream-dwelling larval salamanders, which can comprise greater than 90% of predator biomass in some locations (Murphy and Hall, 1981), influence assemblages of macroinvertebrates through size-selective predation. Interspecific predation and cannibalism among larval salamanders also can determine the species composition of salamander assemblages.

This chapter is divided into 10 sections, each with a focal topic. For each topic, subsection A discusses fishes and subsection B discusses amphibians. In each section, we address one or two organizing questions. The questions covered are as follows: II. Terminology: What is a larva? What is metamorphosis? III. Vertebrates with a Larval Period: Which vertebrates having free-living and feeding larvae will we cover? IV. Life-History Evolution and Larval Feeding: Why is feeding central to the life-history strategy of having a larva? V. Sources of Larval Nutrition: How do larvae make the transition from reliance on endogenous food supplied within eggs to ingestion of exogenous foods from the aquatic environment? VI. Diet: How general or specific are the foods larvae consume? VII. Reynolds Number: How does larval size influence the mechanics of feeding in water? VIII. Ontogeny of Feeding Structure and Function: How do feeding structures and methods change after hatching? IX. Effects of Metamorphosis on Feeding Mechanisms: How do feeding structures and methods change at metamorphosis? X. Evolution of Larval Feeding Mechanisms: To what extent are the morphology and function of larval feeding a function of ancestry? How does heterochrony (the truncation or acceleration of larval development) contribute to the derivation of novel feeding modes? XI. Future Research: Which topics present interesting new questions?

II. TERMINOLOGY

A. FISHES

The term "metamorphosis" refers to a dramatic and abrupt change in an animal's form or structure during postembryonic development (Youson, 1988). However, during ontogeny, all fish species experience a period of postembryonic change in structure, with the degree of change varying between species (Youson, 1988). Because the degree of change that constitutes metamorphosis has not been specified (Williamson, 1992), confusion and inconsistency have arisen in the application of the term "metamorphosis" to patterns of postembryonic change in fishes.

Similar confusion extends to the term "larva," which often is defined in the context of metamorphosis. We will define a fish larva as the immature life-history stage that differs morphologically from the juvenile and adult (Kendall *et al.,* 1984; Osse and van den Boogaart, 1995).

What ontogenetic event marks the onset of the larval period? Unfortunately, there is disagreement about whether the larval period in fishes begins at hatching or at the time of first feeding (e.g., Balon, 1984; Blaxter, 1988; Noakes and Godin, 1988). For the purposes of this chapter, the larval period is considered to begin at hatching, when the embryo is freed from the egg envelope, as this is usually the time when growth and metabolism increase dramatically [review in Kamler (1992)].

The time between hatching and metamorphosis may range from a few days in tropical fish species, to a few weeks or months in most temperate fishes, to years in the eel *Anguilla* (Blaxter, 1969). The mean larval period duration for 94 marine and freshwater teleost species is approximately 33 days (Houde and Zastrow, 1993). Typical changes that occur during metamorphosis to produce juvenile/adult characters include scale formation, pigmentation, fin migration, completion of meristic characters such as fin rays, and attainment of adult body shape (Blaxter, 1988; Kendall *et al.,* 1984).

B. Amphibians

The general characteristics that define the larval stage for amphibians are similar to those described for fish. The larval stage begins at hatching, when an embryo escapes from its egg capsule. Larvae are morphologically distinct from adults and are not reproductive (Duellman and Trueb, 1986).

Specifically, anuran larvae are defined by the presence of a tail and the absence of (functional) fore- and hindlimbs. Metamorphosis is abrupt and dramatic in anurans, affecting almost every tissue, but most obvious are reabsorption of tail, repatterning of skull, shortening and transformation of digestive tract, eruption of limbs, and change of mode of locomotion.

Salamander larvae have external gills, early development of limbs, a tail fin, which in some species extends up the back, distinctive larval dentition, and only a rudimentary tongue. Metamorphic changes include resorption of the caudal fin and gills, closing of the gill slits, and repatterning of the skull (Dodd and Dodd, 1976).

Caecilian larvae are distinct from adults in that they lack certain bones, have an open gill slit, have a tail fin, and lack scales. Caecilian larvae hatch at advanced stages relative to anurans and salamanders. At metamorphosis elements of the bony palate transform (Reiss, 1996), gill slits close, and the dorsal caudal fin degenerates. The skin thickens and develops glands and scales.

III. VERTEBRATES WITH A LARVAL PERIOD

Vertebrates that have indirect development from a larva to a juvenile, involving a metamorphic phase, belong to the classes Cephalaspidomorpha (lampreys), Osteichthyes, and Amphibia (Youson, 1988). The aquatic lifestyle and close phylogenetic relationship of these groups may be factors in their possession of a larval period and metamorphosis (Youson, 1988).

A. FISHES

In dipnoans (lungfishes; Kemp, 1996) and in the actinopterygian families Polypteridae (bichir), Acipenseridae (sturgeon), Lepisosteidae (gar), and Amiidae (bowfin), some structures that are present after hatching but not in the adult stage have been described as larval (see Norman and Greenwood, 1975; Youson, 1988). However, because most of these fishes change form gradually rather than undergoing an abrupt transformation into a juvenile at metamorphosis (Youson, 1988), and because limited information is available on early feeding mechanisms in these fishes, we have excluded them from our discussion. We will focus on teleost fishes (e.g., Fig. 2), which are the most abundant group in the class Osteichthyes.

A larval period that culminates in metamorphosis has been a "highly successful strategy among fishes" (Youson, 1988, p. 183). Most fish species in the class Osteichthyes have free-swimming, feeding larvae (Blaxter, 1969; Noakes and Godin, 1988). However, the vast majority of fish larvae have not yet been identified to the level of species (Kendall and Matarese, 1994). Some oviparous freshwater fishes have unusually large, precocial hatchlings that can be considered to lack a larval stage (Balon, 1984; Houde, 1994). Ovoviviparous and viviparous osteichthyan species also may be essentially juveniles at birth (Blaxter, 1988; Kendall et al., 1984).

Youson (1988) states that there are no fish species in the class Chondrichthyes that have a larval period terminating in true metamorphosis. According to Blaxter (1988), the young of many oviparous elasmobranchs effectively are juveniles at hatching, although they possess a yolk sac.

B. AMPHIBIANS

All three orders of the class Amphibia have species with free-living aquatic larvae: Caudata, salamanders; Gymnophiona, caecilians; and Anura, frogs (Wassersug, 1975; Collins et al., 1993; Wake, 1993). However, it is among

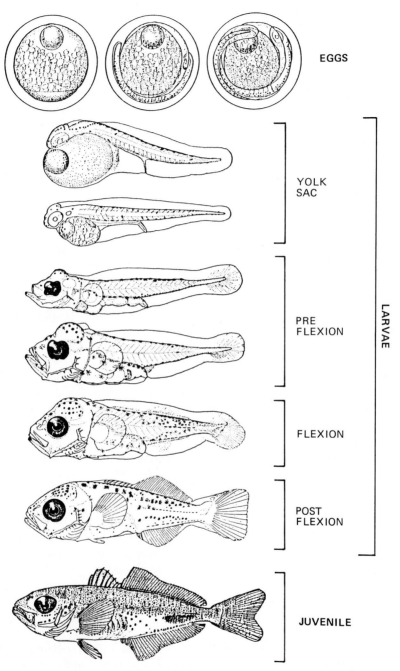

EGGS

YOLK
SAC

PRE
FLEXION

FLEXION

POST
FLEXION

JUVENILE

LARVAE

FIGURE 2 Generalized early life-history stages of teleost fishes, as exemplified by the jack mackerel. From the original drawings of Ahlstrom and Ball in Kendall *et al.* (1984, Fig. 5, p. 12), with permission.

the Anura in which a feeding larva is numerically the dominant life-history strategy. Approximately 90% of 250 anuran genera have free-living and feeding tadpoles. Among the approximately 400 species of salamanders, 113 have distinct larval forms. Probably more than 50% of the approximately 170 species of caecilians have derived reproductive modes without a free-living larval stage. Gymnophiones will not be discussed in depth as very little is known about feeding in free-living caecilian larvae.

IV. LIFE-HISTORY EVOLUTION AND LARVAL FEEDING

A. FISHES

The sizes, morphologies, and ontogenies of teleost larvae vary tremendously, such that teleost larvae are as diverse as adult teleosts (Houde, 1994; Moser, 1981). Due to this variability, consumption, growth, and mortality rates differ substantially among species (e.g., Letcher et al., 1997; Pepin, 1991). For example, almost all freshwater fish species are demersal spawners that produce relatively few but large eggs, whereas almost all marine fish species are pelagic spawners that release many small eggs. On average, marine teleost fish larvae are 10 times smaller than freshwater teleost fish larvae at hatching, have higher mortality rates, have higher weight-specific respiration rates, and remain in the larval period approximately 75% longer (15 days longer) than freshwater larvae (Houde, 1994). All of these differences between marine and freshwater larvae are statistically significant (Houde and Zastrow, 1993).

On average, only 0.12% of the individuals in a cohort of marine fish survive through the larval stage, whereas 5.30% of a typical cohort of freshwater fish survive (Houde, 1994). Given the low survival of fish larvae, the functional consequences of form changes during larval fish ontogeny are expected to be of particular importance in the evolution of the life history (Osse, 1990). Weight-specific ingestion rates calculated for 22 species of marine and freshwater teleost fish larvae indicate that, on average, larvae must consume more than 50% of their body weight daily to grow at mean reported rates (Houde and Zastrow, 1993). Consequently, interactions among prey consumption rates, body size, and degree of development may play an important role in the evolution of teleost life-history parameters, particularly in marine fish species. Hunter (1981) suggested that food size preferences may be linked closely to life-history strategies.

B. AMPHIBIANS

The life-history strategy of having a free-living aquatic stage reflects an evolutionary history in which the ability to exploit ephemeral resources was a driving force in maintaining the larval stage (Wassersug, 1975; Wilbur, 1980). Indeed, many manipulative field experiments (too many to cite fully) have shown that one of the major selective forces acting on larval amphibians is intra- and interspecific competition for food resources (Brockelman, 1969; Wilbur, 1972; DeBenedictus, 1974; Hairston, 1987; Fauth et al., 1990; Skelly, 1995; Smith and van Buskirk, 1995; Kupferberg, 1997b). The general conclusion from this body of literature is that competition for food is an important process structuring multispecies assemblages of amphibians and that population regulation in many amphibians occurs during the aquatic larval stage. The mode of feeding is important to food resource partitioning and contributes to determining how many species can co-occur in a given habitat. In the case of trophic polymorphisms, where some tadpoles become carnivorous and others remain herbivorous, variation in feeding mode may decrease intraspecific competition as well (Pfennig, 1992a).

Although the primary role of amphibian larvae appears to be growth, it is important to note that most species transform at a small fraction of full adult size. The largest proportion of somatic growth (90–99.9% for the majority of 68 species of *Rana, Hyla,* and *Bufo* analyzed) occurs postmetamorphosis (Werner, 1986, for a few exceptions to this pattern, see Emerson (1988)]. This pattern, along with low survival rates of larval cohorts, ranging from 1 to 5% [Licht, 1974; and other references cited in Werner (1986)], prompted Werner to develop the hypothesis that the complex amphibian life history evolved not only to maximize growth but to minimize the risk of predation.

V. SOURCES OF LARVAL NUTRITION

A. FISHES

Fish larvae undergo trophic ontogeny as the diet changes markedly with growth (Govoni et al., 1986). Four types of nutrient acquisition are recognized in fish larvae: endogenous, absorptive, mixed, and exogenous (Balon, 1986).

1. Endogenous Feeding

Yolk platelets and oil globules serve as the primary source of nutrition for the embryos of oviparous and ovoviviparous fishes (Heming and Buddington, 1988). Although there is considerable variability among species in the duration of endogenous feeding, the yolk-sac larvae of most fish species rely on these endogenous energy supplies for approximately the first 3–5 days after hatching (Miller et al., 1988). During the time after hatching but prior to exogenous feeding, the digestive, sensory, muscular, circulatory, and respiratory systems develop in preparation for the functional demands of first feeding on plankton.

2. Absorptive Feeding

Many marine invertebrate larvae are capable of supplementing their nutrition by the uptake of dissolved organic matter (DOM) via transepidermal transport [reviews in Boidron-Métairon (1995) and Levin and Bridges (1995)]. Few data are available on the acquisition of nutrients from the external medium via uptake at the body surface or at specialized external gut and finfold structures in oviparous fishes (Balon, 1986). According to Heming and Buddington (1988), Pütter's theory of the assimilation of dissolved organic matter from the external medium by aquatic animals from a number of taxonomic groups has been largely discounted. However, the issue remains unresolved, given that embryonic and juvenile fishes of a number of species have been shown to absorb, from the external environment, such organic compounds as pyruvate, glycine, and glucose [review in Heming and Buddington (1988)].

Morris (1955) observed mucus cells in the roof and floor of the oral cavity in three larval fish species belonging to different families and proposed that muscle activity moves mucus posteriorly toward the esophagus. He hypothesized adsorption and simple adhesion of "dissolved" organic matter (e.g., 40 Å) to this mucus and suggested that such a source of nutrition could be important.

The most intriguing potential example of absorptive feeding in fishes is the leptocephalus larva of the elopomorph teleosts (Fig. 3). This larva is characterized by a nearly transparent, laterally compressed body that is composed primarily of acellular gelatinous material covered by a thin layer of myomeres (Castle, 1984; Smith, 1984). During a radical metamorphosis the leptocephalus resorbs the gelatinous matrix, often resulting in a dramatic decrease in the body size of the juvenile relative to that of the larva. The juvenile has a typical body shape and composition and feeds orally (Pfeiler, 1986).

Although yolk appears to be exhausted when leptocephali are less than a centimeter long, leptocephali commonly are 5–10 cm at metamorphosis and can reach 1.8 m in some species (Pfeiler, 1986). Although the dentition is well-developed and the jaws appear to be functional, food has been found only rarely in the guts of leptocephali collected in the field. In fact, Hulet (1978) described the leptocephalus gut of one eel species as being imperforate, without a discernible lumen. Pfeiler (1986) concluded that the poorly differentiated gut, the equilibrium of body ionic composition with seawater, and the high levels of essential amino acids in leptocephali support the hypothesis that epidermal uptake of DOM supplies at least some nutrition to leptocephali.

Otake et al. (1993) found that detrital aggregates (<20 μm diameter) and zooplankton fecal pellets were common in the guts of three species of eel

S. Laurie Sanderson and Sarah J. Kupferberg

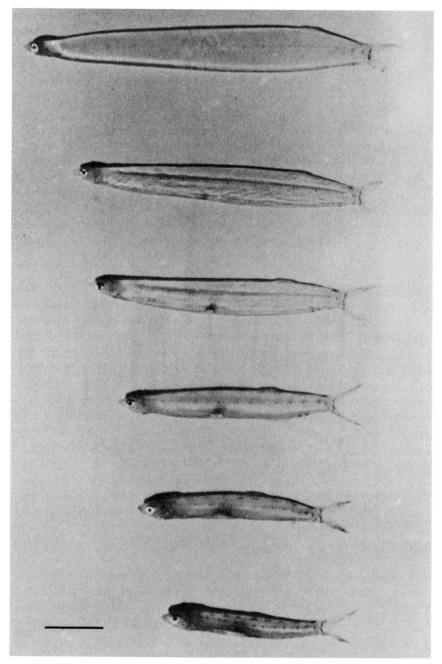

FIGURE 3 The leptocephalus larva of bonefish (top) metamorphoses over a period of 8–10 days to become a juvenile (bottom). Scale bar = 1 cm. From Pfeiler and Luna (1984, Fig. 1, p. 244), with kind permission from Kluwer Academic Publishers.

leptocephali collected off Japan. The ultrastructure of the midgut mucosal epithelium was suggestive of active seawater absorption. On the basis of these observations, they concluded that these leptocephali selected detrital particles using vision and olfaction, as well as ingested DOM by actively swallowing sea-water.

In his discussion of the evolution of the leptocephalus developmental strat-egy, Pfeiler (1986) proposed that the utilization of DOM reduced competition for food, resulting in the prolongation of the larval period from months to years. He suggested, however, that an increase in the duration of the vulnerable larval period in fact was not a strong selective advantage and noted that the leptocephalus is not a trait of the more derived teleosts.

There is no evidence to support Hulet's (1978) suggestion that leptocephali might use their forward-directed, fanglike teeth for puncturing organisms to ingest the body fluids. However, Mochioka et al. (1993) reported that premetamorphosing leptocephali of two species of eel in captivity tore mouth-fuls from lumps of a paste made by grinding squid with a mortar and pestle. The ingested paste was defecated completely within approximately 1 hr.

3. Mixed Feeding

Balon (1986) used the term "mixed feeding" to refer to any combination of endogenous, absorptive, and exogenous feeding. Generally, mixed feeding occurs as concurrent endogenous and exogenous feeding during the time between first exogenous feeding and complete resorption of the yolk sac (Kamler, 1992). When the duration of the mixed feeding period has been compared to the duration of the endogenous feeding that occurs between hatching and first exogenous feeding, the duration of mixed feeding has been found to vary widely among species, ranging from as short as one-half to as long as six times the duration of endogenous feeding [review in Kamler (1992)].

Yolk-sac larvae of most species that have been tested in the laboratory are capable of mixed feeding during a short interval prior to complete yolk absorp-tion (Heming and Buddington, 1988). Mixed feeding has been proposed to increase growth and survival by (1) offsetting any potential metabolic deficit prior to complete yolk absorption (Heming and Buddington, 1988) and (2) allowing a reliance on endogenous food reserves during a "learning period" in preparation for obligate exogenous feeding (e.g., Coughlin, 1991; Hunt von Herbing et al., 1996b). However, the extent to which mixed feeding occurs naturally in the field is unclear (Heming and Buddington, 1988).

4. Exogenous Feeding

a. First Feeding

Unlike the case for amphibians, a trend toward the loss of exogenous feeding in fish larvae has not been identified. Apparently, all fish larvae engage in mixed and/or exogenous feeding at some time during development (Just *et al.*, 1981; Kamler, 1992).

First feeding has been defined as the initiation of the capture of mobile prey (Hunt von Herbing *et al.*, 1996c), the first oral ingestion and intestinal digestion of food (Balon, 1984), or simply the first occurrence of feeding as evidenced in many cases by the presence of material in the alimentary tract rather than by the observation of prey capture (Miller *et al.*, 1988). The first definition is intended to exclude apparent filter-feeding on phytoplankton by larvae. The second definition excludes the ingestion of food that cannot yet be digested.

There are many unanswered questions regarding the digestive and absorptive mechanisms and the assimilation efficiencies of exogenously feeding larvae (Govoni *et al.*, 1986; Kjørsvik *et al.*, 1991; Walther and Fyhn, 1993). Kjørsvik and Reiersen (1992) noted that both halibut and cod larvae showed only limited food absorption by the gut epithelium during the first days of feeding. They proposed that the first ingestion and absorption of food may trigger an increase in digestive capacity and suggested that the role of phytoplankton in this process could be important, but is not understood.

b. Critical Period

For almost a century, considerable attention has been focused on determining whether there are "critical periods" during larval fish development when mortality is exceptionally high [reviews in Braum (1978) and Gerking (1994)]. Some researchers have proposed that metamorphosis is a critical period, perhaps due to the small size of the developing stomach in some fishes (e.g., Thorisson, 1994). However, most investigations of critical periods have emphasized the threat of starvation during the transition from endogenous or mixed feeding to exogenous feeding, primarily due to the requirement at this time for prey of specific sizes, types, and concentrations (Heath, 1992; May, 1974). From a different perspective, the larval period can be hypothesized to have evolved to enable fish to pass through the critical period at the transition from endogenous to exogenous feeding (Youson, 1988).

The existence of a critical period at the time of first feeding in fishes has been debated extensively in the literature, but the issue has not been resolved (see Blaxter, 1988). Starved marine invertebrate larvae are known to reach a fatal point of no return, when development cannot continue even with food (Boidron-Métairon, 1995). There is evidence that starved fish larvae of a number of species also reach a point of no return at the time of first feeding, when they become less active and are unable to capture and/or digest prey even if adequate food becomes available [reviews in Gerking (1994); Kjørsvik *et al.*, 1991; Kamler, 1992]. The point of no return has been reported to occur as early as one or two days following complete yolk absorption in some species (see Gerking, 1994). Large numbers of pelagic marine larvae are collected with empty guts (Blaxter, 1988), a finding that has fueled the controversy regarding the existence of a critical period. However, a number of factors now are known to lead to empty guts in larvae sampled from the field, including defecation and regurgitation during capture and preservation, rapid digestion, diurnal feeding rhythms, and the preferential capture of weak larvae in nets (Blaxter, 1969; Kjelson *et al.*, 1975).

c. Filter Feeding

As hagfish have large, yolky eggs and direct development, lampreys are the only extant agnathans with a larval stage (Youson, 1988). Although adult lampreys are either parasitic or nonfeeding, larval lampreys (ammocoetes) are filter feeders that use mucus entrapment and ciliary transport to consume minute prey such as diatoms, bacteria, and organic detritus (Sutton and Bowen, 1994). Goblet cells in the pharynx secrete mucus, which is moved medially across the gill pouches by cilia and water currents (Mallatt, 1981; Moore and Mallatt, 1980). At metamorphosis in parasitic species, the blind ammocoete that is virtually sedentary in a burrow transforms into an active juvenile with well-developed eyes and a rasping tongue in a suctorial mouth (Fig. 4; Youson, 1988).

The larvae of many teleost fish species collected in the field as well as reared in the laboratory have been reported to have guts containing phytoplankton (e.g., dinoflagellates and diatoms), generally described as "green material" [review in Gerking (1994)]. The ingestion of phytoplankton has been referred to as "filter feeding" by some authors (e.g., Hunt von Herbing *et al.*, 1996c; Reitan *et al.*, 1994; van der Meeren, 1991). Gerking (1994) summarized the available data by concluding that only the youngest larval stages of particular species and those larval species that are very small after hatching appear to ingest phytoplankton. He noted that these larvae generally consume phytoplankton for only a short period of time before switching to zooplankton.

Houde (1973) reported that the growth and survival of cultured marine fish larvae were enhanced by the presence of a dense phytoplankton bloom in the rearing tanks. He stated that the larvae that he had reared in this manner did not consume the phytoplankton and suggested that the phytoplankton "conditioned" the water by removing metabolites and possibly by supplying oxygen (see Section XI.B).

Van der Meeren (1991) added phytoplankton (1–10 μm) at natural spring bloom concentrations (1×10^6 to 10×10^6 cells per liter) to tanks housing cod yolk-sac larvae that were 1–6 days old. He found several phytoplankton species in the larval guts and determined that the larger phytoplankters (6–

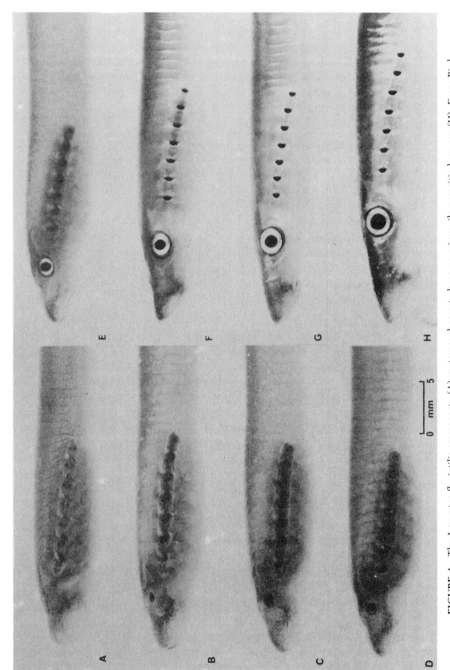

FIGURE 4 The *Lampetra fluviatilis* ammocoete (A) metamorphoses to become a juvenile parasitic lamprey (H). From Bird and Potter (1979, Fig. 4, p. 138).

10 μm) were concentrated in the guts at rates from approximately 500 to 7000 times the drinking rate. These ingestion rates corresponded to a carbon intake of 1–2 small copepod nauplii per day per larva, which is insufficient to supply all of the energy needed for growth. However, van der Meeren (1991) suggested that the phytoplankton could provide essential nutrients such as free amino acids. Reitan *et al.* (1994) concluded that the assimilation efficiency of Atlantic herring larvae that had been exposed to ^{14}C-labeled phytoplankton was in the range of 1–5%. Moffatt's (1981) analyses of pheophytin *a*, a metabolic breakdown product of chlorophyll, indicated that first-feeding Northern anchovy larvae may gain nutrients from phytoplankton directly, as well as indirectly via the ingestion of zooplankton that contain phytoplankton in their guts.

The feeding mechanisms that are used by larvae to engulf and retain phytoplankton are unknown. Adult filter-feeding fish engulf massive volumes of water by swimming forward with an open mouth (ram filter feeders) or by generating suction to draw water into the mouth (pump filter feeders) [review in Sanderson and Wassersug (1993)]. Depending on the developmental state of larval feeding structures, larvae could, in theory, use either or both methods of water transport past the filtering apparatus. Although the filtering apparatus has been identified in the adults of a few fish species (e.g., Hoogenboezem *et al.*, 1991; Sanderson *et al.*, 1996), the sites and methods of food particle entrapment and retention in filter-feeding larvae remain unstudied.

De Ciechomski (1967) reported that phytoplankton was common in wild-caught Argentine anchovy larvae $\geq$38 mm in length, but was absent from the guts of smaller larvae. However, the criteria used to determine the transition between larva and juvenile were not stated, and specimens from 38 to 40 mm were referred to alternately as larvae and as juveniles. We include this study here because de Ciechomski is one of the few authors to mention gill raker development in relation to apparent larval filter feeding. Noting that the gill rakers on the first branchial arch appeared as small protuberances in larvae that were 13.5 mm in length but were well-developed in larvae of 31.0 mm, she concluded that the filtering apparatus began to function at the time that phytoplankton appeared in the diet. She suggested that the absence of phytoplankton from the adult diet was due to the coarser mesh of the adult branchial filter.

Reitan *et al.* (1994) reported a negative correlation between the clearance rate (volume of medium cleared of phytoplankton per unit time) and the distance between the branchial arches with growth in 28- to 56-day-old Atlantic halibut larvae. However, the extent of branchial arch abduction in the scanning electron micrographs was not reported. In addition, they pointed out that fixation artifacts may have affected the measurements.

By using interference contrast microscopy, van der Meeren (1991) observed single phytoplankton cells (6–10 μm) or rows of cells along the slits (8–

12 μm wide) between the branchial arches of 2- to 6-day-old larval cod. Groups
of cells also were noted in the posterior part of the pharynx. In morphological
studies of cod larvae at comparable temperatures (Hunt von Herbing et al.,
1996b,c), the mouths of 2- to 3-day-old larvae were open, and 4- to 6-day-
old larvae were able to make infrequent and uncoordinated jaw movements.
However, gill rakers and gill filaments were absent from the branchial arches
of 2- to 6-day-old cod larvae (Hunt von Herbing et al., 1996b). Because cod
larvae of this age probably are incapable of generating the negative pressure
necessary to suck substantial volumes of water into the mouth (Hunt von
Herbing et al., 1996c), the phytoplankton may have entered while the larva
swam forward with an open mouth (van der Meeren, 1991).

Ellertsen et al. (1980) observed live 1- to 5-day-old cod larvae in plastic
chambers under a low-power binocular microscope. They noted that, at 1–100
times the natural spring bloom concentration, phytoplankton (7–9 μm) were
swallowed after incidentally entering the immovable larval mouth and clogging
the slits between the branchial arches. When the jaw became functional and
the larvae began to capture Artemia nauplii, microscopic observations showed
that the larvae spat the phytoplankton out of their mouths rather than swallowing the cells, and there was a concomitant substantial reduction in the percent
of larvae with green material in their guts. They concluded that this small
phytoplankton (7–9 μm) was ingested passively in larvae that had not developed sufficiently to expel the phytoplankton from their mouths whereas a
larger phytoplankton species (50–80 μm) was consumed actively by first-
feeding larvae.

d. Suction Feeding

In general, fish larvae are diurnal, particulate planktivores that ingest their
prey whole (Govoni et al., 1986; Hunter, 1980) by using suction and/or ram
feeding (see Section VIII.A.3.a).

B. AMPHIBIANS

To facilitate comparisons between fishes and amphibians, we classify nutrient
acquisition by free-living larvae into the same four types discussed earlier:
endogenous, absorptive, mixed, and exogenous. However, alternative terminology has been established for amphibians, e.g., endotrophic and exotrophic
instead of endogenous and exogenous (Altig and Johnston, 1989). These categories (excluding absorptive feeding) represent an ontogenetic sequence for
many taxa, but an evolutionary trend toward increased reliance on yolk reserve

has occurred to the extent that some taxa do not require exogenous food to complete metamorphosis.

1. Endogenous Feeding

The ancestral condition of the amphibian life history is to have a free-swimming, feeding larva, with a period between hatching and feeding when larvae rely on yolk reserves. Yet a large number of reproductive innovations have evolved in which exogenous feeding is lost. An evolutionary trend toward increasing terrestriality is indicated by the phylogenetic distribution of direct development and viviparity in caecilians, urodeles, and anurans (Duellman and Trueb, 1986; Wake, 1989, 1993). Interesting parallels are found among marine invertebrates in which the taxonomic distribution of larval traits also indicates a trend toward the loss of larval feeding (Strathmann, 1978; Emlet, 1990; Havenhand, 1995).

Increased reliance on yolk by free-living larvae appears to be an important evolutionary step toward direct development. Eggs must have enough yolk for larvae to metamorphose successfully in the absence of feeding. Unlike taxa with feeding larvae, which lay large numbers of very small eggs, these species lay a small number of large eggs. There are examples from at least five families and twelve anuran genera with species that have free-living tadpoles, but that can develop successfully into frogs via yolk absorption alone. Bufonidae: *Pelophryne,* the Costa Rican golden toad; *Bufo periglenes* (Crump, 1989) and *Bufo haematiticus* from Ecuador (McDairmid and Altig, 1990). Ranidae: *Rana tagoi* (Kusano and Fukuyama, 1989) and *R. sakuraii* (Maeda and Matsui, 1989). Leptodactylidae: *Eupsophusa* and *Cycloramphus stejnegeri.* Rhacophoridae: *Rhacophorus gauni* from Borneo (Inger, 1992) and Microhylidae: *Anodonthyla, Kalophrynus, Platypelis,* and *Plethodontyla* (Duellman and Trueb, 1986).

The consequences of endogenous feeding are manifest in trade-offs between growth and development. Apparently, when yolk is the nutrient source, differentiation and development take precedence over growth. Experiments with the facultatively endotrophic *B. periglenes* demonstrate that unfed individuals metamorphose faster, but at a smaller size, than fed individuals (Crump, 1989). During the period between hatching and feeding, spadefoot toad larvae *(Scaphiopus intermontanus)* use yolk reserves to continue differentiating, but growth is restricted to increases in tail length (Hall *et al.,* 1997).

2. Absorptive Feeding

Absorption of DOM, as observed in larval invertebrates, has not been reported in the wild for recently hatched amphibians. In a laboratory experiment, however, *Bufo americanus* tadpoles (developmental stage not specified) successfully filtered precipitates of DOM (Ahlgren and Bowen, 1991). The nutrition derived from the precipitates was insufficient for growth, however. There is a suggestion from a laboratory practice used to prevent tadpoles from producing feces that tadpoles can absorb dissolved nutrients in a commercial growth medium containing essential fatty acids, glucose, oligosaccharides, minerals, and vitamins (Viertel, 1990). Whether larvae might encounter such nutrient soups under natural conditions is a question open for speculation.

3. Mixed Feeding

The simultaneous reliance on endogenous and exogenous food supplies has been documented for a brief period following hatching in *Rana temporaria, B. calamita,* and *B. bufo* (Viertel, 1991), and presumably it occurs in many other anuran genera and species. First feeding in these species precedes the exhaustion of yolk in the gut, the differentiation of the filter plates, and the mucus produced by the filter plates and occurs prior to the differentiation of keratinized beaks and oral tooth rows. Hence, the first foods of these species are likely to be small particles from plankton or sediment and, thus, very different from the later foods that will be scraped from substrates. How common mixed feeding is among amphibians is not known because, in contrast to mixed feeding for larval fishes, this feeding mode has not received explicit attention from herpetologists. There is some evidence, however, of selective pressure against continued reliance on yolk after hatching to the extent that the inert mass of yolk can contribute to decreased locomotory performance and increased susceptibility to predation in young anuran larvae (Kaplan, 1992).

4. Exogenous Feeding

a. First Feeding

Although there are some larvae with sufficient yolk reserves to complete metamorphosis without feeding (see endogenous feeding discussed earlier), most anurans experience a short period (up to a few days) before yolk supplies are exhausted and ingestion of exogenous food begins. Among anurans, the developmental stage at first feeding varies among species (Viertel, 1991). There is much earlier commencement of ingestion in *Rana temporaria* [Gosner (G) stage 23], *B. calamita* (G stage 23), and *B. bufo* (G Stage 24) relative to *Xenopus laevis*. Whereas fish have a yolk sac, anuran yolk reserves at hatching are found in the digestive tract, primarily the coiled gut. In *Xenopus*, suspension feeding does not begin until the yolk supply located within the buccal cavity and esophagus is exhausted.

Within a single species, time to first feeding is not fixed. The amount of yolk and size at hatching can vary widely due to complex interactions between maternal investment (egg size) and temperature (Kaplan, 1992). Additionally, variation in time to first feeding may be determined by factors other than quantity of yolk. Williamson and Bull (1989) found no consistent relationship between egg size and time to hatching or first feeding within the myobatrachid frog *Ranidella signifera*. Significant differences among clutches suggested that some other maternal factors were important. For the red-eyed treefrog (*Agalychnis callidryas*), time to hatching and the delay until feeding are influenced by predation (Warkentin, 1995). When clutches of eggs laid on arboreal leaves were attacked by snakes, hatching age was reduced significantly (5 vs 8–10 days). The younger hatchlings had longer delays before feeding than larvae with long embryonic periods and suffered greater losses to shrimp and fish predators. The overall fitness consequences of the variability in age to first feeding that occurs in these, and presumably many other species, thus is difficult to assess because of such trade-offs.

Compared to anurans, the period between hatching and feeding is much longer in salamanders. A period of 6–10 weeks has been reported before first feeding in larvae of the stream-dwelling plethodontid salamander *Pseudotriton ruber* (Gordon, 1966). During this period, pond-dwelling salamanders in the families Hynobiidae, Salamandridae, and Ambystomatidae use balancers, a pair of mucus-secreting rodlike structures on either side of the head, to keep from sinking into the sediment (Duellman and Trueb, 1986). Both environmental and maternal factors determine the length of time between hatching and first feeding. Warmer temperatures speed the progression from hatching [stage 40 of Harrison (1969)] to feeding (Harrison stage 46). For *Ambystoma maculatum* the difference is 50.4 ± 1.6 days at 10°C vs 11.1 ± 1.5 days at 20°C and for *A. tigrinum* it is 32.5 ± 12.6 days at 10°C vs 7.3 ± 1.2 days at 20°C (Kaplan, 1980a). Larvae from large eggs develop quickly from hatching to feeding (Kaplan, 1980b). Larvae from large eggs also have more energy reserves than small hatchlings because embryos use 1.45 cal to hatch, regardless of egg size (which ranges from 16.5 to 29 cal/egg among different females), temperature, or species. Large hatchlings then develop into feedlings more quickly because they have high respiratory rates associated with their greater surface area. This leads to efficient conversion of yolk into tissue. These feedlings continue to retain a greater portion of their initial energy content and have higher size specific growth rates when fed *ad libitum*. Thus, egg-size-related growth advantages accrue after exogenous feeding has commenced. It would be very interesting to investigate whether these advantages derive from the continuing conversion of stored energy (i.e., mixed feeding) or whether first-feeding capability, such as capture rate, is enhanced.

b. Critical Period

Undoubtedly there is a length of time beyond which if a hatchling does not eat, death by starvation will ensue. Such a "critical period" has not been a focus of research by herpetologists as it has for ichthyologists. For anurans, even those that oviposit in highly oligotrophic waters, there probably is always some particulate matter in the water or substrate that can be consumed. Perhaps salamander larvae could face the paucity of correctly sized prey that some fish larvae experience.

c. Filter Feeding

Most tadpoles are microphagous and feed on suspended particles or can facultatively create their own suspensions by scraping larger items with external keratinized beaks and denticles. The particles are trapped on the surface of the gill filters and other buccopharyngeal structures (Fig. 5). Tadpoles are continuous suction feeders [as defined by Sanderson and Wassersug (1993)], meaning that while tadpoles are feeding, the pumping of water created by depressions of the buccal floor occurs continuously to maintain the flow of suspended food over the surfaces where particles are trapped.

There is an incredible diversity of morphological adaptations among tadpoles to consume a wide variety of foods via filter feeding (Duellman and Trueb, 1986; Altig and Johnston, 1989). In lentic, pond habitats there is a

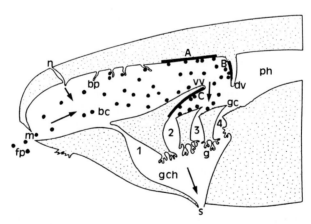

FIGURE 5   Parasagittal section through the anterior portion of a tadpole showing the flow of water (arrows) and food particles (fp) through the mouth (m), into the buccal cavity (bc), past the ventral velum (vv), and past the gill filter plates (gc). Large food particles can be trapped on the buccal papillae (bp), and smaller particles can adhere to mucus secreted on the roof of the buccal cavity (region A), on the dorsal velum (dv) (region B), and on the ventral velum (region C). Other abbreviations: g, gill; gch, branchial cavity; n, nare; ph, pharyngeal cavity; s, spiracle; 1–4, gill arches. From Hourdry et al. (1996, Fig. 2, p. 222), with permission from The University of Chicago Press.

generalized pond type of tadpole (Fig. 6A). Its characteristics include an ovoid body, a tail about two times as long as the body, an anteriorly directed mouth bordered laterally and ventrally by 1–2 rows of small papillae, 2–3 rows of denticles on the upper lip, and 3–4 rows on the bottom lip. The jaws have keratinized beaks. Among pond-dwellers, there are distinct morphologies associated with feeding in different microhabitats. Surface feeders can have funnel-shaped mouths, and midwater filter feeders can have terminal mouths that completely lack keratinized denticles and beaks as in *Xenopus* and *Rhinophrynus*. Taxa occupying lotic habitats can have morphological feeding specializations that include a ventral mouth, an increased number of denticle rows, and a suctorial disk for attaching to substrates (Fig. 6B). Some arboreal tadpoles are filter feeders of detritus and have retained many of the internal oral features of generalized pond tadpoles (e.g., large gill filters) while having evolved other morphological traits specific to hypoxic conditions (e.g., reduced aquatic respiratory surfaces, large, anteriorly directed glottis) (Lannoo *et al.*, 1987). Many arboreal detritivores have rows of denticles greater than the two rows above and three rows below the lips typical of the generalized pond form. Species found in extremely small bodies of water, such as leaf axils and bromeli-

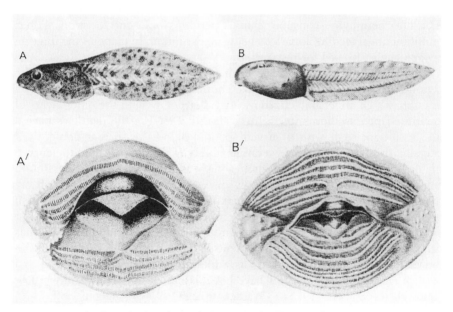

FIGURE 6   Body shape (top) and mouth (bottom) of (A) *Rana palmipes*, a generalized pond type tadpole, and (B) *Hyla lindae*, a stream-dwelling tadpole. From Duellman and Trueb (1986, Figs. 6–12a, 6–13a, 6–15b, 6–16b, pp. 158–161).

ads, typically have elongate bodies. Those found in larger tanks, e.g., treeholes, have more normal body proportions and short tails (Lannoo *et al.*, 1987). For a complete treatment of the filter-feeding and other ecomorphological types of tadpoles, see Altig and Johnston (1989).

In contrast to the importance of plankton to fish larvae, attached algae, and diatoms in particular, are important tadpole foods. Because the most common tadpole feeding mode is benthivorous, we first focus on the morphological structures that have evolved for consuming benthic resources. We then discuss the morphological innovations that allow other feeding modes, such as the midwater suspension feeding that occurs in the families Pipidae and Microhylidae and the macrophagous carnivory that occurs in a number of arboreal breeding species (Lannoo *et al.*, 1987).

In scanning electron micrographic studies comparing the mouthparts of diverse periphyton consumers, Arens (1989, 1994) has identified four functional problems that periphyton grazers must solve. Algae must be detached from the substrate, collected, and crushed, and in lotic habitats the water current must be shielded to prevent the food from being washed away. The taxa compared, including five different orders of aquatic insect larvae and one tadpole (the Brazilian hylid frog, *Ololygon heyeri*), have modifications of different structures to form the various "tools" used for scraping, transporting to mouth, crushing, and shielding the current. In most of the arthropod cases, each of a periphyton consumer's tools performs only one function, with a multifunctional feeding apparatus being an exception. In tadpoles the filtering apparatus also is used for respiration.

In tadpoles the most common form of oral apparatus for scraping is the keratinized denticles borne on the lips. The denticles fit Arens' (1989) definition of rasps, which are characterized by a rigid structure with the outer layers of keratin shaped as thorns, teeth, or scales. In *Ololygon* the denticles have a nearly straight shaft with a curved spoonlike tip. The shaft is compressed laterally, thus decreasing the overall flexibility of the denticle, and the curved tip has dentate edges. Use of such a rasp as a grazing tool is not found among the arthropods. For scraping, insect larvae have brushes, which are modifications of the mandibles and maxillae and both pairs of palps. A brush is equivalent to a rasp, but consists of many slim bristle-like components.

The problem of transporting food through the mouth and into the gut has been solved in tadpoles by using elements of the cartilaginous branchial skeleton to form a pump and the gills as a filtering organ. Water transport and particle entrapment in tadpoles have been described in detail (Kenny, 1969; Gradwell, 1968, 1971, 1975a,b; Wassersug, 1972; Wassersug and Hoff, 1979; Wassersug and Rosenberg, 1979) and compared to suspension feeding mechanisms in a diversity of vertebrates by Sanderson and Wassersug (1993). The following summary is based on their overview. A current of water containing suspended particles flows into the mouth by suction when the buccal cavity expands. This expansion occurs when the paired ceratohyal cartilages, which form a plate in the floor of the buccal cavity, are depressed, thus acting like the piston of a pump (Fig. 7). The ceratohyal cartilages articulate laterally with the horizontally oriented palatoquadrate (Fig. 8). When one set of muscles connected to the palatoquadrate (the orbitohyoideus in particular) contracts, the ceratohyal is depressed. When another muscle connected to the ventral surface of the ceratohyals contracts, the ceratohyal plate is lifted. This muscle, the interhyoideus, is positioned like a sling under the ceratohyal cartilages. The point of articulation between the palatoquadrate and the ceratohyal cartilage and the length of the ceratohyal past that pivot point determine the length

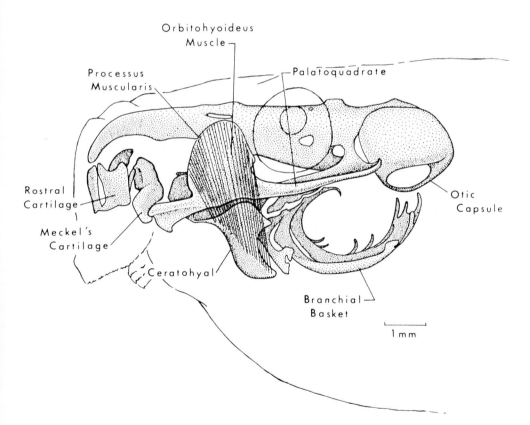

FIGURE 7 Lateral view of the chondrocranium of a tadpole *(Alytes obstetricans)*, with major structures labeled and the orbitohyoideus depressor muscle drawn in place. The point of articulation between the palatoquadrate and the ceratohyal is the fulcrum around which the ceratohyal rotates when the orbitohyoideus contracts. From Wassersug and Hoff (1979, Fig. 1, p. 227).

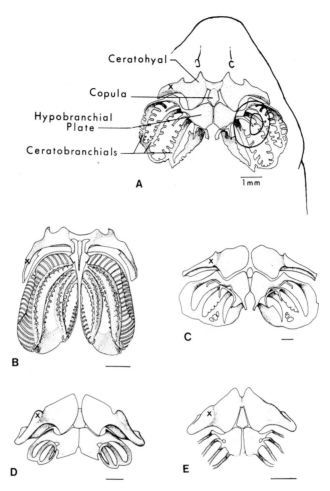

FIGURE 8   Branchial skeletons in dorsal view for tadpoles from five species, selected to illustrate basic structures and diversity. Anterior is toward the top of the page. The "x" on each drawing indicates the point of articulation of the ceratohyal with the palatoquadrate bar on one side (not shown). The ceratobranchials form the branchial baskets and the gill filters arise from the ceratobranchials. (A) *Rana pipiens,* with major structures labeled. This larva is a dietary generalist, feeding on a coarse suspension of particles generated through the action of its keratinized beaks and denticles. (B) *Gastrophryne carolinensis,* an obligate, midwater suspension feeder. (C) *Heleophryne natalensis,* a benthic tadpole adapted to fast-flowing water. This larva has a large suctorial mouth and grazes on periphyton. (D) *Anotheca spinosa,* an arboreal, macrophagous, carnivorous larva. (E) *Hyla microcephala,* a pond tadpole that ingests large filamentous plant fragments. From Wassersug and Hoff (1979, Fig. 2, p. 228).

of the lever arm, i.e., the mechanical advantage that the depressor muscles have. Variation in the length of the ceratohyal lever arm and the amount of deflection of the buccal floor clearly are correlated to feeding mode (Fig. 8). Carnivorous macrophagous tadpoles create a large buccal volume by depressing a large buccal floor area a short distance by using a long lever arm. This is analogous to a large-bore, short-stroke piston. The action of a midwater obligate suspension feeder, in which a short lever arm deflects a relatively smaller buccal floor area over a relatively longer distance, is analogous to a small-bore, long-stroke piston. Stream-dwelling larvae, which also need to generate a large amount of negative pressure to adhere to substrates, have a long lever arm but only moderate buccal volume. Their diet of periphyton does not require that large volumes of water be ingested. The negative pressure created by the pump along with the structure of the suctorial disk probably also contribute to shielding the current, thus preventing the loss of food once it is detached from the substrate.

Water flow in tadpoles is unidirectional. Opening and closing of the mouth and flaps on the nares serve as external valves. Internally, there is the ventral velum, an epithelial flap that arises from the floor of the buccal cavity and covers the branchial baskets (Fig. 9). "Branchial basket" is the collective term referring to the set of gill filter plates that arise from the pharyngeal arches or ceratobranchials. When the buccal floor is depressed, the posterior free edge of the ventral velum is pushed against the roof of the buccal cavity, thus separating the buccal cavity from the pharyngeal cavity. When the buccal floor rises, the ventral velum falls and water is pushed into the branchial baskets. A second flap attached to the roof of the buccal cavity, the dorsal velum, serves to guide the flow of water to the gill filters. Pipids lack a ventral velum so that unidirectional flow of water is maintained by flaplike covers of the paired opercula. The opercular chamber encloses the gills and opens externally through spiracles.

Food particles can be retained on several surfaces within the buccopharyngeal cavities (Fig. 5). Large particles can be trapped by simple sieving on papillae in the buccal cavity or on ruffled epithelium occurring on gill filters. The branchial baskets are reduced or absent in carnivorous larvae, whereas microphagous larvae have extremely large branchial baskets. To trap very small particles simply by passing water through a mesh or strainer, the mesh size would have to be smaller than the particles and an excessively large amount of pressure would be required to push water through. Instead, small particles are collected by adhering to mucus-coated surfaces. Mucus is secreted by cells on the surfaces of the ventral velum, the dorsal velum, and the anterior end of the filter cavity. These secretory areas are termed the branchial food traps. Once particles are aggregated on mucus strands they are transported to the

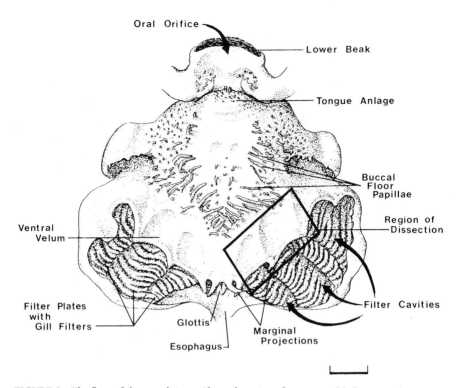

FIGURE 9   The floor of the mouth in an *Alytes obstetricans* larva, stage 36. Structures important to the feeding process and some common anatomical landmarks are labeled. The ventral velum is the valvular flap attached to the dorsal margin of the filter plate. The branchial food traps lie on its ventral surface in all suspension-feeding tadpoles except pipids, which lack a valvular velum. Scale bar = 1 mm. From Wassersug and Rosenberg (1979, Fig. 1, p. 394). Surface anatomy of branchial food traps of tadpoles: A comparative study. *Journal of Morphology,* Copyright © 1979. Reprinted by permission of Wiley-Liss, Inc., a subsidiary of John Wiley & Sons, Inc.

esophagus by cilia. In contrast to this method, freshwater arthropod larvae use combs, brooms, and hairy pads for trapping and transporting particles.

The arrangement of the entrapping surfaces on the ventral velum differs between families recognized by Ford and Cannatella (1993) as recent (Ranidae, Hylidae, and Bufonidae) and more primitive (Ascaphidae, Discoglossidae, and Pelobatidae). The derived taxa have secretory pits and ridges with secretory cells arranged in rows at the tops of the ridges (Fig. 10) (Wassersug and Rosenberg, 1979). Most of the primitive taxa lack the organization of pits and ridges, and when ridges are present (as in *Rhinophrynus*) the apices of the secretory cells are not in rows. The arrangement into ridges does not appear

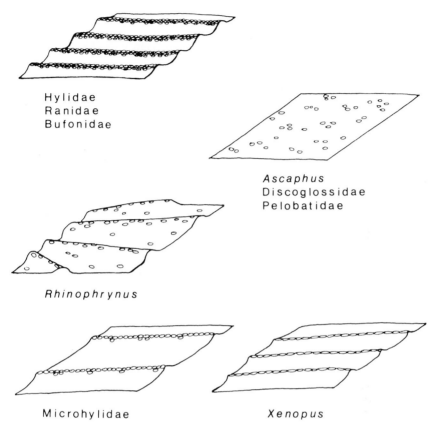

Hylidae
Ranidae
Bufonidae

Ascaphus
Discoglossidae
Pelobatidae

Rhinophrynus

Microhylidae   Xenopus

FIGURE 10   Schematic drawing of the surface anatomy of branchial food traps of tadpoles. Only the apices of the secretory cells are shown. The typical neobatrachian pattern, with secretory cells in pits arranged in rows at the tops of ridges, is illustrated in the upper left. From Wassersug and Rosenberg (1979, Fig. 30, p. 407). Surface anatomy of branchial food traps of tadpoles: A comparative study. *Journal of Morphology,* Copyright © 1979. Reprinted by permission of Wiley-Liss, Inc., a subsidiary of John Wiley & Sons, Inc.

to be essential for trapping small particles, but having a dense field of columnar mucus-secreting cells with apices protruding through the epithelium is essential. Regardless of familial relationship, obligate macrophagous tadpoles have reduced branchial traps, lack ridges and pits, or lack mucus-secreting cells altogether, whereas obligate midwater suspension feeders display parallel ridges.

For crushing algal cells, aquatic insect larvae have masticators or break cells by the rubbing action of a scraping apparatus, whereas in larval anurans, cells are digested without previous breakup. Neither the denticles nor the keratinized beak can crush algal cells. Observation of the diatoms in tadpole feces (Kupferberg et al., 1994) showed that many cells passed through the gut with frustules intact, but with cell contents gone. Tadpoles use digestive enzymes that are produced in the pancreas and liver and delivered directly to the coiled midgut, where extracellular digestion takes place. In general, herbivorous tadpoles rely on the large surface area of very long coiled intestines to maximize nutrient absorption (Burggren and Just, 1992).

The relative sizes of the gut, pancreas, and liver are constrained by adult morphology in addition to factors directly relating to their role in larval digestion. Because frogs have the shortest spinal column among all vertebrates, tadpoles face severe packing constraints for the internal viscera. Among three sympatric species of Bornean stream-dwelling pelobatid tadpoles, variation in these organs relates more to spatial microhabitat distribution rather than directly corresponding to the type of food consumed (Nodzenski et al., 1989). Of the three species, the one with the largest gut/liver volume (24% of total volume for *Leptobatrachium montanum*) and the one with the smallest gut/ liver volume (10% of total volume for *Leptolalax gracilis*) have similar diets, consisting of relatively large tracheoid plant fragments (Inger, 1986). The former species uses a wide array of habitats, whereas the latter is very slender, living in rock crevices on the bottom of riffles.

d. Suction Feeding

Rather than consuming small particles in suspension, obligately carnivorous tadpoles ingest live prey whole. As indicated by the modifications of the buccal pump and mucus entrapment surfaces described earlier, predaceous tadpoles use suction rather than chasing and swimming with an open mouth (i.e., ram feeding) to engulf prey. Tadpoles of the pipid *Hymenochirus* scull slowly with their tails to obtain a position near enough to zooplankton prey to catch prey by suction (Wassersug, 1989). Another type of obligate macrophagy is the consumption of anuran eggs among arboreal breeding frog species. Tadpoles rely on trophic eggs deposited regularly (5-day intervals for *Osteocephalus oophagus*) by the parents and starve if parents fail to return (Lannoo et al., 1987; Jungfer and Schiesari, 1995).

Philautus carinensis, a rhacophorid frog that breeds in treeholes in Thailand, typifies the unusual morphologies associated with egg eating. It has an anteriorly directed mouth, a shortened gut, and an enlarged stomach and is exclusively air-breathing (Wassersug et al., 1981). *Osteocephalus oophagus*, a hylid species from the central Amazon, shows a reduction in the number of denticle rows and an absence of beak serrations compared to other members of the genus (Jungfer and Schiesari, 1995). This reduction of denticle rows that normally are used for rasping is common among arboreal tadpoles (Lannoo *et al.*, 1987) and may be an adaptation for oophagy in which scraping is not necessary.

Salamander larvae are exclusively suction feeders. Similar to suspension feeding in anurans, suction feeding in larval salamanders is achieved by expansion of the oral cavity to create negative pressure relative to the surrounding water and thus cause a flow of water into the mouth that brings prey within range of the jaws. The functional morphology of suction feeding has been described in detail for the genus *Ambystoma* (Lauder and Shaffer, 1985). Salamanders have two different sets of musculoskeletal apparatus for opening the mouth (Fig. 11): one primitive and retained throughout other lower vertebrate lineages, and the second an innovation acquired by adult lungfishes and salamanders independently (Lauder and Shaffer, 1993). The first set involves depression of the hyoid arch, which places tension on the ligament connecting the hyoid and the mandible, causing the mouth to open. This movement is achieved by contraction of the rectus cervicis muscle. The second mechanism for mandibular depression is contraction of a muscle that inserts directly onto the lower jaw, the depressor mandibulae. The timing of skullbone movements during initial prey capture also is conserved across all lower vertebrates. First, peak gape is reached and then maximal hyoid depression, which in turn is followed by expansion of the posterior portion of the skull and water exiting through the gills. In salamander larvae, as in adult fish, the flow of water is unidirectional, from anterior to posterior. The gill bars act as a valve to prevent water flow in the opposite direction (Fig. 12), as maximum gape coincides with maximal adduction of the gill bars. Transport of food items to the esophagus is achieved by water flow produced from rapid jaw movements similar to those observed during initial prey capture, except that mouth opening and hyoid retraction occur more slowly (Reilly, 1996).

Cranial kinematics do not differ with prey type (Reilly and Lauder, 1989), but the kinematics and muscle patterns of suction feeding are but one part of the predation cycle. Prey detection and the decision to attack once an encounter occurs appear to be the aspects of salamander feeding behavior that modulate in response to environmental factors. Parker (1993a) made behavioral observations of *Dicamptodon* larvae (44–63 mm in snout–vent length, SVL) feeding on large and small mayflies (5.8 or 3.3 mm total length, TL) and stoneflies (20 or 8.8 mm TL). Encounter rates largely depended on prey movement and size, with larger prey being detected from a greater distance than small prey. Both the probability of attack given an encounter and handling time increased with prey size. Despite longer handling times for large prey, the ratio of prey biomass to handling time increased with prey size, indicating that large prey provide a better energetic return.

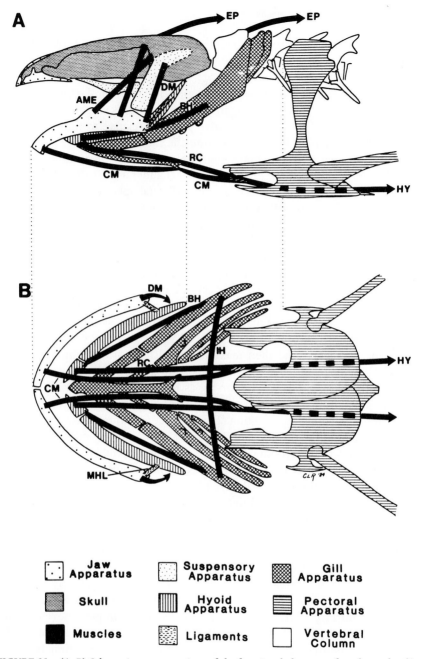

FIGURE 11 (A, B) Schematic representations of the functional elements of a salamander (*Ambystoma mexicanum*) head involved in aquatic suction feeding. Opening and closing of the mouth is achieved by depression of the hyoid apparatus via contraction of the depressor mandibulae muscle (DM) and by elevation of the head by the epaxial muscles (EP). Other abbrevia-

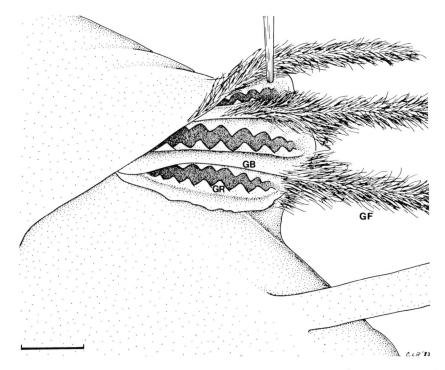

FIGURE 12 Ventral view of the branchial area of the salamander *Ambystoma mexicanum*, exposed with forceps. When the gill rakers (GR) interlock with the gill bars (ceratobranchials, GB) during suction feeding, a high resistance to water flow is achieved. GF = gill filaments; scale bar = 0.5 cm. From Lauder and Schaffer (1985, Fig. 6, p. 307). Functional morphology of the feeding mechanism in aquatic ambystomatid salamanders. *Journal of Morphology*, Copyright © 1985. Reprinted by permission of Wiley-Liss, Inc., a subsidiary of John Wiley & Sons, Inc.

Caecilian larvae also are suction feeders. Because several aspects of their morphology [e.g., dentition (Wake, 1980), cranial and hyoid structure and musculature] are similar to those of adults at birth or hatching, it has been assumed that the feeding mechanism is similar to that of adults (Burggren and

tions: AME, adductor mandibulae externus muscle; BH, branchiohyoideus muscle; CM, coracomandibularis muscle; HY, hypaxial muscles; IH, interhyoideus muscle; MHL, mandibulohyoid ligament; RC, rectus cervicis muscle. From Lauder and Shaffer (1985, Fig. 1, p. 303). Functional morphology of the feeding mechanism in aquatic ambystomatid salamanders. *Journal of Morphology*, Copyright © 1985. Reprinted by permission of Wiley-Liss, Inc., a subsidiary of John Wiley & Sons, Inc.

Just, 1992). Caecilians are carnivorous, using heavily fanged jaws with single rows of recurved teeth to capture elongate prey such as worms.

VI. DIET

A. Fishes

Studies of freshwater teleost larvae show that many species consume cladocerans and/or calanoid and cyclopoid copepods (e.g., Faurot and White, 1994; Wahl *et al.,* 1993). At first feeding, most marine fish larvae consume diverse prey that are approximately 50–100 μm in width, including tintinnids, phytoplankton, mollusk larvae, ciliates, and copepods [citations in Hunter (1980)]. With growth, marine fish larvae tend to specialize on the naupliar through adult stages of copepods (Hunter, 1980; Leis, 1991), even in species that primarily are herbivorous as adults (Gerking, 1994). The occurrence of carnivorous larvae in herbivorous marine and freshwater species may be due to the relatively undeveloped morphology and enzyme systems of the larval alimentary tract (Blaxter, 1988).

In his compilation of the diets of tropical marine shorefish larvae, Leis (1991) noted that some species show extreme feeding specialization on certain types of plankton (e.g., larvaceans). Some fish larvae specialize in piscivory on other larval fishes and have a large, well-developed caecum similar to the adult stomach (e.g., Jenkins *et al.,* 1984). In general, as larvae approach metamorphosis, their feeding habits diverge to become more species-specific (Hunter, 1980).

Simultaneous quantification of prey available in the environment and prey in gut contents indicates that larvae can be selective [review in Heath (1992)]. For example, the calculation of electivity values for larval walleye in a freshwater lake demonstrated that a diet change during early growth was not associated with a change in the composition of the zooplankton in the lake (Graham and Sprules, 1992). Johnston and Mathias (1994) reported that walleye larvae selected most strongly for prey types and sizes that were relatively uncommon.

As fish larvae grow, the minimum size of prey that is consumed often increases more gradually than the maximum prey size. Consequently, larger larvae tend to consume a larger range of prey sizes (see Hunter, 1980; Wahl *et al.,* 1993). Reports of such an increase in the range of prey sizes with increasing predator size are not uncommon in the ecological literature (e.g., Werner and Gilliam, 1984). Experiments indicate that the maximum prey size consumed is related to larval mouth width (Hunter, 1980) and gape (e.g., Schael *et al.,* 1991). Hunter (1980) concluded from the literature that, at least during the first few weeks of feeding, the minimum size of prey consumed appears to be determined by metabolic constraints, whereas the maximum size of prey is controlled by mouth size.

Larval diet is determined by a large number of factors, including larval morphology, foraging behavior, and prey capture mechanisms, as well as prey morphology, behavior, and abundance. For example, in laboratory experiments on larval bluegill, Bremigan and Stein (1994) found that the average size of zooplankton ingested increased significantly as both available zooplankton size and larval gape increased. However, because gizzard shad larvae and juveniles selected smaller prey consistently, the prey size of gizzard shad larvae and juveniles was unrelated to gape. The juvenile gizzard shad may have ingested the zooplankton using filter feeding, which does not favor the capture of the larger, more evasive prey (Bremigan and Stein, 1994).

B. AMPHIBIANS

Anuran larvae eat an amazingly wide variety of foods. These foods represent the entire spectrum of primary and secondary, autochthonous and allochthonous, production observed in freshwater aquatic habitats and span an incredible range of particle sizes. "Herbivorous" diets include vascular plants (Storre, 1994), vascular plant-based detritus (Diaz-Paniagua, 1985), periphyton which can include filamentous green algae, cyanobacteria, diatoms, bacteria, and desmids (Dickman, 1968; Kupferberg et al., 1994), planktonic green algae (Seale, 1980; Battish and Sandhu, 1988), cyanobacteria (Seale and Beckvar, 1980), ultraplankton (Wassersug, 1972), and pollen (Wagner, 1986; Britson and Kissell, 1996).

Two lines of evidence support the interpretation that tadpoles feed indiscriminately on this wide array of foods. Algae and/or zooplankton in the guts of tadpoles from pond environments reflect the species composition in the environment (Farlowe, 1928; Jenssen, 1967; Heyer, 1972, 1976; Nathan and James, 1972; Seale, 1980; Diaz-Paniagua, 1985). Similarly, laboratory feeding trials on suspended particles do not indicate selectivity (Wassersug, 1972; Seale and Beckvar, 1980; Viertel, 1992). Interspecific differences in diet usually have been interpreted as a reflection of segregation of the species among microhabitats (Heyer, 1976; Waringer-Loschenkohl, 1988; Skelly, 1995) and the morphological adaptations allowing them to occupy these habitats (Diaz-Paniagua, 1985; Nodzenski et al., 1989). Some larvae that feed on patchily distributed attached algae have displayed the ability to distinguish and differentially consume high-quality taxa that are rich in protein and lipids and that enhance growth and developmental rates (Kupferberg, 1997c).

Some tadpoles are "carnivorous," many opportunistically so, with no special morphological adaptations for macrophagy. The diversity of animal prey consumed facultatively includes protozoans (Nathan and James, 1972), mosquito larvae (Blaustein and Kotler, 1993), fairy shrimp (Pfennig, 1990), eggs–embryos of other amphibians (Petranka *et al.*, 1994), and conspecific tadpoles (Crump, 1983, 1986; Polis and Meyers, 1985). Obligate carnivores, such as those living in plant-held waters, or phytotelmata, consume conspecific unfertilized (Lannoo *et al.*, 1987) or fertilized (Jungfer and Schiersari, 1995) eggs and aquatic insect larvae (Pimm and Kitching, 1987). There also is the special case of spadefoot toad tadpoles in which a cannibal morph is induced environmentally (Pfennig, 1990) (see discussion of resource polymorphisms in Section VIII.B.3).

By comparison, salamanders eat a less diverse array of foods and generally consume only animal material. Larval salamanders are gape-limited predators that expand the breadth of their diet, taking larger prey as they grow. Salamanders consume zooplankton–microcrustacea (Dodson, 1970; Morin, 1987), benthic macroinvertebrates–insects (McWilliams and Bachmann, 1989a,b), terrestrial insects falling into the water (Parker, 1994), mollusks (Collins and Holomuzki, 1984), tadpoles (Loeb *et al.*, 1994), other salamanders (Collins and Holomuzki, 1984; Gustafson, 1993), and fish (Parker, 1993b).

VII. REYNOLDS NUMBER

The Reynolds number *(Re)* is a dimensionless index that provides qualitative insights into the nature of the hydrodynamic regime in which aquatic larvae must feed, respire, and locomote. Low Reynolds numbers ($Re \ll 1$) indicate that viscous or "fluid stickiness" forces predominate, whereas high Reynolds numbers ($Re \gg 1$) indicate that inertial or "fluid momentum" forces predominate. Reynolds numbers for biological activities span more than 14 orders of magnitude (Vogel, 1994). To calculate the Reynolds number for a swimming larva, the swimming velocity is multiplied by the larval body length, and this amount is divided by the kinematic viscosity of seawater or freshwater at the given temperature.

A. FISHES

Weihs (1980) suggested that the Reynolds numbers that are relevant to swimming larval fishes can be divided into three distinct hydrodynamic regimes: a viscous regime ($Re < 10$), an intermediate or transitional regime ($10 < Re < 200$), and an inertial regime ($Re > 200$). As larvae increase in size, they grow through the viscous and the intermediate regimes (Weihs, 1980). Müller and

Videler (1996) hypothesized that the larvae of at least some fish species grow in length as fast as possible to reach the inertial flow regime.

This change in Re during ontogeny may be responsible for ontogenetic changes in the following aspects of routine (voluntary, spontaneous) swimming: (1) the proportion of the body that undulates (e.g., Osse, 1990); (2) the amplitude of undulation at locations near the head vs locations near the tail (e.g., Batty, 1984); (3) the use of continuous swimming vs beat-and-glide intermittent swimming (e.g., Weihs, 1980); and (4) the proportion of turns, made to change direction during swimming, that are a wide angle from the initial direction of travel vs a shallow angle (Fuiman and Webb, 1988).

Many fish species have larvae that are in the range of 3–8 mm in length at hatching (Osse, 1990). These yolk-sac larvae operate in the viscous regime (Weihs, 1980), where resistive forces are more important than inertial forces. Therefore, these small larvae often use the entire length of their eel-like bodies to push against the water during swimming (e.g., Dabrowski, 1989; Osse, 1990; Osse and van den Boogaart, 1995) and bend their bodies continuously during swimming rather than gliding or coasting periodically (Osse and van den Boogaart, 1995; Weihs, 1980). The amplitude of side-to-side body movement during swimming increases linearly toward the tail, which is indicative of the importance of resistive forces (Batty, 1984).

As larvae grow and the Re increases, there is a gradual transition to a beat-and-glide mode of swimming in anguilliform swimmers (Weihs, 1980). The energetic advantages of this intermittent swimming increase with increasing Re, as reactive (inertial) forces predominate (Weihs, 1980). The amplitude of side-to-side body movement during swimming increases more rapidly than linearly toward the tail, as the posterior portion of the body contributes more to thrust than the anterior portion (Batty, 1984). Above an Re of 200, the larvae maintain a comparatively rigid anterior half of the body and primarily use the posterior one-half to one-third of the body, including the tail, for propulsion (Batty, 1984; Dabrowski, 1989; Osse, 1990; Osse and van den Boogaart, 1995).

The preceding model of the effects of Re on locomotor strategy during ontogeny has important implications for larval feeding. Webb and Weihs (1986) predicted that there should be a minimum length at which yolk-sac larvae progress to first feeding, in order to minimize the costs of searching for food by using beat-and-glide swimming above an Re of 10. At a typical foraging speed of two body lengths/s, they calculated this minimum length to be approximately 3 mm. They noted that most larval fishes are larger than 3 mm at first feeding and commented that the exceptions to this pattern warranted further study to discover whether such smaller larvae utilized a faster swimming speed to increase the Re above 10.

Other important implications of the ontogenetic changes in Re involve the increasing rigidity of the anterior body with a gradual transition to beat-and-glide swimming. Drost et al. (1988b) suggested that premaxillary protrusion, which causes a highly directed flow of water into the oral cavity, does not develop until carp larvae reach a size at which reduction in the yaw of the head during swimming causes an increase in the aiming accuracy of prey capture. In addition, Osse (1990) hypothesized that the release of the anterior body from the need to undulate during swimming allows rapid development of the abdominal organs. When carp larvae that are 10 mm in length swim at 2–3 body lengths/s, they are in the inertia-dominated regime above an Re of 200 and therefore can rely on their posterior body for propulsion. The anterior body remains relatively stiff during swimming, and there is sudden growth in the length of the intestine. At this stage of development, the stiffness of the anterior body may contribute to a functional separation between food transport in the alimentary tract and swimming. The slow early growth of the intestine in larvae less than 10 mm long is thought to facilitate flexion of the anterior body during undulatory locomotion at transitional Re values (Osse, 1990).

Osse (1990) suggested that this relationship between Reynolds number and the development and functioning of abdominal organs may apply to fish larvae in general. This hypothesis should be tested, particularly because Govoni et al. (1986) noted that development of the larval fish alimentary canal occurs in a rapid burst, in contrast to the gradual changes that have been reported for other organ systems (O'Connell, 1981). A rapid change occurs just before first feeding as the alimentary tract develops from a straight, undifferentiated incipient gut to a segmented, differentiated fore-, mid-, and hindgut that are separated by muscular valves (Govoni et al., 1986). As mentioned previously, Webb and Weihs (1986) predicted a change in hydrodynamic regime at first feeding, which results in a change in swimming mode. Information on the timing of larval gut development, swimming mode, and calculations of Re for larval fishes should be examined for possible correlations between Re and gut development.

Changes in Re during ontogeny also may have implications for larval foraging behavior. For example, Fuiman and Webb (1988) noted that larval zebra danio turned through larger angles more often when they were swimming rapidly than when they were swimming slowly. Below Re = 23, the larvae turned through small angles (<63°) exclusively (Fuiman and Webb, 1988). Building on this observation, Coughlin et al., (1992) suggested that clownfish larvae increase their swimming speed when they encounter a patch of dense food due to constraints imposed by viscosity. They calculated that first-feeding clownfish larvae swim at a low Re (~38) that is not conducive to gliding through turns of a large angle. In order to stay within a patch of dense food, clownfish larvae may increase their inertia by increasing their velocity, thereby improving their ability to turn through wide angles.

The end of the viscous regime at $Re = 10$ and the beginning of the inertial regime at $Re = 200$ were proposed for larval fishes on the basis of both empirical data on flow around spheres and cylinders and hydrodynamic theory (Weihs, 1980; Webb and Weihs, 1986). By quantifying the swimming speed and kinematics of larval Atlantic herring as the viscosity of seawater was increased using methyl cellulose, Fuiman and Batty (1997) obtained data suggesting that the viscous regime for this species extends to at least $Re = 300$. Because changes in hydrodynamic regime with growth have important implications for larval development, similar experimental manipulations of the viscosity of the medium are needed to determine the hydrodynamic regime for additional species.

B. AMPHIBIANS

Amphibian larvae, by virtue of their size, are not operating at the same low Re as fish larvae with regard to swimming. Maximum tadpole length for 60 species ranges from approximately 7 to 220 mm TL, with most falling between 30 and 60 mm (Emerson, 1988). Tadpoles swim at Re values within a range of 100–10,000 (Liu et al., 1996), well into the inertial flow regime experienced by the largest fish larvae. Despite their globose bodies and wobbling snouts, tadpoles are efficient undulatory swimmers. Unlike the problems caused by anterior yaw in larval fish, oscillations at the snout are not detrimental to tadpole feeding because most are grazers and do not need to aim their mouths accurately at prey. Rather, the yaw contributes to forward thrust, and estimates of propeller Froude efficiency (an index based on swimming speed and speed of the propulsive wave) for bullfrog tadpoles (Rana catesbeiana) are high. Two-dimensional models estimate efficiency at 82–83% (Wassersug and Hoff, 1985; Liu et al., 1996), but a three-dimensional model shows a lower efficiency of 45% (Liu et al., 1997). Tadpoles are not constrained to low Re by the mechanics of fluid flow around the body and tail. Through the use of an elegant computational fluid dynamics model, Liu et al. (1996) showed that at Re above 10,000, as if tadpoles were bigger and faster than they really are, propeller efficiency would increase. The effects of low Re on tadpole locomotion may be more important to predator avoidance than to food intake. For example, viscous forces allow tadpoles to stop simply by ceasing tail undulation and glide into vegetation or sediment, often without a trace (Wassersug, 1989).

In contrast to tadpoles, larval salamanders have elongate bodies with drag-creating external gills and limbs that render them relatively inefficient anguilliform swimmers (Hoff et al., 1989). For a given velocity, tadpoles have a lower tail beat frequency than salamander larvae. Ambystoma larvae in the size range of 20–90 mm exhibit sustained swimming velocities (1.26–18.7 L/s, L = body

length) similar to those of comparably sized *Bufo, Rana,* and *Xenopus* tadpoles
(0.5–14.2 L/s) and, thus, operate in the same range of *Re* as tadpoles. Hydrodynamics plays an important role in salamander foraging mode. The high-
amplitude (>30% of body length), short-wavelength movements of *Ambystoma*
larvae are better suited for acceleration than steady swimming. This is consistent with the "sit and wait" predatory behavior observed in other salamander
taxa (Petranka, 1984; Parker, 1994), which involves rapid acceleration over
short distances to lunge at prey.

With regard to the fluid dynamics directly relating to feeding mechanism,
there is no literature directly assessing whether viscous forces affect the filter
feeding of tadpoles or suction feeding of salamander larvae. For tadpoles this
is probably because "the flow into the tadpole pharynx is pulsatory and the
irregular shape of the ventral velum and the branchial food traps do not lend
themselves to a simple hydrodynamic analysis" (Wassersug and Rosenberg,
1979, p. 409).

VIII. ONTOGENY OF FEEDING STRUCTURE AND FUNCTION

A. FISHES

A multitude of fundamental morphological changes occur during the larval
period, including the development of structures involved in feeding, respiration, digestion, sensory systems, and locomotion. Not surprisingly, these morphological changes are correlated among systems and are associated with
functional and behavioral changes.

Extensive morphological data are available on teleost larval development,
including a number of detailed studies of the viscerocranium. Our review
summarizes data on the development of structures involved in feeding. Because
feeding and respiration are linked closely in adult fishes, we include mention
of respiratory structures and functions.

Hunt von Herbing *et al.* (1996c) examined the effects of temperature on
the development of larval cod from two genetically discrete populations. They
concluded that, during the yolk-sac stage, the development of structures associated with prey capture and respiration primarily was under genetic control.
However, on the basis of the size- and age-dependent variation that they
observed in the intestinal development of these larval cod raised at two different
temperatures, they suggested that environmental factors could be particularly
important in producing phenotypic variability during the exogenous feeding
stage.

Webb and Weihs (1986) examined the body and fin shapes of larvae belonging to species that, as adults, differ dramatically in body and fin shape and in locomotor mode (thunniform, esociform, and chaetodontiform). The limited available data indicate that, at hatching, all of the larvae studied were similar in body and fin shape. These larvae began to diverge toward the adult form at a length of approximately 1.5 times hatching length, near the size at which first feeding occurs. These findings should stimulate further investigation, because similarity among species in body and fin shape during the early larval period is suggestive of common phylogenetic and/or functional constraints that may operate in yolk-sac larvae. In addition, the widespread occurrence of a U-shaped growth gradient in fish larvae, with intense growth in width and depth at the head and tail regions (Fuiman, 1983), deserves further study (Osse and van den Boogaart, 1995).

1. Yolk-Sac Larvae

In general, the mouth and jaws have been reported as not formed or not functional at hatching (Blaxter, 1988; Kendall *et al.*, 1984). The mouth then becomes functional at some time during the yolk-sac period. In contrast, larvae from the demersal eggs of coastal marine and freshwater fishes may be capable of feeding at hatching (Kendall *et al.*, 1984).

In Atlantic cod at hatching, an oropharyngeal membrane seals the mouth completely (Hunt von Herbing *et al.*, 1996b). After 2–3 days at 5–7°C, this membrane is ruptured and the mouth opens. However, at this time, there are no distinct articulations between the upper and lower jaws nor between the elements of the suspensory apparatus, and no mouth movements are observed (Fig. 13). In some species, the jaws of early larval stages have been reported to move or "quiver" even before the tissue over the mouth has ruptured (e.g., McElnan and Balon, 1979; Paine and Balon, 1984a).

Most cartilaginous elements of the branchial skeleton and the mandibular arches appear prior to hatching in the Japanese medaka, and ossification of most of these structures occurs in early Japanese medaka larvae that feed exogenously (Langille and Hall, 1987). The formation of cartilaginous structures such as Meckel's cartilage also has been reported before hatching in Northern logperch and rainbow darter (Paine and Balon, 1984a,b). However, the timing of chondrification and ossification varies among species. For example, little cartilage is observed in the head of a white sucker until after hatching (McElman and Balon, 1980). Ossification of elements of the head skeleton begins prior to first feeding in walleye (McElman and Balon, 1979) and North American landlocked Arctic charr (Balon, 1980), but well after first feeding in Northern logperch (Paine and Balon, 1984a) and European landlocked Arctic charr (Balon, 1980). The number of species for which both developmental

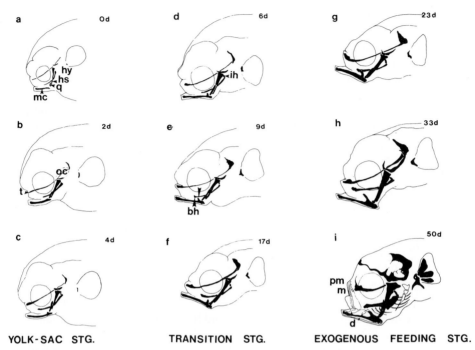

FIGURE 13   Cranial skeletal elements in Atlantic cod larvae. (a–c) yolk-sac stage, 0–4 days after hatching; (d–f) transition (mixed feeding) stage, 6–17 days after hatching; (g–i) exogenous feeding stage, 23–50 days after hatching. Abbreviations: bh, basihyal; d, dentary; hs, hyomandibulo-symplectic; hy, hyoid; ih, interhyal; m, maxilla; mc, Meckel's cartilage; oc, otic capsule; pm, premaxilla; q, quadrate; t, trabeculum cranii. Arrow in (e) points to the anterior insertion point of the adductor mandibulae muscle on mc. Stippled elements represent cartilage; solid elements represent bone. From Hunt von Herbing *et al.* (1996b, Fig. 1, p. 18). Ontogeny of feeding and respiration in larval Atlantic cod *Gadus morhua* (Teleostei, Gadiformes): I. Morphology. *Journal of Morphology,* Copyright © 1996. Reprinted by permission of Wiley-Liss, Inc., a subsidiary of John Wiley & Sons, Inc.

and functional data are available is insufficient to determine the relationship between environmental variables (e.g., temperature), ecological and functional aspects of exogenous feeding, and the timing of chondrification and ossification of feeding structures.

In cod, the upper jaw does not become protrusible until approximately 4 weeks after food is observed in the intestine (Hunt von Herbing *et al.,* 1996b). Similarly, premaxillary protrusion was not observed in first-feeding carp larvae, but was 3.8% of the length of the head in older larvae (Fig. 14; Drost and van den Boogaart, 1986).

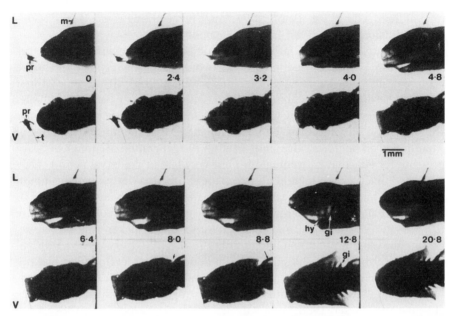

FIGURE 14   Selected frames of a shadowgraphic film (1250 frames/s) showing capture of prey (pr, *Artemia* nauplius) by a carp larva (9.5 mm standard length) in a lateral (L) and a ventral (V) view. Time is indicated in milliseconds. The *Artemia* is glued to a 12-μm nylon thread (t, faintly visible) connected to a micromanipulator (m). The upper jaw protrudes 3.8% of the length of the head between *t* = 4.8 and 8 ms. The opercular valves open at *t* = 8 ms (arrow). Abbreviations: pf, pectoral fin; gi, branchial arches; hy, hyoid. From Drost and van den Boogaart (1986, Fig. 3, p. 375).

The development of gill rakers is of interest due to their potential importance in the filtration of food particles from water that has entered the oral cavity. However, there are relatively few reports of gill raker development in early larvae. Paine and Balon (1984b) noted that gill rakers appear on all ceratobranchials in early rainbow darter larvae that are feeding exogenously. In Atlantic cod, however, gill rakers do not form on all four branchial arches until late in the larval period, approximately 5 weeks after the beginning of exogenous feeding (Hunt von Herbing *et al.*, 1996b). The relationship between diet, feeding mechanism, and the timing of gill-raker development deserves further study (see Mullaney and Gale, 1996).

In general, most cartilaginous elements of the branchial arches are formed before first feeding, and some branchial circulation has been established (e.g., Balon, 1977; McElman and Balon, 1979; Paine and Balon, 1984a,b). Blood generally flows through developing gill filaments and pseudobranchs by the time of first feeding (e.g., Balon, 1977; Hunt von Herbing et al., 1996b; McElman and Balon, 1979, 1980; Paine and Balon, 1984b).

Newly hatched yolk-sac larvae of Atlantic cod lack an opercular apparatus and therefore have an open opercular cavity or canal in the region of the future opercular slit (Fig. 15; Hunt von Herbing et al., 1996b). At the time of first feeding, the opercular apparatus is composed primarily of connective tissue, and an epithelium almost encloses the opercular cavity to create an opercular slit. Hunt von Herbing et al. (1996b) concluded that the opercular bones of Atlantic cod do not form sufficiently to allow the generation of enough negative pressure for effective gill ventilation until very late in the larval period. They suggested that the pseudobranchs, which are exposed to water when the mouth opens, may be important in larval gas exchange. McElman and Balon (1980) suggested that the gill filaments may exchange gases as a result of their exposure to the water via the open opercular cavity in white sucker yolk-sac larvae, despite the closed, nonfunctional mouth.

The contribution of the gills to respiration during development has not been well-documented (Osse, 1989). In general, structures for branchial respiration do not develop fully until sometime during the exogenous feeding stage (El-Fiky and Wieser, 1988; Hunt von Herbing et al., 1996b; Osse, 1989). Osse (1989) calculated that the thickness of the oxygen diffusion boundary layer in the oral and opercular cavities of carp larvae early in the exogenous feeding stage (6–7 mm SL) prevented adequate gas exchange. Consequently, early larvae rely on cutaneous respiration. The function of the median finfold that is characteristic of teleost yolk-sac larvae has not been established, but may involve an enhancement of gas exchange via an increase in the surface area to volume ratio (Osse and van den Boogaart, 1995). Given that the opercular structures that are considered essential for adequate gill ventilation do not form until a late stage of development in larval cod, Hunt von Herbing et al. (1996b) concluded that initial viscerocranial structures are specialized for feeding in cod and perhaps in most fish larvae (also see Osse, 1990).

2. First-Feeding Larvae

From data compiled on the larvae of 63 marine, freshwater, and anadromous fish species, Miller et al. (1988) noted that the time to first feeding increased significantly with larval length at hatching. The time to yolk depletion also increased significantly with hatching length, and the time to first possible feeding, defined as the time at which a larva is first physically or anatomically capable of feeding, decreased significantly with hatching length. They concluded that larvae that are large at hatching have a greater "window of opportunity" for first feeding than do larvae that are small at hatching. For larvae that

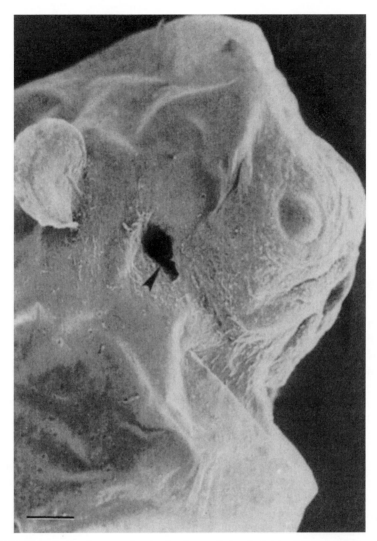

FIGURE 15 Scanning electron micrograph of a newly hatched Atlantic cod larva showing an open opercular cavity (arrow). Cod larvae at this stage have four branchial arches but lack gill filaments. Scale bar = 100 μm. From Hunt von Herbing *et al.* (1996b, Fig. 2b, p. 19). Ontogeny of feeding and respiration in larval Atlantic cod *Gadus morhua* (Teleostei, Gadiformes): I. Morphology. *Journal of Morphology,* Copyright © 1996. Reprinted by permission of Wiley-Liss, Inc., a subsidiary of John Wiley & Sons, Inc.

hatch at large sizes, the period of time during which first feeding must occur to avoid starvation is longer than that for larvae that hatch at small sizes.

Fish larvae possess branchial arches, a functional lower jaw, and opercular slits by the time at which they begin to capture mobile prey (Blaxter, 1988; Hunt von Herbing et al., 1996c). In 4- to 5-day-old yolk-sac larvae of Atlantic cod, the formation of the interhyal cartilage, the articulation between Meckel's cartilage and the quadrate, and the articulation between the hyomandibulosymplectic and the cranium lead to the first jaw movements (Hunt von Herbing et al., 1996b). Hunt von Herbing et al. (1996b) described these "gulping" movements as infrequent and uncoordinated. Because these initial movements occur prior to first feeding, they are thought to provide the larvae with time to improve the coordination of viscerocranial structures with changes in locomotion and growth. In addition, they proposed that these erratic movements, which probably generate very little negative pressure in the oral cavity, facilitate the expansion and growth of the oral and pharyngeal cavities.

3. Exogenous Feeding

Exogenous feeding begins after complete resorption of the yolk sac. In general, the exogenous feeding stage involves increases in the size of larval structures that differentiated during the endogenous and mixed feeding stages (e.g., Hunt von Herbing et al., 1996c; McElman and Balon, 1980).

Fine-scale studies of the relationships between structural, functional, and behavioral changes in feeding during larval fish ontogeny are rare. Numerous difficulties are encountered in conducting such studies, including the following. (1) Larvae must be reared in a laboratory environment that does not alter substantially their development and behavior, or a field sampling program must be designed that provides functional and behavioral data. (2) Although temperature has a profound effect on development, little is known about the extent to which the relative timing of different developmental events varies with temperature (Hunt von Herbing et al., 1996a). (3) The observation and quantification of the functional morphology and behavior of feeding in small fishes are challenging.

Ontogenetic changes in prey capture mechanisms, which could be related to ontogenetic changes in diet or could be independent of diet, might be manifested at a number of interrelated levels: (1) behavioral aspects of prey capture, such as attack kinematics (e.g., the swimming velocity and acceleration of the predator during the attack, the distance from the prey at which the predator intiates the attack) and the duration of the attack; (2) cranial kinematics of prey capture, such as the timing, velocity, and extent of mouth opening, premaxillary protrusion, hyoid depression, cranial elevation, and opercular abduction; (3) cranial muscle activity patterns associated with prey capture;

(4) timing and magnitude of oral and opercular pressures; and (5) use of suction feeding vs ram feeding. Due to the small size of fish larvae, empirical data on cranial muscle activity patterns and oral and opercular pressures cannot be obtained with the equipment presently available. However, Drost *et al.* (1988a) have developed a quantitative hydrodynamic model of larval suction feeding that includes frictional effects.

In theory, all of the preceding variables could be affected by the body size of the predator. Variables that have a component that is dependent on predator body size are expected to change with growth. For example, body size had significant effects on kinematic timing and velocity variables in a study of feeding of juvenile and adult largemouth bass (Richard and Wainwright, 1995).

Possible methods of adjusting for predator body size effects include (1) expressing distance measurements (e.g., distance between predator and prey when the attack is initiated, maximum premaxillary protrusion) as a proportion of predator body length or structure length (e.g., premaxillary length), (2) expressing acceleration and velocity measurements in units of predator body length, and (3) expressing timing variables (e.g., duration of the expansive phase of prey capture, time to maximum premaxillary protrusion) as a proportion of the duration of the complete prey capture event.

One purpose of calculating variables relative to predator body length rather than in absolute terms is to determine whether larval prey capture is functionally equivalent to the capture of prey by a miniature adult (e.g., Cook, 1996). If sufficient data are available from conspecific fish of a range of sizes, an allometric analysis can be conducted. If the goal is to compare data from the larvae of a number of species that differ in size, an analysis of covariance can be performed with body size as a covariate.

a. Suction Feeding vs Ram Feeding

Suction feeding and ram feeding are best viewed as opposite ends of a continuum (e.g., Norton, 1991; Norton and Brainerd, 1993). To capture planktonic prey, a larval fish could use suction feeding, ram feeding, or a combination of the two. During pure suction feeding, the predator remains stationary as the prey is engulfed. Suction of the prey item into the predator's oral cavity results from the generation of negative pressure as oral cavity volume is increased rapidly. In contrast, during pure ram feeding, the prey is engulfed due to the forward body movement of the predator rather than suction. The predator swims its open mouth forward around the prey as water exits from the predator's oral and opercular cavities posteriorly via the opercular slit created by the abducted opercular valves. In general, adult fish use some combination of suction feeding and ram feeding as they generate negative pressure and open their mouths while approaching the prey (see Norton and Brainerd, 1993). The ram–suction index developed by Norton and Brainerd (1993) is a quantitative measure of the relative importance of ram feeding vs suction feeding during an individual prey capture event.

The importance of adjusting for body size was demonstrated in a study of prey capture by conspecific juvenile cottids over a range of sizes (Cook, 1996). During prey capture, the juvenile cottids functioned essentially as small adults, with the exception of a gradual decrease in the ram–suction index, indicating a decrease in the importance of the ram component of the attack and an increase in the suction component. Consequently, early juveniles of this cottid species used a ram-dominated feeding mode that was inconsistent with the suction-feeding characteristics related to their morphology (small gape, large buccal volume) and attack kinematics (short predator–prey distance, low attack velocity). Although this conclusion should be verified using a variety of prey types and sizes, these results illustrate a potential mismatch between the ontogeny of morphology, kinematics, and feeding mode and stimulate intriguing questions about ontogenetic changes in function.

Given that larval and adult fish of the same species differ in a number of features, such as head morphology, hydrodynamic regime, and diet, larval fish should not be expected to use the same feeding mechanisms as adults. Ontogenetic changes in feeding mechanisms are of interest because such changes could provide insight into morphological and biomechanical constraints that operate during ontogeny, the role of morphology in inducing ontogenetic niche shifts, and developmental constraints on function (e.g., Galis, 1993; Galis et al., 1994). Unfortunately, there have been few studies that compare the functional morphology of feeding in larval vs adult fish. Liem (1991) reported that an increase in the ratio of head depth to head length in the larvae of species belonging to three families was associated with a change from a cylindrical oropharyngeal cavity at hatching to the shape of a truncated cone in later larvae (Fig. 16). The relatively small mouth at the apex of the cone, combined with a greater head depth in the region of the suspensory apparatus, is thought to result in improved suction feeding. High-speed video analysis showed that these morphological changes were accompanied by a switch from ram feeding in early larvae to suction feeding in later larvae.

Using a video recorder (60 fields/s) with a darkfield modified-Schlieren optical system, Coughlin (1991) concluded that young Atlantic salmon during the first few days of feeding used a ram-dominated feeding mode by opening the opercular valves early while lunging toward the prey, but incorporated suction feeding by generating a highly directed flow into the mouth. With growth, there was a gradual transition to a suction-dominated feeding mode, during which forward body movement of the predator became less important and the opercular valves opened later in the attack.

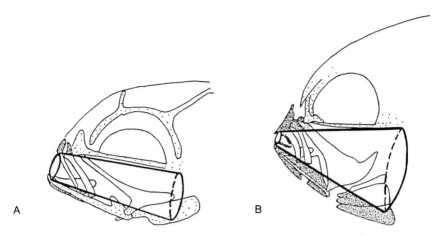

FIGURE 16 (A) Early pomacentrid larva with a cylindrical oropharyngeal cavity and a large gape. (B) Late larva of the same species with a conical oropharyngeal cavity and small gape. From Dilling (1989) in Liem (1991, Fig. 11, p. 240).

Although the data reported by Liem (1991) and Coughlin (1991) are suggestive of an ontogenetic pattern involving a switch from ram feeding to suction feeding, Coughlin (1994) used the modified-Schlieren optical system to determine that larvae of a species in the pomacentrid genus studied by Liem generated suction as well as swam forward during prey capture on the first or second day of feeding. In addition, in high-speed films (200–1250 frames/s) of carp larvae (5.8–15 mm standard length, SL), Drost and van den Boogaart (1986) observed that prey capture involved some forward body movement but was dominated by suction (Fig. 14). The 5.8-mm larvae were referred to as first-feeding larvae (Drost and van den Boogaart, 1986). However, Coughlin (1994) referred to these carp larvae as older larvae. Pike larvae (12–15 mm SL) that were referred to as first-feeding larvae were reported to use a combination of ram feeding and suction feeding during prey capture (Drost, 1987). Given the results described earlier, the available data are insufficient to determine whether there are patterns in the use of ram feeding vs suction feeding during larval fish ontogeny and whether the use of ram vs suction feeding is related to the development of cranial structures that are involved in the generation of suction (e.g., hyoid apparatus, opercular apparatus).

On the basis of a decrease in the ratio of head depth to head length with growth in larval cod and the proposed relationship between this ratio, oropharyngeal shape (cylindrical vs conical), and the use of ram feeding vs suction feeding (Liem, 1991), Hunt von Herbing *et al.* (1996c) predicted that cod switch from a suction-dominated feeding mode in early larvae to a ram-dominated feeding mode with growth. However, a quantitative morphological analysis of the relationship between the shape of the oropharyngeal cavity and the ratio of head depth to head length is not available for fish larvae. In addition, the functional significance of oropharyngeal shape has not been investigated empirically in fish larvae.

b. Spatial Relationships and Connections between Functional Elements

Two particularly promising areas for research on the feeding mechanisms of larval fishes are the functional significance of (1) ontogenetic changes in the proportions and spatial relationships of cranial structures and (2) the development of connections (e.g., muscles, ligaments, and/or articulations) between functional elements. For example, important ontogenetic changes in mouth-opening mechanisms have been described for the few larval fish species that have been examined. Interestingly, these ontogenetic changes are similar in many respects to phylogenetic changes in mouth-opening mechanisms.

In early larvae belonging to three families in the suborder Labroidei (Cichlidae, Pomacentridae, and Embiotocidae), the mandible is depressed by contraction of the geniohyoideus muscle, which connects the hyoid to the mandible, while the hyoid is held in place or depressed by contraction of the sternohyoideus muscle (Liem, 1991; Otten, 1982). As the larvae grow, however, the jaw joint between the quadrate and the mandible (=quadratal joint) moves rostrally, causing the working line of the geniohyoideus muscle to shift dorsal to the quadratal joint. In later larvae, therefore, the geniohyoideus muscle can no longer function as a mouth-opening mechanism (Liem, 1991). Coincident with the loss of this mechanism, the levator operculi muscle differentiates and bones of the opercular apparatus form and become linked by ligaments (Otten, 1982). Contraction of the levator operculi causes rotation and posterior movement of opercular elements, which results in depression of the lower jaw via the newly formed interoperculomandibular ligament. The transition between reliance on the geniohyoideus to open the mouth and reliance on the opercular mouth-opening mechanism occurs extremely rapidly in the cichlid *Haplochromis elegans,* within one day (Otten, 1982).

In early cichlid and pomacentrid larvae that rely on the geniohyoideus mouth-opening mechanism, the hyoid was observed in the three species studied to retract (depress) synchronously with the lower jaw (Liem, 1991; Otten, 1982). In the later larvae of these species, which rely on the opercular mouth-opening mechanism, hyoid depression lagged behind lower jaw depression, such that the hyoid actually protracted as the lower jaw began to depress during respiration.

Liem (1991) examined the ontogeny of mouth-opening mechanisms in a species in the viviparous family Embiotocidae for comparison with the closely related oviparous families Cichlidae and Pomacentridae. He observed the same switch from a geniohyoideus to an opercular mouth-opening mechanism as the working line of the geniohyoideus muscle moved dorsal to the jaw joint before birth, although he noted that the transition occurred over a greater length of time in the embiotocid relative to the cichlids and pomacentrids. Since embiotocid larvae are released from the female parent at an advanced stage of development, the switch in mouth-opening mechanisms occurs in the ovary under the influence of very different functional demands for respiration and feeding from those encountered by free-living cichlid and pomacentrid larvae. Liem suggested that the presence of the shift in embiotocid larvae was indicative of internal programmatic rules of the assembly of form that may have been inherited from plesiomorphous oviparous ancestors.

In the larvae of cod (Hunt von Herbing *et al.*, 1996c) and rainbow trout (Verraes, 1977), there is also a shift from a hyoid to an opercular mouth-opening mechanism, although the musculoskeletal linkage that is involved in the hyoid mouth-opening mechanism in these two species differs from that that operates in the cichlids, pomacentrids, and embiotocids. Mouth opening in the early stages of cod and rainbow trout larvae is accomplished by contraction of the sternohyoideus muscle, which causes hyoid retraction and subsequent mandibular depression via the mandibulohyoid ligament. In later cod larvae and juvenile rainbow trout, a levator operculi coupling similar to that that serves as the opercular mouth-opening mechanism in cichlids, pomacentrids, and embiotocids becomes functional. Unlike the labroid species, however, cod and rainbow trout retain a functional hyoid coupling throughout life.

Thus, all of the few species that have been studied pass through an early larval stage involving one of two hyoid mouth-opening mechanisms and then develop the same opercular mouth-opening mechanism. This is similar to the phylogenetic pattern of mouth-opening mechanisms in adult actinopterygian fishes. Primitive actinopterygians (e.g., *Polypterus* and *Lepisosteus*) effect mandibular depression via the mandibulohyoid ligament (Lauder, 1980). At the halecostome level, a second biomechanical pathway for mouth-opening, the levator operculi coupling, evolved (Lauder, 1979, 1980; Lauder and Liem, 1989). Hunt von Herbing *et al.* (1996c) discussed the potential developmental and evolutionary significance of these two independent biomechanical pathways to effect mouth opening. They pointed out that species which, as adults, use similar mouth-opening mechanisms can develop the mechanisms very differently in early life-history stages. The converse can also occur. For example, species with larvae that possess similar hyoid depression mechanisms early in ontogeny can have very different mechanisms as adults (Adriaens and Verraes, 1994).

The preceding discussion of mouth-opening mechanisms illustrates the potential importance of ontogenetic changes in the spatial relationships and connections between functional elements. For the most part, such changes involved in the development of premaxillary protrusion, cranial elevation, and suspensorial and opercular abduction during feeding, and the functional consequences of such changes, have not been investigated in larval fishes (but see Hernández, 1995). Such studies will be challenging because they will necessitate detailed morphological examination as well as high-speed motion analysis. Another approach to the study of spatial relationships and functional connections was provided by Otten's (1983) computer model of kinematics and force equilibria in a juvenile cichlid, which was based on anatomical points and articulation surfaces measured from serial sections.

B. AMPHIBIANS

1. Developmental Sequences

The developmental sequences of the oral structures among tadpole species with four different kinds of mouthparts have been compared by Thibadeau and Altig (1988). The development of the most common suite of oral structures, including keratinized jaw sheaths, labial teeth set on a fleshy oral disk surrounded by papillae (*Hyla chrysoscelis* and *Rana sphenocephala*), was compared to (1) jaw sheaths with a reduced oral disk and no teeth, (2) the absence of all keratinized mouthparts (pipids and most microhylids), and (3) no keratinized mouthparts but fleshy structures like barbels or oral flaps (*Rhinophrynus*). The general order of development is as follows: stomodeal invagination, oral pad development, jaw sheath delimitation, tooth row ridge formation, and keratinization of jaw sheaths followed by labial tooth keratinization. Thibadeau and Altig (1988) found that the "interspecific differences [in shape of homologous structures] result from the truncation of developmental sequences such that structures are terminally omitted rather than any kind of alterations of more proximal steps." With regard to intraspecific variation in mouthpart development, ontogenetic sequence and eventual total number of labial tooth rows appear to be phenotypically plastic. Rearing temperature and geographic origin play roles in determining the numbers of tooth rows [Hall *et al.* (1997) and references cited therein].

Once anuran larvae have commenced suspension feeding, developmental stage has a significant effect on filtering rate, ingestion rate, and retention efficiency of suspended particles. Filtering rate is the volume of water ingested per unit time, whereas ingestion rate is the number of particles captured per individual or per unit biomass over a given time period. The retention efficiency is the proportion of food particles actually removed from the ingested water. Viertel (1992) compared these measures of ingestion among premetamorphic

(Gosner stages 28 and 32) and prometamorphic (stage 40) tadpoles of five species with differing feeding modes and filtering apparatus. In general, stage 28 had the most effective ingestion followed by stages 32 and 40. This observation is interesting because the surface anatomy of the branchial food traps is not likely to change through these stages (Wassersug and Rosenberg, 1979), and the differences likely were caused by ontogenetic changes in buccal pumping rate.

In addition to the components of the filtering structures in anuran larvae, other organ systems that relate to feeding mode display ontogenetic variation. These organs develop either early or late during ontogeny, depending on the type of trophic specialization. For example, species that are benthic feeders (e.g., the Bornean pelobatid frog *Leptolalax gracilis*) show delayed development of the lungs in order to maintain negative bouyancy throughout the larval period (Nodzenski *et al.,* 1989).

Developmental changes in larval salamander feeding have been investigated in the context of ontogenetic niche shifts by comparing the behavior and diet of small and large size classes. Comparisons of gross motor movements associated with feeding (swimming, lunging, crawling, etc.) vary significantly among size classes of pond-dwelling ambystomatid larvae (Leff and Bachmann, 1986; McWilliams and Bachmann, 1989a). Because early feeding on zooplankton by very small larvae precedes full development of limbs, swimming in the water column to catch prey is the most common mode. Larger prey in the benthos are selected after larvae develop the ability to crawl on the bottom. In the laboratory, electivity for small zooplankton prey, such as *Daphnia,* appears to decline with increasing larval size whereas preference for macroinvertebrates, e.g., water boatmen, increases (Leff and Bachmann, 1986). Although salamander larvae tend to incorporate larger prey into their diets as they grow (Collins and Holomuzki, 1984; Petranka, 1984), field studies have indicated that large larvae can continue to eat large numbers of very small zooplankton in ponds (Dodson and Dodson, 1971) and insect prey in streams (Parker, 1994). Ontogenetic changes in feeding behavior are driven more by microhabitat selection than by gape limitation (McWilliams and Bachmann, 1989b). Changes in feeding behavior also may be driven by hydrodynamic constraints on swimming. Hoff *et al.* (1989) found that small larvae are capable of much higher length-specific swimming velocities than large larvae. Independent of current body size, however, feeding rate can depend on the level of prey abundance during the first and early feeding stages. Larvae that have experienced a period of early food deprivation reject fewer prey than larvae fed at a richer resource level (Maurer, 1996), as if to compensate for lost growth opportunities.

Despite these ontogenetic differences in feeding, the basic mechanisms of suction feeding are present in posthatching salamanders and are not different in later stages. Reilly (1995) filmed the feeding kinematics of *Salamandra salamandra* (which emerge from the maternal oviduct at a diversity of developmental stages) capturing *Tubifex* worms. He found that the head movements involved in initial prey capture and capture success were the same at the time of birth and just prior to metamorphosis. Because intraoviductal cannibalism is known to occur in this species, it is not clear whether these were first feedings.

2. Resource Polymorphisms

One type of ontogenetic change with very interesting ecological and evolutionary significance is trophic polymorphism, in which conditions early during larval ontogeny trigger the development of a phenotype with an alternative feeding mode. Reviews of resource polymorphisms in a diversity of vertebrates (Skúlason and Smith, 1995; Smith and Skúlason, 1996) have stressed the potential role that these polymorphisms can play in population divergence and speciation.

There are three species of amphibians displaying polymorphisms in which the variant phenotype is carnivorous or cannibalistic. For the New Mexico spadefoot toad (Pelobatidae: *Scaphiopus multiplicatus*), diet early during larval ontogeny triggers the development of the carnivorous and/or cannibalistic morph (Pfennig, 1990, 1992b). Tadpoles in ponds with high relative abundances of fairy shrimp develop the carnivorous morphology (Pfennig, 1990, 1992b), which is characterized by hypertrophied jaw musculature, fewer teeth, shorter intestines, and less pigmentation than omnivores (Orton, 1954). The proximate mechanism is that these crustacea contain thyroid hormone, T_2 (diiodotyrosine). Tadpoles given thyroxine (T_4, tetraiodotyrosine) have the same induction of the phenotype as those consuming the fairy shrimp. For tiger salamanders (Ambystomatidae: *Ambystoma tigrinum*), the cue does not directly appear to be dietary (Loeb *et al.*, 1994). Rather the cue is the density of conspecifics (Collins and Cheek, 1983) and the presence of nonsiblings (Pfennig and Collins, 1993). For the Eastern long-toed salamander, *Ambystoma macrodactylum columbianum*, diet appears to influence head shape and size but does not induce the complete suite of carnivorous traits (Walls *et al.*, 1993).

Although cannibalism can occur in some *Ambystoma* species without a specialized cannibal morph (*Ambystoma annulatum*; Nyman *et al.*, 1993), the morphological change from a typical phenotype to a specialized one can influence the mechanism/performance of feeding. For *Ambystoma tigrinum*, the main morphological differences between the typical and cannibal phenotypes are that cannibals have greater head widths and hypertrophy of the vomer bone and vomerine tooth patches (Fig. 17). Of these traits, head width alone does not significantly affect aquatic prey capture (Lauder and Shaffer, 1985; Lauder and Reilly, 1988), whereas the vomerine teeth, which project ventrally

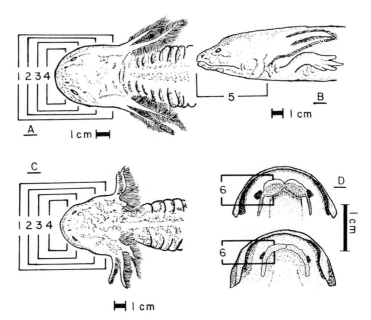

FIGURE 17   Illustrations of six characters used in a multivariate analysis of head and vomer shape in *Ambystoma tigrinum nebulosum*. (A) Dorsal view of a cannibal morph: 1 = width at jaws, 2 = width at midparietal, 3 = ocular width of head, 4 = nasal width of head. (B) Lateral view of cannibal, 5 = length of upper jaw. (C) Dorsal view of a typical larva. (D) View of vomerine ridge in cannibalistic (upper) and typical (lower) morphs; 6 = width of vomerine ridge. From Collins *et al.* (1993, Fig. 3, p. 171), with kind permission from Kluwer Academic Publishers.

from the palate and oppose the larval tongue, are important for prey manipulation performance (Reilly *et al.*, 1992). Compared to similarly sized (i.e., equivalent SVL) typical morphs, the vomerine tooth patches of cannibal morphs are about 1 mm longer, 4.5 times wider, and 2.5 times greater in area. Gape size and mass of the hyoid detractor muscles (rectus cervicis) do not differ. These are the muscles that create negative pressure in the buccal cavity during suction feeding (Lauder and Shaffer, 1985; Reilly and Lauder, 1989). In feeding trials, ability to suck in equivalent volumes of water by the two morphs was indicated by the lack of difference in capturing elusive prey such as guppies and small conspecifics. The main functional advantage that the cannibals had was that when they grasped large prey they were able to hold onto it and eventually swallow it, whereas typical morphs eventually lost prey that they held in their mouths. This advantage derives from the hypertrophied vomerine ridge and teeth, which enable them to manipulate and hold large prey that are not directly sucked in head first.

These polymorphisms are examples of intraspecific heterochrony (*sensu* Reilly *et al.*, 1997) in that the rate, onset, or offset of development diverges from the normal pattern observed within a species. Two different types of alteration of the growth pattern of typical morphs (presumed to be ancestral) contribute to the origin of the cannibal morph in tiger salamanders (Collins *et al.*, 1993). First, the wide head of cannibals appears to have evolved as a result of delayed development (Fig. 18). The relatively U-shaped head of cannibal larvae is most similar to the shape of hatchlings. Second, the hypertrophied vomerine tooth patches that characterize the cannibal phenotype display a pattern of accelerated growth relative to typical morphs. In the case of spadefoot toad tadpoles, it appears as if some traits precociously undergo transformations resembling metamorphosis, e.g., gut shortening, whereas other characters remain in their larval state, e.g., complete tailfin. In light of advances in understanding the roles of thyroid hormones and regulatory genes involved in anuran morphogenesis (Shi, 1996; Stolow *et al.*, 1997), answers to questions regarding the mechanisms and induction of these resource polymorphisms seem achievable. For example, do carnivorous spadefoot toad tadpoles digest animal protein using pepsin? Are the genes that code for production of pepsin up-regulated by the ingestion of exogenous thyroid hormone?

IX. EFFECTS OF METAMORPHOSIS ON FEEDING MECHANISMS

A. FISHES

Whereas the effects of metamorphosis on diet and dentition have been documented in fishes (e.g., Govoni, 1987; Mullaney and Gale, 1996), changes in the functional morphology of feeding due to metamorphosis remain relatively unstudied (but see Hernández, 1995).

B. AMPHIBIANS

The effects of metamorphosis on structures related to feeding are far more dramatic in the Anura than in urodeles and gymnophiones. Atrophy of the external mouthparts during metamorphosis generally follows the reverse pattern of formation but is much less ordered. Tooth rows initiate atrophy in reverse order, jaw sheaths detach when the tooth rows have almost disappeared, tooth ridges decline in reversed order of appearance, and then what remains of the oral disk atrophies (Thibadeau and Altig, 1988). The cartilaginous elements of the skull repattern and the buccal pump is lost (Wassersug and

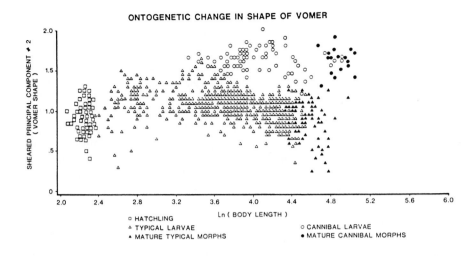

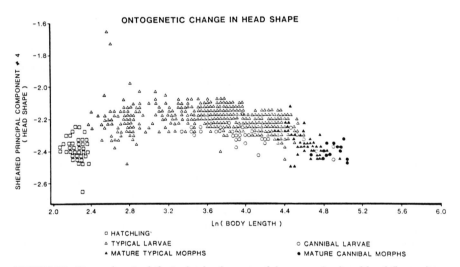

FIGURE 18 Heterochronic shifts in the development of the vomer (top) and head (bottom) in polymorphic tiger salamanders (*Ambystoma tigrinum nebulosum*). Dependent variables are principal component scores, which refer to overall shape and were derived from the morphometric variables illustrated in Fig. 17. From Collins *et al.* (1993, Fig. 4, p. 175), with kind permission from Kluwer Academic Publishers.

Hoff, 1982). Specifically, the palatoquadrate shortens and changes from a horizontal plane of orientation to a more vertical one. The degree of reorientation has a phylogenetic component, in that the anuran families recognized as more primitive have the least change at metamorphosis. Recall from the description of the buccal pump mechanism earlier (Section V.B.4.c.) that the orientation of the palatoquadrate is horizontal. This is necessary to create an up and down, rather than a back and forth, movement of the ceratohyal when the muscles, which connect the anterior palatoquadrate to the ceratohyal, contract. Meckel's cartilage also elongates to allow a larger mouth opening. As tadpoles metamorphose, the pharyngeal filtering apparatus also disappears.

The digestive tract is totally restructured as tadpoles become frogs [reviewed by Hourdry et al. (1996)]. In the foregut of the tadpole, the glandular sheath regresses and no longer protrudes from the gut wall. A large permanent stomach is formed to replace the sheath. The muscle layer of the stomach thickens and the gastric glands develop. The intestine shortens and the coils become rearranged. At a cellular level, the larval epithelium undergoes a process of programmed cell death, or apoptosis, whereas stem cells proliferate leading to the production of a new folded epithelium. All but one of the approximately 20 genes involved in apoptosis and morphogenesis of adult intestinal epithelium are up-regulated by thyroid hormone [reviewed by Stolow et al. (1997)]. These changes in the epithelium during metamorphosis make absorption of nutrients impossible. The endocrine cells of the digestive tract also are redistributed, and the production of proteolytic enzymes begins with metamorphosis. The pancreas makes trypsin and the stomach secretes pepsin and chitinases. These enzymes enable adult frogs to digest animal proteins and the exoskeletons of chitinous insects. Pepsin is absent in larvae, with the exception of the obligately carnivorous tadpole of *Lepidobatrachus laevis* (Carroll *et al.*, 1991).

Although there may be a brief period during metamorphosis when salamanders do not feed (Kuzmin, 1991), and gut length can shorten up to 25% (Tilley, 1964), the change in the alimentary canal is relatively small in salamanders compared to anurans. A pepsin-secreting stomach, which is necessary for the digestion of animal protein, already is present in larvae (Burggren and Just, 1992).

At metamorphosis salamander larvae show a decrease in prey capture success (Reilly and Lauder, 1988, 1992). Larvae lose their gills and gill slits, thereby changing the hydrodynamics of aquatic prey capture. Instead of unidirectional flow from mouth to gills, water must reverse direction and also leave through the mouth. The negative and positive pressure waveforms in the buccal cavity created by the reversal of flow are sufficient to decrease feeding performance (Lauder and Reilly, 1988). Metamorphic changes in the hyobranchial apparatus and a decrease in the size of muscles that create suction into the buccal cavity also tend to decrease performance (Lauder and Reilly, 1990).

Despite these differences, the kinematics of larval and transformed individuals capturing and transporting prey in water are similar (Shaffer and Lauder, 1988). The muscle activity patterns, revealed by electromyography, of transformed salamanders during terrestrial prey transport also appear similar to the motor pattern during larval aquatic prey transport (Reilly and Lauder, 1991).

X. EVOLUTION OF LARVAL FEEDING MECHANISMS

A. FISHES

1. Phylogenetic Patterns

At present, the information on the functional morphology of larval fishes is inadequate to identify phylogenetic patterns in feeding morphology and mechanisms.

2. Heterochrony

The role of heterochrony in the evolution of fish feeding mechanisms has not been studied extensively. The available research indicates that heterochrony may lead to phenotypic plasticity in trophic morphology and the development of trophic morphs. Meyer (1987) suggested that dietary differences and possible differences in feeding mode were responsible for differential timing of developmental events in a cichlid species, resulting in phenotypic differences in trophic morphology. Although a number of trophic polymorphisms have been reported in fishes [reviews in Skúlason and Smith (1995) and Smith and Skúlason (1996)], the ontogeny of trophic morphology is not known in most cases (Skúlason *et al.,* 1989). On the basis of laboratory-rearing experiments, Skúlason *et al.* (1989) proposed that the morphological variability among four coexisting trophic morphs of Arctic charr may result from variable regulatory genes that influence the timing and rate of morphological differentiation.

B. AMPHIBIANS

1. Phylogenetic Patterns

At the level of interfamilial comparisons among anurans, there are some general phylogenetic patterns in feeding morphology. Hylids, ranids, and bufonids, families in the Neobatrachia, have similar particle entrapment surfaces on the ventral velum. The mucus-secreting cells are arranged in pits and rows at the tops of the ridges. Taxa recognized as more primitive (Ford and Cannatella, 1993), *Ascaphus, Rhinophrynus,* Discoglossidae, and Pelobatidae, lack that organization of secretory cells (Wassersug and Rosenberg, 1979). These primitive taxa also display a smaller degree of reorientation in the angle of the cartilaginous elements of the skull associated with jaw suspension than Neobatrachians (Wassersug and Hoff, 1982).

Alternatively, evolution of morphological features associated with larval feeding in the Anura often can be characterized by convergence and parallelism. Carr and Altig (1992) compared the anatomy of the rectus abdominis muscle in 60 species representing 13 families. The rectus abdominis–rectus cervicis muscles participate in moving the ceratobranchials and thus are likely to be important in feeding. In some cases, configuration corresponded to ecomorphological guild. In the benthic feeders, the anterior terminus of r. abdominis is a straight or fan-shaped array loosely associated to the r. cervicis, whereas in macrophagous suction feeders and bromeliad inhabitants the r. abdominis is contiguous with the r. cervicis. In other cases configuration corresponded to family. In suspension feeders representing the Pipidae and Rhinophrynidae the two muscles are contiguous, but not in the microhylid suspension feeders. Within a family, there can be such a good correlation between internal larval anatomy and habitat and/or feeding mode that it is difficult to detect finer levels of phylogenetic pattern. For the Leptodactylidae, Wassersug and Heyer (1988) found it difficult to identify larval character states that clearly were a function of ancestry and helpful in resolving open systematic questions. They concluded that there is a generalized pond form and that the other forms are either elaborations or simplifications of that general plan.

For salamander families, we found no phylogenetic comparisons for larvae, but comparisons made for adults (including branchiate–neotenic taxa) that feed in water can serve as an illustrative substitute. Reilly and Lauder (1992) conducted comparisons of seven measures of cranial movements during suction feeding (gape cycle time, maximum gape angle, maximum head elevation, time to maximum head elevation, maximum hyoid depression, and time to maximum hyoid depression) among species representing six salamander families. Multivariate analyses showed that, with respect to kinematic patterns, four families grouped together (Ambystomatidae, Amphiumidae, Dicamptodontidae, and Proteidae), whereas two families (Sirenidae and Cryptobranchidae) showed divergent patterns of kinematics. However, there was no clear correspondence between interfamilial differences in morphological traits, such as gill opening number, jaw length, and head width, and the differences in kinematics. Electromyographic observations on a subset of four of the taxa revealed that variation in muscle activity contributes to the lack of concordance between morphology and kinematics (Lauder and Reilly, 1996).

2. Heterochrony

When comparisons are made between species using closely related taxa to represent presumed ancestral types, the morphological innovations that result in derived feeding modes appear to be the result of truncation or acceleration of the developmental program in larval amphibians [i.e., interspecific heterochrony *sensu* Reilly *et al.* (1997)]. Carnivory among tadpoles, for example, appears to be associated with adult-like gut morphology. The stomach of the carnivorous leptodactylid frog, *Lepidobatrachus laevis*, is characterized by a distendable fundic portion, tubular glands, and a pyloric sphincter (Ruibal and Thomas, 1988). There is no larval stage in the development of the stomach in comparison to related noncarnivorous taxa (Carroll *et al.*, 1991). The pepsinogen produced by individuals after hatching and the onset of larval feeding is electrophoretically indistinguishable from one of the isozymes produced by postmetamorphic individuals.

In addition to the evolution of derived feeding modes, transition to a nonfeeding larval form also may be achieved by heterochrony. Haas (1996) compared the development of cranial morphology of tadpoles within the neotropical Hemiphractinae (Hylidae), a monophyletic group. These are known as marsupial frogs because females brood developing eggs and larvae on their backs or in brood pouches. There are three developmental modes: (1) tadpoles emerge to live freely and feed; (2) tadpoles live freely but do not feed (*Flectonotus*); or (3) embryos develop directly and emerge as frogs. The absence of feeding structures in *Flectonotus goeldii* represents a truncation of the normal developmental sequence. The soft tissue structures necessary for filtering small particles out of the water are not developed. The branchial basket is small and shallow, branchial food traps, filter rows, and ciliary cushions are lacking, and the branchial muscles are absent or poorly developed. Several skeletal structures of the cranium also stop growing and differentiating precociously in *F. goeldii* compared to larval *Gastrotheca*, a sister taxon that includes species with free-living and feeding larvae as well as species with direct development.

For *Rhacophorus gauni*, which has both feeding and nonfeeding morphs, the morphs found with yolk-filled guts are characterized by few or no gut coils compared to those of equivalent developmental stage observed with exogenous food (sand grains and fine particles) (Inger, 1992). Yolk-filled tadpoles display an overall reduction in the size of the oral disk and have smaller and fewer numbers of labial teeth and jaw sheaths than the feeding morph. Although it is not known how this decreased development–allocation to food-gathering structures is triggered, insight may be gained from manipulations of the quantity of endogenous nutrient supply that have been performed on echinoid invertebrate larvae with the intention of measuring effects on morphogenesis. Removal of cytoplasm from ova causes changes in morphogenesis that mimic an evolutionary change in the amount of parental investment per offspring (Sinervo and McEdward, 1988). Perhaps maternal provisioning contributes to the triggering of the *Rhacophorus* morphs.

Developmental plasticity in response to nutrition may contribute to the evolutionary transition from feeding to nonfeeding larval forms. Exogenous food supplies can have morphogenetic effects on echinoid larvae similar to those caused by the level of endogenous food (Strathmann *et al.*, 1992). When sea urchin larvae are exposed to varying levels of food rations, those raised at high food levels display growth allocation to juvenile structures. In contrast, low food levels induce allocation to structures for food gathering. These morphogenetic effects caused by an increase in the exogenous nutrient supply resemble the morphogenesis of species with greater endogenous nutrient supplies and nonfeeding larvae. Because a heterochronic shift in the timing of development of juvenile structures can be induced by levels of exogenous food, this "developmental plasticity that is adaptive with exogenous provisioning of larvae could be preadaptive for changes in parental provisioning of ova" (Strathmann *et al.*, 1992, p. 984).

Among amphibians, many investigators similarly have manipulated quantity and/or quality of exogenous foods. The effects on growth and development have been interpreted mainly from an ecological perspective within a paradigm concerned with the role of competition, optimal timing of metamorphosis (Wilbur and Collins, 1973), and trade-offs between growth and predation risk (Werner, 1986), rather than the evolution of feeding mechanisms. A study on tadpole competition and phenotypic plasticity by Smith and van Buskirk (1995), however, has attempted to integrate the ecological and evolutionary perspectives. Reciprocal transplants of tadpoles among rock pools varying in the abundance of food relative to the abundance of predators and the risk of desiccation indicated that two *Pseudacris* (Hylidae) species differ with regard to which morphological traits are plastic: those devoted to feeding vs those devoted to predator avoidance. Although they did not measure internal feeding morphology, by using relative body mass as a surrogate and assuming that larger mass connotes increased size of buccal pump and increased volume of food digested, they found that interspecific differences in allocation to feeding traits could explain the species distribution among pools.

Rather than a means for deriving innovations in larval feeding mode, heterochrony in salamanders is the means by which larval feeding mode can be retained by a reproductive individual (Reilly, 1994). Some neotenic species (such as *Ambystoma talpoideum* and the axolotl) permanently delay somatic development and become reproductive while remaining completely larval with respect to the morphology and hydrodynamics of feeding. Other taxa display incomplete metamorphosis with the retention of some larval features, such as external gills (in some morphs of *Notopthalmus viridescens*) or gill openings (as in hellbenders), but have transformed feeding mechanisms in terms of bidirectional flow of water and, thus, do not retain the larval feeding performance associated with unidirectional flow of water from mouth to gills. Because unidirectional flow feeding systems are so much more effective than bidirectional flow systems, and because all suction-feeding lower vertebrates have similar biomechanical features in their skulls despite divergent morphological evolution, the extrinsic hydrodynamic constraints on skull design may limit the innovation of new modes of extracting moving prey from water (Lauder and Shaffer, 1993).

XI. FUTURE RESEARCH

Many areas of fruitful research on vertebrate larval feeding are suggested by differences in the extent of knowledge on the topics covered in this review. For example, intra- and interspecific competition, trophic polymorphisms, the effects of metamorphosis on feeding mechanisms, and the evolution of feeding mechanisms have been studied in greater depth among larval amphibians than in fishes. The effects of hydrodynamics on ontogenetic changes in foraging behavior are better understood in fishes than in salamanders. It is not clear whether such differences are driven by the natural history of the organisms or by the history of what herpetologists and ichthyologists have found interesting to study. In addition to the need for synthesis across the disciplines of development, functional morphology, and trophic ecology, there is a need for collaborative efforts that cross taxonomic boundaries.

A. FISHES

The functional morphology of larval fish feeding is a promising area of research. Are there patterns in the integration of the functional elements during ontogeny? Is larval prey capture functionally equivalent to the capture of prey by a miniature adult? To what extent do changes in the Reynolds number during ontogeny have effects on feeding mechanisms? Answers to these questions should be related to life-history strategy (e.g., oviparous or viviparous) and trophic ecology.

B. AMPHIBIANS

Another promising topic for research is the connection between larval nutrition and morphogenesis. Because expression of many of the genes involved in amphibian morphogenesis is regulated by thyroid hormone (Shi, 1996), it is reasonable to assume that consumption of foods that influence thyroid function could play a significant role. One hypothesis is that high-protein foods have a positive influence on thyroid function in amphibian larvae (Kupferberg, 1997c). Alternatively, the essential fatty acid composition of foods, which has been shown to be important for a wide variety of freshwater and marine organisms (Brett and Müller-Navarra, 1997), may determine a link between diet, endocrine function, and morphogenesis. Foods rich in essential poly- and highly unsaturated fatty acids (PUFA and HUFA) promote efficient, rapid growth and development of aquatic animals. HUFA have key regulatory roles for animal cell membrane fluidity and are precursors to many animal hormones, including prostaglandins.

Brett and Müller-Navarra (1997, p. 487) have hypothesized that "larval organisms may be more dependent on dietary HUFA than adults because their high somatic growth rates cannot be satisfied by their fatty acid conversion capacities, i.e. the ability to synthesize these compounds from related precursors." Animals are able to convert 18-carbon-chain PUFA into HUFA, but 18-carbon-chain PUFA must be obtained in the diet and are made only by plants. PUFA and HUFA are found in diatoms to a much greater degree than in algae or bacteria. Consistent with this pattern, tadpoles fed diatom-rich diets reach metamorphosis sooner and at a larger size than tadpoles fed diets of filamentous green algae, cyanobacteria, or detritus (Kupferberg et al., 1994). There is a vast literature from the aquaculture field showing that hatchery growth, survival, stress resistance, and metamorphosis of fish and invertebrate larvae can be increased either by boosting the fatty acids in artificial feeds or by rearing the initial feed, usually rotifers, on algae rich in long-chain HUFA [Sorgeloos and Leger, 1992; Tuncer et al., 1993; Ako et al., 1994; and references cited in Brett and Müller-Navarra (1997)]. The HUFA that may be important to amphibian morphogenesis are eicosapentanoic and aracadonic acids, which are precursors to eicosanoids, a class of regulatory hormones.

It already has been shown for larval fishes that dietary deficiencies of HUFA cause a range of developmental abnormalities (Kanazawa, 1993; Naess et al., 1995). Given the importance of heterochrony to the derivation of novel feeding modes, investigations of the interactions between development sequences and the foods available in the environment may provide new insights regarding the evolution of vertebrate larval feeding.

ACKNOWLEDGMENTS

We thank D. Gaff, J. Brown, M. Evans, S. Shin, and M. Stebar for assistance with literature searches and manuscript preparation; J. Thomas for figure preparation; B. Malmqvist for assistance in scanning illustrations; L. P. Hernández, K. Liem, and C. Souza for providing helpful information on larval fish feeding biology; and R. Kaplan for discussions about first feeding in Amphibia. S.J.K. thanks the Umeå–Berkeley Exchange Program, funded by the Swedish Foundation for International Cooperation in Research and Higher Education, and M. H. Wake for a postdoctoral position in the Department of Integrative Biology. The preparation of this chapter was supported in part by NSF Grant IBN-9458114 to S.L.S.

REFERENCES

Adriaens, D., and Verraes, W. (1994). On the functional significance of the loss of the interhyal during ontogeny in *Clarias gariepinus* Burchell, 1822 (Teleostei: Siluroidei). *Bel. J. Zool.* **124**, 139–155.

Ahlgren, M. O., and Bowen, S. H. (1991). Growth and survival of tadpoles (*Bufo americanus*) fed amorphous detritus derived from natural waters. *Hydrobiologia* **218**, 49–51.

Ako, H., Tamaru, C. S., Bass, P., and Lee, C. S. (1994). Enhancing the resistance to physical stress in larvae of *Mugil cephalus* by the feeding of enriched *Artemia* nauplii. *Aquaculture* **122**, 81–90.

Altig, R., and Johnston, G. F. (1989). Guilds of anuran larvae: Relationships among developmental modes, morphologies, and habitats. *Herpetol. Monogr.* **3**, 81–109.

Arens, W. (1989). Comparative functional morphology of the mouthparts of stream animals feeding on epilithic algae. *Arch. Hydrobiol., Suppl.* **83**, 253–354.

Arens, W. (1994). Striking convergence in the mouthpart evolution of stream-living algae grazers. *J. Zool. Syst. Evol. Res.* **32**, 319–343.

Balon, E. K. (1977). Early ontogeny of *Labeotropheus* Ahl, 1927 (Mbuna, Cichlidae, Lake Malawi), with a discussion on advanced protective styles in fish reproduction and development. *Environ. Biol. Fishes* **2**, 147–176.

Balon, E. K. (1980). Comparative ontogeny of charrs. In "Charrs: Salmonid Fishes of the Genus *Salvelinus*" (E. K. Balon, ed.), pp. 703–720. Dr. W. Junk Pub., The Hague, The Netherlands.

Balon, E. K. (1984). Reflections on some decisive events in the early life of fishes. *Trans. Am. Fish. Soc.* **113**, 178–185.

Balon, E. K. (1986). Types of feeding in the ontogeny of fishes and the life-history model. *Environ. Biol. Fishes* **16**, 11–24.

Battish, S. K., and Sandhu, J. S. (1988). Food spectrum of the skipper frog *Rana cyanophlyctis* Schneider. *Ann. Biol. (Ludhiana, India)* **4**, 14–19.

Batty, R. S. (1984). Development of swimming movements and musculature of larval herring (*Clupea harengus*). *J. Exp. Biol.* **110**, 217–229.

Bertram, D. F., and Leggett, W. C. (1994). Predation risk during the early life history periods of fishes: Separating the effects of size and age. *Mar. Ecol: Prog. Ser.* **109**, 105–114.

Bird, D. J., and Potter, I. C. (1979). Metamorphosis in the paired species of lampreys, *Lampetra fluviatilis* (L.) and *Lampetra planeri* (Bloch). 1. A description of the timing and stages. *Zool. J. Linn. Soc.* **65**, 127–143.

Blaustein, L., and Kotler, B. P. (1993). Oviposition habitat selection by the mosquito *Culiseta longiareolata*: Effects of conspecifics, food, and green toad tadpoles. *Ecol. Entomol.* **18**, 104–108.

Blaxter, J. H. S. (1969). Development: Eggs and larvae. In "Fish Physiology" (W. S. Hoar and D. J. Randall, eds.), Vol. 3, pp. 177–252. Academic Press, New York.

Blaxter, J. H. S. (1988). Pattern and variety in development. In "Fish Physiology" (W. S. Hoar and D. J. Randall, eds.), Vol. 11A, pp. 1–58. Academic Press, New York.

Boidron-Métairon, I. F. (1995). Larval nutrition. In "Ecology of Marine Invertebrate Larvae" (L. McEdward, ed.), pp. 223–248. CRC Press, Boca Raton, FL.

Braum, E. (1978). Ecological aspects of the survival of fish eggs, embryos and larvae. In "Ecology of Freshwater Fish Production" (S. D. Gerking, ed.), pp. 102–131. Blackwell, Oxford.

Bremigan, M. T., and Stein, R. A. (1994). Gape-dependent larval foraging and zooplankton size: Implications for fish recruitment across systems. Can. J. Fish. Aquat. Sci. 51, 913–922.

Brett, M. T., and Müller-Navarra, D. C. (1997). The role of highly unsaturated fatty acids in aquatic food-web processes. Freshwater Biol. 38, 483–499.

Britson, C. A., and Kissell, R. E., Jr. (1996). Effects of food type on developmental characteristics of an ephemeral pond-breeding anuran, Pseudacris triseriata feriarum. Herpetologica 52, 374–382.

Brockelman, W. Y. (1969). An analysis of density dependent effects and predation in Bufo americanus tadpoles. Ecology 50, 632–644.

Brönmark, C., Rundle, S. D., and Erlandsson, A. (1991). Interactions between freshwater snails and tadpoles: Competition and facilitation. Oecologia 87, 8–18.

Burggren, W. W., and Just, J. J. (1992). Developmental changes in physiological systems. In "Environmental Physiology of the Amphibians" (M. E. Feder and W. W. Burggren, eds.), pp. 467–530. University of Chicago Press, Chicago.

Caldwell, J. P. (1993). Brazil nut fruit capsules as phytotelmata: Interactions among anuran and insect larvae. Can. J. Zool. 71, 1193–1201.

Carr, K. M., and Altig, R. (1992). Configuration of the rectus abdominis muscle of anuran tadpoles. J. Morphol. 214, 351–356.

Carroll, E. J., Jr., Seneviratne, A. M., and Ruibal, R. (1991). Gastric pepsin in anuran larvae. Dev., Growth Differ. 33, 499–507.

Castle, P. H. J. (1984). Notacanthiformes and Anguilliformes: Development. In "Ontogeny and Systematics of Fishes" (H. G. Moser, W. J. Richards, D. M. Cohen, M. P. Fahay, A. W. J. Kendall, and S. L. Richardson, eds.), Am. Soc. Ichthyol. Herpetol. Spec. Publ. No. 1, pp. 62–93. Allen Press, Lawrence, KS.

Collins, J. P., and Cheek, J. E. (1993). Effect of food and density on development of typical and cannibalistic salamander larvae in Ambystoma tigrinum nebulosum. Am. Zool. 23, 77–84.

Collins, J. P., and Holomuzki, J. R. (1984). Intraspecific variation in diet within and between trophic morphs of larval tiger salamanders (Ambystoma tigrinum nebulosum). Can. J. Zool. 62, 168–174.

Collins, J. P., Zerba, K. E., and Sredl, M. J. (1993). Shaping intraspecific variation: Development, ecology, and the evolution of morphology and life history variation in tiger salamanders. Genetica 89, 167–183.

Cook, A. (1996). Ontogeny of feeding morphology and kinematics in juvenile fishes: A case study of the cottid fish Clinocottus analis. J. Exp. Biol. 199, 1961–1971.

Coughlin, D. J. (1991). Ontogeny of feeding behavior of first-feeding Atlantic salmon (Salmo salar). Can. J. Fish. Aquat. Sci. 48, 1896–1904.

Coughlin, D. J. (1994). Suction prey capture by clown-fish larvae (Amphiprion perideraion). Copeia, pp. 242–246.

Coughlin, D. J., Strickler, J. R., and Sanderson, B. (1992). Swimming and search behavior in clownfish, Amphiprion perideraion, larvae. Anim. Behav. 44, 427–440.

Crump, M. L. (1983). Opportunistic cannibalism by amphibian larvae in temporary aquatic environments. Am. Nat. 121, 281–289.

Crump, M. L. (1986). Cannibalism by younger tadpoles: Another hazard of metamorphosis. Copeia, pp. 1007–1009.

Crump, M. L. (1989). Life history consequences of feeding versus non-feeding in a facultatively non-feeding toad larva. Oecologia 78, 486–489.

Dabrowski, K. (1989). Formulation of a bioenergetic model for coregonine early life history. *Trans. Am. Fish. Soc.* **118**, 138–150.

DeBenedictus, P. A. (1974). Interspecific competition between tadpoles of *Rana pipiens* and *Rana sylvatica:* An experimental field study. *Ecol. Monogr.* **44**, 129–151.

de Ciechomski, J. D. (1967). Investigations of food and feeding habits of larvae and juveniles of the Argentine anchovy *Engraulis anchoita. CALCOFI Rep.* **11**, 72–81.

Diaz-Paniagua, C. (1985). Larval diets related to morphological characters of five anuran species in the biological reserve of Doñana (Huelva, Spain). *Amphibia-Reptilia* **6**, 307–322.

Dickman, M. (1968). The effects of grazing by tadpoles on the structure of a periphyton community. *Ecology* **49**, 1188–1190.

Dilling, L. (1989). An ontogenetic study of the jaw mechanism and feeding modes in *Amphiprion frenatus* and *A. polymnus.* BSc. Hon. Thesis, Department of Organismic and Evolution Biology, Harvard University, Cambridge, MA.

Dodd, M. H. I., and Dodd, J. M. (1976). The biology of metamorphosis. *In* "Physiology of the Amphibia" (B. A. Lofts, ed.), pp. 467–599. Academic Press, New York.

Dodson, S. L. (1970). Complementary feeding niches sustained by size-selective predation. *Limnol. Oceanogr.* **15**, 131–137.

Dodson, S. I., and Dodson, V. E. (1971). The diet of *Ambystoma tigrinum* larvae from western Colorado. *Copeia*, pp. 614–624.

Drost, M. R. (1987). Relation between aiming and catch success in larval fishes. *Can. J. Fish. Aquat. Sci.* **44**, 304–315.

Drost, M. R., and van den Boogaart, J. G. M. (1986). The energetics of feeding strikes in larval carp, *Cyprinus carpio. J. Fish Biol.* **29**, 371–379.

Drost, M. R., Muller, M., and Osse, J. W. M. (1988a). A quantitative hydrodynamical model of suction feeding in larval fishes: The role of frictional forces. *Proc. R. Soc. London, Ser. B* **234**, 263–281.

Drost, M. R., Osse, J. W. M., and Muller, M. (1988b). Prey capture by fish larvae, water flow patterns and the effect of escape movements of prey. *Neth. J. Zool.* **38**, 23–45.

Duellman, W. E., and Trueb, L. (1986). "Biology of Amphibians." McGraw-Hill, New York.

El-Fiky, N., and Wieser, W. (1988). Life styles and patterns of development of gills and muscles in larval cyprinids (Cyprinidae; Teleostei). *J. Fish Biol.* **33**, 135–145.

Ellertsen, B., Solemdal, P., Strømme, T., Tilseth, S., Westgard, T., Moksness, E., and Øiestad, V. (1980). Some biological aspects of cod larvae (*Gadus morhua* L.). *Fiskeridir. Skr. Ser. Havunders.* **17**, 29–47.

Emerson, S. B. (1988). The giant tadpole of *Pseudis paradoxa. Biol. J. Linn. Soc.* **34**, 93–104.

Emlet, R. B. (1990). World patterns of developmental mode in echinoid echinoderms. *Adv. Invertebr. Reprod.* **5**, 329–335.

Farlowe, V. (1928). Algae of ponds as determined by the examination of the intestinal contents of tadpoles. *Biol. Bull. (Woods Hole, Mass.)* **55**, 443–448.

Faurot, M. W., and White, R. G. (1994). Feeding ecology of larval fishes in Lake Roosevelt, Washington. *Northwest Sci.* **68**, 189–196.

Fauth, J. E., Resetarits, W. J., and Wilbur, H. M. (1990). Interactions between larval salamanders: A case of competitive equality. *Oikos* **58**, 91–99.

Ford, L. S., and Canatella, D. C. (1993). The major clades of frogs. *Herpetol. Monogr.* **7**, 94–117.

Fuiman, L. A. (1983). Growth gradients in fish larvae. *J. Fish Biol.* **23**, 117–123.

Fuiman, L. A., and Batty, R. S. (1997). What a drag it is getting cold: Partitioning the physical and physiological effects of temperature on fish swimming. *J. Exp. Biol.* **200**, 1745–1755.

Fuiman, L. A., and Webb, P. W. (1988). Ontogeny of routine swimming activity and performance in zebra danios (Teleostei: Cyprinidae). *Anim. Behav.* **36**, 250–261.

Galis, F. (1993). Interactions between the pharyngeal jaw apparatus, feeding behaviour, and ontogeny in the cichlid fish, *Haplochromis piceatus*: A study of morphological constraints in evolutionary ecology. *J. Exp. Zool.* **267**, 137–154.

Galis, F., Terlouw, A., and Osse, J. W. M. (1994). The relation between morphology and behavior during ontogenetic and evolutionary changes. *J. Fish Biol.* **45**, 13–26.

Gerking, S. D. (1994). "Feeding Ecology of Fish." Academic Press, San Diego, CA.

Giguere, L. (1979). An experimental test of Dodson's hypothesis that *Ambystoma* (a salamander) and *Chaoborus* (a phantom midge) have complementary feeding niches. *Can. J. Zool.* **57**, 1091–1097.

Gordon, R. E. (1966). Some observations on the biology of *Pseudotriton ruber schenki*. *J. Ohio Herpetol. Soc.* **5**, 163–164.

Gotceitas, V., Puvanendran, V., Leader, L. L., and Brown, J. A. (1996). An experimental investigation of the 'match/mismatch' hypothesis using larval Atlantic cod. *Mar. Ecol: Prog. Ser.* **130**, 29–37.

Govoni, J. J. (1987). The ontogeny of dentition in *Leiostomus xanthurus*. *Copeia*, pp. 1041–1046.

Govoni, J. J., Boehlert, G. W., and Watanabe, Y. (1986). The physiology of digestion in fish larvae. *Environ. Biol. Fishes* **16**, 59–77.

Gradwell, N. (1968). The jaw and hyoidean mechanism of the bullfrog tadpole during aqueous ventilation. *Can. J. Zool.* **46**, 1041–1052.

Gradwell, N. (1971). *Xenopus* tadpole: On the water pumping mechanism. *Herpetologica* **27**, 107–123.

Gradwell, N. (1975a). The bearing of filter feeding on the water pumping mechanism of *Xenopus* tadpoles (Anura: Pipidae). *Acta Zool.* **56**, 119–128.

Gradwell, N. (1975b). Experiments on oral suction and gill breathing in five species of Australian tadpole (Anura: Hylidae and Leptodactylidae). *J. Zool.* **177**, 81–98.

Graham, D. M., and Sprules, W. G. (1992). Size and species selection of zooplankton by larval and juvenile walleye (*Stizostedion vitreum vitreum*) in Oneida Lake, New York. *Can. J. Zool.* **70**, 2059–2067.

Gustafson, M. P. (1993). Intraguild predation among larval plethodontid salamanders: A field experiment in artificial stream pools. *Oecologia* **96**, 271–275.

Haas, A. (1996). Non-feeding and feeding tadpoles in the hemiphractine frogs: Larval head morphology, heterochrony, and systematics of *Flectonotus goeldii* (Amphibia: Anura: Hylidae). *J. Zool. Syst. Evol. Res.* **34**, 163–171.

Hairston, N. G. (1987). "Community Ecology and Salamander Guilds." Cambridge University Press, New York.

Hall, J. A., Larsen, J. H., Jr., and Fitzner, R. E. (1997). Postembryonic ontogeny of the spadefoot toad, *Scaphiopus intermontanus* (Anura: Pelobatidae): External morphology. *Herpetol. Monogr* **11**, 124–178.

Harrison, R. G. (1969). Harrison stages and description of the normal development of the spotted salamander, *Ambystoma punctatum* (Linn.). *In* "Organization and Development of the Embryo" (R. G. Harrison, ed.), pp. 44–66. Yale University Press, New Haven, CT.

Havenhand, J. N. (1995). Evolutionary ecology of larval types. *In* "Ecology of Marine Invertebrate Larvae" (L. McEdward, ed.), pp. 79–122. CRC Press, Boca Raton, FL.

Heath, M. R. (1992). Field investigations of the early life stages of marine fish. *Adv. Mar. Biol.* **28**, 1–174.

Heming, T. A., and Buddington, R. K. (1988). Yolk absorption in embryonic and larval fishes. *In* "Fish Physiology" (W. S. Hoar and D. J. Randall, eds.), Vol. 11A, pp. 407–446. Academic Press, San Diego, CA.

Hernández, L. P. (1995). The functional morphology of feeding in three ontogenetic stages of the zebrafish, *Danio rerio*. *Am. Zool.* **35**, 104A.

Heyer, W. R. (1972). Food item analysis of mixed tadpole populations from Thailand. *Herpetol. Rev.* **4**, 127.

Heyer, W. R. (1976). Studies in larval amphibian habitat partitioning. *Smithson. Contrib. Zool.* **242**, 1–27.

Hoff, K. von S., Huq, N., King, V. A., and Wassersug, R. J. (1989). The kinematics of larval salamander swimming (Ambystomatidae: Caudata). *Can. J. Zool.* **67**, 2756–2761.

Hoogenboezem, W., van den Boogaart, J. G. M., Sibbing, F. A., Lammens, E. H. R. R., Terlouw, A., and Osse, J. W. M. (1991). A new model of particle retention and branchial sieve adjustment in filter-feeding bream (*Abramis brama,* Cyprinidae). *Can J. Fish. Aquat. Sci.* **48**, 7–18.

Houde, E. D. (1973). Some recent advances and unsolved problems in the culture of marine fish larvae. *Proc. 3rd Annu. Workshop World Maricul. Soc.,* pp. 83–112.

Houde, E. D. (1987). Fish early life dynamics and recruitment variability. *Am. Fish. Soc. Symp.* **2**, 17–29.

Houde, E. D. (1994). Differences between marine and freshwater fish larvae: Implications for recruitment. *ICES J. Mar. Sci.* **51**, 91–97.

Houde, E. D., and Schekter, R. C. (1980). Feeding by marine fish larvae: Developmental and functional responses. *Environ. Biol. Fishes* **5**, 315–334.

Houde, E. D., and Zastrow, C. E. (1993). Ecosystem- and taxon-specific dynamic and energetics properties of larval fish assemblages. *Bull. Mar. Sci.* **53**, 290–335.

Hourdry, J., L'Hermite, A., and Ferrand, R. (1996). Changes in the digestive tract and feeding behavior of anuran amphibians during metamorphosis. *Physiol. Zool.* **69**, 219–251.

Hulet, W. H. (1978). Structure and functional development of the eel leptocephalus *Ariosoma balearicum* (de la Roche, 1809). *Philos. Trans. R. Soc. London, Ser. B* **282**, 107–138.

Hunter, J. R. (1980). The feeding behavior and ecology of marine fish larvae. *In* "Fish Behavior and Its Use in the Capture and Culture of Fishes" (J. E. Bardach, J. J. Magnuson, R. C. May, and J. M. Reinhart, eds.), pp. 287–330. International Center for Living Aquatic Resources Management, Manila.

Hunter, J. R. (1981). Feeding ecology and predation of marine fish larvae. *In* "Marine Fish Larvae: Morphology, Ecology and Relation to Fisheries" (R. Lasker, ed.), pp. 33–77. Washington Sea Grant Program, Seattle.

Hunt von Herbing, I., Boutilier, R. G., Miyake, T., and Hall, B. K. (1996a). Effects of temperature on morphological landmarks critical to growth and survival in larval Atlantic cod (*Gadus morhua*). *Mar. Biol.* **124**, 593–606.

Hunt von Herbing, I., Miyake, T., Hall, B. K., and Boutilier, R. G. (1996b). Ontogeny of feeding and respiration in larval Atlantic cod *Gadus morhua* (Teleostei, Gadiformes): I. Morphology. *J. Morph.* **227**, 15–35.

Hunt von Herbing, I., Miyake, T., Hall, B. K., and Boutilier, R. G. (1996c). Ontogeny of feeding and respiration in larval Atlantic cod *Gadus morhua* (Teleostei, Gadiformes): II. Function. *J. Morph.* **227**, 37–50.

Inger, R. F. (1986). Diets of tadpoles living in a Bornean rain forest. *Alytes* **5**, 153–164.

Inger, R. F. (1992). A bimodal feeding system in a stream-dwelling larva of *Rhacophorus* from Borneo. *Copeia,* pp. 887–890.

Jenkins, B., and Kitching, R. L. (1990). The ecology of water-filled tree holes in Australian rainforests: Food web reassembly as a measure of community recovery after disturbance. *Aust. J. Ecol.* **15**, 199–205.

Jenkins, G. P., Milward, N. E., and Hartwick, R. F. (1984). Food of larvae of Spanish mackerels, genus *Scomberomorus* (Teleostei: Scombridae), in shelf waters of the Great Barrier Reef. *Aust. J. Mar. Freshwater Res.* **35**, 477–482.

Jenssen, T. A. (1967). Food habits of the green frog, *Rana clamitans,* before and during metamorphosis. *Copeia,* pp. 214–218.

Johnston, T. A., and Mathias, J. A. (1994). Feeding ecology of walleye, *Stizostedion vitreum*, larvae: Effects of body size, zooplankton abundance, and zooplankton community composition. *Can. J. Fish. Aquat. Sci.* **51**, 2077–2089.

Jollie, M. (1982). What are the 'Calcichordata'? and the larger question of the origin of chordates. *Zool. J. Linn. Soc.* **75**, 167–188.

Jungfer, K.-H., and Schiesari, L. C. (1995). Description of a central Amazonian and Guianan tree frog, genus *Osteocephalus* (Anura, Hylidae), with oophagous tadpoles. *Alytes* **13**, 1–13.

Just, J. J., Kraus-Just, J., and Check, D. A. (1981). Survey of chordate metamorphosis. *In* "Metamorphosis: A Problem in Developmental Biology" (L. I. Gilbert and E. Frieden, eds.), pp. 265–326. Plenum, New York.

Kamler, E. (1992). "Early Life History of Fish: An Energetics Approach." Chapman & Hall, London.

Kanazawa, A. (1993). Nutritional mechanism involved in the occurrence of abnormal pigmentation in hatchery-reared flatfish. *J. World Aquacult. Soc.* **24**, 275–284.

Kaplan, R. H. (1980a). Ontogenetic energetics in *Ambystoma*. *Physiol. Zool.* **53**, 43–56.

Kaplan, R. H. (1980b). The implications of ovum size variability for offspring fitness and clutch size within several populations of salamanders (*Ambystoma*). *Evolution (Lawrence, Kans.)* **34**, 51–64.

Kaplan, R. H. (1992). Greater maternal investment can decrease offspring survival in the frog *Bombina orientalis*. *Ecology* **73**, 280–288.

Kemp, A. (1996). Role of epidermal cilia in development of the Australian lungfish, *Neoceratodus forsteri* (Osteichythes: Dipnoi). *J. Morph.* **228**, 203–221.

Kendall, A. W. J., and Matarese, A. C. (1994). Status of early life history descriptions of marine teleosts. *Fish. Bull.* **92**, 725–736.

Kendall, A. W. J., Ahlstrom, E. H., and Moser, H. G. (1984). Early life history stages of fishes and their characters. *In* "Ontogeny and Systematics of Fishes" (H. G. Moser, W. J. Richards, D. M. Cohen, M. P. Fahay, A. W. J. Kendall, and S. L. Richardson, eds.), *Am. Soc. Ichthyol. Herpetol.*, Spec. Publ. No. 1, pp. 11–22. Allen Press, Lawrence, KS.

Kenny, J. S. (1969). Feeding mechanisms in anuran larvae. *J. Zool.* **157**, 225–246.

Kjelson, M. A., Peters, D. S., Thayer, G. W., and Johnson, G. N. (1975). The general feeding ecology of postlarval fishes in the Newport River Estuary. *Fish. Bull.* **73**, 137–144.

Kjørsvik, E., and Reiersen, A. L. (1992). Histomorphology of the early yolk-sac larvae of the Atlantic halibut (*Hippoglossus hippoglossus* L.)—an indication of the timing of functionality. *J. Fish Biol.* **41**, 1–19.

Kjørsvik, E., van der Meeren, T., Kryvi, H., Arnfinnson, J., and Kvenseth, P. G. (1991). Early development of the digestive tract of cod larvae, *Gadus morhua* L., during start-feeding and starvation. *J. Fish Biol.* **38**, 1–15.

Kupferberg, S. J. (1997a). Facilitation of periphyton production by tadpole grazing: Functional differences between species. *Freshwater Biol.* **37**, 427–437.

Kupferberg, S. J. (1997b). Bullfrog (*Rana catesbeiana*) invasion of a California river: The role of larval competition. *Ecology* **78**, 1736–1751.

Kupferberg, S. J. (1997c). The role of diet in anuran metamorphosis. *Am. Zool.* **37**, 146–159.

Kupferberg, S. J., Marks, J. C., and Power, M. E. (1994). Effects of variation in natural algal and detrital diets on larval anuran (*Hyla regilla*) life history traits. *Copeia*, pp. 446–457.

Kusano T., and Fukuyama, K. (1989). Breeding activity of a stream-breeding frog (*Rana* sp.). *In* "Current Herpetology in East Asia" (M. Matsui, T. Hikida, and R. Goris, eds.), pp. 314–322. Herpetological Society of Japan, Kyoto.

Kuzmin, S. L. (1991). Feeding of the salamander *Ranodon sibiricus*. *Alytes* **9**, 135–143.

Langille, R. M., and Hall, B. K. (1987). Development of the head skeleton of the Japanese medaka, *Oryzias latipes* (Teleostei). *J. Morphol.* **193**, 135–158.

Lannoo, M. J., Townsend, D. S., and Wassersug, R. J. (1987). Larval life in the leaves: Arboreal tadpole types, with special attention to the morphology, ecology, and behavior of the oophagous *Osteopilus brunneus* (Hylidae) larva. *Fieldiana, Zool.,* [n.s.] **38**, 1–31.

Lauder, G. V., Jr. (1979). Feeding mechanisms in primitive teleosts and in the halecomorph fish *Amia calva. J. Zool.* **187**, 543–578.

Lauder, G. V., Jr. (1980). Evolution of the feeding mechanism in primitive actinopterygian fishes: A functional anatomical analysis of *Polypterus, Lepisosteus,* and *Amia. J. Morphol.* **163**, 283–317.

Lauder, G. V., and Liem, K. F. (1989). The role of historical factors in the evolution of complex organismal functions. *In* "Complex Organismal Functions: Integration and Evolution in Vertebrates" (D. B. Wake and G. Roth, eds.), pp. 63–78. Wiley, Chichester.

Lauder, G. V., and Reilly, S. M. (1988). Functional design of the feeding mechanism in salamanders: Causal bases of ontogenetic changes in function. *J. Exp. Biol.* **143**, 219–233.

Lauder, G. V., and Reilly, S. M. (1990). Metamorphosis of the feeding mechanism in tiger salamanders (*Ambystoma tigrinum*). *J. Zool.* **222**, 59–74.

Lauder, G. V., and Reilly, S. M. (1996). The mechanistic bases of behavioral evolution: A multivariate analysis of musculoskeletal function. *In* "Phylogenies and the Comparative Method in Animal Behavior" (E. P. Martins, ed.), pp. 104–137. Oxford University Press, New York.

Lauder, G. V., and Shaffer, H. B. (1985). Functional morphology of the feeding mechanism in aquatic ambystomatid salamanders. *J. Morphol.* **185**, 297–326.

Lauder, G. V., and Shaffer, H. B. (1993). Design of feeding systems in aquatic vertebrates: Major patterns and their evolutionary interpretations. *In* "The Skull" (J. Hanken and B. K. Hall, eds.), Vol. 3, pp. 113–149. University of Chicago Press, Chicago.

Lauder, G. V., Crompton, A. W., Gans, C., Hanken, J., Liem, K. F., Maier, W. O., Meyer, A., Presley, R., Rieppel, O. C., Roth, G., Schluter, D., and Zweers, G. A. (1989). Group report: How are feeding systems integrated and how have evolutionary innovations been introduced? *In* "Complex Organismal Functions: Integration and Evolution in Vertebrates" (D. B. Wake and G. Roth, eds.), pp. 97–115. Wiley, Chichester.

Leff, L. G., and Bachmann, M. D. (1986). Ontogenetic changes in predatory behavior of larval tiger salamanders (*Ambystoma tigrinum*). *Can. J. Zool.* **64**, 1337–1344.

Leis, J. M. (1991). The pelagic stage of reef fishes: The larval biology of coral reef fishes. *In* "The Ecology of Fishes on Coral Reefs" (P. F. Sale, ed.), pp. 183–230. Academic Press, San Diego, CA.

Letcher, B. H., Rice, J. A., Crowder, L. B., and Binkowski, F. P. (1997). Size- and species-dependent variability in consumption and growth rates of larvae and juveniles of three freshwater fishes. *Can. J. Fish. Aquat. Sci.* **54**, 405–414.

Levin, L. A., and Bridges, T. S. (1995). Pattern and diversity in reproduction and development. *In* "Ecology of Marine Invertebrate Larvae" (L. McEdward, ed.), pp. 1–48. CRC Press, Boca Raton, FL.

Licht, L. E. (1974). Survival of embryos, tadpoles, and adults of the frogs *Rana aurora aurora* and *Rana pretiosa pretiosa* sympatric in southwestern British Columbia. *Can. J. Zool.* **52**, 613–627.

Liebold, M. A., and Wilbur, H. M. (1992). Interactions between food web structure and nutrients on pond organisms. *Nature (London)* **360**, 341–343.

Liem, K. F. (1991). A functional approach to the development of the head of teleosts: Implications on constructional morphology and constraints. *In* "Constructional Morphology and Evolution" (N. Schmidt-Kittler and K. Vogel, eds.), pp. 231–249. Springer-Verlag, Berlin.

Liu, H., Wassersug, R. J., and Kawachi, K. (1996). A computational fluid dynamics study of tadpole swimming. *J. Exp. Biol.* **199**, 1245–1260.

Liu, H., Wassersug, R., and Kawachi, K. (1997). The three dimensional hydrodynamics of tadpole locomotion. *J. Exp. Biol.* **200**, 2807–2819.

Loeb, M. L. G., Collins, J. P., and Maret, T. J. (1994). The role of prey in controlling expression of a trophic polymorphism in *Ambystoma tigrinum nebulosum. Funct. Ecol.* **8**, 151–158.

Maeda, N., and Matsui, M. (1989). "Frogs and Toads of Japan." Bun-Ichi Sogo Shuppan Co, Ltd., Tokyo.

Mallatt, J. (1981). The suspension feeding mechanism of the larval lamprey Petromyzon marinus. J. Zool. 194, 103–142.

Mallatt, J. (1985). Reconstructing the life cycle and the feeding of ancestral vertebrates. In "Evolutionary Biology of Primitive Fishes" (R. E. Foreman, A. Gorbman, J. M. Dodd, and R. Olsson, eds.), pp. 59–68. Plenum Press, New York.

Mark, W., Wieser, W., and Hohenauer, C. (1989). Interactions between developmental processes, growth, and food selection in the larvae and juveniles of Rutilus rutilus (L.) (Cyprinidae). Oecologia 78, 330–337.

Maurer, E. F. (1996). Environmental variation and behavior: Resource availability during ontogeny and feeding performance in salamander larvae (Ambystoma texanum). Freshwater Biol. 35, 35–44.

May, R. C. (1974). Larval mortality in marine fishes and the critical period concept. In "The Early Life History of Fish" (J. H. S. Blaxter, ed.), pp. 3–19. Springer-Verlag, New York.

McDairmid, R. W., and Altig, R. (1990). Description of a bufonid and two hylid tadpoles from western Ecuador. Alytes 8, 51–60.

McElman, J. F., and Balon, E. K. (1979). Early ontogeny of walleye, Stizostedion vitreum, with steps of saltatory development. Environ. Biol. Fishes 4, 309–348.

McElman, J. F., and Balon, E. K. (1980). Early ontogeny of white sucker, Catostomus commersoni, with steps of saltatory development. Environ. Biol. Fishes 5, 191–224.

McWilliams, S. R., and Bachmann, M. (1989a). Predatory behavior of larval small-mouthed salamanders, Ambystoma texanum. Herpetologica 45, 459–467.

McWilliams, S. R., and Bachmann, M. (1989b). Foraging ecology and prey preference of pond-form larval small-mouthed salamanders: Ambystoma texanum. Copeia, pp. 948–961.

Meyer, A. (1987). Phenotypic plasticity and heterochrony in Cichlasoma managuense (Pisces, Cichlidae) and their implications for speciation in cichlid fishes. Evolution (Lawrence, Kans.) 41, 1357–1369.

Miller, T. J., Crowder, L. B., Rice, J. A., and Marschall, E. A. (1988). Larval size and recruitment mechanisms in fishes: Toward a conceptual framework. Can. J. Fish. Aquat. Sci. 45, 1657–1670.

Mochioka, N., Iwamizu, M., and Kanda, T. (1993). Leptocephalus eel larvae will feed in aquaria. Environ. Biol. Fishes 36, 381–384.

Moffatt, N. M. (1981). Survival and growth of northern anchovy larvae on low zooplankton densities as affected by the presence of a Chlorella bloom. Rapp. P.-V. Réun.—Cons. Int. Explor. Sci. Mer Mediterr. 178, 475–480.

Moore, J. W., and Mallatt, J. M. (1980). Feeding of larval lamprey. Can. J. Fish. Aquat. Sci. 37, 1658–1664.

Morin, P. J. (1983). Predation, competition, and the composition of larval anuran guilds. Ecol. Monogr. 53, 119–138.

Morin, P. J. (1987). Salamander predation, prey facilitation, and seasonal succession in microcrustacean communities. In "Predation: Direct and Indirect Impact on Aquatic Communities" (W. C. Kerfoot and A. Sih, eds.), pp. 174–187. University of New England Press, Hanover, NH.

Morin, P. J., Lawler, S. P., and Johnson, E. A. (1988). Competition between aquatic insects and vertebrates: Interaction strength and higher order interaction. Ecology 69, 1401–1409.

Morris, R. W. (1955). Some considerations regarding the nutrition of marine fish larvae. J. Cons., Cons. Int. Explor. Mer 20, 255–265.

Moser, H. G. (1981). Morphological and functional aspects of marine fish larvae. In "Marine Fish Larvae: Morphology, Ecology, and Relation to Fisheries" (R. Lasker, ed.), pp. 89–131. Washington Sea Grant Program, Seattle.

Mullaney, M. D. J., and Gale, L. D. (1996). Ecomorphological relationships in ontogeny: Anatomy and diet in gag, *Mycteroperca microlepis* (Pisces: Serranidae). *Copeia,* pp. 167–180.

Müller, U. K., and Videler, J. J. (1996). Inertia as a 'safe harbor': Do fish larvae increase length growth to escape viscous drag? *Rev. Fish Biol. Fish.* **6**, 353–360.

Murphy, M. L., and Hall, J. D. (1981). Varied effects of clear-cut logging on predators and their habitat in small streams of the Cascade mountains, Oregon. *Can. J. Fish. Aquat. Sci.* **38**, 137–145.

Naess, T., Germain-Henry, M., and Naas, K. E. (1995). First feeding of Atlantic halibut *(Hippoglossus hippoglossus)* using different combinations of *Artemia* and wild zooplankton. *Aquaculture* **130**, 235–250.

Nathan, J. M., and James, V. G. (1972). The role of protozoa in the nutrition of tadpoles. *Copeia,* pp. 669–679.

Noakes, D. L., and Godin, J. J. (1988). Ontogeny of behavior and concurrent developmental changes in sensory systems in teleost fishes. *In* "Fish Physiology" (W. S. Hoar and D. J. Randall, eds.), Vol. 11B, pp. 345–395. Academic Press, San Diego, CA.

Nodzenski, E., Wassersug, R. J., and Inger, R. F. (1989). Developmental differences in visceral morphology of megophryne pelobatid tadpoles in relation to their body form and mode of life. *Biol. J. Linn. Soc.* **38**, 369–388.

Norman, J. R., and Greenwood, P. H. (1975). "A History of Fishes," 3rd ed. Ernest Benn Ltd., London.

Northcutt, R. G., and Gans, C. (1983). The genesis of neural crest and epidermal placodes: A reinterpretation of vertebrate origins. *Q. Rev. Biol.* **58**, 1–28.

Norton, S. F. (1991). Capture success and diet of cottid fishes: The role of predator morphology and attack kinematics. *Ecology* **72**, 1807–1819.

Norton, S. F., and Brainerd, E. L. (1993). Convergence in the feeding mechanics of ecomorphologically similar species in the Centrarchidae and Cichlidae. *J. Exp. Biol.* **176**, 11–29.

Nyman, S., Wilkinson, R. F., and Hutcherson, J. E. (1993). Cannibalism and size relations in a cohort of larval ringed salamanders (*Ambystoma annulatum*). *J. Herpetol.* **27**, 78–84.

O'Connell, C. P. (1981). Development of organ systems in the northern anchovy, *Engraulis mordax,* and other teleosts. *Am. Zool.* **21**, 429–446.

Orton, G. L. (1954). Dimorphism in larval mouthparts in spadefoot toads of the *Scaphiopus hammondi* group. *Copeia,* pp. 97–100.

Osse, J. W. M. (1989). A functional explanation for a sequence of developmental events in the carp. The absence of gills in early larvae. *Acta Morphol. Neerl.-Scand.* **27**, 111–118.

Osse, J. W. M. (1990). Form changes in fish larvae in relation to changing demands of function. *Neth. J. Zool.* **40**, 362–385.

Osse, J. W. M., and van den Boogaart, J. G. M. (1995). Fish larvae, development, allometric growth, and the aquatic environment. *ICES J. Mar. Sci.* **201**, 21–34.

Otake, T., Nogami, K., and Maruyama, K. (1993). Dissolved and particulate organic matter as possible food sources for eel leptocephali. *Mar. Ecol.: Prog. Ser.* **92**, 27–34.

Otten, E. (1982). The development of a mouth-opening mechanism in a generalized *Haplochromis* species: *H. elegans* Trewavas 1933 (Pisces, Cichlidae). *Neth. J. Zool.* **32**, 31–48.

Otten, E. (1983). The jaw mechanism during growth of a generalized *Haplochromis* species: *H. elegans* Trewavas 1933 (Pisces, Cichlidae). *Neth. J. Zool.* **33**, 55–98.

Paine, M. D., and Balon, E. K. (1984a). Early development of the northern logperch, *Percina caprodes semifasciata,* according to the theory of saltatory ontogeny. *Environ. Biol. Fishes* **11**, 173–190.

Paine, M. D., and Balon, E. K. (1984b). Early development of the rainbow darter, *Etheostoma caeruleum,* according to the theory of saltatory ontogeny. *Environ. Biol. Fishes* **11**, 277–299.

Parker, M. S. (1993a). Size selective predation on benthic macroinvertebrates by stream-dwelling salamander larvae. *Arch. Hydrobiol.* **128**, 385–400.

Parker, M. S. (1993b). Predation by Pacific giant salamander larvae on juvenile steelhead trout. *Northwest. Nat.* **74**, 77–81.

Parker, M. S. (1994). Feeding ecology of stream-dwelling Pacific giant salamander larvae (*Dicamptodon tenebrosus*). *Copeia*, pp. 705–718.

Pepin, P. (1991). Effect of temperature and size on development, mortality, and survival rates of the pelagic early life history stages of marine fish. *Can. J. Fish. Aquat. Sci.* **48**, 503–518.

Petranka, J. W. (1984). Ontogeny of the diet and feeding behavior of *Eurycea bislineata* larvae. *J. Herpetol.* **25**, 355–357.

Petranka, J. W., Hopey, M. E., Jennings, B. T., Baird, S. D., and Boone, S. J. (1994). Breeding habitat segregation of wood frogs and American toads: The role of interspecific tadpole predation and adult choice. *Copeia*, pp. 691–697.

Pfeiler, E. (1986). Towards an explanation of the developmental strategy in leptocephalous larvae of marine teleost fishes. *Environ. Biol. Fishes* **15**, 3–13.

Pfeiler, E., and Luna, A. (1984). Changes in biochemical composition and energy utilization during metamorphosis of leptocephalous larvae of the bonefish (*Albula*). *Environ. Biol. Fishes* **10**, 243–251.

Pfennig, D. W. (1990). The adaptive significance of an environmentally cued developmental switch in an anuran tadpole. *Oecologia* **85**, 101–107.

Pfennig, D. W. (1992a). Polyphenism in spadefoot toad tadpoles as a locally adjusted evolutionary stable strategy. *Evolution (Lawrence, Kans.)* **46**, 1408–1420.

Pfennig, D. W. (1992b). Proximate and functional causes of polyphenism in an anuran tadpole. *Funct. Ecol.* **6**, 167–174.

Pfennig, D. W., and Collins, J. P. (1993). Kinship affects morphogenesis in cannibalistic salamanders. *Nature (London)* **362**, 836–838.

Pimm, S. L., and Kitching, R. L. (1987). The determinants of food chain length. *Oikos* **50**, 302–307.

Polis, G. A., and Meyers, C. A. (1985). A survey of intraspecific predation among reptiles and amphibians. *J. Herpetol.* **19**, 99–107.

Reilly, S. M. (1994). The ecological morphology of metamorphosis: Heterochrony and the evolution of feeding mechanisms in salamanders. *In* "Ecological Morphology: Integrative Organismal Biology" (P. C. Wainwright and S. M. Reilly, eds.), pp. 319–338. University of Chicago Press, Chicago.

Reilly, S. M. (1995). The ontogeny of aquatic feeding behavior in *Salamandra salamandra*: stereotypy and isometry in feeding kinematics. *J. Exp. Biol.* **198**, 701–708.

Reilly, S. M. (1996). The metamorphosis of feeding kinematics in *Salamandra salamandra* and the evolution of terrestrial feeding behavior. *J. Exp. Biol.* **199**, 1219–1227.

Reilly, S. M., and Lauder, G. V. (1988). Ontogeny of aquatic feeding performance in the eastern newt *Notophthalmus viridescens* (Salamandridae). *Copeia*, pp. 87–91.

Reilly, S. M., and Lauder, G. V. (1989). Physiological bases of feeding behavior in salamanders: Do motor patterns vary with prey type? *J. Exp. Biol.* **141**, 343–358.

Reilly, S. M., and Lauder, G. V. (1991). Prey transport in the tiger salamander: Quantitative electromyography and muscle function in tetrapods. *J. Exp. Zool.* **260**, 1–17.

Reilly, S. M., and Lauder, G. V. (1992). Morphology, behavior, and evolution: Comparative kinematics of aquatic feeding in salamanders. *Brain, Behav. Evol.* **40**, 182–196.

Reilly, S. M., Lauder, G., and Collins, J. P. (1992). Performance consequences of a trophic polymorphism: Feeding behavior in typical and cannibal phenotypes of *Ambystoma tigrinum*. *Copeia*, pp. 672–679.

Reilly, S. M., Wiley, E. O., and Meinhardt, D. J. (1997). An integrative approach to heterochrony: The distinction between interspecific and intraspecific phenomena. *Biol. J. Linn. Soc.* **60**, 119–143.

Reiss, J. (1996). Palatal metamorphosis in basal caecilians (Amphibia: Gymnophiona) as evidence for Lissamphibian monophyly. *J. Herpetol.* **30**, 27–39.

Reitan, K. I., Bolla, S., and Olsen, Y. (1994). A study of the mechanism of algal uptake in yolk-sac larvae of Atlantic halibut (*Hippoglossus hippoglossus*). *J. Fish Biol.* **44**, 303–310.

Resh, V. H., and Rosenberg, D. M. (1984). "The Ecology of Aquatic Insects." Praeger, New York.

Richard, B. A., and Wainwright, P. C. (1995). Scaling the feeding mechanism of largemouth bass (*Micropterus salmoides*): Kinematics of prey capture. *J. Exp. Biol.* **198**, 419–433.

Ruibal, R., and Thomas, E. (1988). The obligate carnivorous larvae of the frog, *Lepidobatrachus laevis* (Leptodactylidae). *Copeia*, pp. 591–604.

Sanderson, S. L., and Wassersug, R. (1993). Convergent and alternative designs for vertebrate suspension feeding. *In* "The Skull." (J. Hanken and B. K. Hall, eds.), Vol. 3, pp. 37–112. University of Chicago Press, Chicago.

Sanderson, S. L., Stebar, M. C., Ackermann, K. L., Jones, S. H., Batjakas, I. E., and Kaufman, L. (1996). Mucus entrapment of particles by a suspension-feeding tilapia (Pisces: Cichlidae). *J. Exp. Biol.* **199**, 1743–1756.

Schael, D. M., Rudstam, L. G., and Post, J. R. (1991). Gape limitation and prey selection in larval yellow perch (*Perca flavescens*), freshwater drum (*Aplodinotus grunniens*), and black crappie (*Pomoxis nigromaculatus*). *Can. J. Fish. Aquat. Sci.* **48**, 1919–1925.

Seale, D. B. (1980). Influence of amphibian larvae on primary production, nutrient flux, and competition in a pond ecosystem. *Ecology* **61**, 1531–1550.

Seale, D. B., and Beckvar, N. (1980). The comparative ability of anuran larvae (genera: *Hyla, Bufo,* and *Rana*) to ingest suspended blue-green algae. *Copeia*, pp. 495–503.

Semlitsch, R. D., Scott, D. E., and Pechmann, J. H. K. (1988). Time and size at metamorphosis related to adult fitness in *Ambystoma talpoideum*. *Ecology* **69**, 184–192.

Shaffer, H. B., and Lauder, G. V. (1988). The ontogeny of functional design: Metamorphosis of feeding behavior in the tiger salamander (*Ambystoma tigrinum*). *J. Zool.* **216**, 437–454.

Shi, Y.-B. (1996). Thyroid hormone-regulated early and late genes during amphibian metamorphosis. *In* "Metamorphosis: Post-embryonic Reprogramming of Gene Expression in Amphibian and Insect Cells" (L. I. Gilbert, J. R. Tata, and B. G. Atkinson, eds.), pp. 505–538. Academic Press, San Diego, CA.

Sinervo, B., and McEdward, L. R. (1988). Developmental consequences of an evolutionary change in egg size: An experimental test. *Evolution (Lawrence, Kans.)* **42**, 885–899.

Skelly, D. K. (1995). Competition and the distribution of spring peeper larvae. *Oecologia* **103**, 203–207.

Skúlason, S., and Smith, T. B. (1995). Resource polymorphism in vertebrates. *Trends Ecol. Evol.* **10**, 366–370.

Skúlason, S., Noakes, D. L. G., and Snorrason, S. S. (1989). Ontogeny of trophic morphology in four sympatric morphs of arctic charr *Salvelinus alpinus* in Thingvallavatn, Iceland. *Biol. J. Linn. Soc.* **38**, 281–301.

Smith, D. C. (1987). Adult recruitment in chorus frogs: Effects of size and date at metamorphosis. *Ecology* **68**, 344–350.

Smith, D. C., and van Buskirk, J. (1995). Phenotypic design, plasticity, and ecological performance in two tadpole species. *Am. Nat.* **145**, 211–233.

Smith, D. G. (1984). Elopiformes, Notacanthiformes and Anguilliformes: Relationships. *In* "Ontogeny and Systematics of Fishes" (H. G. Moser, W. J. Richards, D. M. Cohen, M. P. Fahay, A. W. J. Kendall, and S. L. Richardson, eds.), Am. Soc. Ichthyol. Herpetol., Spec. Publ. No. 1, pp. 94–102. Allen Press, Lawrence.

Smith, T. B., and Skúlason, S. (1996). Evolutionary significance of resource polymorphism in fishes, amphibians, and birds. *Annu. Rev. Ecol. Syst.* **27**, 111–133.

Sorgeloos, P., and Leger, P. (1992). Improved larviculture outputs of marine fish, shrimp, and prawn. *J. World Aquacult. Soc.* **23**, 252–264.

Sprules, W. G. (1972). Effects of size-selective predation and food competition on high altitude zooplankton communities. *Ecology* **53**, 375–386.

Stolow, M. A., Ishizuya-Oka, A., Su, Y., and Shi, Y.-B. (1997). Gene regulation by thyroid hormone during amphibian metamorphosis: Implications on the role of cell-cell and cell-extracellular matrix interactions. *Am. Zool.* **37**, 195–207.

Storre, R. (1994). Predation of cattail (*Typha latifolia*) seedlings by tadpoles. *Hortus Northwest* **5**, 1.

Strathmann, R. R. (1978). The evolution and loss of feeding larval stages of marine invertebrates. *Evolution (Lawrence, Kans.)* **32**, 907–914.

Strathmann, R. R. (1987). Larval feeding. *In* "Reproduction of Marine Invertebrates" (A. C. Giese, J. S. Pearse, and V. B. Pearse, eds.), Vol. 9, pp. 465–550. Blackwell, Palo Alto, CA.

Strathmann, R. R., Fenaux, L., and Strathmann, M. F. (1992). Heterochronic developmental plasticity in larval sea urchins and its implications for evolution of nonfeeding larvae. *Evolution (Lawrence Kans.)* **46**, 972–986.

Sutton, T. M., and Bowen, S. H. (1994). Significance of organic detritus in the diet of larval lampreys in the Great Lakes basin. *Can. J. Fish. Aquat. Sci.* **51**, 2380–2387.

Thibadeau, D. G., and Altig, R. (1988). Sequence of ontogenetic development and atrophy of the oral apparatus of six anuran tadpoles. *J. Morphol.* **197**, 63–69.

Thorisson, K. (1994). Is metamorphosis a critical interval in the early life of marine fishes? *Environ. Biol. Fishes* **40**, 23–36.

Tilley, S. G. (1964). A quantitative study of the shrinkage in the digestive tract of the tiger salamander (*Ambystoma tigrinum* Green) during metamorphosis. *J. Ohio. Herpetol. Soc.* **4**, 81–85.

Tinsley, R. C., and Tocque, K. (1995). The population dynamics of a desert anuran, *Scaphiopus couchii. Aust. J. Ecol.* **20**, 376–384.

Tuncer, H., Harrell, R. M., and Chai, T.-J. (1993). Beneficial effects of N-3 HUFA enriched *Artemia* as food for larval palmetto bass *Morone saxatilis* x *Morone chrysops. Aquaculture* **110**, 341–359.

van der Meeren, T. (1991). Algae as first food for cod larvae, *Gadus morhua* L.: Filter feeding or ingestion by accident? *J. Fish Biol.* **39**, 225–237.

Verraes, W. (1977). Postembryonic ontogeny and functional anatomy of the ligamentum mandibulo-hyoideum and the ligamentum interoperculo-mandibulare, with notes on the opercular bones and some other cranial elements in *Salmo gairdneri* Richardson, 1836 (Teleostei: Salmonidae). *J. Morphol.* **151**, 111–120.

Viertel, B. (1990). Suspension feeding of anuran larvae at low concentrations of *Chlorella* algae (Amphibia, Anura). *Oecologia* **85**, 167–177.

Viertel, B. (1991). The ontogeny of the filter apparatus of anuran larvae (Amphibia, Anura). *Zoomorphology (Berlin)* **110**, 239–266.

Viertel, B. (1992). Functional response of suspension feeding anuran larvae to different particle sizes at low concentrations (Amphibia). *Hydrobiologia* **234**, 151–173.

Vogel, S. (1994). "Life in Moving Fluids: The Physical Biology of Flow," 2nd ed. Princeton University Press, Princeton, NJ.

Wagner, W. E., Jr. (1986). Tadpoles and pollen: Observations on the feeding behavior of *Hyla regilla* larvae. *Copeia*, pp. 802–804.

Wahl, C. M., Mills, E. L., McFarland, W. N., and DeGisi, J. S. (1993). Ontogenetic changes in prey selection and visual acuity of the yellow perch, *Perca flavescens. Can. J. Fish. Aquat. Sci.* **50**, 743–749.

Wake, D. B., and Roth, G. (1989). The linkage between ontogeny and phylogeny in the evolution of complex systems. *In* "Complex Organismal Functions: Integration and Evolution in Vertebrates" (D.B. Wake and G. Roth, eds.), pp. 361–377. Wiley, Chichester.

Wake, M. H. (1980). Fetal tooth development and adult replacement in *Dermophis mexicanus* (Amphibia: Gymnophiona): Fields vs. clones. *J. Morphol.* **166**, 203–216.

Wake, M. H. (1989). Phylogenesis of direct development and viviparity in vertebrates. In "Complex Organismal Functions: Integration and Evolution in Vertebrates" (D. B. Wake and G. Roth, eds.), pp. 235–250. Wiley, Chichester.

Wake, M. H. (1993). Evolution of oviductal gestation in amphibians. *J. Exp. Zool.* **266**, 394–413.

Walls, S. C., Belanger, S. S., and Blaustein, A. R. (1993). Morphological variation in a larval salamander: dietary induction of plasticity in head shape. *Oecologia* **96**, 162–168.

Walther, B. T., and Fyhn, H. J., eds. (1993). "Physiological and Biochemical Aspects of Fish Development." University of Bergen, Norway.

Waringer-Loschenkohl, A. (1988). An experimental study of micro-habitat selection and microhabitat shifts in European tadpoles. *Amphibia-Reptilia* **9**, 219–236.

Warkentin, K. M. (1995). Adaptive plasticity in hatching age: A response to predation risk tradeoffs. *Proc. Natl. Acad. Sci. U.S.A.* **92**, 3507–3510.

Wassersug, R. J. (1972). The mechanism of ultraplanktonic entrapment in anuran larvae. *J. Morphol.* **137**, 279–288.

Wassersug, R. J. (1975). The adaptive significance of the tadpole stage with comments on the maintenance of complex life cycles in anurans. *Am. Zool.* **15**, 405–417.

Wassersug, R. J. (1989). Locomotion in amphibian larvae (or "Why aren't tadpoles built like fishes?"). *Am. Zool.* **29**, 65–84.

Wassersug, R. J., and Heyer, W. R. (1988). A survey of internal oral features of leptodactyloid larvae (Amphibia: Anura). *Smithson. Contrib. Zool.* **457**.

Wassersug, R. J., and Hoff, K. (1979). A comparative study of the buccal pumping mechanism of tadpoles. *Biol. J. Linn. Soc.* **12**, 225–259.

Wassersug, R. J., and Hoff, K. (1982). Developmental changes in the orientation of the anuran jaw suspension. *Evol. Biol.* **15**, 223–245.

Wassersug, R. J., and Hoff, K. (1985). The kinematics of swimming in anuran larvae. *J. Exp. Biol.* **119**, 1–30.

Wassersug, R. J., and Rosenberg, K. (1979). Surface anatomy of branchial food traps of tadpoles: A comparative study. *J. Morphol.* **159**, 393–426.

Wassersug, R. J., Frogner, K. J., and Inger, R. F. (1981). Adaptations for life in treeholes by rhacophorid tadpoles from Thailand. *J. Herpetol.* **15**, 41–52.

Webb, P. W., and Weihs, D. (1986). Functional locomotor morphology of early life history stages of fishes. *Trans. Am. Fish. Soc.* **115**, 115–127.

Weihs, D. (1980). Energetic significance of changes in swimming modes during growth of larval anchovy, *Engraulis mordax. Fish. Bull.* **77**, 597–604.

Werner, E. E. (1986). Amphibian metamorphosis: Growth rate, predation risk, and the optimal size at transformation. *Am. Nat.* **128**, 319–341.

Werner, E. E., and Gilliam, J. F. (1984). The ontogenetic niche and species interactions in size-structured populations. *Annu. Rev. Ecol. Syst.* **15**, 393–425.

Wilbur, H. M. (1972). Competition, predation and the structure of the *Ambystoma-Rana sylvatica* community. *Ecology* **53**, 3–21.

Wilbur, H. M. (1980). Complex life cycles. *Annu. Rev. Ecol. Syst.* **11**, 67–93.

Wilbur, H. M., and Collins, J. P. (1973). Ecological aspects of amphibian metamorphosis. *Science* **182**, 1305–1314.

Williamson, I., and Bull, C. M. (1989). Life history variation in a population of the Australian frog *Ranidella signifera:* Egg size and early development. *Copeia,* pp. 349–356.

Williamson, D. I. (1992). "Larvae and Evolution." Chapman & Hall, New York.

Youson, J. H. (1988). First metamorphosis. In "Fish Physiology" (W. S. Hoar and D. J. Randall, eds.), Vol. 11B, pp. 135–196. Academic Press, San Diego, CA.

CHAPTER 11

Phenotypic Variation in Larval Development and Evolution: Polymorphism, Polyphenism, and Developmental Reaction Norms

ERICK GREENE
Division of Biological Sciences
The University of Montana
Missoula, Montana

I. OVERVIEW

Virtually every individual varies in some ways from conspecifics at a similar life-history stage. Such variation can range from relatively small, continuous phenotypic differences to large, discrete morphological differences. In extreme

The Origin and Evolution of Larval Forms

cases this has led to conspecific individuals being classified as different species. Phenotypic polymorphism is nearly ubiquitous, having been recorded in all phyla of animals from protozoans to vertebrates (Mayr, 1963), as well as in many plants (Bradshaw, 1965).

As a spectacular example of phenotypic variation within a species, consider two individuals of the southeast Asian ant *Pheidologeton diversus* (Fig. 1). These colony mates, probably sisters, differ by about 500-fold in dry weight! At birth, these ant larvae were pluripotent, with the capacity to develop into either of these morphs, as well as any of three other female morphs described later. The developmental "decisions" that determined each ant's eventual phenotype were made during larval development and depended upon the amount and quality of food received by the larva, as well as on chemical cues that reflected social conditions in the nest.

FIGURE 1   Example of extreme size and shape variation among nest mates of the worker caste of the Asian ant *Pheidologeton diversus*. The minor worker weighs about 1/500 that of the larger ant. At hatching, larvae can develop into either caste. The subsequent developmental trajectory of an individual larva depends upon the stimuli received during its development. Photograph care of Dr. Mark W. Moffett/Minden Pictures.

Understanding the causes and consequences of such phenotypic variation integrates all levels of biological enquiry. From developmental and genetic perspectives, what are the relative roles of genetic versus environmental effects in producing phenotypic variation? In species that use environmental cues as triggers for developmental programs, how are those environmental cues detected and monitored? How are environmental signals originally "captured" in a developmental sense? How does genetic architecture orchestrate the myriad complex developmental events that produce an individual's eventual phenotype?

The significance of phenotypic variation also is of great interest to evolutionary biologists and ecologists. Is phenotypic variation maladaptive, selectively neutral, or adaptive? When and why is phenotypic variation favored? What limits phenotypic variation? What conditions favor genetic polymorphisms versus environmentally cued phenotypic plasticity? How does phenotypic plasticity influence rates and modes of speciation? On a practical and descriptive level, understanding the range of phenotypic variation within a species (and at higher taxonomic levels) is critically important for taxonomists to construct realistic phylogenies and evolutionary histories.

The purpose of this chapter is to provide a broad overview of the evolutionary causes and consequences of phenotypic variation. The focus is phenotypic variation in larval forms, especially invertebrates. However, as in the *Pheidologeton* ant example, phenotypic variants that are expressed in adults typically are the result of complex developmental events that occurred during the larval stage. Hence, a fairly general result is that phenotypic variation, expressed in any life-history stage of an organism, nearly always involves developmental events in the larval stage.

I first will provide a broad overview and comparison of different processes that give rise to phenotypic variation (genetic polymorphisms, polyphenisms, and developmental reaction norms). This historical categorization is somewhat artificial and has retarded a full understanding of the evolution of phenotypic variation: the proximate causes of phenotypic variation actually are not discrete developmental phenomena but occur along a continuum of relative importance of genetic and environmental effects. To illustrate different evolutionary and developmental phenomena, I will provide a smorgasbord of interesting case studies of phenotypic variation. The literature on phenotypic variation is enormous, and so this survey admittedly is idiosyncratic and not exhaustive. The interested reader will be steered to entry points into the large literature. Next, models that address the question of when environmentally cued phenotypic plasticity may be favored over genetic polymorphisms will be reviewed. Finally, I will review the implications of phenotypic plasticity for evolutionary processes. Although mention will be made of developmental and hormonal control of polymorphisms, this is the subject of Nijhout's chapter.

II. TAXONOMY OF TYPES OF
PHENOTYPIC VARIATION

A biological survey reveals a diversity of proximate developmental and genetic mechanisms that produce phenotypic variation. At one extreme, a phenotypic character may be determined entirely (or nearly so) by an individual's genotype. In this case, phenotypic variation in a population arises due to genetic differences among individuals and is called *genetic polymorphism*. At the other extreme, such as the *Pheidologeton* ants, an individual may have the capacity to develop along several discretely different trajectories, and its eventual phenotype depends upon external stimuli received during development (*polyphenism*). At intermediate levels, an individual's developmental trajectory may depend on both genotype and environmental conditions and is described by a *developmental reaction norm*.

The differences between genetic polymorphisms, polyphenisms, and reaction norms are illustrated schematically in Fig. 2. Each curve represents one genotype and maps its developmental response (shown as the phenotype on the y-axis) to some measure of environmental variation (shown along the x-axis). Genetic polymorphisms, illustrated by genotypes A and B, have fixed phenotypes that depend only on their genetic constitution and do not vary in

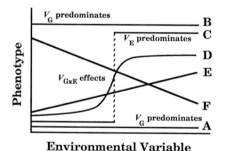

Environmental Variable

FIGURE 2   Schematic representation of the differences between genetic polymorphisms, polyphenisms, and reaction norms. Each curve represents one genotype and shows its developmental response to environmental variation. *Genetic polymorphism:* Genotypes A and B have fixed phenotypes that depend only on their genetic constitution and do not vary in their developmental responses to the environmental variation. *Polyphenism:* Genotype C shows a polyphenism, in which an individual can develop into two discretely different forms, depending upon the environmental conditions experienced during development. This developmental response shows a step function, in which one phenotype is produced below some threshold value of the environmental variable and the other is produced above the threshold value; no intermediate phenotypes are produced. *Reaction norms:* Genotypes D, E, and F show reaction norms, in which an individual's phenotype depends both on its genotype and on environmental conditions. Genotype D shows a sigmoidal reaction norm, whereas genotypes E and F show linear reaction norms.

their developmental responses to the environmental variation. Polyphenism, illustrated by curve C, shows a case in which two discrete morphs can be produced by the same genotype, depending upon the environmental conditions experienced during development. This developmental response shows a step function, in which one morph is produced below some threshold value of the environmental variable. No intermediate phenotypes are produced. Reaction norms show developmental responses to environmental variation that are intermediate between genetic polymorphisms and polyphenisms: genotype D shows a sigmoidal developmental response, whereas genotypes E and F show linear reaction norms. Genetic polymorphisms, polyphenisms, and reaction norms are discussed in more detail later.

Although the prevalence of environmentally induced phenotypic plasticity has been recognized for several centuries, it had been largely neglected in favor of studies of genetic polymorphisms. Similarly, developmental biologists emphasized those aspects of developmental processes that minimize variation among individuals (e.g., canalization) and largely ignored developmental mechanisms that promote phenotypic diversity (West-Eberhard, 1989). This historical bias is due in part to the perception that strictly genetically based sources of variation were "neater" and free of the Lamarckian overtones of environmentally induced phenotypic variation (Mayr, 1963; Shapiro, 1976). Environmental and developmental sources of phenotypic variation were considered embarrassing developmental defects. Indeed, when discussing environmentally cued phenotypic plasticity, Sir R. A. Fisher remarked to Sir V. Wigglesworth, ". . . it is not surprising that such elaborate machinery should sometimes go wrong" (Wigglesworth, 1961, quoted in West-Eberhard, 1989).

The purported distinction between genetic polymorphisms and environmentally induced phenotypic plasticity is somewhat misleading; developmental causes of phenotypic variation actually occur along a gradient of the relative importance of genotypic effects versus environmental effects. Note that it also is inaccurate to characterize environmentally induced polyphenisms as "nongenetic" phenomena: the developmental programs that respond to external, environmental cues have a genetic basis (illustrated by the *Onthophagus* and *Drosophila* examples that follow).

The continuum nature of genetic and environmental effects on phenotypic variation is made explicitly clear by the classic quantitative genetics expression for variance components in phenotypic traits:

$$V_P = V_G + V_E + V_{GE},$$

where total phenotypic variance in a population (V_P) can be partitioned into components due to genetic effects (V_G), environmental effects (V_E), and the interactions between an individual's genotype and its developmental environment (V_{GE}). Phenotypic differences among individuals that arise primarily due to genotypic differences (i.e., $V_G \neq 0$, V_E, $V_{GE} = 0$) are called genetic polymorphisms, variations that arise predominantly due to differences in environmental conditions experienced by individuals (i.e., $V_E \neq 0$, V_G, $V_{GE} = 0$) are called polyphenisms, and variation that involves both genetic and environmental effects (i.e., V_G, V_E, $V_{GE} \neq 0$) is described by reaction norms. This simple formulation explicitly acknowledges the different sources of phenotypic variation and makes it clear that the sources of phenotypic variation lie on a continuum, with different relative effects of genetic, environmental, and G × E interaction effects (Fig. 2).

Phenotypic plasticity is being resurrected as an important developmental, ecological, and evolutionary phenomenon (Shapiro, 1976; Stearns, 1989; West-Eberhard, 1989; Pennisi and Roush, 1997). Indeed, Sharloo (1989) has gone so far as to suggest that biology is emerging from the "Dark Ages of Electrophoresis" into the "Renaissance of the Phenotype!"

A. GENETIC POLYMORPHISMS

The maintenance of genetic variation was a focus of the New Synthesis. Fisher's fundamental theorem of natural selection, which states that the rate of increase of fitness of a population is proportional to the additive genetic variance in a population (Fisher, 1930), suggested an apparent paradox: these simple theoretical models suggested that directional or stabilizing selection should minimize genetic variation rather quickly, yet genetic polymorphisms and genetic variation were found to be surprisingly common in nature. It was realized that spatial and temporal variation in the environment might be critically important in maintaining genetic diversity. In particular, subsequent models investigating equilibrium levels of genetic variation have focused on the importance of local selection regimes in habitat patches versus gene flow among patches with different selection regimes. Two broad classes of genetic models have been developed to investigate the conditions required to maintain genetic polymorphisms: (1) *isolation-by-distance models,* in which patch types are very large relative to gene flow distances (Wright, 1943), and (2) *multiple-niche polymorphism models,* in which habitat patches of different selection regimes are small relative to gene flow distances (Levins, 1968). A review of these models is beyond the scope of this article, but the interested reader is referred to reviews by Hedrick *et al.* (1976), Endler (1986), Hedrick (1986), and Mitchell-Olds (1992).

Classic cases of genetic polymorphisms involve color polymorphisms in which various morphs are cryptic against different backgrounds or microhabitats, such as different color morphs of larvae of the citrus swallowtail *Papilio demodocus* (Clarke *et al.,* 1963) and the pine looper *Bupalus piniarius* (den

Boer, 1971), and industrial melanism in the pepper moth *Biston betularia* and many other species of Lepidoptera (Kettlewell, 1973). It is thought that selection by predators may be important generally in maintaining such color polymorphisms: when predators consume disproportionately fewer of the uncommon morphs (due to search images or other mechanisms), then apostatic selection (frequency-dependent selection) will produce an equilibrium frequency of morphs proportional to their degree of crypsis (Clarke and O'Donald, 1964; Endler, 1978).

1. Walking Sticks

The walking stick *Timema cristinae* (Phasmatodeae: Timemida) is a newly described species that occurs in the chaparral habitat in southern California and eats two shrubs, chamise (*Adenostoma fasciculatum*) or ceanothus (*Ceanothus spinosus*). Simple genetic polymorphisms control body color (three color morphs: green, red, or gray) and pattern (two pattern morphs expressed only in the green morph: striped or unstriped) (Sandoval, 1994a). The vulnerability of the different morphs was examined on different substrates with birds and lizards as predators: strong differential predation indicates that the different morphs are most cryptic on the host plant on which they normally occur (Sandoval 1994a). Morph frequencies in a particular patch of chaparral habitat also are influenced by the balance between strong differential predation on the different morphs and gene flow between patches: the frequency of a morph was high on its preferred host plant when the patch of that food plant was very large, isolated from other patches, or much larger than other adjacent patches (Sandoval, 1994b).

2. Color Polymorphisms in Lepidoptera

Other studies of color polymorphisms in Lepidoptera have revealed that some cases that had been considered to be strict genetic polymorphisms have environmentally induced effects as well. For example, poplar hawkmoth (*Laothoe populi*) caterpillars occur in genetically determined green or red-spotted morphs. However, if a caterpillar is not green, it can develop into a whitish morph if raised on poplar foliage or into a yellowish form if raised on willow foliage. The reflective properties of the leaves trigger the development of these different color morphs (Grayson and Edmunds, 1989). Studies of *Eumorpha* hawkmoth caterpillars also have revealed that the food plants eaten by the larvae influence their color (Fink, 1995), and pupal color morphs in the butterfly *Danaus chrysippus* have environmental components (Smith *et al.*, 1988).

   Color changes in insects involve many different families of pigments, such as ommochromes, melanins, pterins, bile pigments, carotenoids, and quinones.

The chemistry and genetics of insect color and color polymorphisms have been reviewed by Fuzeau-Braesch (1985).

B. POLYPHENISMS

Mayr (1963) defined the term polyphenism to distinguish environmentally induced phenotypic variation from genetic polymorphisms: "the occurrence of several phenotypes in a population, the differences between which are not the result of genetic differences." The term polyphenism embraces a large number of phenomena, including morphological, physiological, and behavioral variation, and the literature is enormous. The reader is referred to Shapiro's (1976) classic review of seasonal polyphenisms. In the following sections, I highlight some interesting examples of polyphenisms.

1. A Diet-Induced Polyphenism in a Caterpillar

The geometrid moth *Nemoria arizonaria* is a bivoltine species, with distinct spring and summer broods of caterpillars. Caterpillars of the spring brood, which hatch when oaks are in flower, develop into superb mimics of oak catkins (Fig. 3, left). Caterpillars of the summer brood emerge after the catkins have fallen. If they developed into catkin morphs, they would be very conspicuous. The summer brood of caterpillars develops instead into superb mimics of oak twigs (Fig. 3, right). In addition to the striking differences in external skin morphology, these two morphs also develop very different jaw structures and hiding behaviors (Greene, 1989).

Rearing experiments showed that only larval diet induces these different developmental responses. Regardless of the photoperiod, temperature, or wavelength of light (Greene, 1989, 1996), all caterpillars that were raised on catkins developed into catkin mimics, whereas all that were raised on leaves developed into twig mimics. Juvenile hormones appear to be involved in mediating the developmental switch: experimental increases in juvenile hormone titres (with application of exogenous methoprene, a synthetic form of juvenile hormone) to caterpillars feeding on catkins completely overrides the dietary cues, and those caterpillars develop into perfect twig morphs. Application of a juvenile hormone block to caterpillars feeding on leaves induces the development of the catkin morphology. The developmental commitment to either morph is made during a brief sensitive period early in larval development. Developmental commitment to either morph is irreversible, and hormone manipulations and diet-switching either before or after the sensitive phase do not affect morph development (E. Greene, unpublished observation).

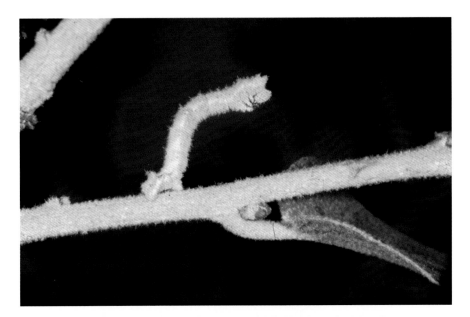

FIGURE 3 A diet-induced seasonal polyphenism in the geometrid caterpillar *Nemoria arizonaria*. The species is bivoltine, with distinct early spring and summer generations. Caterpillars that eat oak catkins develop into superb mimics of the catkins (top), whereas those that eat oak leaves develop into mimics of oak twigs (bottom). These two caterpillars were full sibs; the only difference in rearing conditions was that the caterpillar on the top was reared on oak pollen, whereas the caterpillar on the bottom was reared on oak leaves. Reprinted with permission from Greene, E. (1989). A diet-induced developmental polymorphism in a caterpillar. *Science* **243**, 643–646. Copyright 1989 American Association for the Advancement of Science.

FIGURE 8 Example of a sequential polyphenism, in which an individual passes through different forms during ontogeny. The geometrid caterpillar *Cochisea rigidaria* resembles juniper foliage during its first three larval instars (top); after its final larval moult, it resembles a scaly juniper twig (middle: this is the same caterpillar as that shown in the top panel). A closely related species (bottom) also passes through a foliage-mimicking stage and a scaly-twig-mimicking stage and then develops into a smooth twig form. Photographs by Erick Greene.

2. Alary Polyphenisms

Many insects, such as water striders (Brinkhurst, 1959) and plant hoppers, exhibit wing dimorphism: larvae can develop into either long-winged, flying forms (macropterous morph, Fig. 4, top), or short-winged, flightless forms (brachypterous morph, Fig. 4, bottom).

Wing polyphenisms have been especially well-studied in the plant hopper *Prokelisia marginata* (Denno *et al.,* 1985). This plant hopper feeds on the marsh grass *Spartina alterniflora* along the Atlantic coast, and more than 80% of adults develop into the macropterous form. Plant hoppers spend the winter as nymphs in persistent coastal stands of *Spartina.* During the spring, the nymphs feed on the coastal *Spartina,* and densities can increase rapidly to over 1000 plant hoppers/m^2 (Denno *et al.,* 1985). Simultaneously, the nutritional quality of *Spartina* sap decreases during the summer (Denno *et al.,* 1985). While the quality of the coastal *Spartina* marsh grass deteriorates, patches of

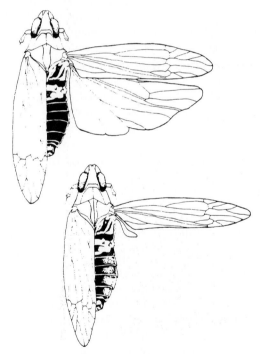

FIGURE 4   The plant hopper *Prokelisia marginata* shows wing dimorphism typical of many insects. The larvae develop into either long-winged macropter (top), which is an excellent flier and colonizer, or short-winged, flightless brachypter (bottom). Reprinted with permission from Denno *et al.* (1985).

high-quality marsh grass grow along inland streams. Development of the
winged forms allows individuals to track the seasonally varying food resource,
migrating between the persistent coastal marsh grass and temporary streamside
patches. Both crowding and host plant nutritional quality directly influence
the fitness of plant hoppers and are used as developmental cues that trigger
the production of appropriate wing forms: crowding during the nymphal stage
induces the development of the winged form in females, but not in males, in
which large percentages of winged males develop even when raised in isolation.
High nutritional quality of marsh grass suppresses the development of wings
(Denno *et al.*, 1986).

3. Caste Systems in Social Insects

Ants, social wasps and bees, and termites can develop different castes. The
two *Pheidologeton* ants shown in Fig. 1 illustrate the spectacular size variation
possible in such caste systems. Caste systems allow efficient division of labor
within colonies (Hölldobler and Wilson, 1990). The small minor worker (Fig.
1, top) conducted many of the routine colony tasks, such as tending the
developing brood, foraging for insects and plant material, and defending the
nest. The major worker (Fig. 1, bottom), acted as a "bulldozer" by packing
the soil with its massive head to create trail systems and hoisting and chewing
large objects. In addition to the two castes shown in this photograph, three
other distinct female phenotypes are produced: medias help the minors with
foraging and trail construction; repletes act as storage vessels by storing food
in their swollen abdomens and sharing it later with other colony workers; and
queen ants produce eggs.

Although different environmental cues trigger caste differentiation in social
insects, all of the underlying developmental programs appear to be mediated
by juvenile hormones (Wheeler, 1986; Nijhout's Chapter 7 in this volume).

4. Seasonal Polyphenism in Butterfly Wing Patterns

Many species of butterflies exhibit different wing patterns during different
times of the year and tend to have darker, more melanized wings during cooler
periods than individuals that emerge during hotter periods (Shapiro, 1976;
Nijhout, 1991). The developmental commitment to wing pattern and melanization is made during larval development, with larvae exposed to longer rearing
photoperiods developing paler wings as adults, and appears to be mediated
by ecdysteroids (Shapiro, 1976; Nijhout's Chapter 7 in this volume).

It has long been assumed that variation in melanization patterns was adaptive, with alternative phenotypes experiencing the highest fitness during the
season in which it normally occurs. A set of ingenious experiments with *Pontia*

butterflies has demonstrated the adaptive nature of phenotypic variation in wing pattern. The darker wings of the early spring forms allow butterflies to heat up more rapidly, whereas the paler wings of the summer forms reduce radiative heating (Kingsolver, 1987). Mark-recapture studies of different wing phenotypes indicated seasonally fluctuating selection on melanization on the base of the dorsal hindwings; darker individuals survived better in the cooler spring, whereas lighter individuals survived better during hot periods (Kingsolver, 1995a). Releases of captive-breed butterflies (Shapiro, 1976; Kingsolver, 1995a,b,c), and butterflies with experimentally altered wings (Kingsolver, 1996) showed that males with darker wings suffered higher mortality rates during hot periods.

5. Phenotypic Plasticity in Aphids

Aphid life cycles can be extremely complex, involving seasonal successions of different morphological forms, seasonal shifts to different host plants, and seasonal shifts between asexual and sexual reproduction (cyclical parthenogenesis).

The complexity of some aphid life cycles is stunning, and the cast of phenotypically distinct characters rivals the complexity of a Dostoevsky novel. This is illustrated by the life cycle of the host-alternating aphid *Aphis fabae* (Fig. 5). The fertilized eggs spend the winter on the primary host, the tree *Evonymus europaeus*. The eggs give rise to wingless females called "stem mothers" or *fundatrices*. A *fundatrix* parthenogenetically produces wingless female offspring called *fundatrigeniae*. Later in the summer, winged migratory females, called *emigrant alatae* develop and then disperse to the secondary summer host plants (mainly herbaceous plants related to beans). The *emigrant alatae* reproduce asexually, producing wingless female *alienicolae*. As these colonies become crowded later in the summer, winged female *alatae viviparae* develop, and then disperse and found new colonies on unoccupied secondary host plants. Later in the summer or early fall, winged female *alate gynoparae* (or *sexuparae*) and winged males migrate back to the primary host plant. The females are viviparous and parthenogenetic and produce wingless sexual females called *oviparae*. The *oviparae* mate with males and lay overwintering eggs on the primary host, and the cycle starts again.

The life cycle of *Aphis fabae* is typical of heteroecious, host-alternating aphid life cycles. Aphid life cycles can be modified in many ways, giving rise to a huge number of life cycle possibilities (Moran, 1988, 1991, 1994).

Phenotypic variation among seasonal forms of aphids can be pronounced, as illustrated by the normal summer larvae and the overwintering morph of the aphid *Periphyllus grandulatus* (Fig. 6). The spring and summer forms have smooth, globular bodies. During the fall, larvae develop into the overwintering

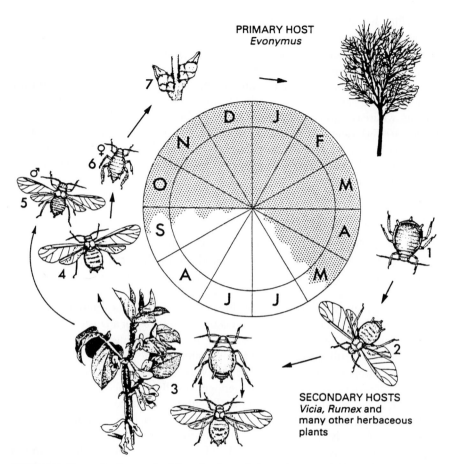

FIGURE 5 Example of a complex life cycle of an aphid, involving seasonal changes in morphology, mode of reproduction, and alternation of host plants. In species such as *Aphis fabae* shown here, the *fundatrix* (1) founds a colony on new growth of the primary host *Evonymus* during the spring. These parthenogentically produced offspring give rise to alatae (2) that emigrate to colonize various secondary host plants, such as beans. As the summer colonies become crowded, *alate viviparae* are produced and these disperse to form new colonies. In response to decreasing day length, *alate gynoparae* (4) and winged males (5) are produced that migrate back to the primary host. The progeny of the *alate gynoparae* are wingless sexual females called *oviparae* (6), which mate with males and lay overwintering eggs (7) in the bud axils of the primary host. Reprinted with permission from Blackman and Eastop (1994).

form: their bodies are flattened, have a thick, waxy coating, and have peripheral hairs that are flattened and veined. These larvae enter a long diapause period, during which the space between their body and appendages is closed over by the fringe of flattened hairs.

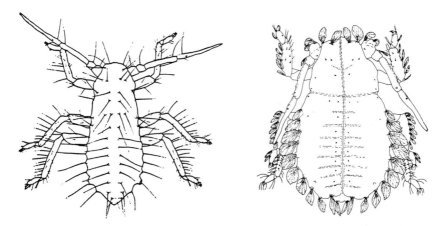

FIGURE 6   Seasonal morphs of the aphid *Periphyllus granulatus*. (A) The normal spring and summer morph. (B) The overwintering morph. Reprinted from Hille Ris Lambers (1966). With permission, from The Annual Review of Entomology, Volume II, © 1966, by Annual Reviews Inc.

The developmental control of an aphid's phenotype is determined during embryonic development and can be influenced by environmental factors such as photoperiod, temperature, humidity, crowding, nutritional quality of sap, water content of the plant tissue, and complex interactions among these factors (Hardie and Lees, 1985). The endocrine system of aphids is similar to that of other hemipterans, and the role of juvenile hormones and ecdysteroids has been investigated most intensively in the control of wing production and sexuality (sexual versus parthenogenetic morphs). The results have been somewhat equivocal [reviewed by Hardie and Lees (1985)], probably because of the large number of environmental cues to which aphids respond, and because of the very early developmental commitment to various pathways (parthenogenetically reproducing aphids can contain many "telescoping" generations of offspring). However, it is clear that juvenile hormones and other hormones control the complex life cycles of aphids.

6. Phenotypic Plasticity in Amphibian Larvae

Many amphibians have aquatic larvae and terrestrial adult forms. For some species of anurans, breeding habitats can be unpredictable and ephemeral. For example, spadefoot toads that live in desert habitats depend upon puddles that form during monsoon rains. In some years, the rains fail and breeding does not occur, in other years puddles are small and dry up rapidly, and in still other years of heavy rains, ponds are more substantial and stable. In addition to high variance in pond duration, the density of predators, food, and conspecifics also is highly variable and unpredictable.

Mexican spadefoot toads (*Scaphiopus multiplicatus*) breed in ephemeral ponds in the southwestern United States. The tadpoles can develop into two morphologically distinct, environmentally induced morphs (Pfennig, 1990): carnivorous morphs have large jaw musculature, short intestines, eat fairy shrimp, grow rapidly, and store little fat (Fig. 7, top), whereas smaller omnivorous morphs have reduced jaw musculature, long intestines, eat detritus, grow slowly, and store more fat (Fig. 7, bottom). Development of the carnivorous morph is induced facultatively when tadpoles eat shrimp (Pfennig, 1990), but tadpoles can revert to the omnivore morph if their diet is switched (Pfennig, 1992a). Carnivores have a higher probability of metamorphosis than omnivores in short-lived ponds, because of their higher growth rates. However, although they grow more slowly, omnivores accumulate more fat reserves, and they survive better in longer lived ponds because of higher postmetamorphic survival rates. Because of competition for food, the performance of a particular morph depends upon the frequency of that morph in the pond: manipulations of morph frequencies in experimental ponds showed that competition for food among the more common morphs made the rarer morph more successful (Pfennig, 1992b). This flexible and reversible developmental program is adaptive, in that it maximizes an individual's chances of successfully metamorphosing in the face of unpredictable larval environments.

Many other species of anuran larvae show plasticity in developmental traits, such as age and size at metamorphosis and growth rates (Newman, 1992; Smith and Skúlason, 1996). In some cases, such developmental plasticity

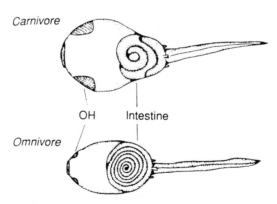

FIGURE 7   Two morphs of tadpoles of the New Mexico spadefoot toad (*Scaphiopus multiplicatus*). The carnivorous tadpoles (top) eat mainly anostracan shrimp, whereas the morphologically distinct omnivores (bottom) eat detritus. Carnivorous larvae have an enlarged orbitohyoideus muscle attached to their jaws that allows them to consume shrimp, as well as a shortened intestine. The two environmentally induced morphs typically occur in the same pond. Reprinted with permission from Pfennig (1992b).

appears to be adaptive. For example, *Bufo americanus* tadpoles metamorphose sooner and at smaller sizes when predators, *Notophthalmus* newts, are in the pond (Wilbur and Fauth, 1990). In other cases, plasticity may not be adaptive: low growth rates of tadpoles growing in crowded ponds may simply reflect the constraint of lower food abundance (Newman, 1992).

7. Sequential or Ontogenetic Polyphenism

All of the previous examples are what could be called *alternative polyphenisms* (Nijhout's Chapter 7 in this volume): a developmental fork commits an individual to one of two different developmental pathways. In many cases, commitment to one trajectory is irreversible (e.g., catkin or twig morphs in *Nemoria* caterpillars; Greene, 1989), whereas in some cases the development of a particular phenotype is reversible (e.g., carnivorous and omnivorous morphs of *Scaphiopus* tadpoles; Pfennig, 1992a). At a very basic level, metamorphosis can be considered a *sequential* or *ontogenetic polyphenism* (Nijhout's Chapter 7 in this volume): during development, an individual passes through different morphological stages, some of which can be extremely different (e.g., caterpillar, pupa, butterfly). Although metamorphosis and polyphenism generally are not thought to be related, Nijhout (Chapter 7 in this volume) points out that in insects these phenomenologically diverse developmental events are extremely similar in their regulatory features. Developmental switches are mediated by the same endocrine mechanisms (juvenile hormones and ecdysteroids), and developmental switches for both are triggered by some token environmental cue. The evolution of developmental events associated with alternative polyphenisms through modification of metamorphic machinery has been poorly studied and would be an extremely fruitful area of research.

The geometrid caterpillar *Cochisea rigidaria* (Fig. 8) provides an interesting example of a sequential color polyphenism that occurs during metamorphosis. These caterpillars eat the foliage of alligator-barked juniper (*Juniperus deppeana*). During the course of their larval life they resemble different microsites within the host plant. During the first several instars, the caterpillar resembles the fine, distal juniper foliage (Fig. 8, left). However, as the caterpillar grows longer and fatter, it is no longer as well-hidden among the thin juniper needles. After the fourth instar molt, the caterpillar's skin resembles the scaly, brown juniper twigs (Fig. 8, middle). These quantum shifts in morphology are accompanied by corresponding changes in behavior: the foliage mimic hides in the distal foliage, and the twig mimic hides among scaly twigs. Another closely related geometrid larva (Fig. 8, right) also passes through foliage mimic and scaly twig mimic forms, but it grows much larger and is no longer well-hidden among the scaly twigs. In its last larval instar, this caterpillar resembles the smooth, pinkish juniper stems found farther down in the juniper trees.

C. REACTION NORMS

Reaction norms explicitly describe the effects of both genotypic and environmental effects on an individual's phenotype. First coined by Woltereck (1909)
while working with seasonal morphological changes in *Daphnia* (Fig. 9), reaction norms are maps of the developmental response (i.e., the phenotype) of
genotypes as a function of the environment. Dobzhansky astutely pointed out
the importance of developmental reaction norms: "What counts in evolution
are the phenotypes. The genes act through the developmental patterns which
the organism shows in each environment. What changes in evolution is the
norm of reaction of the organism to the environment" (1951, quoted in
Stearns, 1989).

Reaction norms are important for genetics and life-history evolution. Reaction norms can be viewed as a developmental mirror that transforms environmental variation into phenotypic variation. The reaction norms of two hypo-

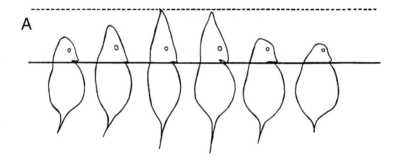

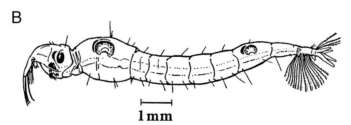

1 mm

FIGURE 9   (A) Example of predator-induced cyclomorphosis in the cladoceran *Daphnia*, in
which the size and shape of the helmet and tail spine vary over the course of the season. The
individuals shown were collected from July (left) through January (right). (B) Aquatic larva of
the midge *Chaoborus americanus*, which is a voracious predator of *Daphnia*. Reprinted with
permission from Dodson (1989), © 1989 American Institute of Biological Sciences.

thetical genotypes in three different environments are represented in Fig. 10. Relatively steep developmental reaction norms will spread out phenotypic variation resulting from environmental variation (Fig. 10, genotype 1), whereas flatter developmental responses will compress phenotypic variation (Fig. 10, genotype 2). Such crossing reaction norms, representing a genetic diversity of developmental responses to environmental conditions, appear to be virtually ubiquitous in sexually reproducing organisms (e.g., Fig. 11; Gebhardt and Stearns, 1993). The extent of crossing of reaction norms in a population is measured in quantitative genetics as the genotype by environment interaction.

The shapes and slopes of reaction norms have two important consequences for the distribution of phenotypic traits and are important in understanding how traits and correlated traits respond to selection (Hillesheim and Stearns, 1991; Stearns, 1989; van Noordwijk, 1989). First, they determine whether or not heritable variation in a trait can be detected in a particular environment. For example, in the region where the reaction norms cross in environment 2 (Fig. 10), genotypes 1 and 2 cannot be distinguished on the basis of their phenotypes. Thus, heritability for the trait measured in environment 2 would be zero. In environments 1 and 3, however, genotypes 1 and 2 have distinctly different nonoverlapping phenotypes; heritability of the trait would be signifi-

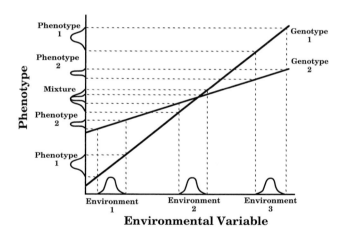

FIGURE 10   Reaction norms convert environmental variation into phenotypic variation. Two hypothetical lines show the reaction norms for two different genotypes. Crossing reaction norms produce genotypes that are distinguishable by their phenotypes in environments 1 and 3. Note that the phenotypic rankings of the genotypes reverse in environments 1 and 3 and that heritabilities measured in environments 1 and 3 would be significant. Where the reaction norms cross in environment 2, the genotypes cannot be separated on the basis of their phenotypes and, thus, heritability measured in environment 2 would be zero. After Stearns (1989), © 1989 American Institute of Biological Sciences.

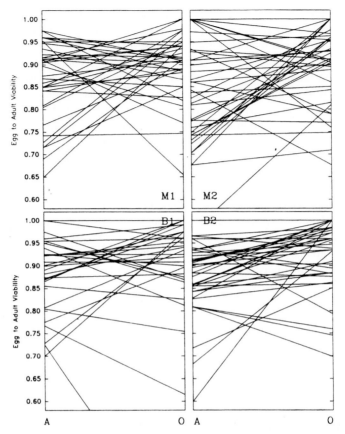

FIGURE 11 Example of reaction norms for mainland populations (M1 and M2) and Baja populations (B1 and B2) of *Drosophila mojavensis*. These plots show how egg to adult viability varied for half-sib families reared on two different larval foods: fermenting tissue of agria cactus (A) or organ pipe cactus (B). These plots show substantial effects of genotype, environment, and genotype by environment interactions. Reprinted with permission from Etges (1993).

cantly nonzero when measured in environments 1 and 3. Second, reaction norms influence the relative rankings of phenotypes, and crossing reaction norms reverse the relative rankings of the genotypes in the extreme environments (Stearns, 1989). For example, genotype 1 has more extreme values of the trait in environment 3, whereas genotype 2 has more extreme values in environment 1. Such crossing reaction norms would influence the results of selection in different environments. If selection favored more extreme values of the trait, genotype 1 would be favored in environment 3, whereas genotype

2 would be at a selective advantage in environment 1; selection would be neutral with respect to genotype in environment 2.

1. Cyclomorphosis

Many organisms show striking phenotypic changes over the course of a season, involving size, shape, and mode of reproduction. Cyclomorphosis is especially common in aquatic and intertidal habitats and occurs in such diverse taxa as algae, rotifers, ciliates, bryozoans, cladocerans, snails, and barnacles (Dodson, 1989). Many of the changes involve the seasonal production of large spines, helmets, spikes, or other protuberances ("exuberant" phenotypes). Figure 9A shows a typical example of cyclomorphosis in the cladoceran *Daphnia:* the size and shape of the helmet and tail spine vary over the course of a season. Helmet and spines are induced by the presence of predators, such as the predaceous dipteran larvae *Chaoborus* (Fig. 9B).

It generally is thought that cyclomorphosis is driven by a trade-off between predation risk and growth and reproduction. By having inducible defenses, individuals can invest resources into growth or reproduction during periods of low predator densities and divert some of these resources into defensive structures only when needed (Dodson, 1989; Zaret, 1969, 1972). Other work has shown that there is substantial genetic variance in *Daphnia* for the induced life-history cost (delayed maturation): in some genotypes, the life-history cost of the induced helmet and spine is great, whereas in other genotypes it is much less (Spitze, 1992).

A large number of environmental cues have been implicated as triggering cyclomorphic changes, including dietary control (Gilbert, 1966; Gilbert and Thompson, 1968), turbulence, temperature (Hazelwood, 1966), and specific organic substances correlated with richness of food supply (α-tocopherol; Gilbert and Thompson, 1968; Gilbert, 1973a,b).

2. Horned Beetles

Many species of male beetles have "horns," analogous to the antlers or horns of some ungulates. In some species of horned beetles, variation in this secondary sexual trait is not a continuous but a dimorphic trait, with males possessing relatively long horns, relatively short horns, or lacking horns (Figs. 12 and 13). Dimorphism in horn length results from nonlinear sigmoidal allometric relationships between body size and horn length (Fig. 14A): the steep slope around the inflection point of the allometric relationship separates males into short-horned or long-horned forms.

Especially thorough studies of this fascinating system have been conducted on *Onthophagus acuminatus,* a beetle that rears its young on the dung of howler

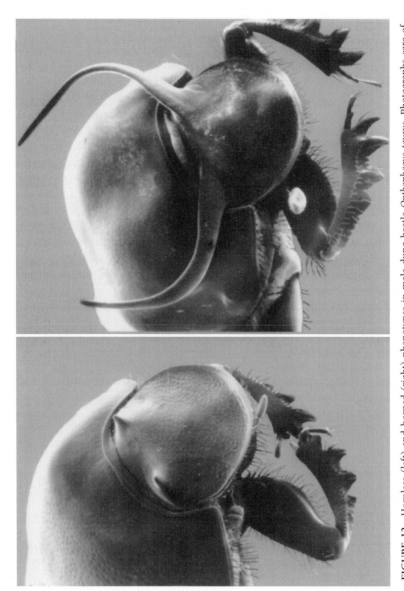

FIGURE 12 Hornless (left) and horned (right) phenotypes in male dung beetle *Onthophagus taurus*. Photographs care of Douglas Emlen [photo of horned beetle reprinted with permission from Nijhout and Emlen (1998), © 1998 National Academy of Sciences, USA].

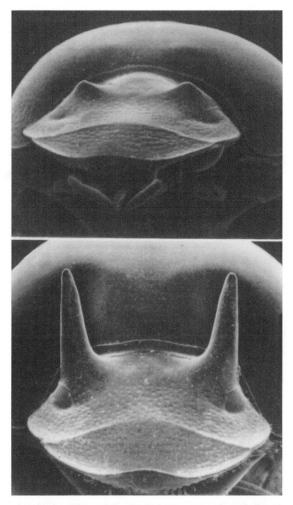

FIGURE 13 Hornless (top) and horned (bottom) phenotypes in the male dung beetle *Onthophagus acuminatus*. Photographs care of Douglas Emlen [reprinted with permission from Figure 1, p. 568, of Emlen (1997), © Royal Society of London.]

monkeys in tropical lowland forests (Emlen, 1994, 1996). Beetles fly to fresh dung shortly after it falls to the ground. Females excavate tunnels under the dung, pack round masses of dung into "brood balls," place a brood ball into an underground chamber, and lay one egg on the brood ball. After the egg hatches, the developing larva consumes its brood ball and metamorphoses in the chamber. After eclosion, the newly hatched adult digs to the surface and the cycle begins again.

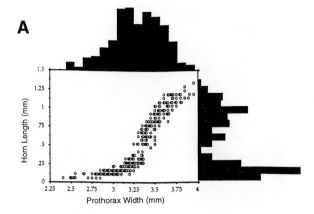

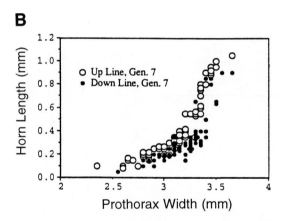

FIGURE 14 (A) Sigmoidal allometric relationship between horn length and body size in male dung beetles *Onthophagus acuminatus* from Panama. The sigmoidal allometry results in a bimodal distribution of male horn lengths (right margin). (B) The developmental switch determining horn production is made during larval development, and the location of the reaction norm can be shifted laterally by artificial selection. Beetle lines were selected for increased relative horn length (open circles) and decreased relative horn length (closed circles) for seven generations. Reprinted with permission from Emlen (1996).

Males appear to use two different behavioral–morphological strategies to compete with other males for access to females. Large males attempt to guard a female's tunnel. Larger horns allow guarding males to more successfully defend the tunnels from other males attempting to climb down. Small males are unable to compete in direct fights with larger horned males, but they can successfully reproduce by digging side tunnels that join the tunnel complex below the guarding male. There seems to be disruptive selection against males with intermediate-sized horns, because they are unable to compete effectively in fights with larger horned males and they are encumbered and slower in the sneaking strategy.

There is no measurable heritable genetic variation for horn length (i.e., the horn lengths of sons are not influenced by the horn morphology of their fathers; Emlen, 1994). Rather, body size and horn size of males depend upon the nutritional conditions experienced by the developing larvae. Because each larva completes its development on the single brood ball, the size and nutritional quality of the dung ball completely determines its final size. Although there is no heritable variation for horn length per se, the shape of the allometric relationship that relates body size and horn length has a genetic component: after seven generations of artificial selection for relatively long horns or relatively short horns (and a control selection line), Emlen (1996) produced lines of beetles with relatively long horns or relatively short horns for their body size (Fig. 14B). This selection experiment did not change the shape of the sigmoidal allometric relationship, but rather shifted the position of the inflection either to the right (short-horned selection lines) or to the left (long-horned selection lines).

These studies are illuminating for several reasons. First, *Onthophagus* beetles are an example of how complex developmental processes and the resulting phenotypes are integrated across life-history stages of an individual: although horns are expressed and used only during the adult life-history stage, the developmental commitment to horn morphology is made during the larval stage. Activation of horn development proceeds only if a larva has surpassed a threshold of body fat prior to metamorphosis. In addition, development of horns is costly and necessitates trade-offs with other adjacent body parts that also are growing during development: the lines of beetles selected for relatively large horns and experimental manipulation of relative horn size with juvenile hormones showed that males with large horns that develop near eye disks have proportionally smaller eyes (Nijhout and Emlen, 1998). This suggests that the allocation of resources during development in this closed system is a zero-sum game and, thus, limits the range of developmental possibilities of beetles. This example also shows that even though horn morphology is not heritable, the developmental reaction norm has a genetic basis and can be shifted rapidly by selection.

3. Models of Evolution of Phenotypic Plasticity

Although it is a relatively straightforward process to plot reaction norms, the extent to which reaction norms are simply the byproduct of selection on trait means in different environments is unclear (Via and Lande, 1985, 1987; Via,

1987, 1993a,b). The alternative view is that reaction norms themselves are the target of selection (Gavrilets and Scheiner, 1993; Scheiner, 1993a). According to the first view, reaction norms are epiphenomena of different selection regimes in different environments, and there is no need to invoke genes that control the shape of reaction norms (Via, 1993b). Others argue that there are environmentally dependent regulatory genes (i.e. plasticity genes) that, in essence control the shape of reaction norms (Scheiner, 1993b; Schlichting and Pigliucci, 1993).

Theoretical models of the evolution of phenotypic plasticity fall into two general classes of models: (1) "character state" models that arise from quantitative genetics approaches, in which reaction norms are described as a set of phenotypic means expressed by a genotype in different environments (Via and Lande, 1985; Van Tienderen, 1991; Gomulkiewicz and Kirkpatrick, 1992) and (2) "polynomial" models, in which reaction norms are described as polynomial functions of an environmental variable. Although De Jong (1995) has shown that in some cases character state and polynomial models are mathematically equivalent, it is unclear how much of the debate about reaction norms and the evolution of phenotypic plasticity is due to the different mathematical formulations used in different models. The literature on models of phenotypic plasticity is technically difficult, but nice overviews of the important conceptual issues and comparison of models can be found in Arnold (1992), Via *et al.* (1995), and Gotthard and Nylin (1995).

III. EVOLUTIONARY MAINTENANCE OF PHENOTYPIC VARIATION

How organisms respond to fluctuating environments in an evolutionary sense depends, in part, on the length of their life relative to the periodicity of the variation. For large, long-lived organisms, environmental variability may be best handled by the ability to adjust physiologically or behaviorally in response to changing conditions (e.g., migratory behavior in response to seasonally changing habitats, seasonal changes in fur thickness, molting in birds). Species whose life span is closer to the characteristic time span of environmental variation may experience natural selection that adjusts gene frequencies associated with morphs, resulting in genetic polymorphisms. In cyclical or seasonally varying environments, however, there is a time lag between the selective environment and the population genetic structure responding to the selective events. In addition, average fitness may be balanced by a substantial genetic load across generations (Levins, 1968). The magnitudes of the genetic load and the time lag depend upon the difference between an organism's generation time and the length of the environmental periodicity. The worst possible fit to changing environmental conditions would be a species with two generations per year occurring in two different selective environments (e.g., a spring environment and a winter environment): in such a system, the population genetic structure would be perpetually out of phase with the selective environment. It is thought that such situations could favor the evolution of phenotypic plasticity (polyphenisms and reaction norms) rather than genetic polymorphisms.

Moran (1992) developed a nongenetic optimality model to investigate the general conditions that would favor the evolutionary maintenance of a developmental switch and polyphenism over two monophenic developmental strategies (i.e., genetic polymorphism). These general models suggest that the most important factors favoring adaptive phenotypic plasticity are (i) the accuracy with which current environmental cues can be used to predict a future selective environment; (ii) the degree to which environmental variation is temporal versus spatial; (iii) the relative frequencies of different selective environments; and (iv) the relative fitness benefits of alternative phenotypes in different environments. Moran (1992) models a situation with two selective environmental states and three different morphological strategies: two alternative monophenic strategies (i.e., a genetic polymorphism, in which an individual can develop into only one of two morphs, depending on their genotype) or a polyphenic strategy, in which an individual can develop into either morph. It is assumed that individuals experience only one environment during their life, with the probability of encountering that environment equal to the frequency of that environment. First of all, the maintenance of phenotypic diversity requires a fitness trade-off: if one phenotype has the highest fitness in both environments, then that phenotype will be favored over the other monophenic strategy and the polyphenic strategy.

Figure 15 shows the combinations of conditions in Moran's model under which the polyphenic strategy is favored over monophenic genetic polymorphism for temporal environmental variation. This model shows that the polyphenic strategy may be favored under a broad range of conditions, especially when the correlation between present and future environmental conditions is high (e.g., decreasing day length is a good predictor that winter is approaching), both environmental types occur in intermediate frequencies, or the asymmetries in the fitness advantages of different phenotypes in their "right" and "wrong" environments are large (i.e., large z values). Furthermore, Moran's model suggests that polyphenism is more difficult to maintain when environmental variation is purely spatial.

These are precisely the conditions typically experienced by the organisms in which polyphenisms are most common, such as short-lived, multivoltine, or bivoltine animals in seasonal environments.

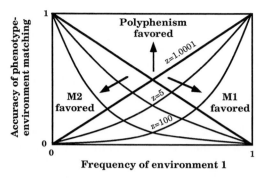

FIGURE 15 Moran's model (1992) showing the combination of conditions under which polyphenism is favored over monomorphic strategies for temporally varying environments. This particular model assumes that only two discrete environmental states exist (environments 1 and 2); a polyphenic strategy allows individual organisms to develop into one of two morphs, whereas two monophenic strategies allow individuals to develop only one phenotype (M1 or M2, respectively). The phenotype M1 is assumed to be better suited to environment 1, whereas phenotype M2 does best in environment 2. The parameter z gives a measure of the fitness advantage of the "right" over the "wrong" phenotype in either environment. These general models show that the polyphenic developmental strategy is advantageous over the monophenic strategies under a broad range of conditions, especially when the the accuracy of predicting future environments from present conditions is high, when the fitness cost of being in the "wrong" environment is high (i.e., high z value), and when the different selective environments occur in roughly the same proportion. After Moran (1992), with permission.

IV. IMPLICATIONS FOR SPECIATION AND EVOLUTION

Because evolution generally is defined as changes in gene frequencies over time, it has been argued that phenotypic variants produced by flexible developmental responses to environmental conditions are outside the domain of evolutionary change. This view is flawed for several reasons. First, environmental conditions, by influencing developmental processes, determine the range of expressed phenotypes that are exposed to natural selection. Stated differently, developmental processes may act as filters by selectively exposing genotypes to natural selection. Second, developmental characteristics of phenotypic plasticity may be heritable and respond quickly to selection (Bradshaw, 1965; Gupta and Lewontin, 1982; Stearns, 1983; Emlen, 1996).

West-Eberhard (1986, 1989) has argued that phenotypic plasticity may play an important and underappreciated role in evolution. She proposes that phenotypic plasticity may facilitate and accelerate the evolutionary origin of novelty, the rates and mode of speciation, and macroevolutionary patterns. Once a developmental switch arises, allowing the facultative expression of traits, selection can act independently upon different forms. This uncoupling of selection on alternate forms may allow morphological divergence of stable alternative phenotypes prior to reproductive isolation and speciation. Such bifurcation of developmental programs and uncoupling of different phenotypes may allow extensive phenotypic divergence of different forms, which may promote sympatric speciation (West-Eberhard, 1989).

Possible examples of plasticity leading to speciation include sympatric host shifts and timing of emergence or diapause among herbivorous insects. The most persuasive cases of host race formation have been studied in herbivorous sawfly larvae (Knerer and Atwood, 1973) and apple maggot flies (*Rhagoletis pomonella;* Feder *et al.,* 1990a,b). These fruit flies oviposited on native haw-thorns, but also have developed the ability to use domestic apples during the last 150 years. The seasonal availability of ripe fruits suitable for larvae differs, because domestic apples ripen before hawthorn fruits. Timing of eclosion is important for adults to match reproduction to availability of suitable host fruits. Genetic analyses have revealed complex patterns of sympatric interhost divergence superimposed on north–south clinical patterns (Feder *et al.,* 1990a,b). Such partial allochronic isolation may allow sympatric speciation. The general importance of host use plasticity promoting speciation remains unknown. For example, host specialization by polyphytophagous fall canker-worm caterpillars (*Alsophila pometaria*) does not seem to favor sympatric incipient speciation (Futuyma *et al.,* 1984; Futuyma and Philippi, 1987).

Such a speciation process may lead to "phenotypic fixation" (West-Eberhard, 1989), when one morph of a polyphenic species may be lost: such a process would produce species complexes in which one species resembles one morph of another closely related polyphenic or polymorphic species. Candidates for examples of speciation arising by phenotype fixation may found in closely related species with different life cycles. Many taxa of aphids contain species in which one species shows host alternation in its life cycle whereas a closely related species only occurs on one host (Moran and Whitham, 1988). Similarly, 13-year periodical cicadas may have evolved from a polyphenic cicada capable of switching between a 13- or 17-year life cycle (Martin and Simon, 1988). There is evidence that some monophenic, univoltine pierid butterflies evolved from multivoltine and polyphenic species (Shapiro, 1971).

In addition, development of phenotypic plasticity may interact with other selective processes, such as sexual selection premating isolation mechanisms (Liou and Price, 1994) or selection acting on spatially structured or geographi-cally separated populations (Thompson, 1994), to synergistically increase spe-ciation rates. Explosive speciation in African cichlids and Hawaiian *Drosophila* (Dominey, 1984) may have arisen by an interaction of polyphenism and sex-ual selection.

V. CONCLUSIONS

Phenotypic variation within a species is a nearly universal phenomenon. We are now beginning to appreciate the wide diversity of developmental and genetic pathways that produce phenotypic variation, as well as the diversity of ecological reasons favoring variation. Although many examples of phenotypic plasticity have been well-described, and in some cases the endocrine and genetic mechanisms underlying them have been elucidated, overall we have a very incomplete understanding of phenotypic evolution.

Understanding the causes and consequences of phenotypic variation is emerging as one of the most exciting fields in biology. For a full understanding of phenotypic variation, this emerging discipline of "evo–devo" requires an integration across all levels of biological organization (Pennisi and Roush, 1997; Arnold, 1992). Some of the outstanding questions are the following: How do developmental events, and the timing of those events, influence the range of possible phenotypes? How do developmental trade-offs in the allocation of resources influence the universe of possible phenotypes and developmental outcomes? How do developmental processes impose genetic constraints? Developmental events typically precede, by a considerable time, the selective environments for which they are required. How does this uncoupling of developmental events and selective environment come about? How do ecological interactions between organisms and their environments influence phenotypic evolution? How do patterns of phenotypic variation influences patterns and modes of speciation and macroevolutionary patterns (West-Eberhard, 1989)? These issues will be at the forefront of developmental and evolutionary biology for some time to come.

REFERENCES

Arnold, S. J. (1992). Constraints on phenotypic evolution. *Am. Nat.* **140**, S84–S107.

Blackman, R. L., and Eastop, V. F. (1994). "Aphids of the World's Trees: An Identification and Information Guide." CAB International in association with the Natural History Museum, Cambridge, UK.

Bradshaw, A. D. (1965). Evolutionary significance of phenotypic plasticity in plants. *Adv. Genet.* **13**, 115–155.

Brinkhurst, R. O. (1959). Alary polymorphism in the Gerroidea (Hemiptera–Heteroptera). *J. Anim. Ecol.* **28**, 211–230.

Clarke, B. C., and O'Donald, P. (1964). Frequency-dependent selection. *Heredity* **19**, 201–206.

Clarke, B. C., Dickson, G. C., and Sheppard, P. M. (1963). Larval color pattern in *Papilio demodocus*. *Evolution (Lawrence, Kans.)* **17**, 130–137.

de Jong, G. (1995). Phenotypic plasticity as a product of selection in a variable environment. *Am. Nat.* **145**, 493–512.

den Boer, M. H. (1971). A colour polymorphism in caterpillars of *Bupalus piniarius* (L.) (Lepidoptera: Geometridae). *Neth. J. Zool.* **21**, 61–116.

Denno, R. F., Douglass, L. W., and Jacobs, D. (1985). Crowding and host plant nutrition: Environmental determinants of wing-form in *Prokelisia marginata. Ecology* **66**, 1588–1596.

Denno, R. F., Douglass, L. W., and Jacobs, D. (1986). Effects of crowding and host plant nutrition on a wing-dimorphic planthopper. *Ecology* **67**, 116–123.

Dobzhansky, T. (1951). "Genetics and the Origin of Species." Columbia University Press, New York.

Dodson, S. (1989). Predator-induced reaction norms. *Bioscience* **39**, 447–452.

Dominey, W. J. (1984). Effects of sexual selection and life history on speciation: Species flocks in African Cichlids and Hawaiian *Drosophila. In* "Evolution of Fish Species Flocks" (A. A. Echelle and I. Kornfield, eds.), pp. 231–249. University of Maine Press, Orono.

Emlen, D. J. (1994). Environmental control of horn length dimorphism in the beetle *Onthophagus acuminatus* (Coleoptera: Scarabaeidae). *Proc. R. Soc. London, Ses. B* **256**, 131–136.

Emlen, D. J. (1996). Artificial selection on horn length-body size allometry in the horned beetle *Onthophagus acuminatus* (Coleoptera: Scarabaeidae). *Evolution (Lawrence, Kans.)* **50**, 1219–1230.

Emlen, D. J. (1997). Diet alters male horn allometry in the beetle *Onthophagus acuminatus* (Coleoptera: Scarabaeidae). *Proc. R. Soc. London, Ses. B* **264**, 567–574.

Endler, J. A. (1978). A predator's view of animal color patterns. *Evol. Biol.* **11**, 319–364.

Endler, J. A. (1986). "Natural Selection in the Wild." Princeton University Press, Princeton, NJ.

Etges, W. J. (1993). Genetics of host-cactus response and life-history evolution among ancestral and derived populations of cactophilic *Drosophila mojavensis. Evolution (Lawrence, Kans.)* **47**, 750–767.

Feder, J. L., Chilcote, C. A., and Bush, G. L. (1990a). The geographic pattern of genetic differentiation between host associated populations of *Rhagoletis pomonella* (Diptera: Tephritidae) in the eastern United States and Canada. *Evolution (Lawrence, Kans.)* **44**, 570–594.

Feder, J. L., Chilcote, C. A., and Bush, G. L. (1990b). Regional, local and microgeographic allele frequency variation between apple and hawthorn populations of *Rhagoletis pomonella* in western Michigan. *Evolution (Lawrence, Kans.)* **44**, 595–608.

Fink, L. S. (1995). Foodplant effects on colour morphs of *Eumorpha fasciata* caterpillars (Lepidoptera: Sphingidae). *Biol. J. Linn. Soc.* **56**, 423–437.

Fisher, R. A. (1930). "The Genetical Theory of Natural Selection." Clarendon Press, Oxford.

Futuyma, D. J., and Philippi, T. E. (1987). Genetic variation and covariation in responses to host plants by *Alsophila pometaria* (Lepidoptera: Geometridae). *Evolution (Lawrence, Kans.)* **41**, 269–279.

Futuyma, D. J., Cort, R. P., and van Noordwijk, I. (1984). Adaptation to host plants in the fall cankerworm (*Alsophila pometaria*) and its bearing on the evolution of host affiliation in phytophagous insects. *Am. Nat.* **123**, 287–296.

Fuzeau-Braesch, S. (1985). Colour changes. *In* "Comprehensive Insect Physiology, Biochemistry, and Pharmacology" (G. A. Kerkut and L. I. Gilbert, eds.), Vol. 9, pp. 549–589. Pergamon, New York.

Gavrilets, S., and Scheiner, S. M. (1993). The genetics of phenotypic plasticity. V. Evolution of reaction norm shape. *J. Evol. Biol.* **6**, 31–48.

Gebhardt, M. D., and Stearns, S. C. (1993). Phenotypic plasticity for life history traits in *Drosophila melanogaster.* I. Effect on phenotypic and environmental correlations. *J. Evol. Biol.* **6**, 1–16.

Gilbert, J. J. (1966). Rotifer ecology and embryological induction. *Science* **151**, 1234–1237.

Gilbert, J. J. (1973a). The adaptive significance of polymorphism in the rotifer *Asplanchna.* Humps in males and females. *Oecologia* **13**, 135–146.

Gilbert, J. J. (1973b). Induction and ecological significance of gigantism in the rotifer *Asplanchna sieboldi. Science* **181**, 63–66.

Gilbert, J. J., and Thompson, G. A., Jr. (1968). Alpha tocopherol control of sexuality and polymorphism in the rotifer *Asplanchna. Science* **159**, 734–736.

408 Erick Greene

Gomulkiewicz, R., and Kirkpatrick, M. (1992). Quantitative genetics and the evolution of reaction norms. *Evolution (Lawrence, Kans.)* **46**, 390–411.

Gotthard, K., and Nylin, S. (1995). Adaptive plasticity and plasticity as an adaptation: A selective review of plasticity in animal morphology and life history. *Oikos* **74**, 3–17.

Grayson, J., and Edmunds, M. (1989). The causes of colour and colour change in caterpillars of the poplar and eyed hawkmoths (*Laothoe populi* and *Smerinthus ocellata*). *Biol. J. Linn. Soc.* **37**, 263–279.

Greene, E. (1989). A diet-induced developmental polymorphism in a caterpillar. *Science* **243**, 643–646.

Greene, E. (1996). Effect of light quality and larval diet on morph induction in the polymorphic caterpillar *Nemoria arizonaria* (Lepidoptera: Geometridae). *Biol. J. Linn. Soc.* **58**, 277–285.

Gupta, A. P., and Lewontin, R. C. (1982). A study of reaction norms in natural populations of *Drosophila pseudoobscura*. *Evolution (Lawrence, Kans.)* **36**, 934–948.

Hardie, J., and Lees, A. D. (1985). Endocrine control of polymorphism and polyphenism. *In* "Comprehensive Insect Physiology, Biochemistry and Pharmacology" (G. A. Kerkut and L. I. Gilbert, eds.), Vol. 8, pp. 441–490. Pergamon, New York.

Hazelwood, D. H. (1966). Illumination and turbulence effects of relative growth in *Daphnia*. *Limnol. Oceanogr.* **11**, 212–216.

Hedrick, P. W. (1986). Genetic polymorphisms in a heterogeneous environment: A decade later. *Annu. Rev. Ecol. Syst.* **17**, 535–566.

Hedrick, P. W., Ginevan, M. E., and Ewing, E. P. (1976). Genetic polymorphisms in heterogeneous environments. *Annu. Rev. Ecol. Syst.* **7**, 1–32.

Hille Ris Lambers, D. (1966). Polymorphism in Aphididae. *Annu. Rev. Entomol.* **11**, 47–78.

Hillesheim, E., and Stearns, S. C. (1991). The responses of *Drosophila melanogaster* to artificial selection on body weight and its phenotypic plasticity in two larval food environments. *Evolution (Lawrence, Kans.)* **45**, 1909–1923.

Hölldobler, B., and Wilson, E. O. (1990). "The Ants." Belknap Press, Cambridge, MA.

Kettlewell, H. B. D. (1973). "The Evolution of Melanism: A Study of Recurring Necessity." Oxford University Press, Oxford.

Kingsolver, J. (1987). Evolution and coadaptation of thermoregulatory behavior and wing pigmentation pattern in pierid butterflies. *Evolution (Lawrence, Kans.)* **41**, 427–490.

Kingsolver, J. (1995a). Estimating selection on quantitative traits using capture–recapture data. *Evolution (Lawrence, Kans.)* **49**, 384–388.

Kingsolver, J. (1995b). Viability selection on seasonally polyphenic traits: Wing melanin pattern in western white butterflies. *Evolution (Lawrence, Kans.)* **49**, 932–941.

Kingsolver, J. (1995c). Fitness consequences of seasonal polyphenism in western white butterflies. *Evolution (Lawrence, Kans.)* **49**, 942–954.

Kingsolver, J. (1996). Experimental manipulation of wing pigment pattern and survival in western white butterflies. *Am. Nat.* **146**, 296–306.

Knerer, G., and Atwood, C. E. (1973). Diprionid sawflies: Polymorphism and speciation. *Science* **179**, 1090–1099.

Levins, R. (1968). "Evolution in Changing Environments." Princeton University Press, Princeton, NJ.

Liou, L. W., and Price, T. D. (1994). Speciation by reinforcement of premating isolation. *Evolution (Lawrence, Kans.)* **48**, 1451–1459.

Martin, A. P., and Simon, C. (1988). Anomalous distribution of nuclear and mitochondrial DNA markers in periodical cicadas. *Nature (London)* **336**, 237–239.

Mayr, E. (1963). "Animal Species and Evolution." Harvard University Press, Cambridge, MA.

Mitchell-Olds, T. (1992). Does environmental variation maintain genetic variation? A question of scale. *Trends Ecol. Evol.* **7**, 397–398.

Moran, N. A. (1988). The evolution of host-plant alternation in aphids: Evidence for specialization as a dead end. *Am. Nat.* **132**, 681–706.

Moran, N. A. (1991). Phenotype fixation and genotypic diversity in the complex life cycle of the aphid *Pemphigus betae*. *Evolution* (*Lawrence, Kans.*) **45**, 957–970.

Moran, N. A. (1992). The evolutionary maintenance of alternative phenotypes. *Am. Nat.* **139**, 971–989.

Moran, N. A. (1994). Adaptation and constraint in the complex life cycles of animals. *Annu. Rev. Ecol. Syst.* **25**, 573–600.

Moran, N. A., and Whitham, T. G. (1988). Evolutionary reduction of complex life cycles: Loss of host-alternation in *Pemphigus* (Homoptera: Aphididae). *Evolution* (*Lawrence, Kans.*) **42**, 717–728.

Newman, R. A. (1992). Adaptive plasticity in amphibian metamorphosis. *BioScience* **42**, 671–678.

Nijhout, H. F. (1991). "The Development and Evolution of Butterfly Wing Patterns." Smithsonian Institution Press, Washington, DC.

Nijhout, H. F., and Emlen, D. J. (1998). Competition in body parts in the development and evolution of insect morphology. *Proc. Natl. Acad. Sci.* **95**, 3685–3689.

Pennisi, E., and Roush, W. (1997). Developing a new view of evolution. *Science* **227**, 34–37.

Pfennig, D. W. (1990). The adaptive significance of an environmentally-cued developmental switch in an anuran tadpole. *Oecologia* **85**, 101–107.

Pfennig, D. W. (1992a). Proximate and functional causes of polyphenism in an anuran tadpole. *Func. Ecol.* **6**, 167–174.

Pfennig, D. W. (1992b). Polyphenism in spadefoot toad tadpoles as a locally adjusted evolutionarily stable strategy. *Evolution* (*Lawrence, Kans.*) **46**, 1408–1420.

Sandoval, C. P. (1994a). Differential visual predation on morphs of *Timema cristinae* (Phasmatodeae: Timemidae) and its consequences for host range. *Biol. J. Linn. Soc.* **52**, 341–356.

Sandoval, C. P. (1994b). The effects of the relative geographic scales of gene flow and selection on morph frequencies in the walking-stick *Timema cristinae*. *Evolution* (*Lawrence, Kans.*) **48**, 1866–1879.

Scharloo, W. (1989). Developmental and physiological aspects of reaction norms. *BioScience* **39**, 465–471.

Scheiner, S. M. (1993a). Genetics and evolution of phenotypic plasticity. *Annu. Rev. Ecol. Syst.* **24**, 35–68.

Scheiner, S. M. (1993b). Plasticity as a selectable trait: Reply to Via. *Am. Nat.* **142**, 371–373.

Schlichting, C. D., and Pigliucci, M. (1993). Control of phenotypic plasticity via regulatory genes. *Am. Nat.* **142**, 366–370.

Shapiro, A. M. (1971). Occurrence of a latent polyphenism in *Pieris virginiensis* (Lepidoptera: Pieridae). *Entomol. News* **82**, 13–16.

Shapiro, A. M. (1976). Seasonal polyphenism. *Evol. Biol.* **9**, 259–333.

Smith, D. A. S., Shoesmith, E. A., and Smith, A. G. (1988). Pupal polymorphism in the butterfly *Danaus chrysippus* (L.): Environmental, seasonal and genetic influences. *Biol. J. Linn. Soc.* **33**, 17–50.

Smith, T. B., and Skúlason, S. (1996). Evolutionary significance of resource polymorphisms in fishes, amphibians, and birds. *Annu. Rev. Ecol. Syst.* **27**, 111–133.

Spitze, K. (1992). Predator-mediated plasticity of prey life history morphology: *Chaoborus americanus* predation on *Daphnia pulex*. *Am. Nat.* **139**, 229–247.

Stearns, S. C. (1983). The evolution of life-history traits in mosquitofish since their introduction to Hawaii in 1905: Rates of evolution, heritabilities, and developmental plasticity. *Am. Zool.* **23**, 65–75.

Stearns, S. C. (1989). The evolutionary significance of phenotypic plasticity. *BioScience* **39**, 436–445.

Thompson, J. N. (1994). "The Coevolutionary Process." University of Chicago Press, Chicago.

van Noordwijk, A. J. (1989). Reaction norms in genetical ecology. *BioScience* **39**, 453–458.

Van Tienderen, P. H. (1991). Evolution of generalists and specialists in spatially heterogeneous environments. *Evolution (Lawrence, Kans.)* **45**, 1317–1331.

Via, S. (1987). Genetic constraints on the evolution of phenotypic plasticity. *In* "Genetic Constraints on Adaptive Evolution" (V. Loeschcke, ed.), pp. 47–71. Springer, Berlin.

Via, S. (1993a). Adaptive phenotypic plasticity: Target or byproduct of selection in a variable environment? *Am. Nat.* **142**, 352–365.

Via, S. (1993b). Regulatory genes and reaction norms. *Am. Nat.* **142**, 374–378.

Via, S., and Lande, R. (1985). Genotype-environment interaction and the evolution of phenotypic plasticity. *Evolution (Lawrence, Kans.)* **39**, 505–523.

Via, S., and Lande, R. (1987). Evolution of genetic variability in a spatially variable environment: Effects of genotype-environment interaction. *Genet. Res.* **49**, 147–156.

Via, S., Gomulkiewicz, R., De Jong, G., Scheiner, S. M., Schlichting, C. D., and Van Tienderen, P. H. (1995). Adaptive phenotypic plasticity: Consensus and controversy. *Trends Ecol. Evol.* **10**, 212–217.

West-Eberhard, M. J. (1986). Alternative adaptations, speciation, and phylogeny. *Proc. Natl. Acad. Sci. U.S.A.* **83**, 1388–1392.

West-Eberhard, M. J. (1989). Phenotypic plasticity and the origins of diversity. *Annu. Rev. Ecol. Syst.* **20**, 249–278.

Wheeler, D. E. (1986). Developmental and physiological determinants of caste in social Hymenoptera: Evolutionary implications. *Am. Nat.* **128**, 13–34.

Wigglesworth, V. B. (1961). Insect polymorphism—A tentative synthesis. *R. Entomol. Soc. London Symp.* **1**, 103–113.

Wilbur, H. M., and Fauth, J. E. (1990). Experimental aquatic food webs: Interactions between two predators and two prey. *Am. Nat.* **135**, 176–204.

Woltereck, R. (1909). Weitere experimentelle Untersuchungen über Arveränderung, speziell über das Wesen Quantitativer Artunterschiede bei Daphniden. *Verh. Dtsch. Zool. Ges.*, pp. 110–172.

Wright, S. (1943). Isolation by distance. *Genetics* **28**, 114–138.

Zaret, T. M. (1969). Predation-balanced polymorphism on *Ceriodaphnia cornuta* Sars. *Limnol. Oceanogr.* **14**, 301–303.

Zaret, T. M. (1972). Predators, invisible prey, and the nature of polymorphism in the Cladocera (class Crustacea). *Limnol. Oceanogr.* **17**, 171–184.

CHAPTER 12

Epilogue: Prospects for Research on the Origin and Evolution of Larval Forms

MARVALEE H. WAKE* AND BRIAN K. HALL†

*Department of Integrative Biology and Museum of Vertebrate Zoology, University of California, Berkeley, California, †Department of Biology, Dalhousie University, Halifax, Nova Scotia, Canada

Why are there larvae, and how did they come to be? What, indeed, is a larva? How do larvae function—physiologically, ecologically, and as a component of life history? How does larval development differ from and/or contribute to adult form and function? What is the mechanistic basis of the development of larval form and function? What is the interplay of developmental pattern and process with evolutionary pattern and process? The contributors to this volume have clarified and increased our understanding of larval biology, while pointing out a number of areas that remain unresolved. They variously have defined larvae, assumed definitions, or defied definitions. In general, two types of definitions of larvae emerge: one is characterized by morphological attributes that alter during the lifetime of the animal, often dramatically, whereas the other features ecological change during the life history of the animal. Some definitions combine these features. At the same time, several authors point out that the definitions are general ones and that there are exceptions to those generalizations in all major groups of animals. This terminological incongruity is a clue to the variation in larval biology that contributes to the wide range of patterns of evolution discussed in these chapters.

The complexity of larvae, their contribution to a species' life history, some of the controls over their development, maintenance, and metamorphosis, their function and behavior, and their ecologies are discussed in various frameworks by the contributors. We appear to be reaching a better understanding of larval biology at several levels, but the investigations that established our understanding have opened many new questions. In fact, the realization and elucidation of new questions, and the reconsideration of some "ancient and honorable" ones in the light of new information, are important contributions of our work, which we hope will stimulate some new research venues as a consequence of what is known and, more importantly, what is not known about the biology of larvae.

An "ancient and honorable" question that is not fully answered, nor even fully grasped, is why larvae evolved at all. The facile answer usually provided is that larvae contribute to the survivorship of species by diversifying both larval and adult life forms and thereby the habitat and resources available. A more interesting question might be why adults evolved. In other words, which came first? Was the reproductive form sexual or not? Did organisms with "larval" features acquire sexual reproduction, or did larvae evolve secondarily to diversify species? Different groups of animals might provide different insights into an answer to these questions. As noted in our Introduction (Chapter 1), Garstang recognized such questions as whether larval and adult evolution occur in parallel, and whether larval evolution is an "escape from specialization" or the achievement of specialization. These and other major questions are not yet answered, but new tools and techniques, data, and analytical and theoretical frameworks provide means of approaching problems of the interplay of development and evolution with new insights. Our contributors provide several examples of new contributions through synthetic approaches.

At the same time, each author either implies or directly states a series of areas that need study or poses questions that are not yet resolved. Virtually all contributors decry the limited numbers of species that have been studied and the relative absence of comparative analysis. Several point out that comparative assessment done in a phylogenetic context is essential to understand the pattern and direction of evolution. Many animal groups for which there are excellent descriptive and even mechanistic studies for individual species are not represented by well-corroborated hypotheses of phylogenetic relationships; understanding of the evolution of such systems remains speculative or intuitive at best. Most authors have found that mechanistic bases for developmental and evolutionary change are poorly known, studied in only a few species, and often of limited generality.

Hickman wrestles with the enormous diversity of invertebrate larvae, especially problems of analysis of their biology, origins, potential contribution to understanding of phylogenetic relationships, and the emerging evidence in the fossil record that is changing the ways that paleontologists, systematists, and evolutionary biologists view the evolution of major invertebrate groups. She resolves the three conflicting definitions of "larva" (one is based on structure, another on ecology, and the third on morphogenetic mechanisms) in favor of the structural one. Given that a larva is a structural state or series of states—which are transitory and lost at metamorphosis but reflect adaptations for larval life—the structural features allow recognition of larvae and their biology, including ecology, development, and evolution. Further, they permit analysis of the diversity of larval form and structure in terms of characters for phylogenetic analysis. Hickman contends that currently available terminology (and there is a lot of it) and classification vastly underestimate the structural diversity of invertebrate larvae, both at higher taxonomic levels and at those of the individual, population, and species. She points out that inducible traits and phenotypic plasticity can be better addressed in a structural framework. She addresses the question of the origin of larvae by first insisting that questions be framed clearly and then identifying what must be done to effect resolution—largely dependent upon robust hypotheses of relationships. Similarly, the evidence from the fossil record, including first appearances of adults and larval skeletal remains and larval soft anatomy, has new meaning as additional characters are included in phylogenetic analysis. Hickman challenges developmental biologists to incorporate the paleontological evidence into research on developmental mechanisms in order to shed more light on evolutionary history. Her probing questions open several new areas for research on invertebrate larval biology, cast in a structural framework and facilitated by a well-supported phylogenetic hypothesis.

Hanken, in his discussion of amphibian larvae in development and evolution, emphasizes that homology of larvae remains unresolved. Further, variable and inconsistent terminology hinders syntheses of information about development, including interpretations of heterochrony. The embryonic derivation of larval and adult features is not well-studied for any species. Conservatism of cranial development generally is accepted, but there are huge gaps in our knowledge—the assumption of conservatism is based on the study of a handful of frogs, information about salamanders is limited, and there is less about caecilians. For example, there is no detailed information for any species of amphibian about the contribution of neural crest to cranial development, comparable to that available for certain amniotes. Hanken notes that the evolution of larvae through the modification of developmental patterns, especially the timing of events, is well-known, but that the mechanistic bases of the modifications are poorly known. He states that more attention should be paid to early embryology, for perturbations of early ontogeny relevant to later development are not well-documented. Finally, the loss of larvae in lineages for which a biphasic life history is ancestral deserves more attention in terms of developmental patterns, especially mechanisms, as does the recruitment of larval features to perform new functions in embryos and fetuses.

Webb asks what drives larval specializations and how specializations are selected. She notes the need for synthesis of data about morphology, growth, development, and ecology. The data are distributed in a number of pure and applied subfields of biology, often with nonoverlapping participants. Fish larvae are highly diverse—they can be extremely different from adults or rather similar, and they can metamorphose gradually or abruptly. The mechanistic basis for these differences is not understood. Webb stresses the multiple sensory systems possessed by larval fishes and their interaction to effect the ontogeny of behavior, which is another area of study often discussed, but for which there are few data about mechanisms of interaction and ontogenetic change. She strongly states the case for the need for more experimental study of development in comparative and phylogenetic contexts, asking for an interdisciplinary approach to larval biology that ranges from shipboard collaborations to that among laboratories and museums with different perspectives.

Hart and Wray extend a challenge to those who consider heterochrony an "explanation" of pattern and process in development and evolution. They find that heterochrony is *not* an important generative process in the origin and evolution of larvae, but that it arises as a result of epigenetic interactions and, therefore, is a consequence, not a process. They suggest that an alternative research program to that of larval ecology might be feasible, but that it would be limited to "smaller scale variation in morphological development of species with invariant order of developmental events."

Chris Rose's thesis is that complex life histories *produce* specialized larval forms. He, too, recognizes that few species (of amphibians, but this applies to all groups) have been studied in terms of mechanistic versus descriptive development and that there is a significant need for mechanistic research in a phylogenetic context. He develops his concern by discussing the pattern of endocrine control of amphibian development and the apparent conservatism of the few hormones involved, coupled with the essential modification of timing of events. The mechanistic bases for differentiation of patterns among species are not known. He posits that only a phylogenetic approach will allow resolution of the causation of developmental and evolutionary change and allow an understanding of the evolution of diversity of form and function.

In contrast, Fred Nijhout illustrates that whereas postembryonic development in insects is controlled by hormones and developmental transformation involves only a few hormones that have broad activity, embryonic development involves cascades of genetic activation, with no extrinsic or hormonal control. He suggests that morphogenetic field size is a key to this distinction. As the organism increases in size, the fields must operate under "local control;" hormones are a mechanism that provides long-range signals that coordinate development. Nijhout poses the contrast between postembryonic development, in which hormones operate in an all-or-none matter, and adult metabolism, in which hormones operate in terms of threshold levels and control. He recommends two major research arenas: (1) a better understanding of the ways that environmental cues are transduced by hormones during short windows of sensitivity and (2) an examination of the "modular control" of development, whereby body parts can develop, and potentially evolve, independently, thereby generating great organismal diversity.

Raff's examination of cell lineages in the development and evolution of larvae also reveals gaps in our knowledge and new questions to be explored. If relatively few cleavages are needed to generate the number of cells that establish lineages, what generates diversity? Is pattern conservation due to developmental constraints, or are features relatively unconstrained but limited for other reasons? The causal basis of cell lineage modification begs for comparative studies. One approach that Raff recommends is an examination of the way that selection maintains suites of larval characters (e.g., for feeding). Rapid evolution of cell lineages and gene expression can occur when selection is relaxed. Such studies, in a phylogenetic context, have the potential to elucidate mechanism.

Nagy and Grbić provide an innovative approach to the analysis of cell lineages in invertebrates, using insects as their model. They note that the concept of invariant cell lineage "can sometimes be a red herring;" developmental mechanisms can evolve while the cell lineage pattern does not change. The syncytial early development of insects and their patterning via gradients that then establish domains of gene expression as the embryo cellularizes suggest that at least some embryonic parts are lineage-independent. Nagy and Grbić make a conceptual leap as they realize that the linkage of gene expression domains and developmental fate is well-established in several model animals and recommend that "groups of cells under similar genetic control" be substituted for "differentiating cell lineages" in order to bring new coherence to the study of gene networks and cellular differentiation. They recommend that comparison of groups of cells that share common gene expression patterns be analyzed comparatively and in a phylogenetic framework in order to understand the origin and evolution of insect larvae. They provide lucid examples of this approach, including analyses of life-history evolution, origins of larvae, and modification of adult form, and find that such phenomena as insertions of novel stages into life histories, dissociation of developmental features, and the rearrangement of gene circuits that generate new morphologies have occurred repeatedly in the evolution of insect larvae. The concept of "groups of cells that express the same gene networks" as an analogue for the standard concept of cell lineage promises a new approach to comparative analysis of mechanisms and new principles underlying morphogenesis.

Sanderson and Kupferberg's discussion of aquatic larval feeding in fishes and amphibians further illustrates the need for synthesis of developmental, functional, and ecological information in an evolutionary framework. They propose that several research directions would be profitable. For larval fishes, examination of the function of ontogenetic changes in proportions and spatial relations of cranial structures, and the connections between functional elements, would yield a better understanding of larval development and diversification within and across lineages. For larval amphibians, the role of diet quality in morphogenesis and the ability of larvae to select foods that induce changes in morphogenesis are virtually unexplored; some data exist for frogs, but effectively there are none for salamanders or caecilians. The more general questions of the ontogenetic changes in sources of nutrition and means of its acquisition and the role of habitat segregation in feeding require more study as well.

Greene, too, advocates the integration of studies of genetics, development, and environmental effects in order to understand the evolutionary causes and consequences of phenotypic variation. He points out that we have an incomplete understanding of phenotypic evolution and asks a series of questions: How do developmental events and timing influence the range of possible phenotypes? How do developmental trade-offs in resource allocation influence possible phenotypic and developmental outcomes? Do developmental processes impose genetic constraints? Developmental events precede selection events for which they are required—how does the decoupling of development and selection occur? How do ecological interactions between the organism and its environment influence phenotypic evolution? How do patterns of phenotypic variation influence patterns and modes of speciation and macroevolutionary patterns?

In summary, questions and problems for new research on the origin, function, and evolution of larvae abound. These questions and problems pertain to the biology of larvae in general and specifically for different lineages of animals, to different processes of regulation of development, to different structural and functional attributes, and to different responses to environments. Old questions remain unresolved, and new questions constantly arise as consequences of investigation. New techniques and new theoretical constructs permit new insights. More species must be studied, within and across lineages of animals, in order to assess the generality of current conclusions. Cross-disciplinary syntheses of data and approaches will provide more profound answers to questions of larval biology. A comparative approach in a phylogenetic context is most likely to provide answers to questions of mechanism and causation of the evolution of larvae and their development, structure, and function. The research presented by the contributors to this volume and their new options for study indicate that investigation of the origin and evolution of larval forms is a field of study as lively and engaging today as it was for Balfour and Garstang nearly a century ago.

INDEX

hemimetabolous, 226, 232–233,
282–285
heteromorphosis, 241, 248–249
holometabolous, 247–248, 281–283,
288–291
cell lineages in, 275–296
evolution of, 281–285
hormone sensitive periods, 222–223
larval → adult molt, 226
larval → pupal molt, 226
life histories, 281–285
novel stages in, 285–287
metamorphosis, 228–233
size and, 235–240
modular mechanism of developmental
control, 234–235
molting, 225–227
parasitic wasps, 286
polyembryony, 286
polyphenism, 241–248
prolegs, 289–296
convergence of within dipteran insects,
292–294
regulation of development, 221
seasonal polyphenism in butterflies,
245–248
spatial heterogeneity of response to
hormones, 224–225
temporal heterogeneity of response to
hormones, 225
theories for the origin of, 283–284
tissue sensitivity to hormones, 223–224;
see also Juvenile hormone
Invertebrate larvae, 21–55
adaptations for feeding, 42–45
adaptations for swimming, 42–45
antipredatory adaptations, 46–47, 50
classification of, 33
defensive structures, 46
developmental sequences, 52–54
ecological definition, 25–26, 32-33
fossil record, 39–42
intracapsular adaptations, 45–46
larval types, 27–33
morphogenetic regulatory definition, 26
origins of, 36–39
and phylogeny reconstruction, 50–54
stages, 27, 413
structural definition, 24–25, 52

J
Juvenile hormone
control of molting, 225–227
and diet-induced polyphenism, 386
and insect larval metamorphosis, 218–219,
228–233, 284
neurosecretory control of, 219
and polyphenism, 242–248
queen induction, 222, 224
and seasonal morphs in aphids, 391
sensitive periods, 222–223, 229–233
soldier termites, 245; see also Insect larval
development

L
Lampreys, 12, 316
Larvae
definitions 5, 24–27, 305–307, 411, 413;
see also Invertebrate larvae
evolution of, 10, 67, 90, 181–200,
262–268, 281–295, 412
feeding, 12–13
kinds of, 7–9
phylogeny of, 10–12; see also Primary
larvae; Secondary larvae
Larval adaptations, 4–8, 42–50, 138, 256
polyphenism as, 241–248
Larval constraints, 68, 90, 94–95
Larval developmental programs, 7
Larval dispersal, 16
of invertebrate larvae, 23–24, 32–33
Larval dormancy, 15
Larval ecology, 14–16, 32–33, 163, 303
Larval feeding, 37–39, 301–364
absorptive feeding, 311–313, 319–320
aquatic larvae, 301–364, 416
critical periods, 314–315, 321–322
diet, 334–336
endogenous feeding, 310–311, 319
exogenous feeding, 313–318, 320–334,
346–352
evolution of, 359–363
filter feeding, 315–318, 322–330
mixed, 320
ontogeny of feeding structures, 340–356
predation and, 303–304
Ram feeding, 347–350
suction feeding, 318, 330–334, 347–350
suspension feeding, 304